ShipbuildingTechnology

Shipbuilding Technology

First Edition: July 2015

Printed in Korea by Jisungsa, 2F. 4, Jinheung-ro 1-gil,
Eunpyeong-gu, Seoul, Korea.
www. jisungsa.co.kr

* This book is translation of the original Korean edition published
by the Society of Naval Architects of Korean in 2011.

ISBN 978-89-7889-304-6 (93550)

A CIP catalogue record of the National Library of Korea for this book is available at the homepage of CIP(http://seoji.nl.go.kr)
and Korean Library Information System Network(http://www.nl.go.kr/kolisnet). (CIP2015016473)

Shipbuilding Technology

조선 기술

The Society of Naval Architects of Korea

Jisungsa

A Word from the President

Founded in 1952, The Society of Naval Architects of Korea (SNAK) has become a premier institution for practitioners and scholars of the shipbuilding sciences and technology. SNAK has played an essential role in not only stimulating career development but also imparting consequential research on the shipbuilding industry. Korea's ascent to international shipbuilding power since the start of the 21st century can be attributed to key technological innovations in the field. Despite these notable gains, it is difficult to find English-language academic literature chronicling these far-reaching achievements and contributions as well as the on-going status of the Korean shipbuilding industry. The publication of *Shipbuilding Technology* marks the first comprehensive effort in English-language academic literature.

The present volume, *Shipbuilding Technology*, is a revised and updated edition of the 2011 cognational Korean-language volume, which introduced the science and technology of the latest ships and ocean engineering to the domestic public. Likewise, the English-language volume is written in accessible and concise prose for the interested laymen. It presents all of the technology used in Korean shipbuilding ranging from fundamental theories and discourse to first-rate practices and technical expertise. The contents encompass main topics such as contract, design, resistance and propulsion, ship motion and maneuvering, engines, general arrangement, ship structure, outfitting, construction, project management, warships, leisure boats, and offshore structures, and expectations for technology. Readers are free to begin from the chapter or section of their interest without respect to chronological order. This literature can also be used as an introductory textbook for university students.

The arduous task of producing and directing *Shipbuilding Technology* from development to publication is due to the tireless efforts of many experts including Professor Hyochul Kim of Seoul National University, and Dr. Byungsuk Lee of the University of Strathclyde. I would also like to take this time to thank the dedication of other Committee members.

Last but not least, it is my hope that this publication will bring to light a concrete understanding of the aspects of Korean shipbuilding industry and its state-of-the-art technology to the international audience.

July, 2015

Jong Gye Shin, Ph.D.

President (2014-2015)

Foreword

The Society of Naval Architects of Korea (SNAK) held her inaugural meeting at the training center of Korea Shipbuilding and Engineering Corporation (currently Hanjin Heavy Industries) in Busan on 9th November, 1952 during the Korean War. After infancy SNAK commenced various activities as an academic society including publication of text books in Korean and the participation of ITTC and ISSC and the involvement of PRADS founding from the beginning of the 1970's.

Korean shipbuilding industries also made their debut on the international market from the beginning of the 1970's. The forty years after the debut brought the new comer of the market to one of the most competitive countries. During the period, the Korean shipbuilders have achieved great progresses based upon the technologies developed by themselves and those transferred from foreign countries.

SNAK held the 60th anniversary ceremony on 15th November, 2012 in Changwon, Korea and published 'The Society of Naval Architects of Korea 2002 ~2012'. SNAK Directors discussed what could be done beside the ceremony and the history book to commemorate her 60th anniversary. Finally, the 326th Board of Directors meeting held on 8th March, 2013 decided to translate *Shipbuilding Technology* into English. The contents of *Shipbuilding Technology* cover from 'Sales and Purchase of Ship' to 'Production Management.' Not only merchant ships but also naval ships and offshore structures are dealt with in the book. The special ship construction techniques developed in Korea are included as well.

Prof. Hyochul Kim, former SNAK President (1994–1995) and Professor Emeritus of Seoul National University, was proposed to initiate and coordinate the translation work and that proposal was accepted by him. Later he invited Dr. B. S. Lee of University of Strathclyde, Glasgow to improve the English not only literarily but also technologically.

I firmly believe that the introduction of recently developed Korean shipbuilding technologies to the world could be a kind of reward for the foreign financial and technological supports during the 1970's and a small contribution to the further development of marine technology.

July, 2015
Sang-Rai Cho
President (2012-2013)

Foreword from Korean edition

This book is intended for use as a bird's eye view of general concepts and technologies in the field of naval architecture and ocean engineering to guide newcomers to the field or non-naval architects already practicing in shipyards. Publication of this book was possible by constant efforts of many contributors involved in diverse ways, and therefore, it would be my obligation to take this occasion to appreciate their efforts and to bring out the motive of the publication as well.

It seems to mystify outsiders that Korea, one of the poorest countries until the 1960's asserted to be absolutely impossible to escape from poverty, emerged from next to nothing and has grown into one of the major economic powers in the world in only a few decades. A well-known answer to the puzzle is simply the education that enthusiastic parents provide their children at any cost. Our shipbuilding industry shares its voyage with and contributed much toward the drastic growth of the national economy and it is not just fortunate enough for the industry to retain a leading position in the world market amid recurring economic crises and fierce competitions prevailing. Likewise for the nation, the success of the Korean shipbuilding industry may not be possible without valuable manpower and technological knowledge accumulated over a long time through devotion and sacrifice of the members of the society of the naval architects of Korea (SNAK) endeavor to promote the industry in yards, research institutes, and schools.

SNAK, an organization for promotion of practical and scientific knowledge in naval architecture and ocean engineering and fellowship among its members, has been confidently dedicated to bringing up outstanding technical manpower in these fields. Publication of various books for technical education is one of such activities and was possible owing to full support of the members. However, it is disappointing to discover that the books upon which new students just entering the field or non-naval architects practicing in shipyards can easily rely in order to understand general concepts and technologies in the field of naval architecture and ocean engineering are still scarce even though almost 60 years have passed since the society was established. Therefore, it is a privilege of SNAK to have the timely proposal from senior fellows of the society to publish this book to make up for the deficiency and to finally realize the proposal at this moment. The members who played important roles in the process of publication include Jae-Wook Lee, Suk Chang, Chang-Sup Lee, Sung-Hyuk Hwang as well as Hyochul Kim, among others, and their contribution would be highly appreciated. Gratitude should be expressed also to Dong Sik Shin, the president of Korea maritime consultant co. Ltd., for his encouragement and financial support which made the publication of this book conceivable.

Many members of the society take part in writing of this book and many thanks should be given. Among them, Hyochul Kim, the editor-in-chief, and other members of the editorial board, and major research groups of the society have contributed so much in the publication process and would also be highly appreciated. Needless to say, the successful publication of this book is indebted to the members of the publication committee of the society under the chairmanship of Hyun-Kyoung Shin and the commitment of the publishing company, Jisungsa, was very helpful in perfecting the publication. It may be an interesting reminder that the first thought to publish this book came across during the discussion on translation of 'Naval Architecture for Non-naval Architects' published by SNAME in 2006 into Korean.

I anticipate that the successful publication of this book will be an opportunity for the shipbuilding industries to take a step further by assisting cultivation of excellent naval architects and ocean engineers and that more books of purpose continue to emerge in the future.

December, 2011
Seung-Hee Lee
President (2010-2011)

Editor's Preface

Entering midsummer of 2010, spirited discussions came about at unexpected occasions among the naval architects including the president of the Society of Naval Architects of Korea (SNAK), Prof. Seung-Hee Lee, on the strategies for effective accomplishment of research tasks. The major consensus of opinion was appeared that the domestic education on naval architecture has been particularly dependent on foreign literature and such absurd situation should be improved urgently. It is also thought that consulting a specific technical problem can be a very important contribution to domestic shipbuilding industry; but more important is the provision of proper guidelines to assist individuals engaged in the industry to understand naval architecture at least in a broad sense. Such a critical opinion has been arose from the current abnormal educational environment depending on old foreign literature dealing with the past knowledge; instead of the newest technology developed by own industry.

Coincidentally, Emeritus Prof. Jong Heul Hwang, a honorary member of the SNAK and also a member of the National Academy of Sciences, was sincerely considering a research proposal on the compilation of domestic shipbuilding technology to the academy; Prof. Shin Hyung Rhee, a member of the SNAK, had just started to translate the book titled 'Naval Architecture for Non-naval Architects,' which was published by the Society of Naval Architects and Marine Engineers of America. It is obvious that both of the works could play an important role in the propagation of shipbuilding technology. But it is also recognized that the publication of a new textbook comprised of more recent and extensive shipbuilding technology over which shipyard personnel with a rudimentary engineering knowledge can easily read is necessary to further strengthen competitiveness of domestic shipbuilding industry which has been leading the world shipbuilding industry for decade.

Subsequently, it was decided to make an effort to study feasibility of writing and publishing the new book which can provide an overview of core shipbuilding technology for engineers determined to serve the industry regardless of his major, and to materialize the specific publication plan. On the line of context, free discussions came up in friendly meeting of some senior members of the SNAK in July of 2010 and the meeting evolved into a pre-editorial committee and sustained three meetings until October 2010. The committee has figured out the nature of the text book, editorial scope, contents, recommendation of writer, support from industry, frame of budget and etc. and members were:

- Chang, Suk : Ex-President, KRISO
- Kim, Hyochul : Emeritus Professor, Seoul National University
- Kim, Jung Ho : Executive Director, Korea Marine Equipment Association
- Lee, Chang-Sup : Professor, Chungnam National University
- Lee, Jae Wook : Emeritus Professor, Inha University
- Lee, Seung-Hee : President, Society of Naval Architects of Korea
  Professor, Inha University

Among them Hyochul Kim played a role as a meeting coordinator.

The annual meeting of the SNAK which was held in Changwon on 21st October, 2010 has provided a momentum to talk about the constitution of the editorial board, and 22 board members have been elected from the research panels and major member organizations of the Society. At the same time the annual meeting could gain wide encouragement from the industry and nationwide participation of researchers and scholars. The editorial board could be activated by the financial support of CEO Dong Shik Shin of Korea Maritime Consultant Co. and the mobilization of writing work could follow. The publishing company, Jisungsa, has willingly joined the publication of the text book and comprehensively supported the publication.

The editorial board of '*Shipbuilding Technology*' was made official on the 17th of November 2010 with 16 additional board members invited from all of the major fields of naval architecture and various relevant organizations, as listed below.

- Bong, Hyon Soo : Chairman, Board of Technology Council, KOSHIPA Vice-President, Hanjin Heavy Industry Co.
- Cho, Tae Ik : Chairman, Research Panel on Ship Design, SNAK Managing Director, Daewoo Shipbuilding and Marine Engineering
- Chon, Young Kee : Chief Engineer, Technology Assistant Center, KR
- Chun, Ho Hwan : Chairman, Research Panel on Naval Vessel, SNAK

- Hwang, Sung Hyuk : President, Hwanghwa Company
- Hwangbo, Sueng Myun : Managing Director, Samsung Heavy Industry
- Kang, Sa Joon : Managing Director, Korea Shipbuilders Association
- Kim, Choon Gon : Chairman, Research Panel on Ship Production Technology, SNAK
  Vice-President, Hyundai Mipo Dockyard Co. Ltd
- Lee, Chun Ju : Chief Researcher, Marine Transportation research team, KRISO
- Lee, Hong Gi : Chairman, Korea Towing Tank Conference, SNAK
  Managing Director, Hyundai Heavy Industry, Co.
- Leem, Hyo Kwan : Executive Director, STX Offshore and Shipbuilding
- Shin, Hyun Kyoung : Director, Board of Publication, SNAK
  Professor, University of Ulsan
- Shin, Jong Gye : Chairman, Research Panel on Marine leisure Boat, SNAK
  Professor, Seoul National University
- Won, Youn Sang : Chairman, Research Panel on Ocean Engineering, SNAK
- Yoon, Bum Sang : Chairman, Research Panel on Ship Hydrodynamics
  Professor, University of Ulsan
- Hong, Suk Yoon : Chairman, Research Panel on Ship Structure, SNAK
  Professor, Seoul National University

At the kick-off meeting of the editorial board on 17th November 2010, '*Shipbuilding Technology*' has been decomposed into 16 chapters in accordance with major fields of the technology and also a responsible editor for each chapter have been nominated simultaneously. Every editor should have an obligation not only in analyzing the technology tree of the given major field but also coordinating co-authors of consisting sub articles. Nomination of authors for the sub articles have been made since the 14th of January 2011 and accompanied distribution of sample manuscripts to keep the formality. Due to the variety of detailed technical elements, every editorial board member should play a role not only a coordinator among co-authors but also a leading author of typical chapter under his obligation.

Depending on the technical constitution of specified chapters, a few authors are sufficient for preparing the manuscript for a typical chapter while some other chapters have inevitably required a group of authors. Some of the primary manuscripts provided by authors have to be supplemented with the aid of an expert; or in other cases, valuable manuscripts which were provided by individual authors should be merged together to avoid duplication. In addition, refining and unifying the manuscripts proceeded with help of the expert knowledge and therefore the authors of *Shipbuilding Technology* have been continuously increased to reach 66 persons at the final completion of the manuscript as shown in following list of authors

Kang, Sa Joon	Kang Young Jin	Kim, Kyung Rae	Kim, No Shig
Kim, Deok Joon	Kim, Myung Hyun	Kim, Sang Hyun	Kim, Sung Hyun
Kim, Seong Hwan	Kim, Jeong Je	Kim, Jung Ho	Kim, Choon Gon
Kim, Hyochul	Kim, Hee Jung	Nho, In Sik	Lew, Jae Moon
Moon, Jong Leen	Moon, Joon Sik	Park, No Joon	Park, Sang Hyun
Park, Hyoung Gil	Bae, Sang Eun	Bong, Hyon Soo	Seo, Dae Won
Seo, Young Soo	Suh, Jung Chun	Seo, Hyung Kyun	Sung, Jung Hyun
Song, Chun Tae	Shin, Jeong Do	Shin, Jong Gye	Shin, Hyun Kyoung
Sim, Sang Mog	Yang, Yeong Jun	Oh, Jung Keun	Won, Youn Sang
Yoon, Bum Sang	Lee, Sang Su	Lee, Sun Taek	Lee, Soo Geun
Lee, Seung-Hee	Lee, Young Gill	Lee, In Gyu	Lee, Jae Wook
Lee, Jae Ung	Lee, Chang-Sup	Lee, Chun Ju	Lee, Hyun Yup
Lee, Hong Gi	Yim, Sang Heun	Lim, Seung Joon	Leem, Hyo Kwan
Chang, Suk	Chon, Young Kee	Chun, Ho Hwan	Jeong, Byeong Cheol
Chung, Tae Young	Cho, Kwon Hae	Cho, Dae Seung	Cho, Tae Ik
Ha, Tae Bum	Huh, Don	Hong, Suk Yoon	Hwangbo, Sueng Myun
Hwang, Sung Hyuk	Hwang, Jung Ho		

The manuscripts were collected by the due date on the 30th of April 2011 as agreed upon by the meeting of authors. The proof-reading of the manuscripts required painstaking devotion from editorial board members, Jung Ho Kim, Hyochul Kim, Seung-Hee Lee, Jae Wook Lee, Chang-Sup Lee, Suk Chang those who initiated the publication work. Not only for unifying and synthesizing of the manuscript, they should prepare supplement articles for unconsidered elements with the aid of related experts. The editors of the publishing company, Jisungsa, have focused their efforts on simplifying complicated sentences into simple form, unifying tense and mode of sentences, including careful spelling.

The manuscripts have been revised through the six editorial board meetings held during the period from May 12 to December 16, 2011. In this procedure the editorial board members should take a share in proof reading and manuscript correction with the aid of specialists in related topics. Final revised manuscripts were synthesized and handed over to the publisher step by step starting on August 30, 2011. Preparation work for publication such as adjustment of figures and photos also has been carried out concurrently with the confirmation of references and copyrights by the staff of the publishing company. The printing of *Shipbuilding Technology* was realized near the end of the December 2011 in considering the termination of the current term of the SNAK.

It is believed that *Shipbuilding Technology*, which has been edited with the consensus of the naval architects active in the shipbuilding industry and published by the SNAK, could be utilized as a valuable reference in promoting the global understandings of naval architecture to younger generation who are interested in the industry. It is also believed that this book, may serve as an overview reference to the naval architects or a principal reference to guide non-naval architects in advance before joining to the shipbuilding community. Finally, it should be acknowledged that the publication could not be realized without the devotion of Director Je Yang Lew and Assistant Director Mi Kyung Kim of the Society. The appreciation should be extended to the numerous organizations and individuals who devoted time and effort into this publication of the *Shipbuilding Technology*.

Publication of English version

Shipbuilding Technology was finally published in January 2012 and since then it has served as a favorite primer for understanding naval architecture and ocean engineering. The first edition was sold out so fast and additional copies were printed in September 2012. Such a reputation made the book to be chosen as one of the best academic publication of the year by the Ministry of the Culture and Tourism in July 2012. In November, SNAK sounded Prof. Hyochul Kim, chairman of the editorial committee, out on translation of *Shipbuilding Technology* into English.

The translation was a challenging task to the committee but it couldn't simply ignore the proposal. Therefore, the chairman convened a meeting of the authors of *Shipbuilding Technology*, mostly work in the southern part of the country, at a conference room of DSEC Co., Ltd. in Busan. It was decided to go ahead with the translation project and various topics such as an estimate and work scope were discussed along with plans for publication and distribution of the final product. Based on the results of the meeting, the committee for translation of *Shipbuilding Technology* into English was formed on December 14th, 2012 to give shape to the action plan. An additional authors' meeting was held in Geoje region on January 24th, 2013 and subsequently some leading members of the committee visited Hyundai Heavy Industries Co. and Mipo Dockyard Co., Ltd. to ask for their supports.

The participants in the translation was finalized at the committee meeting on February 4th, 2013 held at the Department of Naval Architecture and Ocean Engineering, Chungnam National University and they gathered together for the first time on February 28th, 2013. SNAK's board of directors officially decided on March 8th, 2013 to translate the book into English to commemorate the 60th anniversary of SNAK.

At the committee meeting held on May 23th, 2013 in Jeju Island, it was decided to collect and integrate translated manuscripts until July 30th. However, some of the participants were too busy to meet the schedule and made the whole project stalled until professional expert take over their manuscripts.

The original intention of the committee was to conclude the project no later than end of October 2013 but it actually took more than 18 months to complete the final English manuscript of *Shipbuilding Technology* on February 2015. It should be mentioned here that Dr. B. S. Lee who served at University of Strathclyde, Glasgow contributed greatly to polishing up the translated manuscripts and made the manuscripts become a step closer to familiar English. The committee extends its appreciation to Dr. Lee, and also to Prof. Jong Ho Nam, Mr. Soon Gil Park and Mr. Seung Kyun Park who joined the translation project later to back up.

The committee expects that English version of *Shipbuilding Technology* can be published in September, 2015 and that the book would be a small return to the global shipbuilding community from Korea, the country initiated its own shipbuilding industry with help of Western technology. The committee hopes that this book could be guidance to the new joiner having interest in Naval Architecture and/or shipbuilding industries and wishes this book may be a tiny help to the development of latecomers in the industry.

Hyochul Kim

Editor-in-Chief

Contents

CHAPTER 3

—

Ship Design

CHAPTER 4

—

Hull Form Design

CHAPTER 5

Resistance and Propulsion

CHAPTER 6

Ship Motions and Maneuverability

CHAPTER 7

Propulsion Plant

CHAPTER 8

General Arrangement

CHAPTER 9

Ship Structures

CHAPTER 10

Outfitting

CHAPTER 11

Ship Production Technology

CHAPTER 12

Production Management

CHAPTER 13

—

Naval Ships

CHAPTER 14

—

Marine Leisure Craft

CHAPTER 15

—

Offshore Structures

CHAPTER 16

Life of Ships and Prospects of Ship Technology

CHAPTER 1

Introduction to Ships and Shipbuilding

1. Definitions of Ships and Their Significance
2. Ships and Shipbuilding
3. Types of Ships
4. Origins of Korean Shipbuilding Technologies
5. Outline History of the World Shipbuilding Industry
6. Marine Equipment Industry

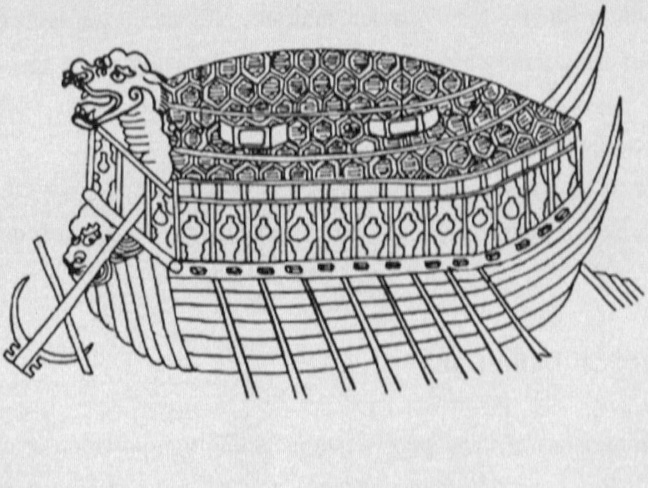

1. Definitions of Ships and Their Significance

1.1 Lexical Definitions

'배 (*bae*)' or '선박 (*seonbak*)' are the Korean equivalents of the English word 'ship.' It is defined in the Korean dictionary as an object made of wood, metal or some other material for navigating on the water to transport people or cargo. Ships can be categorized by their shape or purpose into boats, ferries, warships, cargo ships, passenger ships, oil tankers, etc. The origin of *seonbak* is the Chinese word '船舶,' the first character of which literally means river boat according to a Chinese character dictionary, and the second means a large ocean-going ship. The Chinese character '艇,' which is also commonly used in Korea, indicates a narrow and long small boat, while the character '艦' refers to a battleship furnished with planks to provide defense against arrows and stones. Meanwhile, a navy has fleets, 艦隊, and fleets are composed of warships, 軍艦. Similarly, more than 172 Chinese characters that indicate ship types, equipment and performance are each affixed with another Chinese character '舟' on the left-side radical to indicate that they are associated with ships.

In English dictionaries a 'ship' is usually large while a boat is small, although there is no clear cut criterion distinguishing ships from boats. Large ships are occasionally called boats as exemplified by 'ore boats' at the Great Lakes which are more than 300 meters long ships carrying ores. Koreans, however, use a distinct word '단정(*danjeong*)' for small ships. On the other hand, the word 'vessel' is used regardless of size and thus it could mean any ship, *bae* or *seonbak* in the Korean language; the word 'craft' usually refers to small boats, or speedboats. While the word 'carrier' indicates a means of transport or a shipping agent, it also means an aircraft carrier. The word 'yacht' usually refers to a small sail boat but sometimes indicates a ship large enough to hold onboard meetings for hundreds of people, like the erstwhile Royal Yacht *Britannia*.

All in all, a ship can be defined as an object operating on a body of water, such as sea, river and lake for the purpose of carrying passengers or cargo or performing certain tasks.

1.2 Practical Definitions

The main missions of ships are diverse, including transportation of cargo or passengers, military activities, scientific explorations, industrial support, vessel support services and sports and leisure activities. From engineering perspective it is important to identify the main purpose the ship is to serve before it is designed. A ship must have some basic properties as follows:

Firstly, it must be possible to construct the ship, and when constructed, it must have sufficient buoyancy to carry the designated cargo. In addition, the ship must be easily operable and be able to protect the crew and provide good working environment. It must be fit for purpose, i.e., if it is a merchant ship, it must be profitable to operate.

Secondly, the ship must have appropriate seakeeping performance to ensure safe and comfortable navigation on route to its destination. The performance can be assessed in terms of vertical and transverse accelerations, roll, slamming and deck wetness among others, and are closely related to comfort, workability, schedule reliability and fuel consumption which often place conflicting demands on the design parameters. In order to improve the transverse stability of a ship, for example, the breadth of the ship can be increased, but this will lead to an increase in the resistance of the ship so as to reduce its economic efficiency. A ship is required to have structural strength sufficient to maintain its integrity in rough seas, but higher scantling inevitably increases the hull weight which leads to a reduction of pay load and consequently loss of earning capacity. It is clear, therefore, that ship design is essentially a process of compromising between various conflicting demands.

Thirdly, the ship must be self-sufficient for a required length of time without additional supplies such as fuel, provisions and fresh water during a voyage at sea, although some naval ships do get refueled at sea when required. Durations of voyages vary on the various kinds of ship. For instance, ferries may take only a few minutes to cross a river, whereas naval vessels sometimes maneuver for several months at a time. Machinery and component parts of ships out at sea for an extended period of time must have high reliability to operate uninterrupted during the voyage.

Fourthly, the ship must provide comfortable conditions for its crew so that they can live and perform their duties effectively in confined spaces during a long voyage. Therefore, a ship must be fully furnished not only for daily life but also for the relaxation of the crew. The working and living spaces must be arranged so as to improve the operational efficiency and it is also important to isolate these places from noise and vibration. Ship motions should be limited to a tolerable range and appropriate devices are necessary to prevent extreme temperature variations in inhabited spaces.

Lastly, the safety of the ship must be ensured, and especially for passenger ships very high level of safety is of paramount importance. Therefore, the probability of damage to the ship due to fire, explosion, storm, collision, grounding and so on should be minimized. Furthermore, ships must be equipped with various safety devices that prevent the ship from completely losing their functions, should such an accident occur. Various lifesaving appliances for passengers and crew have to be provided for extreme events when the ship has to be abandoned.

Even if a ship is built as intended at the lowest possible cost, it will not be able to function properly unless above-mentioned conditions are met.

1.3 Legal Definitions

The Korean laws regulating ships and their activities include Ships Act, Law for Safety of Vessels, Fishing Vessels Law, Ship Registry Act, Water Leisure Safety Law, Fisheries Act, Installation and Utilization of Sports Facilities Act, Sea Pollution Prevention Law, Sea Traffic Safety Act, Construction Machinery Management Act, Bicycle and Motorboat Racing Act, Commerce Act, Marine Transportation Act and Harbor Act. The scope of applicable ships varies depending on the law, and some ships are exempted from some of the laws.

The Maritime Commerce Law has been enacted for ships engaged in commercial transportation. Thus non-profit making public vessels are exempted from restrictions of this legislation, and small boats and rowboats are similarly excluded. The primary purpose of fishing vessels is in fishing itself instead of direct commercial activities, but they are included in the category of ships regulated by the current law. In addition, ships which exclusively operate in inland waters such as rivers, lakes and canals are not considered as a ship under this law. Besides, ships that are under construction are not subject to the law either, but in some cases, they may be legally considered as vessels.

According to the Ships Act, ships are defined as floating or submerged vessels or similar structures which can be utilized for navigation; they must have sufficient buoyancy to transport passengers or cargo. For a long time barges and dredgers had not been considered to be ships but simple floating structures, as they usually have no self-propelling ability and must rely on other ships to move by being towed or pushed; but the law currently regards them as ships. According to the law, ships classified as steamships are propelled by engines, sailing ships by sails, and barges are unable to move by themselves.

The Fishing Vessels Law regulates ships used for fishing and related activities such as transporting or processing of fishery products. Official ships engaged in research, survey, guidance, control and training in support of fisheries industries are also considered fishing vessels by the law. A certificate of ship's nationality is issued to each registered fishing vessel of 20 gross tons and over, a certificate of ship registration for ships of 5 to 20 gross tons, and a certificate of registration for ships less than 5 gross tons.

The Water Leisure Safety Law and the Installation and Utilization of Sports Facilities Act control not only canoes and kayaks used for water sports and leisure activities but also simple sports goods, such as surfboards, by placing them under the category of 'ships.' Under the Bicycle and Motorboat Racing Act, a portion of the proceeds earned by the boat racing business must be spent in fostering the motor boat industry. On the other hand, wing-in-ground (WIG) effect ships have been designed to utilize augmented air lift produced by the interaction between their wings and the water surface when flying close to the surface, and are called flying ships. In general, commercial airliners often fly at high altitude because thin air there reduces the air resistance, whereas WIGs can realize energy saving only when flying in close proximity to the water surface. Their structure must therefore be reinforced to withstand unexpected impacts of water slamming. For this reason, WIGs are classified as ships by both the International Maritime Organization (IMO) and the International Civil Aviation Organization (ICAO). The Sea Traffic Safety Act has also classified the WIG and the seaplane (hydroplane) in the category of ships.

Meanwhile, the Sea Pollution Prevention Law classifies drillships and platforms as ships, regardless of whether they are fixed or floating. A floatel (floating hotel) is considered a ship in accordance with the Law for Safety of Vessels; however the same floatel or floating auditorium is not considered a ship if the structure has been authorized by the River Supervision Act or Public Water Surface Management and Reclamation Act.

2. Ships and Shipbuilding

A landmark event that heralded the modern Korean shipbuilding occurred in the early 1970s: Hyundai Heavy Industries Co. Ltd, a company that had nothing but a sandy beach and a huge ambition at the time, was awarded an order to build two very large crude oil carriers (VLCCs). Many people, even some industry 'experts' mistakenly thought that it was easy to build a ship given a few drawings. This was a misunderstanding, to put it mildly, considering the fact that HHI at the time had neither experience in building such large ships nor appropriate facilities. Most people simply did not appreciate the scale and complexity of a VLCC.

Take a 300K DWT (deadweight tons) class VLCC Venus Glory (331m×58m×31m, 15.1 knot) for example. Its depth is equivalent to the height of a 10-story building; the length is about 1.5 times the height of '63 Building,' one of the best-known landmarks in Seoul, and it is as wide as a soccer field. It can even accommodate three soccer fields on the deck. If its 300,000 tons of cargo oil were to be carried by 20-ton class tank lorries, the necessary trucks, when placed in line at 100 meter intervals, will occupy the entire 428km length of the highway from Seoul to Busan almost four times over.

Some consider shipbuilding industry a labor-intensive one. However, consider this: ships built by Korean shipbuilding industry in 2010 amounted to 26.4 million GT (gross tons) and it employed about 132,700 people in the same period. It can be said, therefore, that each person-year built roughly 199 GT on average. Since a passenger car has a carrying capacity of about 0.7 GT and weighs about 1.4 tons, it is not unreasonable to say that each person employed by the Korean shipbuilding industry produced the carrying capacity equivalent to that of about 284 cars a year. Therefore, treating the industry as a merely labor-intensive one is not fair. By the early 1970s the shipbuilding industry was already sufficiently advanced in technology to construct large oil tankers longer than 300 meters entirely by welding. However, all-welding construction of the rail systems was not accomplished until well into the 21st century, more than 100 years after the building of the first domestic railroad from Seoul to Incheon in 1899. The market price of a newly built tanker of 300K DWT was recorded at 146 million US dollars(USD) at the end of 2007. This price is comparable to that of about 10 thousand small passenger cars, thus the shipbuilding industry should perhaps be treated as a major capital-intensive industry instead.

When a ship is built outdoors aligned in the north-south direction, the centerline of the hull is liable to be bent due to the expansion of steel plates on the sunny side. The amount and direction of the hull bending varies from morning to evening with the progress of the sun.

Nevertheless, the centerline of the propeller shaft must remain straight after completion of the hull construction. Therefore, the welding and construction process must be meticulously planned and carried out with high precision, especially if the shell plates – about 60 meters wide – are to be assembled together smoothly. Misalignment at the joints must not exceed 1 mm and this requires an accuracy of 1/60,000 of the beam. This is comparable to the tolerances allowed for manufacturing high precision machineries. With such accuracy, blocks weighing 1,000 tons are routinely manufactured and assembled, each aligned flawlessly to the neighboring blocks together with all the pre-outfitted items, such as pipes and machinery. It is apparent that such precision of this magnitude within the shipbuilding process would never be achieved without skilled labor force.

VLCCs are often referred to as low speed full-bodied ships and are usually designed to travel at the speed of around 15 knots. However, the speed of a VLCC is never as low as such a number appear to indicate, since one knot is equivalent to a speed of 1,852 meters per hour and, therefore, 10 knots corresponds to the speed of a marathoner with a record of 2 hours and 16 minutes. By the way, the current Sea Traffic Safety Act still considers those passenger ships faster than 15 knots as high-speed ships.

The Venus Glory is equipped with a diesel engine of 35,000 horsepower and a propeller approximately 10 meters in diameter, rotating at a rate of 1.5 revolutions per second. The blade tips of the propeller continuously move through the water with a circumferential speed of about 47m/sec. The thrust generated by the propeller is then transmitted to the hull through a single propeller shaft to propel this enormous ship. This ship is a floating structure of over 400,000 tons displacement travelling at a very respectable speed of over 15 knots with sufficient agility to negotiate the many obstacles encountered at sea.

Ship-owners need to invest heavily to acquire ships and they place orders for new ships which preferably have performance characteristics necessary to realize more profit than the similar ships currently in service. However, ship designers have to consider total economic efficiency extending over the entire expected life span of the ship. Since ships are built to order, ship-owners almost invariably request an improved design when placing the order. As a result, for example, the thickness of the shell plate of a tanker has gradually been reduced to 1/15,000 of its length in an effort to drive down the ship's weight. Considering the fact that the thickness of an egg shell is about 1/100 of its largest dimension, difficulties involved in reducing the shell thickness to this extent can be easily imagined. The structure of a ship is designed so as to result in a minimum structural weight, but above all else, the structure must be able to withstand challenges of harsh operating conditions safely and without impairment to its functionality for 20~30 years of its life span at sea. Thus, ship design requires highly advanced design technologies.

When a 320-meter long ship is pitching even in moderate seas with a small angular velocity of 1 degree per second, the relative speed of the bow and stern against the water surface can reach around 6 meters per second, which is comparable to the speed of freely falling objects after descending 1.8m. Therefore, the equipment and machinery on board a ship must have an extremely high level of performance throughout its lengthy and continuous operation at sea under severe sea loads, unthinkable for stationary machinery installed on land. The expected life span of a motor car is far less than 20 years, even though the products are rolled out only after a long-term durability test with prototypes, while a ship is required to serve for 20~30 years without any preliminary durability test. This fact, once more, exemplifies the degree of difficulty facing the shipbuilding professionals.

The number of crew on board a ship has been decreasing significantly through automation over the years, but the voyage time remains virtually unchanged. This means that a ship is usually operated by a relatively small crew in a shift system on a lengthy voyage. Moreover, since crewmen on duty have little chance of enjoying social contact with each other, crew's psychological wellbeing must be fostered by providing comfortable and pleasant accommodation while living on board. In addition, to minimize crew fatigue in daily operations, the ship motion, vibration and noise should be kept at an acceptable level. This is especially true for cruise ships which must provide convenience and comfort to the many passengers who are unfamiliar with sea voyages. A pleasant living condition will add to their tour experience and they are more likely to return.

A ship advancing through wind and waves inevitably experiences motions. Therefore, the ship is required to possess sufficient stability against roll motion and be prepared for slamming and deck wetness as well. Even with 400 kilotons of displacement in full load conditions, the ship must be able to safely maneuver in congested sea routes without mishap. It must also be able to operate as scheduled on long-distance sea routes, crossing the oceans in variable weather conditions. Such ability gives cargo owners and shipping companies the ability to plan their business in advance without too much anxiety.

It has been reported that 376 ships were launched in Korea in 2010, and of these 361 ships were delivered to foreign owners. The value of the ships exported reached 49.1 billion USD, which amounted to approximately 10.5% of Korea's total export that year. This statistic reaffirms that ships built in Korea are recognized for their high quality by owners all around the world, considering the fact that the build quality can vary significantly even though all ships operating in international waters must comply with the relevant rules of the IMO and the classification societies.

A ship must be registered in a country and it acquires the nationality of that country, irrespective of the nationality of the owner. The ship is then recognized as a territory of the flagged country, even while berthing in a foreign port. For this reason, the Korean Navy could arrest some Somali pirates from a hijacked Korean merchant ship in international waters and put them on trial in a Korean court.

3. Types of Ships

Ships can be classified into diverse groups according to their purpose, the way they are used, service area, hull shape and size, speed and propulsion method of the ship.

When a ship is exclusively designed for an operation in a specific region, the performance of that ship in other regions may not be up to the expected standard. In such a case, the operating region of the ship decides the type of the ship. For example, a fishing vessel, designed for fishing in cold heavy seas where icing on deck is a normal occurrence, may be less efficient for fishing operations in warm and mild climate, since the ship may possibly be sub-optimal in such conditions.

Ships may also be classified into tramps or liners depending on the type of shipping service they are engaged in. Ships can also be classified into displacement ships, planing ships, air cushion vehicles, catamarans, trimarans, etc. based on their configurations. Permissible size in entering a particular harbor or waterways, such as canals, is also used as a factor for classifying ships; examples include Panamax, Suezmax and Malaccamax. Naturally, loading capacity, speed and propulsion system can be used as factors for classifying ships as well. In some cases two or more different criteria are applied concurrently for classifying ships.

For the sake of brevity, here we shall focus on grouping in accordance with the purpose of services. A ship is a typical product built to order and consequently each ship is often designed according to its own design concepts which best meet the ship-owner's requirements, although sometimes a series of ships are built to a similar design for economic or operational reasons, in which case they are known as sister ships The purpose of use can be manifest in many ways. However, ships are mostly used for transport of cargo or passengers, military activities, scientific observation or exploration, industrial activities, support activities, sports activities and so on.

Ships used for cargo or passenger transportation are collectively referred to as merchant ships under the Ships Act. Ships used for sports activities are subject to the Water Leisure Safety Law. Naval ships are used for military operations and various research vessels are employed in exploration or scientific studies of ocean. Fishing vessels are typical examples of ships built for industrial activities at sea. There are also various support vessels the main missions of which are to back up the operation of other vessels. It is noteworthy that, in recent years, offshore oil and gas production has become a major marine industry, and offshore structures have gradually become the center of some shipyards' interests as a consequence.

3.1 Merchant Ships

Ships engaged in transporting passengers or cargos are generally known as merchant ships. They are classified into general cargo ships, container ships, bulk carriers, oil tankers, ore and oil combined carriers, liquefied gas carriers, chemical tankers, car carriers and so on, according to the cargos they carry. Merchant ships often specialize in carrying certain types of cargo and, therefore, they must be designed to suit the specific characteristics of the cargo. The major cargos in world trade are summarized in **Table 1-1**.

Table 1-1 Quantity of major sea-borne goods transported in 2008 (unit: Billion ton-miles)

Passengers	Dry cargos			Liquid cargos		Special cargos
	Unitized/Break bulk cargos	Bulk cargos				
Round trip Specified routes Planned routes Passengers with vehicles	Containers Cars Heavy cargos Refrigerated cargos Miscellaneous	Iron ore	4,617	Crude oil	10,117	Livestock Timber products Pressurized gas Cement
		Coal	3,703	Petroleum products	2,860	
		Grains	1,551	Chemicals		
		Bauxite /Alumina	272	LPG/LNG		
		Other ores	9,858	Liquid food products		

3.2 Leisure and Sports Ships

Unlike merchant ships, small craft for leisure sport activities are often mass produced to meet the public demand for affordable craft. In some cases they are custom designed and constructed at higher costs. In extreme cases astronomical amount of money would be spent to produce high-performance vessels for participating in prestigious events, such as the America's Cup yacht race. The world market of leisure craft is huge, and international boat shows are held every year in major cities all around the world. Many yacht races of varying degree of prestige are held regularly in which many yachtsmen and women take part. The crafts for leisure and sports are controlled by the 'Water Leisure Safety Law' in Korea. The law controls not only the crafts but also simple appliances such as inflatable boats and surfboards by placing them in the same category as ships.

3.3 Naval Ships

Naval ships are significantly different from merchant ships. Ordinarily naval ships are designed for operational performance rather than economic efficiency, and a series of ships are mass produced after carefully assessing the performance of a prototype. Naval ships are usually equipped with weapons systems and carry a large number of personnel to operate the systems, whereas merchant ships carry cargo and but a small crew is sufficient for their operation naval ships tend to have heavy structures, and early retirement of naval ships is often due to the on-board weapons systems becoming obsolete rather than the limitations of the physical life of the ships. The make-up of a naval fleet is determined by striking a strategic balance between defensive and offensive policy of the nation.

The main missions of offensive surface ships are in the warfare against hostile surface ships, submarines and aircraft. In recent years naval battles have been shifting to littoral areas and, as a consequence, missions of large naval ships have been expanding to land battles. Major sea powers have aircraft carriers as their main war ships and this has allowed them to expand their operational range considerably. Mine layers/sweepers and escort vessels are deployed for the protection of areas or fleets and there are vessels which are used for landing operations.

Submarines are employed for their stealthy nature to perform reconnaissance, surveillance, intelligence and offensive missions. They are also used for securing sea passages for friendly fleets and protecting battle fleets as well as for blockading enemy bases or cutting off passages of hostile forces. Recent development in weapons systems has enabled submarines to undertake tactical missions and precision strikes. They are also used in support of special operations forces.

Table 1-2 Classification of naval ships

Surface ships		Submarines (designation of US Navy)
Offensive	**Defensive**	
Aircraft carriers	Frigates	Nuclear submarines for ballistic missiles (SSBN)
Battle ships	Corvettes	Nuclear submarines for guided missiles (SSGN)
Cruisers	High speed craft	Nuclear general purpose attack submarines (SSN)
Destroyers	Mine warfare ships	Diesel-electric submarines (SS) with or without AIP
Amphibious war ships	Auxiliary ships	Submersibles

3.4 Fishing Vessels

The Fishing Vessels Law defines ships engaged in fishing, transport of catches and processing of marine products collectively as fishing vessels. Ships used for experiments, investigation, guidance, control or training in fishing industry are also included in this category. The Fisheries Act applies to fishing in oceans, coastal seas and inland artificial seawater ponds, but the inland fisheries are separately controlled by the Inland Water Fisheries Act. Ships engaged in catching or collecting animals or plants outside the areas of East Sea, Yellow Sea, East China Sea and the Pacific Ocean areas west of 140° longitude and north of 25° latitude are treated separately and are controlled by the Pelagic Fisheries Development Act of Korea.

Fishing vessels are classified in many ways according to the fishing implements, methods, target fish species and so on. With the development of aqua culture technologies, vessels for management and support of farming sea vegetables, fish, shellfish, etc. have been developed and also there are some vessels specializing in processing and transporting of marine products.

Table 1-3 Types of fishing vessels

Classification by fishing implement and method	Trawlers, purse seiners, long liners, gillnetters, drift gillnetters, stow netters, creel boats, net setters, diving support vessels
Fish farm support vessels	Management vessels, feeding vessels, seaweed gatherers, shellfish gatherers
Processing vessels	Salting carriers, anchovy processing ships, reefer vessels, factory ships
Carriers	Catch carriers, bait carriers, fisheries supply vessels
Support ships	Ships for fisheries experiment, inspection, guidance, patrol and training

3.5 Other Special Ships

3.5.1 Ships for Scientific Research and Exploration

Ships for scientific research and exploration are used for a variety of tasks, but are designed and constructed with optimum performance in their main missions in mind. Representative types of ships in this category are ocean research vessels, meteorological observation vessels, resource exploration vessels, arctic research vessels, hydro-graphical survey vessels, fisheries resource research vessels and deep sea exploration vessels.

3.5.2 Offshore Structures

The rising demand for hydrocarbon has taken men to offshore in its search. Since its tentative first steps in 1950s offshore industry has matured in the harsh but relatively shallow waters of the North Sea. It has now blossomed into one of the major industries of the world and spread to all corners of the globe. These expanding offshore activities in more and more hostile environment have given rise to an increasing demand for offshore structures of high technical complexity. Many of these structures are produced by shipyards. Even though clear definitions do not appear in the Ships Act, offshore structures furnished with self-propulsive capability should naturally be categorized as ships, since a ship is interpreted to be a vessel which can navigate on or under the water surface. In addition, if a floating structure, which cannot move on its own but is towed or pushed by other ships, is considered to be a ship, offshore structures with self-propulsion capability can also be categorized as ships by the same token. In fact, the Sea Pollution Prevention Law explicitly classifies the floating or stationary drillships and platforms into the categories of ships.

The types of offshore structures are summarized in **Table 1-4**.

Table 1-4 Classification of offshore structures

Fixed offshore structures	Jacketed, gravity-based and tension-leg platforms, guyed towers
Floating offshore structures	Drillships, semisubmersibles, FPSOs, spars
Support/work vessels	Pipe-laying vessels, deep sea oil exploration ships, deep sea submersibles, diving and engineering support vessels, anchor handling boats, ocean-going tugs, supply vessels

4. Origins of Korean Shipbuilding Technologies

4.1 Ships in Ancient Times

The first recorded shipbuilding in Korea is mentioned in '*Dangun Segi* (The History of Dangun Dynasty)' as having taken place in 2131 B.C. The book also records that the Dynasty conquered Japan twice in the years 723 B.C. and 667 B.C., which implies that Koreans had ships sophisticated enough to transport soldiers and war supplies across the strait between Korea and Japan. Bronze age rock engravings on Bangudae in Ulju-gun, Gyeongsangnam-do province in the southeastern part of Korean Peninsula show three boats one of which appears to be carrying 18 people who have succeeded in capturing a large whale (**Figure 1-1**). It can be deduced from this that Koreans built large well-found ships even in those ancient times.

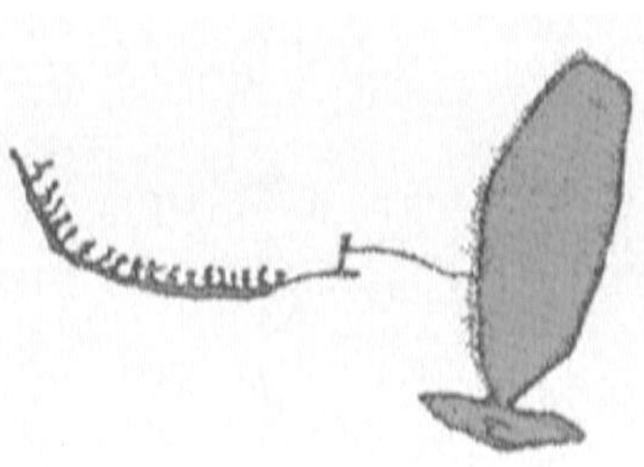

Figure 1-1 Whale boat in petroglyphs of Bangudae

4.2 Ships in the Three Kingdoms and Unified Silla Period (57 B.C. ~ 936 A.D.)

At the time of Three Kingdoms (Goguryeo to the north, Baekje in the south-west and Silla in the south–east of the Korean Peninsula), the main industry was agriculture and fishing was secondary. Nevertheless fishing was brisk around Youngheung Bay, and was carried out from moored vessels utilizing the tidal differences on the coast. Ships including carrier boats, merchant boats, lighters, ferry boats and so on began to be built as fishing flourished.

A ship-shaped ceremonial earthenware vessel used in Silla and Gaya is shown in **Figure 1-2**. It is evidence that shipbuilding technology had already developed to a significant level by as early as the 5th or 6th century A.D. People of Gaya, a confederation of small nations located between Silla and Baekje, were known to have made frequent voyages to Japan well before the Three Kingdoms period. From the fact that similar clay figures of wooden ships of the same period were also excavated in some tumuli in Japan, it can be conjectured that such boats must have been used to cross the strait from the mouth of the Nakdong River to the northern Kyushu via Tsushima Island.

Figure 1-2 Ship-shaped clay figure of Gaya

After unification of the Three Kingdoms, the Silla dynasty reinforced the management and operation of naval vessels to enhance sea transportation and coastal defenses. Jang Bogo, an admiral of Silla, was ordered to establish a naval base of Cheonghaejin in Wando Island at the southern coast and to eradicate the pirates. He was also given the task of promoting trades with Tang and Japan by building large ships and securing the sea routes.

4.3 Ships of Goryeo Dynasty (918 A.D. ~ 1392 A.D.)

The historical fact that the first king of the Goryeo dynasty, Wang Geon, led his fleet to assault the Later Baekje's naval base near Naju demonstrates that Goryeo possessed excellent navigation and shipbuilding skills. Eventually the later Baekje fell at his hand and the later Three Kingdoms period drew to a close. In the *Goryeosa sega*, a book of the history of Goryeo dynasty, the naval vessels of King Wang Geon are described as follows: "a dozen big ships among the fleet could carry 30 seamen, were 90 to 96 feet long, sufficient even to ride a horse on them, and had a deck house with shields all around."

A shipwreck was discovered offshore of Wando Island in 1985 and the vessel was dated to around 1000 A.D. ~ 1100 A.D., soon after the reunification of the Peninsula and about 170~270 years later than the merchant ships of Admiral Jang Bogo. **Figure 1-3** shows a photograph of the restored ship through which the shipbuilding skills of the unified Silla and Goryeo can be deduced.

Figure 1-3 Restored ship of Goryeo dynasty (Wando ship)

The Goryeo dynasty established a government-operated shipping system to transport grains, collected as a tax at that time, to the royal capital. Twelve Jochangs (granaries) were established all over the country and Jounseons (grain carriers) operated to and from 60 ports along the coast and major rivers. The carrier ships were also called Chomaseon and known to have been capable of carrying 1,000 'seom' of collected grains each (a 'seom' is a bale of grain equivalent to 180 liters). It is believed that their appearance and structure resembled the Wando ship.

Another historical record states that, upon the request of Yüan(元), the Mongol dynasty in China, to join their military expedition to Japan, Goryeo established a Bureau of Shipbuilding to improve and expand its capability of naval vessel construction. Some of the naval vessels thus built were exported to Yüan and were used by the allied naval force of Goryeo and Yüan. Although the conquest failed twice due to typhoons (understandably, Japanese started calling them Kamikaze or 'God's wind'), the veritable fact is that the ships of Goryeo were built strong and had good seakeeping qualities, since they could survive the typhoons without sustaining any severe damage. Without doubt the shipbuilding skills of Goryeo period were highly developed.

4.4 Ships in Joseon Dynasty (1392 A.D. ~ 1910 A.D.)

Japanese pirates were very active all around the coast throughout the first half of Joseon period, and much effort was expended to eradicate this plague. Joseon dynasty inherited the strong maritime traditions and shipbuilding technology from the Goryeo dynasty, but it was essential to improve the war ships to wage war against the pirates.

At the time of King Sejong the Great, the Bureau of Naval Ships investigated and analyzed the characteristics of foreign ships in order to develop new hull forms by adopting their strong points. As a result of this endeavor some multipurpose vessels were developed. In war times, a covered deck was added to these ships and they were used for sea battles; in

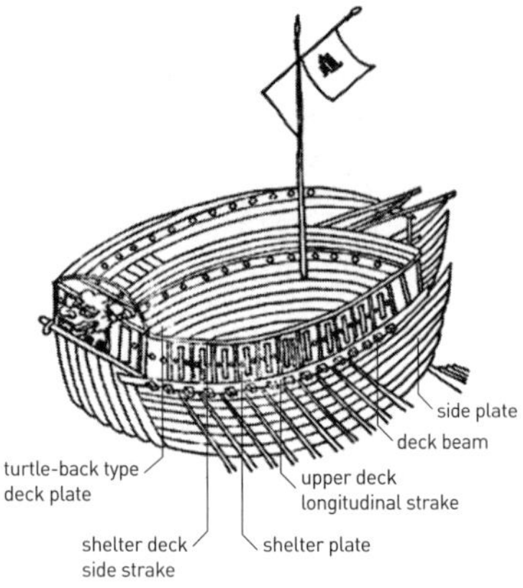

Figure 1-4 Tongjeyoung Geobukseon, the turtle ship

this mode they were called Maengseon. In peace times, they were used as tax-grain carriers under the name of Jounseon, and in this mode they operated with the deck removed.

After the riots of Japanese residents at Sampo (three ports near present-day Busan, Jinhae and Ulsan) in early 16th century, Japanese improved their ships and made them longer, higher and stronger. In response, Joseon had to develop Panokseon as a counter measure. These battle ships were large enough to accommodate 160-strong crew each and had a superstructure supported by pillars erected all over the upper deck to protect the rowers and crews from enemy attack. A deck house was erected on top of the superstructure as a command post. In the process of developing Panokseon from Pyungjeonseon, a battle ship with a flat deck, the builders encountered problems related to balance, stability and strength of the ship, arising from adding superstructure and moving cannons to higher locations. All of these problems presented the shipbuilders a good opportunity for enhancing their understanding and experience and they must have learned a great deal in meeting these challenges.

In 1591 Admiral Yi Sun-shin was appointed the commander-in-chief of the naval force defending the southwest sea. Soon after taking up the office he set about designing a special battle ship, eventually named Geobukseon, the turtle ship, to prepare for the invasion of Japanese pirates. The invention of the turtle ship was an epoch-making event after the development of Panokseon in the annals of Korean shipbuilding. One can say that, if Panokseon was evolved from Pyungjeonseon by providing a protective superstructure, the turtle ship was developed by covering the weather deck of Panokseon with an armored shelter deck. The turtle ship was constructed with sufficient structural strength to endure the installation and firing of 14 big naval guns. The specification and appearance of the turtle ships vary, and the one belonging to Tongjeyoung, the Naval command headquarters, is shown in **Figure 1-4**.

Among the collection of the Royal Library, Gyujanggak, which is now maintained by the Seoul National University, there is an old document called *Junbyeonggakseondo*. This document details the outward appearance of fighting ships, drawn in 45-degree isometric views, and dimensional and other detailed information necessary for their construction. Any ship's carpenter could draw the plans and shell expansion drawing of a ship from the information provided in the document based on the traditional scale and building methods. It is also possible to build replica of these old battle ships using the drawings, when any one of the principal dimensions, length, breadth or depth, is known.

Ships at the time were generally built of pines, which were abundant and provided good strong timber. Therefore, shipyards at that time were usually built on sandy beaches near a pine forest. A lumber mill, blacksmith shops and warehouses for storage of the equipment and material were built in the vicinity. Obviously, the shipbuilding carpenters must have lodged near the yards during the building of any ship. Tidal differences along the complex west and south coastlines of the Korean Peninsula are very large, and therefore shipbuilders must have been able to predict the exact date for the completion of a ship and the maximum height of the tide occurring at that time to place the building berth right at the water's edge. Upon the completion, the ship was launched bow-first at high tide by removing transverse holding beams and pushing the ship until it slid down longitudinal wooden supports, which were 'greased' with soybean paste. After launching, the mast was erected and the equipment installed before the ship's stability was tested. Permanent rock ballasts placed in the hold were used to improve stability and to adjust the longitudinal trim. It can be seen from this that Koreans in Joseon period had great skills and knowledge of ship design and building.

5. Outline History of the World Shipbuilding Industry

5.1 Evolution of Shipbuilding Industry

The factors that have influenced the development of shipbuilding industry include demand for marine transport, navigation technology, shipbuilding materials, propulsion systems, shipbuilding technology and shipbuilding systems. The continuous evolution of these elements has contributed to the growth of shipbuilding industry all over the world.

In ancient Egypt, the need for transport on the Nile gave rise to the advent of reed and wooden river boats propelled by sails or oars. Later, in the time of ancient Greece, galleys dominated the Mediterranean Sea and continued to be used by the Romans.

Medieval ships included Viking ships in Scandinavia and cogs or cog-built ships of the Hanseatic League in northern Germany, while galleys continued to evolve in the Mediterranean region. In the Far East at similar time, Admiral Jang Bogo's Silla merchant ships from Cheonghaejin traded with China and Japan. In the 15th century three- or four-masted carracks were developed by the Portuguese and widely used by European maritime powers, especially for the discovery of Americas and for the exploration of Asia. Between the 17th and 19th centuries, sailing ships increased in size, leading to the development of clippers which had three or more masts and a square rig. Until the early 1900's, all ocean-going ships were sailing vessels built of wood and the United States, with the plentiful supply of high-quality timber from abundant forest resources, led the world shipbuilding industry.

However, mechanization in the wake of the Industrial Revolution and the growth of the steel industry, together with the evolution of shipbuilding materials and engines, allowed Europe to take the lead in developing shipbuilding industry. Britain in particular spearheaded the emergence of steam ships built of steel and became the world leader in shipbuilding, and remained so until the mid-1950's. Korea, on the other hand, tried to modernize the Navy in the final days of Joseon, which had declared itself an Empire, by buying battle ships Yangmu and Gwangje but all that effort was for nothing with the Japanese annexation of Korea. Later on, under Imperial Japanese rule, some of the shipyards in Korea were turned into Japan's armament factories. This gave a few Korean engineers an opportunity to come into contact with Japanese steel shipbuilding technology for the first time.

Due to advancement in welding technology and demand for transporting huge quantities of military supplies during World War II, the United States of America produced a large number of simple bulk carriers and tankers, and in the process became a pioneer in technological innovation and led the world in moving from riveted to welded ships. After the war America's welding technology was introduced to Japan and Japan soon took over the leading position in ship production from the late 1950's, with the development of novel shipbuilding technology named the 'block construction method' applied to the construction of large ships. With the increase in ship size, Japanese shipbuilding industry invested heavily in large building docks and continued to enhance the welding and block construction technology. Due largely to these efforts, Japan dominated the world shipbuilding market until the early 2000's.

During the Korean War, the Korea Shipbuilding and Engineering Corporation (KSEC) played a central role in repairing naval vessels and learned modern shipbuilding methods, thereby laying the foundation for its entry into shipbuilding market and subsequent world domination in shipbuilding. At the turn of the century Korean shipyards began to introduce their own computerized systems into ship design and production processes. These systems have enabled faster and more accurate designs to meet the owners' ever more complex requirements and enhanced precision in hull fabrication. Furthermore, the development of large three-dimensional block assembly method has shortened the construction period and helped Korea to lead the global shipbuilding industry.

5.2 A Brief History of Korean Shipbuilding Companies

5.2.1 Hanjin Heavy Industries Co., Ltd.

The history of KSEC, the predecessor of Hanjin Heavy Industries, represents that of the modern shipbuilding industry in Korea. The company started out as Joseon Heavy Industries Co., Ltd. in July 1937, the first modern shipyard in Korea, and prepared for its business by constructing two 3,000-ton building berths and one 6,000-ton building dock. Once liberated from Japanese occupation, employees at Joseon Heavy Industries took over management of the company for a little while until the US military government handed over its management to Sinhan Corporation, a state-owned company, in December 1945.

After the establishment of the Korean government in 1948, KSEC was founded as a partly state-owned corporation in January 1950, in accordance with the KSEC Act enacted for promotion of the Korean shipbuilding industry. But the company had a difficult time due to lack of orders, high dependency on imported equipment and low productivity. The company was wholly nationalized as a result, and, in accordance with the first national five-year economic development plan, its facilities were expanded with a loan from Japan in December 1965 and an additional domestic funding. By the time the expansion was completed, KSEC had a capacity to build up to 12,000-ton class ships with an annual building capacity of 66,000 tons and repair up to 20,000-ton class ships. Under government management, KSEC properly began to build steel ships, and The Sinyang, a 1,600-ton cargo ship constructed in 1964, was the first from the yard to receive an international certification (from the American Bureau of Shipping). In 1968, the company received an order to build twenty 250-ton fishing vessels from the Taiwanese government and became the first company in Korea to export ships.

In the late 1960's, the company ran into financial difficulties again and was eventually sold to Far Eastern Shipping Co., Ltd. It was reorganized as Korea Shipbuilding and Engineering Corporation Co., Ltd. in November 1968. After the privatization the company gradually returned to normalcy through management overhaul and rationalization in the 1970's, and expanded the second dock to increase its capacity to 60,000 tons. It set up new facilities including a hull assembly shop and ship assembly yard, and extended the second pier. Later, the third dock with a capacity of 150,000 tons was constructed and a variety of mechanical equipment was added.

In 1983 the company received an order from a Norwegian company to build six multipurpose cargo ships (product-oil-bulk-ore carriers – PROBOs). These ships were of a new concept, and first of its kind in the world that could carry various types of cargo ranging from refined oil products to ore and other dry bulk cargo. They were equipped with an advanced satellite navigation system, a computer-based vessel monitoring and managing system, and a cargo containment system. In 1985, however, the global shipping market experienced a serious slump, and the owner requested a postponement of the delivery of the ships. The ships were eventually delivered, the price having been reduced by 40 million US dollar. However, due to accumulated debt and prolonged recession of the shipping market, KSEC experienced severe difficulties and was taken over by the Hanjin Group soon after.

As a result of the Hanjin Group's effort to return the yard to normal management, the company was able to build ships named among the world's best for a number of consecutive years since 1992. These included a membrane type LNG carrier, the world's fastest container ship at the time, deep-sea cable laying ships equipped with state-of-the-art automatic systems, and a 6,000 TEU container ship. In 2006, the company began to build a shipyard on a site of 2.3 million square meters in Subic, the Philippines, and delivered a 4,300-TEU container ship. Since then, the company has consistently tried to extend its business to offshore industry, based on its global production system in collaboration with the Subic shipyard, while focusing on very large container ships, LNG carriers and tankers.

5.2.2 Hyundai Heavy Industries Co., Ltd.

In March 1970 Hyundai Engineering & Construction Co., Ltd. prepared a site in Jeonha Bay, Ulsan, to construct a shipyard capable of building 150K class ships, but the plan was revised later to set up a new company and increase the building capacity funded by a commercial loan from abroad. The shipyard planned to build a dock and workshop buildings of over 124,000 m^2 floor area on a site of about 580,000 m^2 with a capacity of producing five 500K DWT class VLCCs annually.

The shipyard held a groundbreaking ceremony in March 1972 and in April signed a contract with Rivanos of Greece to construct two 260K DWT class crude oil tankers; the company was going to build the ships while the shipyard construction proceeded. After commencing the construction project in 1972, the company made changes to the plan again: the building capacity was increased to 700K DWT, the site was expanded to 2,000,000 m^2 and buildings to 141,300 m^2.

In 1973, upon completion of the hull assembly shop and installment of a Goliath crane, the company was reborn in December that year as Hyundai Shipbuilding and Heavy Industries Co., Ltd. In June 1974 the company held a naming ceremony for the first and second vessels it built, concurrently with a completion ceremony for the shipyard. With growing demand for VLCCs, the company increased the capacity to 1 million DWT ships and expanded the length of the third dock to 560 meters as well as creating a 450-meter outfitting quay. By the time the first expansion was completed in May 1975, annual building capacity of the shipyard reached 3,760,000 DWT with the steel processing capacity of 530,000 tons. The company constructed the 4th and 5th docks in 1977, the 6th dock in 1978, and the 7th dock in 1979.

The 1973 oil crisis, however, caused the global shipping and shipbuilding boom to cool off, and by late 1974, ship-owners were prone to cancel orders, pushing Hyundai Heavy Industries (HHI) into a financial difficulty as well. Beginning in mid-1974, when orders for VLCCs had ceased worldwide, the company diversified its products from large ships to medium-sized ships such as multipurpose cargo ships, bulk carriers and timber carriers, to cope with the crisis and expanded its business by creating a steel structure division and Hyundai Mipo Dockyard within the shipyard.

In 1983, the company opened a research institute for welding technology and, in 1984, established a maritime research institute, and extended its areas of business to ship construction, ocean development, plants, engines, robots and heavy machinery. In September 1991 the company completed a hull assembly shop for LNG carriers and, in June 1994, became the first Korean yard to construct an LNG carrier. In 1996 the company completed the 8th and 9th docks, specifically for VLCCs, and created a VLCC hull assembly shop to relieve the overloaded production line. The company designated each dock to build its own type of ships only, so as to improve quality of the ships and productivity.

In October 2004 HHI successfully built and launched a ship without a dock through the innovative 'ground building method.' In early 2006 the company's cumulative production surpassed 100 million DWT, a record comparable to the world's best shipyards in the UK and Japan with over 100 years' history.

In the 2000s, HHI adopted the strategy of exclusively building high value-added ships, such as very large container ships, passenger ships, LNG carriers and offshore plants. In early 2005 the company received an order from Cosco of China, to construct 10,000-TEU class container ships and in 2007 signed a contract to build eight 13,000-TEU container ships which were the world's largest at the time. In September 2007 HHI got an order from Global Santa Fe of the United States for a drillship, thus launching itself into the drillship market.

Furthermore, the company constructed an ultra-large LNG carrier of 216,000 m^3 capacity propelled by diesel engines in November 2007, and thus became capable of building LNG carriers propelled by any of steam turbines, electric motors or diesel engines. In January 2009 the company completed a dock specializing in FPSOs and successfully reduced construction periods and costs. In March 2010 HHI completed a 1.3 million-ton class dock equipped with a 1,650-ton Goliath crane on a 1,782,000 m^2 site within the Gunjang Industrial Complex, Jeollabuk-do province, and began building large ships.

5.2.3 Daewoo Shipbuilding & Marine Engineering Co., Ltd.

Korea Shipbuilding and Engineering Corporation confirmed a plan to build a shipyard in Okpo in May 1973, and in October of the same year, commenced the shipyard project focusing on a one million DWT class dock, a 150K DWT class berth, and a 500K DWT class repair dock. However, the global financial recession and the slow growth of the shipbuilding market brought the Okpo shipyard project to a halt, and eventually the Daewoo Group acquired the shipyard and formed Daewoo Shipbuilding Industries Co., Ltd. in September 1978.

The shipyard received an order from a Norwegian company, Skibs A/S Stolt, to build four 22,500-ton chemical carriers in September 1979, and delivered the Bow Pioneer, the first vessel that the yard ever constructed, in May 1982. This was a valuable experience to the shipyard of building technically demanding chemical carriers. The yard completed its first one million-ton class dock in April 1981, installed a 900-ton class Goliath crane in September, and held a shipyard completion ceremony in October of that year. In 1985 the company was improved its competitive position by increasing productivity and cutting costs. The company made a profit for the first time in 1991, and, by May that year, the cumulative ship delivery exceeded 10 million tons.

In addition to VLCCs which are their main strength, the company builds diverse ship types including medium-sized crude carriers, product carriers, bulk carriers, car carriers and LNG carriers. The shipbuilding facilities of the company include a one million-ton dock, 900-ton Goliath crane, a 350,000-ton dock and other large equipment like 450-ton jib cranes, with an annual merchant shipbuilding capacity of 2.2 million GT.

In 1983 the company began a special ships business by building and delivering a barge to a ship-owner from the United Arab Emirates, which extended to special purpose ships, such as high-speed patrol boats, destroyers, submarines and deep-sea exploration vessels. The company constructed and delivered the first Korean submarine named 'Yicheon' to the Korean Navy in April 1993, and later delivered four more vessels of the same type to the Navy. In February 1996, the company collaborated with the Far East Marine Research Institute in Russia to develop 'Okpo 6000,' a deep-sea submersible allowing ocean exploration in waters up to 6,000 meters deep.

During the 1997 IMF crisis Daewoo Shipbuilding was named among the 'workout' companies, but its superior technological and cost competitiveness enabled it to return to normal operations in a short period of time. In November 2000 the company became independent of Daewoo Heavy Industries. It was renamed Daewoo Shipbuilding & Marine Engineering Co.(DSME), Ltd. in 2002, and reorganized into a business specializing in shipbuilding and marine engineering.

Recently, the company has built 450K ULCCs, LNG carriers, LPG carriers, FPSOs and FSRUs, and many of them have been noted for their excellent quality. In 2005, in particular, the company became the first shipbuilder in the world to deliver the LNG-RU, an LNG carrier equipped with regasification unit that does not require a shore terminal facility. In 2011, the company constructed and delivered an FPSO, 325 meters long, 61 meters broad, and 32 meters high having lightship displacement of 120,000 tons.

Currently the company owns two building docks, including one 530m long, 131m wide and 14.5m deep, four floating docks and two slipways, and builds about 70 large vessels each year. Starting in 2009, its medium- and long-term goal is to expand its business into the green and renewable energy sector, and become an all-round heavy industry company.

5.2.4 Samsung Heavy Industries Co., Ltd.

It was in May 1973 that the Samsung Group decided to start a shipbuilding business and chose a temporary site for its shipyard at Anjeong-ri, Gwangdo-myeon, Tongyeong-gun Gyeongsangnam-do province. The group formed a joint venture with IHI of Japan, named 'Samsung Heavy Industries Co., Ltd.' in August 1974. The oil crisis, however, caused a delay in the commencement of the shipyard project, until the company acquired Woojin Shipbuilding Co., Ltd., in April 1977 and resumed the project. Woojin Shipbuilding was formed as a successor to Goryeo Shipbuilding Co., Ltd., with the goal of building four 100,000-ton class ships each year in one 100,000-ton class dry dock within the industrial complex located in Jangpyeong-ri, Sinhyeon-myeon, Geoje-gun, Gyeongsangnam-do province.

Samsung changed its name to Samsung Shipbuilding Co., Ltd. in April 1977, and established an independent, new construction plan. Based on the short- and long-term prospects for the global shipbuilding industry, the company decided to revise its initial plan while it was in the middle of the first phase, and a master plan was developed for building a large shipyard during the second phase. The No.1 dock completed in October 1978 was 240 meters long, 46 meters wide and 11 meters deep with a maximum building capacity of 65,000 tons. The first phase was completed by the end of 1979 and the company came to have a building capacity of 150,000 tons.

In March 1980 the company decided to implement the second-phase expansion of the Geoje shipyard centering on the construction of the second dock, together with the expansion of an outfitting shop and the hull assembly shop and the installation of other facilities. By the time the construction work was completed in February 1983, the company established itself as a shipyard with 3,300 employees, a maximum capability of 250,000 DWT class vessels and an annual capacity of 450,000 GT, and began to optimize its production line. In January

1983, Samsung Shipbuilding merged with Daesung Heavy Industries to form Samsung Heavy Industries Co., Ltd.

Between 1992 and 1993, Samsung Heavy Industries completed the development of high value-added ships such as LNG carriers and high-speed ships, while building the third dock in expectation of the booming global shipbuilding market. Upon completion of the third dock, 640 meters long, 97.5 meters wide and 12.7 meters deep, in late October 1994, the company finally became capable of building VLCCs. In December 1996 the company delivered a 138,000 m^3 class LNG carrier and completed a ship block plant in Ningbo, China. The company entered cruiser business by delivering a large passenger ship of 28,000 tons in August 2000. In November 2001 it became the first company in the industry to use a floating dock for ship construction and, in January 2002, completed a 73,000-ton class FPSO only 44 days after starting the erection in the dock. Since then, the company has used the floating dock for building new ships.

In December 2004, the company developed a 12,000-TEU container ship and led the world in building large ships. In July 2005 it introduced GSCAD, an automatic ship design system, to become capable of three-dimensional ship design. Throughout 2005, the company undertook projects worth 7.2 billion dollars in shipbuilding and marine engineering, including the construction of an Arctic tanker in November, and increased the proportion of high value-added vessels to 80%. In January 2010, the company held a ceremony proclaiming 'green management' under the motto of eco-friendly ships, reducing greenhouse gas emissions by 30%, since when the company has developed various green products and technologies. By 2010, the company had evolved into a shipbuilder equipped with an annual capacity of 3.8 million tons (CGT).

5.2.5 Hyundai Mipo Dockyard Co., Ltd.

The Hyundai Mipo shipyard Co., Ltd. evolved from the ship repair division of Hyundai Shipbuilding & Heavy Industries, and became a separate company in April 1975 specializing in ship repairs. Since then the company continued to grow, repairing 48 vessels in 1975, 148 vessels in 1976 and 196 vessels in 1977. Back then, the Hyundai Mipo shipyard had a limited repair capacity of 800K DWT and could not compete against overseas repair yards, let alone meeting the domestic and foreign demands. Therefore, in November 1982, the company expanded its second yard and equipped itself with seven docks and 2.15 million DWT repair capacity.

The company successfully completed a modification project of an underwater pipe-laying ship, which was technically very exacting and demanding, and cemented its position as the world's largest repair yard by executing a variety of repair and remodeling projects from all over the world. The company then expanded its facilities, including the expansion of the fourth dock and the extension of the first quay. The company endeavored to differentiate itself from emerging repair yards and diversify its business by completing a hatch cover factory in December 1994.

With some new ship repair companies arriving on the scene, however, the company decided on a diversification strategy and ventured into the new ship building business. In December 1997 the company delivered its first ship, the Ramform Banff, a Norwegian crude oil production and storage vessel. Based on this success, the company proceeded to sign new ship building contracts with some erstwhile repair clients. In this way Hyundai Mipo Dockyard successfully transformed itself from a simple repair yard into one of the world's best builders of medium-sized vessels.

Hyundai Mipo Dockyard grew by concentrating on medium-sized ships, instead of large ships which are the main business for major domestic and international shipbuilders, and thus avoiding competition with them. It specializes in building medium container ships and petrochemical product carriers.

The petrochemical product carriers built by Hyundai Mipo Dockyard are now called 'Mipo Tankers' by international ship-owners and are renowned for its outstanding quality and technology. Medium-sized product carriers of 35K, 37K and 46K DWT have been developed and the company builds more than half of the world's new ships in this range. Furthermore, in addition to product carriers and container ships which are its key ship types, the company has entered the markets for LPG carriers, PCTCs, ROROs and open-hatch bulk carriers, in order to respond rapidly to the global demand. The company has also secured additional production facilities at Daebul, Jangsaengpo and Onsan in preparation for increased production requirement due to expected increase in orders.

5.2.6 Hyundai Samho Heavy Industries Co., Ltd.

Hyundai International predicted a boom in the large ship market and created Samho Shipyard in Yeongam, Jeollanam-do province, on a 2,900,000 m^2 site in 1992. Within four years of starting the shipyard project, the company completed a dock for new ships and another for new building and repairs, and installed three 600-ton Goliath cranes. In February 1996, Samho Shipyard successfully constructed the first vessel, a 2,500-TEU container ship and went on to build about 30 large ships.

In December 1997, however, it filed for bankruptcy in the aftermath of the foreign exchange crisis, and Hyundai Heavy Industries took over its management in 1999 and changed the name to Samho Heavy Industries. With its experience in shipyard operations and international sales network, the company's temporary management group made successful efforts to get additional orders and quickly restored their ship prices to normal levels.

In 2001, for the first time, Samho Heavy Industries made a net profit worth 800 million US dollars and in July 2002, officially became an affiliate of the Hyundai Heavy Industries Group. The company changed its name to Hyundai Samho Heavy Industries to enhance its brand value in January 2003. By December of the same year, it achieved a cumulative shipbuilding output of 10 million DWT, noteworthy for its fast growth.

In June 2005 the company signed a contract with BP of Britain to build a 155,000-CBM (m^3) LNG carrier of the membrane type. In November of the same year, a contract to build a 321K DWT FPSO was signed with the French company, Total. As a result, Hyundai Samho Heavy Industries gained experience in building VLCCs, bulk carriers, very large container ships, LNG carriers, LPG carriers, car carriers and FPSOs, as well as marine equipment, thereby emerging as a highly competitive shipbuilder in the global market.

In May 2008 and in September 2009, the company completed a ground construction site capable of building 165K DWT class ships each year and a 543-meter dolphin quay while expanding the docks and establishing a new block plant. As a result, Hyundai Samho Heavy Industries became a shipbuilder equipped with two docks, one ground construction site, and 6 Goliath cranes with capacity ranging from 600 to 1,200 tons on a 3,300,000 m^2 site.

5.2.7 STX Offshore & Shipbuilding Co., Ltd.

The company started out as Daehan Shipbuilding Iron Works in Yeongdo, Busan in January 1962, and in April 1967, changed its name to Dongyang Shipbuilding Industrial Co., Ltd. Later, it merged with Anjeon Shipbuilding and Gukje Warehousing, and started anew as Daedong Shipbuilding Co., Ltd.

In August 1973 the company built a 2,000 GT class cellular container ship and acquired a deserved recognition for its technical ability. In December of the same year, the company installed two 7,000-ton slipways and a 30-ton travelling crane. The company focused on building timber carriers in the early 1980's, following which they built tuna long liners, container ships and chemical carriers throughout the rest of the decade.

The company started building a shipyard in Jinhae Bay in 1993 aiming for the market for medium-sized ships. By the time the shipyard was completed in June 1996, the company possessed a dock (320m×74m×11m), two 300-ton Goliath cranes, a hull shop, a pre-outfitting shop and a painting shop on a 726,000 m^2 site in Wonpo-dong, Jinhae-si. Daedong Shipbuilding specialized in the construction of 46K class chemical carriers, container ships and Panamax class oil tankers, but due to a financial crisis, it was taken over by STX Co., Ltd in September 2001.

The company renamed itself STX Shipbuilding Co., Ltd in January 2002, and increased its global market share for medium-sized oil tankers as well as 23,000-CBM LPG tankers. Starting from 2004, STX Shipbuilding achieved an average annual growth of about 40% in both number of ships and tonnage constructed by utilizing the experience in ship construction to develop its proprietary construction methods. As a result, the company won the Three Billion Dollar Export Tower Award in 2009.

In 2004, the company developed a ground construction method called Skid Launching System (SLS). While building several ships in one dock through the semi-tandem process, STX combined innovative methods such as the ROSE (Rendezvous On the Sea for Erection) method by which blocks are erected using module transporters, and the PA/PF method by which very large blocks of the bow and stern of a ship are constructed before being erected, which increased the company's productivity and helped it cement its position as a best-in-class shipyard.

In October 2007 the company acquired Aker Yards of Norway, the world's largest builder of cruise ships operating 18 shipyards in eight countries around the world. Through this acquisition, STX looked to create synergy by entering the high value-added cruise ship market and securing a variety of shipbuilding technologies. Another growth engine that STX Shipbuilding had at its disposal was its plant in Dalian, China. As the first overseas shipyard built by STX, the yard began its operation in 2008 and has since become a key plant driving the growth of the company towards 'Global STX.'

STX's shipbuilding facilities specialize in the construction of bulk carriers, car carriers, oil tankers, container ships and VLCCs, and its offshore plant facility is currently engaged in the construction of drillships, underwater pipe-laying ships and floating storage units (FSUs). The company's engine production base is manufacturing propellers, crank shafts, and main engines and the components, and supplying main engines to major shipyards throughout the world.

Furthermore, the company works hard to gain a competitive advantage in the 'green' business, a new growth engine of the STX Group, by undertaking projects in renewable energy areas such as solar and wind energy.

6. Marine Equipment Industry

6.1 Outfitting as the Key Factor in Ship Performance

6.1.1 Overview of Marine Equipment

For a ship to work properly, all necessary component equipment appropriate for their functions must be installed. Such equipment can be broadly divided into engines, generators, power transmission devices, navigation and steering systems, mooring systems, cargo handling systems for loading and unloading of cargo, lifesaving appliances, fire protection systems and living quarters for passengers and crew. These systems constitute the numerous ship outfitting items and are supplied by the marine equipment industry. The marine equipment is characterized by high-mix, small quantity batch products, related to many other industries ranging from the basic material industries, such as steel and chemical products, to the mechanical, electrical and electronics industries. Furthermore, on-board equipment has to meet the exacting requirements of each ship, and therefore must be ordered from specialist marine equipment manufacturers.

Table 1-5 Categorization of marine equipment by their use and function

Category	Subcategory	Sub-subcategory
Hull parts	Metal products Welding materials Cast steel Chemicals	Mild steel plates, high tensile steel plates, zinc plates, beams, etc. Electric welding rods, oxygen, nitrogen, acetylene, etc. Rudder stocks, rudder pintles, stern tubes, etc. Paint, plastic resin, rubber, glue, etc.
Engine parts	Propulsion machinery Auxiliary machinery	Diesel engines, steam turbines, boilers, propellers, shafts, etc. Generators, air-compressors, fresh water generators, pumps, chillers, fans, etc.
Outfitting	Steering systems Navigation systems Mooring systems Cargo handling systems Fisheries systems Safety systems Living systems Piping systems	Steering gears, rudders, autopilot systems, etc. Radar systems, direction finders, gyrocompass, etc. Anchors, windlasses, capstans, fair leads, Panama chocks, etc. Cranes, winches, derricks, hoists, etc. Fish finders, luring lights, wire drums, trawl winches, etc. Lifeboats, life vests, inert gas systems, etc. Sanitation systems, air-conditioning systems, kitchen supplies, water-tight sliding doors, etc. Valves, flanges, elbows, pipes, etc.

Electrical & electronic parts	Power units Wiring equipment Lighting equipment Communication equipment Control systems Instruments	Motors, batteries, transformers, electric heaters, etc. Main distribution frames, wiring devices, marine cables, etc. Lights, searchlights, etc. Wireless transmitters/receivers, frequency changers, telephones, etc. Control panels Pressure gauges, speedometers, etc.

6.1.2 Required Characteristics of Marine Equipment

1) Anti-vibration property

A ship's entire voyage, from departure to arrival at a port, involves continuous operation of its engine and propeller. The vibratory forces induced by the workings of the engine or propeller can not only cause damage to the fixture of the engine foundation but are also causes of noise and vibration propagated throughout the vessel. Vibration induced by the propeller may also cause excessive wear in the stern tube bearing.

Pumps and hydraulic systems produce internal fluid flows and they also produce noise and vibration. Expansion and contraction caused by temperature variations can damage the ship's piping and wiring. Marine equipment, therefore, must have good anti-vibratory property and care must be taken to avoid vibration resonance.

2) Adaptability to ship motions

Ships are liable to move in all six degrees of freedom due to environmental disturbances such as wind and waves. Slamming due to encountered waves and consequent pitching motions can cause excessive impact at the bow. As a consequence, the equipment located at the bow and stern may be damaged by excessive acceleration and, in some cases they themselves can place excessive loads onto the ship structure. Equipment placed higher up in the ship, such as mast-top radars, must be designed to withstand the acceleration due to roll motion.

3) Anti-corrosion property

Many of ship's equipment and parts are exposed to sea water and damp air, and consequently are vulnerable to corrosion. Rotating shafts, bearings and precision devices, such as electrical and electronic units, that are installed on the ship's exposed weather deck are liable to be affected by salt. Therefore, these parts must be protected from coming into contact with seawater. A better solution is to develop equipment that are unaffected by green water conditions.

4) Weatherproofing

The equipment installed on a weather deck cannot be protected completely. It is exposed to strong sunlight and UV radiation. During summer heat waves, in particular, the temperature of the deck surface can soar up to 70℃. At night, on the other hand, high humidity can cause dew condensation on inner and outer surfaces of the equipment, which also requires attention. The shielding material for deck machineries must provide sufficient weather protection, taking into consideration the effect of condensation.

5) Maintainability and reliability

When a piece of ship's equipment breaks down mid-voyage, the crew is usually responsible for the repairs. If essential equipment for the operation of the ship cannot be repaired by the crew at sea, the ship should at least be able to reach the nearest port. Therefore, it is highly desirable to design equipment simply so that even a non-expert member of the crew can undertake emergency repairs using devices available on board.

Nevertheless, there are many technology-intensive systems which use special materials and technically complex components, including automatic navigation systems, radars and engine monitoring systems, malfunction of which will be difficult for the crew to repair. These systems therefore have to have high efficiency and reliability.

6) Compliance with international quality standards

Selection of marine equipment is heavily dependent on the ship-owner's preference but it must meet the international quality control standards, ship classification criteria and regulations set forth by international conventions. It must also comply with all applicable domestic laws such as the Law for Safety of Vessels.

7) Required time of supply

In general, the time that a shipyard requires a piece of equipment varies depending on the progress of the ship construction. In particular, steel and some other raw material are required at early stages of construction. In a typical ship construction schedule, steel cutting will take place more than thirteen months before the delivery of the ship. Assembly work occurs seven and a half months and launching about five months before the delivery. That is why thick steel plates for hull construction, pipes and equipment for pre-outfitting are required from eight to twelve months before the ship's delivery.

Since some machines such as engines are required at the time of erection, they have to be delivered about six to seven months before the delivery. Demand for equipment used for general outfitting and living quarters occurs six months prior to the ship delivery. Since many of equipment are manufactured to order, items that take longer to manufacture must be ordered with sufficient lead time.

6.2 Korean Marine Equipment Industry

6.2.1 Contribution to the National Economy

The marine equipment cost constitutes roughly 60 to 70% of the total ship price. Machinery and instruments, excluding materials such as hull plates, cast steel, paint and welding materials, account for between 25 and 35%. Of all machinery and instruments, auxiliary machinery in the engine room occupies 15 to 20%, auxiliary instruments 10 to 20%, and the main engine 60 to 75%. Therefore, the price of the main engine takes a significant portion of the total ship price.

Table 1-6 Approximate portion of various equipment in total price of various ship types (unit: %)

Items	Steel	En-gines	Equip-ment	Piping mate-rial	Mate-rials for cabins	Elec-tric outfit-ting	Steel outfit-ting	Others	Total
VLCC	20.0	9.5	16.4	3.8	0.9	0.7	4.8	6.1	62.3
Bulk carriers	26.1	9.0	13.8	2.3	1.0	0.7	3.6	7.0	63.5
Container ships	15.0	12.0	19.5	1.9	1.0	2.3	3.6	8.5	63.7
LNG carriers	10.3	6.2	24.3	5.4	0.6	1.4	2.3	11.5	62.1

The tonnage of ships built in Korea represents about 35% of the total world output, and the percentage of domestic equipment installed on ships constructed in Korea is very high: in the case of VLCCs, bulk carriers and container ships the proportion of domestic equipment is estimated to be approximately 90%. In addition, since the percentage of domestic equipment mounted on LNG carriers is as high as 80%, it can be said that the scale of Korean marine equipment industry is very large. Sales of domestic marine equipment have more than tripled to 10 billion US dollars recently, with an annual average growth rate of over 20%, due to the boom of the world's shipbuilding market after the year 2000 and the increased number of ships constructed by Korean shipyards. The total value of marine equipment exported grew to 1,240 million dollars in 2008, with engines and components accounting for the majority with 990 million dollars.

More than 90% of Korean marine equipment manufacturers are operating in the

Gyeongsangnam-do region including Busan and Ulsan where most shipyards are located. The shipbuilding and marine equipment industries have formed a cluster, creating a synergy which helps maintain the competitiveness of both industries and becoming key industries driving the economy of the southeastern region of Korea.

6.2.2 Major Marine Equipment Sectors

Both ordinary and high-tensile steel plates, thicker than 6 mm and certified by the classification societies are the major material for the construction of hull and machinery. There is a constant pressure to make ships lighter to improve its economic efficiency, and this trend calls for higher quality steel with higher tensile strength, more ductility at low temperature and good high temperature weldability.

Since 1975, POSCO has made a tremendous contribution to the growth of the Korean shipbuilding industry by supplying high-quality steel plates. They provide about four million tons of shipbuilding plates, while Dongkuk Steel provides 1.8 million tons, and Hyundai Steel 1.9 million tons a year. Meanwhile, large shipyards in Korea have worked with Japanese manufacturers to diversify steel supply sources, and recently they have been using some Chinese suppliers carefully selected for the quality of their products.

The power machinery can be classified into two categories: propulsion power sources, such as diesel engines and turbines; and auxiliary power sources such as boilers, generators, pumps, air compressors, etc. Automatic engine control systems are now commonplace, and with them the control system have become highly complex and are often connected to high-temperature and high-pressure piping systems. The engine, which supplies power necessary for ship's operation, is essential machinery, regardless of the ship type. The market share of low- and medium-speed engines produced in Korea is estimated to be about 60% and 20~25% of the global demand, respectively.

Major engines with the brand names such as MAN of Germany and Wartsila NSD of Switzerland are manufactured in Korea on license, and Hyundai Heavy Industries Co., Ltd. developed its own "HiMSEN" engine range since 2001 and is supplying them worldwide. Large diesel engines for ships are produced and supplied in Korea by Hyundai Heavy Industries Co., Ltd. Doosan Engine Co., Ltd. and STX (Engines and Heavy Industries).

6.3 Korean Marine Equipment – 'World-Class Products'

Since 2001 the Ministry of Knowledge and Economy has been implementing a program of government support for technological development and global marketing for 'World-Class

Products' made by Korean companies if their worldwide market share is within top five. Among the products most frequently selected until 2012 are engines and their components, and auxiliary machinery in the engine room, followed by safety systems, cabin facilities and electrical products in the outfitting category as shown in **Table 1-7**.

Table 1-7 List of marine equipment selected as 'World-Class Products'

Year of Item selection	Item Description	Year of certification	Company name
2002	Heavy-duty diesel engines (at least 6,000 horsepower)	2002	Doosan Engine Co., Ltd.
		2002	Hyundai Heavy Industries Co., Ltd.
		2006	STX Heavy Industries Co., Ltd.
	High-pressure carbon dioxide(CO_2) fire suppression systems	2002	NK Co., Ltd.
	Marine cables for ships	2002	Kukdong Electric Wire Co., Ltd.
		2002	LS Cable & System Co., Ltd.
		2002	JS Cable Co., Ltd.
		2007	TMC Co., Ltd.
	Marine boilers	2002	Kangrim Heavy Industries Co., Ltd.
2003	Steering gears	2003	Yoowon Industries Ltd.
	Wall panels for offshore living quarters	2003	BN BIP Co., Ltd.
		2003	STACO Co., Ltd.
	Crankshafts for heavy-duty engines	2003	Hyundai Heavy Industries Co., Ltd.
		2003	Doosan Heavy Industries & Construction Co., Ltd.
		2009	STX Metal Co., Ltd.
2004	Synchronous generators for ships	2004	Hyundai Heavy Industries Co., Ltd.
	Marine diesel engines (4-stroke)	2004	STX Engine Co., Ltd.
		2004	Hyundai Heavy Industries Co., Ltd.
	Fixed pitch propellers for ships	2004	Hyundai Heavy Industries Co., Ltd.
2005	Turbochargers for marine medium-duty diesel engines (not more than 5,000 kw)	2005	STX Metal Co., Ltd.
2006	Crankshafts for marine medium-duty diesel engines (4-stroke)	2006	STX Metal Co., Ltd.
		2006	Hyundai Heavy Industries Co., Ltd.
	Main bearing supports for marine medium- and heavy-duty diesel engines	2006	Daechang Metal Co., Ltd.
		2011	Samyoung M-Tek Co., Ltd.
	Exhaust valves for heavy-duty marine diesel engines	2006	Keumyong Machinery Co., Ltd.
	Economizers	2006	Kangrim Heavy Industries Co., Ltd.

2007	Marine switchboards	2007	Hyundai Heavy Industries Co., Ltd.
	Cargo oil pumps for crude oil tankers	2007	Hyundai Heavy Industries Co., Ltd.
	Guide shoes for heavy-duty marine diesel engines	2007	Shina Metaltech Co., Ltd.
	Cylinder liners for heavy-duty marine diesel engines	2007	Hyundai Heavy Industries Co., Ltd.
		2007	Cape Industries Ltd.
		2007	STX Metal Co., Ltd.
	Air coolers for marine engines	2007	Donghwa Entec Co., Ltd.
2008	Cylinder frames for heavy-duty engines	2008	Hyundai Heavy Industries Co., Ltd.
	Turbochargers for heavy-duty engine lathes	2008	Hyundai Heavy Industries Co., Ltd.
		2010	STX Metal Co., Ltd.
	Inert gas generators	2008	Kangrim Heavy Industries Co., Ltd.
	Electronic fuel injection systems for marine engines	2008	Whayoung Co., Ltd.
	Motors for high-pressure marine thrusters	2008	Hyundai Heavy Industries Co., Ltd.
	Generators for mid-speed diesel engines	2008	Hyundai Heavy Industries Co., Ltd.
2009	Marine propeller shafts	2009	Hyundai Heavy Industries Co., Ltd.
	Side thrusters	2009	Hyundai Heavy Industries Co., Ltd.
	Common rail units	2009	Wonil Co., Ltd.
	Pistons for mid-speed marine diesel engines	2009	Samyoung Machinery Co., Ltd.
	Water mist fire-extinguishing systems for engine rooms	2009	Tanktech Co., Ltd.
	Water desalination plants	2009	DongHwa Entec Co., Ltd.
	Ceiling panels for offshore living quarters	2009	BN BIP Co., Ltd.
		2010	STACO Co., Ltd.
	Bathroom units for offshore living quarters	2009	BN BIP Co., Ltd.
	Ship engine monitoring and control systems	2009	Hyundai Heavy Industries Co., Ltd.
2010	145kV gas-insulated switchgears	2010	Hyundai Heavy Industries Co., Ltd.
	Power supply systems for marine refrigerated containers	2010	Hyundai Heavy Industries Co., Ltd.
	Dual fuel diesel engines	2010	Wärtsilä-Hyundai Engine Company Ltd
	Lifeboats	2010	Hyundai Lifeboats Co., Ltd
	Cylinder heads for medium-speed marine diesel engines	2010	Samyoung Machinery Co., Ltd.
	Prefabricated bathrooms	2010	STACO Co., Ltd.
	Marine fire doors	2010	Cosmo Co., Ltd.
	Cabin units for offshore living quarters	2010	BN BIP Co., Ltd.
	Automatic liquid-solid separators	2010	Yoowon Industries Ltd.

2011	Marine cylinder covers	2011	Hansco Co., Ltd.
2012	Marine ballast water treatment systems	2012	PANASIA Co., Ltd.
	Marine emergency power systems	2012	G.P.C. Co., Ltd.
	Crankshafts for medium-duty marine engines	2012	Sunwoo CS Co., Ltd.
	Cylinder covesr for marine engines	2012	Daechang Metal Co., Ltd.

• Source: Vol. 28 Korea Marine Equipment, 2013.02., Korea Marine Equipment Association

References

1. Chang, S. & Van, S.H. ed., 2002. *Understanding ships* (in Korean). Daejeon: Korea Research Institute of Ship and Ocean.

2. Inha University, 2006. *Ship outfitting* (in Korean). Seoul: The Ministry of Education and Human Resources.

3. Kim, H. et al., 2006. *Advanced Ships of Korea* (in Korean). Seoul: Jisungsa.

4. Kim, Z.G., 1980. *The History of ships* (in Korean). Seoul: Jeongwoosa.

5. Korea Marine Equipment Association, 2010. *The History of the Korean marine equipment industry and the 30-year history of Korea Marine Equipment Association* (in Korean).Seoul: Korea Marine Equipment Association.

6. The Society of Naval Architects of Korea, 1974. *Introduction to shipbuilding engineering* (in Korean). Seoul: Dongmyungsa.

7. The Society of Naval Architects of Korea, 2000. *Ship outfitting* (in Korean). Seoul: Dongmyungsa.

CHAPTER 2

Sale and Purchase of Ships

1. Introduction

1.1 Scope of Shipbuilding Sales

The shipbuilding sales activities try to secure (or even generate at times) orders for new-buildings (NBs) for the shipyard so that it remains profitable. Prosperity or ruin of a shipyard depends on how well its sales operations are conducted across the board. The sales operation manages all activities from receiving new-building orders, assuring that the yard is ready to build the vessel meeting all operational requirements as intended by the ship-owner, and delivering the ship to the owner in the best possible state.

The owner invests a huge amount of money in buying the ship with the aim of making a good profit, and therefore demands a vessel capable of giving the highest performance leading to the best profitability and competitiveness in the competitive shipping market. It follows that the shipyard must be able to design and build the most efficient, high-performance vessels that it can, based on its own technical expertise accumulated over the years.

Before a new-building contract is signed the yard and the owner negotiate the commercial terms and conditions continuously and repeatedly over a considerable span of time over such matters as the vessel's delivery date, price, payment terms and the legal terms and conditions. If the owner wants specifications (i.e., quality, composition and building methods) higher than the normal standards of the shipyard, the owner is liable for the extra costs arising from them. If the owner desires to take delivery of the vessel earlier than as contracted, he/she must meet the costs for the necessary extra or overtime work. Any change in the payment terms is subject to the mutual agreement in good faith between the yard and the owner in respect of the ensuing financial costs.

Once agreement is reached on the vessel's technical specifications, commercial terms and conditions and the legal interpretations through negotiations, the two parties sign the shipbuilding contract. Signing a contract is like breathing life into the shipyard. It triggers activities throughout the whole organization and facilities of the shipyard, encompassing management, design and production. It is the ship sales department that conducts all contract negotiations from the moment of receiving initial inquiries up to the contract signing.

Once the contract is signed, though, all the signed documents are handed over to the contract management department. When the owner pays the first installment to the yard in accordance with the terms of the contract, the yard is then legally liable to commence construction of the vessel. It is the job of the contract management department to follow up on all production process of the vessel, and to maintain communication with the owner's resident representative regarding all production issues on a daily basis. In addition, all milestone events such as work commencement, keel-laying, launching and naming ceremonies are organized by this department. Finally, it is also the task of the contract management department to deliver the vessel to the owner upon confirmation that all contractual obligations have been discharged by the owner as set out in the contract.

After delivery, all post-delivery matters are transferred to and dealt with by the after-delivery service department. It is this department that verifies whether or not the vessel performs as set forth in the contract, and administers all claims submitted by the owner under the relevant clauses of the contract.

In short, the shipbuilding sales operation embraces all these activities relating to ship sales, contract administration and after-sales services.

1.2 Prospects of the Market

Ship's economic life is one of the major factors determining the demand and the supply of ships. The nominal lifespan used to be taken as 20 years, but it was changed to 25 years by the Common Structural Rules (CSR), which came into force in 2006. Henceforth, the maximum durability of the ships against hull corrosion, hull fatigue and the onboard machinery's wear and tear has been set as 25 years. If nothing extraordinary happens, a ship is assumed to be scrapped after 25 years of service. Based on this assumption, the annual demand for replacement is estimated to be around 4% of total world merchant fleet size. This means that, if the total fleet size is assumed to be 700 million GT, about 28 million GT needs to be replaced every year. However, during a period of shipping boom ships well past their standard economic life are often kept in service, while in a depressed market even some young ships are scrapped. Shipyards can forecast the demands for new ships more precisely from the scrapped tonnage.

Figure 2-1 shows that the world seaborne trade grew by about 42% in 10 years from 5.7 billion tons in 1998 to 8.1 billion tons in 2008, which represents an average annual growth at a compound rate of 3.6%. This trade growth can be considered to create a shipping demand about 25.2million tons (3.6% of 700 million GT) directly. If the average annual replacement rate of 4% is added to this figure, the annual new building demand in the short term can be estimated to be 53.2 million GT. **Figure 2-2** shows that the 31.7 million GT of new ships were constructed in 2000. Applying 7.6% of compound annual growth rate for 7 years, the construction of new ships should have been 52.9 million GT by the end of 2006. The actual record of that year was 52.1 million GT. This method forecasts the new building tonnage in 2009 to be 66.8 million GT, but it started increasing steeply in 2007 to reach 77.1 million GT in 2009. It can be surmised from this that the world fleet tonnage increased by far in excess of what the actual demand called for during these three years.

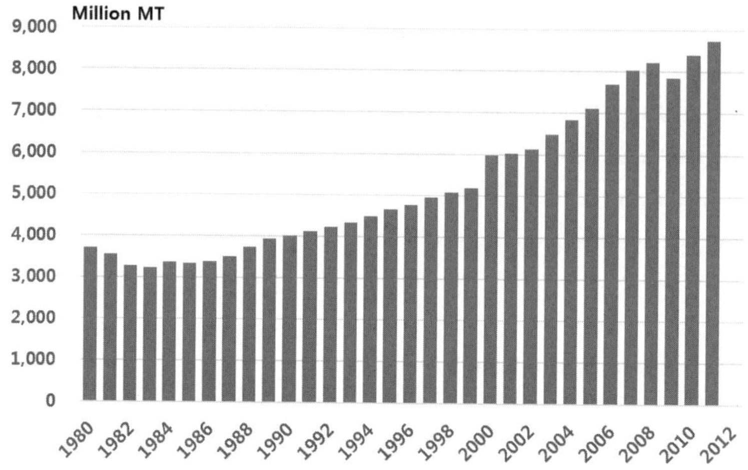

Figure 2-1 Development of international seaborne trade
(Source: Review of Maritime Transport, UNCTAD, 1997-2012)

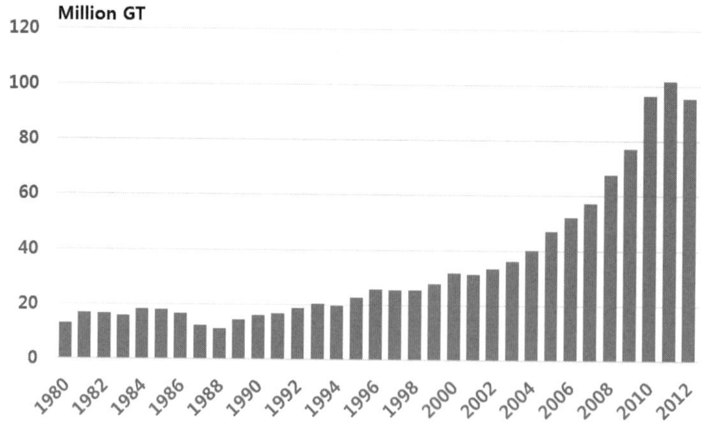

Figure 2-2 Deliveries of new buildings
(Source : Review of Maritime Transport, UNCTAD, 1997-2012)

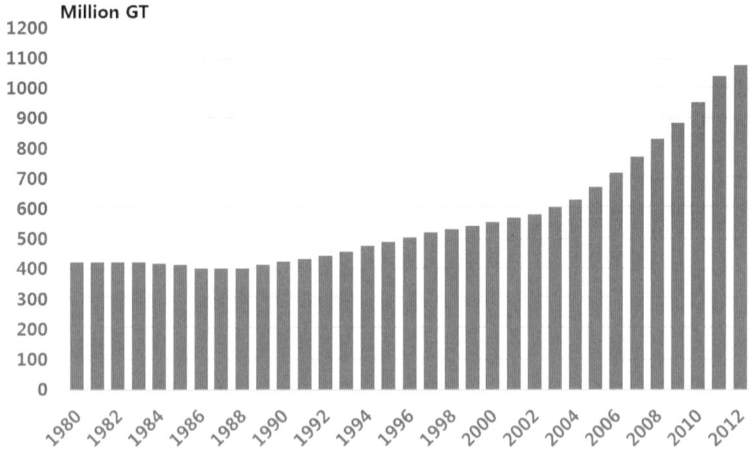

Figure 2-3 World fleet size
(Source : Review of Maritime Transport, UNCTAD, 1997-2012)

Figure 2-3 shows the changes in world fleet size. It can be seen that the tonnage fell slightly between 1982 and 1986 during which time shipbuilding and shipping industries suffered the worst recession in history, but increased steadily ever since. It grew from about 404 million GT in 1987 to 586 million GT in 2002, representing an average increase of 12.1 million GT per year which is about 2.4% of the average world tonnage during that period of 486 million tons. This is lower than the increase of seaborne trade volume of 4.9% as mentioned previously, probably because the shipping industry suffered so much during the most extreme recession between 1982 and 1986.

However, the market situation between 2004 and 2011 was completely different. The total world fleet expanded explosively by on average 62.6 million GT per year from about 605 million to 1,043 million GT, representing about 7.7% of the average tonnage during that period of 815 million GT. This is the fastest growth in shipping history, reflecting the increase in the trade of iron ore, coal and crude oil bound for India and China with steadily growing economy and the consequent expected growth of shipping demand. The ships in service at this time included old tonnage well past their retirement age. Ship-owners analyze the record of world tonnage carefully in this way and forecast its future trend to plan their fleet replacement and manage the size of the fleet.

2. Preparation for Ordering

New-building order(s) are placed by the ship-owners after studying various factors: global economic climate, world shipping market outlook, fleet replacement plan according to operational economy, comprehensive fleet operational plan taking into account new routes and new cargoes, the number of ships to be newly ordered and their delivery time, budget availability and so on.

2.1 Replacing Ships on Operational Policy

Ship-owners replace ships not only when they are old, but also when there are changes in classification rules or national and international regulations. The plan to update the fleet has to be carefully made to ensure that they can respond rapidly to changing circumstances. Moreover, there are other pressures on the owners to update their fleet and these include technological advances resulting in improved fuel economy, more stringent environmental regulations on exhaust gas emissions, and increasing number of automated and eco-friendly systems becoming available.

Shipping companies in developed countries usually try to maintain high operational efficiency by keeping the average age of the fleet low. The ships aged 5 to 7 years are often sold to companies in developing countries, who can deploy these relatively cheap ships in shipping sectors which can render reasonable operational efficiency for around 5 to 10 years. Finally, ships aged over 10 years are acquired and operated by shipping companies of less-developed or emerging countries. Thus, shipping firms from advanced countries tend to be the leaders of maximizing the ship's efficiency and developing the state-of-the-art technology in ship operation. It is the world's top ten shipbuilders who lead the continuous technological development in close cooperation with these 'advanced' shipping companies.

2.2 Changes in Cargoes and Trade Routes

Growth in global economy leads to higher demands for energy resources and raw materials like iron ore, resulting in longer haulage distance as well as higher trade volume. China and India used to produce iron ore, coal and oil, sufficient to meet the domestic demand and even export some surplus until fairly recently when the demand for these commodities were low. Currently, however, both countries import most of the raw materials they require, reversing the trade direction. Furthermore, China and India used to import industrial and consumer goods, but now they export them in huge quantities, thereby drastically increasing demands for container ships. In the meantime, the center of gravity of shipbuilding industry, once dominated by some European countries and USA, has shifted to a few countries in the Far East. Some countries promote shipbuilding and shipping industry under the policy of 'transporting our products on our ships built by ourselves.' China, for example, is actively engaged in building and operating LNG carriers, large container ships, offshore drillships and other offshore structures as a national policy.

3. Ship-owner's Decision to Place Orders

Ordering a new ship offers the owner a great opportunity for future profit-making, but it comes with a high financial risk. A new ocean-going merchant ship costs anything from tens of millions up to billions of US dollars. If successful, such a large-scale investment can bring about huge wealth, but if things go wrong even a little bit, it can bring down the company.

Given this, the ship-owner 'leaves no stones unturned' in carefully and cautiously examining all conceivable factors and scenarios from all angles before placing an order. This is a painstaking process and the owners must exercise utmost prudence in identifying all potential risks prior to ordering.

3.1 Ship-owner's New-building Program

The type and number of ships that the owner should deploy on the intended trading routes for the best competitive advantage must be decided before a plan for financing the project is worked out. It is imperative that the newly built ship is more competitive than the best existing ships already operating on the targeted routes, and this is why owners nearly always pursue ships with the best performance characteristics even for higher prices. At the same time, the owner will want to have his/her ships built by a 'quality' shipyard capable of designing and constructing better-performing ships.

To achieve this, owners ceaselessly collect and analyze information on the shipping market movements and the prices of new and used ships. Based on such information, the owners plan building of new ships best suited to respond to the market need for the most competitive prices. Once the requirements for the ships are settled, the owners need to negotiate and agree on the detailed technical specifications with the shipyard. To make demands more reasonable and realistic, the owners may obtain technical advice from appropriate ship brokers or classification societies.

3.1.1 Determining Type and Size of Ships

The owner's new-building program starts with determining the ship's size. The size of the ship is governed mainly by the type and characteristics of the cargoes the ship is to carry. For ships to carry cargoes of large quantities consistently available at the loading port, such as crude oil and iron ore, shippers' requirements are the most important factor in determining the size. If the trade volume is predetermined, the cargo capacity and speed of the ship is determined to suit the conditions of the loading and unloading ports and the trade routes, so that the ship can be operated at high economic efficiency over a long period. If the loading and discharging ports are already determined, the specific conditions of the two ports as well as the sailing routes are the critical factors in determining the ship's dimensions and operational speed.

The size of ships engaged in trading between the Atlantic and the Pacific oceans is largely determined by whether the ship is to pass through the Panama Canal or not. If passage through the Panama Canal is possible, a merchant ship can save about 20 days against routing through Cape Horn. However, the maximum size of bulk carriers suited for crossing the Panama Canal, known as Panamax, is about 70,000 DWT. There is not much advantage in using the Panama Canal for carrying iron ore or coal from Brazil to China, and consequently ships engaged in this trade are not constrained by Panamax. They are built to a Cape Size, or about 170,000 DWT, but recently a few 400,000 DWT-class VLOCs (very large ore carriers) have been built in Korean and Chinese shipyards.

The same can be said of the Suez Canal. The passage through the Suez Canal connecting the Mediterranean Sea with the Indian Ocean via Red Sea can save about 15 days against rounding the Cape of Good Hope, but it imposes a size constraint to Suezmax of about 150,000 DWT. Should the Suez Canal be closed down, as has happened many times in the past primarily due to conflicts in the area, or should there be a need for carrying more cargoes, it is advantageous to make the ship very large to benefit as much as possible from the economy of scale. This is how VLCCs (very large crude carriers) or ULCCs (ultra large crude carriers) in the range between 250,000 and 500,000 DWT were developed. To match the sailing route or the port conditions specifically, some maximum size vessels have been developed, including 110,000 DWT Aframax oil tankers and 80,000 DWT Kamsarmax bulk carriers (the maximum berthing size at the West African port of Kamsar).

Taking all this into account, the owner decides the number, size and speed of the ships to order after a careful economic evaluation. In the case of ships to carry cargoes that occur at regular intervals in predictable quantities, such as industrial or agro-fishery products, it is relatively easy for the owners to determine the new building plan. In the case of container ships to be employed in an advanced container transport system, their size and speed will be determined depending on in what part of such a system they are to be deployed.

3.1.2 Financing Plan

Once the owner's viable new-build plan is established, the first thing that needs to be resolved prior to contacting shipyards is securing sufficient finance. As mentioned above, constructing a new ship involves an investment of huge scale. Consequently, even a cash-rich owner may well resort to external finance on top of his/her own equity. Once the potential financing plan is determined, the owner makes a cash flow forecast based on estimated gains from the sale or demolition of its retiring tonnage and future revenues arising from the new ships' operations. When the shipyard makes an attractive finance deal, the owners often find it irresistible. It is mainly thanks to the most favorable finance packages that the Japan is able to continue to be one of the major shipbuilding nations of the world despite its lower competitiveness and the pressure from rapidly expanding Chinese shipbuilding industry.

3.1.3 Collecting Information

1) Information sources

A growing number of media dealing with shipping matters and their widening scope of interests have helped the ship-owners collect information more easily. The world's leading shipping and shipbuilding media include 'Llyod's List,' 'Trade Winds,' 'Fair Play' and 'Sea Trade.' Also available are specialized research reports produced by various maritime research institutes as well as information analysts specializing in gas carriers, container ships and offshore structures. Ship broking companies also publish their own market research reports, allowing the owners to have easy access to information on current market condition. It is up to the owners to interpret the bewildering variety of data and information available to them, and good judgement in this regard calls for experience, knowledge and decisiveness.

2) Trade fairs and shipping exhibitions

Shipping trade fairs or maritime exhibitions are one of the most important information sources for the ship-owners. At these exhibitions shipyards and equipment suppliers introduce their latest products and showcase their new technologies. In the conferences often held simultaneously, papers are presented and information on new technologies is released, and they also provide opportunities for the people in the industry to get together and exchange views. Some ship-owners visit such shipping exhibitions to examine all the components on display with their own eyes before starting discussion with shipyards. Major international trade fairs/shipping exhibitions are as follows:

- Poseidonia: World's largest exhibition on maritime technology, shipbuilding, and maintenance and repair, held biennially in Athens, Greece in even numbered years.
- Norshipping: Shipbuilding and maritime exhibition, held biennially in Oslo, Norway, in odd numbered years.
- OTC (Offshore Technology Conference): A festival for ocean engineering industry, held every year in Houston, Texas, USA. It is the world's largest exhibition on drilling, exploration and production of offshore hydrocarbon and environmental protection, and it provides a wonderful forum for discussion.
- SMM (Shipbuilding, Machinery and Marine Technology): A shipbuilding, maritime and equipment exhibition, held biennially in even numbered years in Hamburg, Germany.
- KOMARINE: One of the world's best shipbuilding and maritime exhibitions, held biennially in odd numbered years in Busan, attended by about 1,200 companies from 45 countries, reflecting Korea's global status as a major shipbuilding country.
- Marintec China: A shipbuilding exhibition, held biennially in odd numbered years in Shanghai, China, attended by about 1,200 companies from 87 countries.

3.1.4 Owner's Choices

1) Choosing a shipyard

At an early stage of the process the prospective owner picks out a few shipyards to any one of which, in the owner's judgement, the new ship construction project can be entrusted. In order to do this effectively the owner needs to keep abreast of such information as the operational status of docks/berths and the available time slots in the trustworthy shipyards, as well as the prevailing new and used ship prices. This information is crucial to the owner, because, once the preferred shipyard is finally chosen, the specific information of that shipyard allows the owner to make rational decisions on when to dispose of his/her retiring tonnage and when to sign the contract for its new-build project.

2) Choosing a ship broker

There are cases where neither the shipyard nor the owner desires the intervention of a ship broker, especially if they have built up a strong working relationship based on a good track record of having built several ships to mutual satisfaction.

Generally speaking, however, the owners tend to rely more on their trusted ship brokers. The broker negotiates vis-à-vis with the shipyard personnel on behalf of the owner, find out any sensitive issues and identify things for the owner to prepare in advance. The ship broker must keep abreast of the movement of shipping market, the shipyard's pricing trends and prospects and availability of building berths in order to assist the owner from the first planning stage. It is plain, therefore, that the ship broker must maintain a good trusting relationship with the shipyard as well as the owner. Furthermore, ship brokers must be conversant with maritime law and capable of arranging finance deals.

3) Choosing a classification society

From the very beginning of new-build planning, the owner decides on a classification society to register the ship with. Selection of the right classification is vital, because it affects the insurance premiums and the build quality and the estimation of residual value of the ship after years of service.

If an owner uses one classification society for many ships in the fleet, the crew may become used to it and tend to dislike alternative societies. The shipyard may already have designs which are approved by a society, but if the owner prefers another society, the yard will have to work with the owner's preferred society. In this case the yard may request the owner to meet any additional expenses arising from the design modifications, additional steelwork and modifications of equipment.

4) Ship's nationality or flag

A merchant ship must be registered with a sovereign state, and deciding the ship's nationality is also very important. The owner may register his/her ship under a flag of convenience, such as Panama or Liberia, in order to save taxes or wages, or to avoid stringent government regulations. Ships flying flags of convenience also benefit from freedom to use crew of any nationality, no requirement to disclose ownership or publish annual accounts. In countries like the British Virgin Islands, Cyprus and the Marshall Islands tax rates are so low that they are often called 'tax havens.' In response to this, an increasing number of the major maritime nations of the world lower their shipping tax rates to encourage their own national owners to register their ships with them.

3.2 Invitation for Bids

When he/she is ready, the ship-owner invites some selected shipyards to enter into negotiations with him/her for new ship building. Occasionally the invitation is issued to a single shipyard (sole bidding), but generally an open competitive bidding process is used.

1) Sole bidding

If the owner has complete confidence in a shipyard or both parties have jointly developed the project from the very beginning, the owner may invite that one shipyard to bid as a sole bidder. Especially if the owner cooperated with one shipyard to carry out basic design, budget planning and ship operational planning from the very beginning of the project, both parties can enter directly into negotiations for a shipbuilding contract without the formality of bidding. Of course, absolute mutual trust is the prerequisite for such a practice. However, some owners prefer to formalize the matters by going for competitive bidding despite their good working relationship. Of course, even if the bidding is open to a few shipyards, the yard which has cooperated will enjoy many advantages including the first refusal right.

2) Open competitive bidding

Open competitive bidding is the most common approach to new build projects. This process starts with the owner releasing the basic requirements of the ship to all potential bidders. The shipyards respond by picking out a likely design from their existing design base, and modify it to meet the owner's concrete requirements. They estimate the cost of construction and the earliest possible delivery date for the ship type based on the workload of the shipyard. When financial arrangement is required, the shipyard studies the feasibility of offering an attractive plan at the same time. In some cases the owners may provide their own drawings and specifications produced by their own technical experts or supplied by external consultants. Such being the case, the shipyard needs to calculate the construction cost only.

4. Shipyard's Bid Preparations

Upon receipt of basic requirements or detailed specifications from the owner, the shipyard responds by intimating its 'basic interest' to the prospective owner along with the best possible delivery on a 'without commitment' basis, while preparing a detailed proposal.

4.1 Cost Calculations

The shipbuilding costs can be subdivided into material cost, labor cost, expenses and general management cost. A typical costing structure is shown on the left in **Figure 2-4**, although it varies for different ship type. The proportion of each segment against the total costing is represented as a pie chart on the right. Calculating the costs accurately is a critical factor in producing the most competitive ship price, which may eventually lead the shipbuilder to win the order.

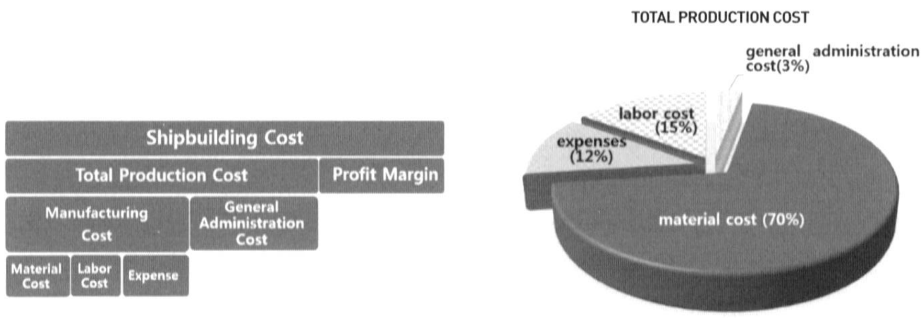

Figure 2-4 Shipbuilding cost structure

In the same context, shipyards need to estimate the quantities of materials required at each stage of construction and the ensuing labor costs accurately. In calculating the total construction cost, shipyards sometimes utilize their own database of similar vessels built by them in the past. They may also resort to their own established standards or criteria for forecasting the total inputs of materials in building a specific ship. On the other hand, general management cost differs from shipyard to shipyard.

Not counting the time needed for negotiations, it takes two to three years to complete a ship from contract-signing to its delivery. Given this, the total production cost must reflect the expected increase in material costs, rises in labor cost, fluctuating exchange rates and oil price changes. They are notoriously difficult to predict with any degree of confidence because of their sensitivity to economic climates home and abroad and international political situation. Shipyards usually bid on fixed basis, but they sometimes insert a 'subject to' escalation clause in its price proposal because of these difficulties.

4.1.1 Material Costs

Material costs cover steel plates, engine room equipment & machinery, on-deck equipment & machinery, electrical equipment, navigation equipment, and accommodation equipment & facilities. Their composition differs for different ship types, as illustrated in **Figure 2-5**. On calculating material costs, the cost estimates need to be updated on the basis of prices quoted each time by the manufacturers, especially when it comes to high-value equipment/machinery, such as main engines and generators. However, if such data are not readily available, their latest price quotes or their last contract prices may sometimes be used.

As for raw materials, steel plates and pipes, it is most important to estimate the precise quantities required. The quantity estimation is based on the actual data from the similar-sized ships of similar types built most recently. Ships are designed these days using specialized ship design computer programs, and they can give an accurate estimation of materials required, thus reducing the overall errors and miscalculations to a great extent. Meanwhile, shipbuilding materials are sourced either domestically or from overseas. In either case their prices are linked to changes in currency exchange rates and it is possible the price may be increased on expiry of their price quotes. In view of all this, maximum care and attention must be given by the yard to all changeable factors in estimating the production cost.

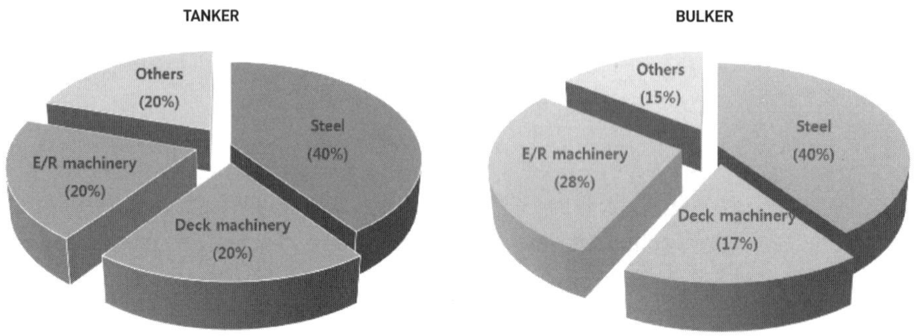

Figure 2-5 Composition of material costs

4.1.2 Labor Cost

Labor cost represents the total cost of human resources employed in constructing the ship, covering steel cutting, welding, assembly, erection and onboard installation of equipment and machinery. In general, labor cost is estimated by multiplying the total man-hours with the unit cost per man-hour. Factors such as annual wage increases and expected savings in man-hours resulting from improved productivity must also be reflected in calculating the labor cost. Some yards treat design cost as a part of direct cost, while others treat it as an indirect cost item or overhead.

4.1.3 General Management Cost

General management cost includes depreciation cost, projected profit, interest on loans and classification fees. Interest on loans has to reflect the prevalent rate in the global financial market.

4.1.4 Exchange Rate

Movement of interest rates and exchange rates is an important factor to consider in calculating the ship's construction cost. Even subtle fluctuations in exchange rates can affect the building cost most seriously. This is because of the considerable lapse of time between the point at which the yard estimates the production cost and when the instalments are paid by the owner. Thus, fluctuations in exchange rates have a critical impact on the yard's overall management environment. If the value of US Dollar goes up after the contract is signed, the yard's profit, when converted and paid into Korean won, will rise. The shipyard can also suffer great losses in the opposite case.

In a bid to minimize the adverse effects of the exchange rate fluctuations, shipbuilding industry has taken to setting up an exchange rate hedging system. The exchange rate, however, does not always move in one upward direction, it sometimes falls heavily all of a sudden. If a shipyard sets up a hedge in preparation for low exchange rate, it has to be prepared to risk any damage arising from the reverse movement. During the world financial crisis in 2008, a number of small Korean shipyards went under due to insurmountable losses inflicted on them by Knock-In, Knock-Out (KIKO) options, a hedge derivative financial product. These barrier options were taken out to guard against the value of Korean won rising unexpectedly, but contrary to expectations the exchange rate rose over the Knock-In Barrier, drastically increasing the exchange losses and consequently driving the small shipyards to bankruptcy. If they were not linked to KIKO to guard themselves from the risk of exchange rate movement, they would have gained high profits instead.

4.2 Setting up Delivery Date

Above all, the shipyard must check whether the owner wants an early or a late delivery. In a boom period the owner will wish to respond quickly to the favorable market conditions by taking an earliest possible delivery of the ship. In hard times, on the other hand, it is to the benefit of the owner to take the latest possible delivery, while trying to adhere to the original contract price. If an early delivery is required, the shipyard may have to consider the possibility of modifying the operational schedule of the docks/berths by switching with another ship or using a gap between the schedules of two ships. The delivery date has important bearings on the management cost, loan interests, man-hours required, exchange rate applicable and rising material prices, and therefore must be considered carefully.

4.3 Payment Methods

Payment method is important for both the owner and the yard, as it has a significant effect on the company's and the owner's cash flow, and therefore must be reflected in the calculation of the ship's price. Cash payment, deferred payment and vessel-lease are the methods of payment most commonly used in shipbuilding contracts.

4.3.1 Cash Payment

In this method paying by five equal installments is the most common. An installment of 20% of the contract price is paid at each of the five occasions: contract-signing, first steel-cutting or 6 months after the contract-signing, keel-laying, launching and delivery.

However, depending on their cash flow situation, the owner may prefer a top-heavy payment scheme through which a greater part of the contract price is paid up front immediately after contract-signing, or tail-heavy payment scheme where a greater proportion of the contract price is paid on the ship's delivery. If the owner's cash flow allows it and the yard is experiencing a cash flow difficulty, the yard may request top-heavy payment in exchange for a discount in the ship price more than sufficient to cover the interest. If the owner wishes tail-heavy payment, the shipyard may include interest for the relevant amount in the ship's price and ask the owner to provide a contract performance guarantee as well.

4.3.2 Deferred Payment

Many countries try to promote shipbuilding industry by providing ship export finance, through which the owners are allowed to make deferred payments. In Korea, it is the KOEXIM Bank which provides this financing scheme. The terms of such financing differ depending on the owner's credit rating. Normally, however, the financing package allows a grace period of two or three years with repayment spread over 6 to 12 years thereafter, with the interest rate being set according to the movement of international financial market. The total annual repayment (capital and interest) is made in two equal installments a year for an agreed period of time. When the commercial interest rate is high, the KOEXIM financing is highly popular amongst international ship-owners. However, when the finance market is buoyant with lower interest rates, this type of financing becomes less attractive.

Nevertheless, it is worth noting that the world shipbuilding was once revived by Japanese financing deals, and the aggressive financial supports by China once distorted the global ship market.

4.3.3 Vessel-lease

It is a tripartite scheme and connects the shipbuilders, ship-owners and an industry requiring continual supply of raw materials which the ship-owner can provide. For example, when a small cash-strapped fishing company wishes to order a shrimp-fishing boat, the boat-builder links the company to a seafood processor who wants a steady supply of shrimps to process. The boat-builder, while maintaining the ownership of the boat, leases the boat to the fishing company who pays back the boat price with the revenue obtained by supplying the seafood processor with shrimps throughout the lease period. When the ship's price and interest has been paid off, the ownership of the boat is then transferred to the fishing company. Such a scheme operated once in the west coast of Korea. A similar ship-lease tripartite scheme can be established between a shipyard, a ship-owner and a steel mill or an oil refinery. A link with financiers can also be introduced in this scheme.

4.4 Establishing Ship Price

Ship price is the most important part of a shipbuilding proposal. The yard arrives at a construction cost by estimating all costs and expenses normally expected to be incurred during the ship's construction. However, the final ship price is often determined by adjusting the total construction cost seemingly arbitrarily but based on the management's experience and the executive policy. Factors such as the latest market trends, the competitors' activities and the relationship with the owner play a significant role in this process. When the market is booming, the yard can add a high profit margin to the total construction cost, whereas in a stagnant market the yard may be forced to offer a considerable reduction from the construction cost and quote a 'survival price.'

4.4.1 Market Trends

A free market is basically governed by the rule of 'balance between supply and demand.' As outlined in Section 1.2 'Prospects of Market' of this chapter, during the period stretching from 1982 to 1986 it was a buyer's market. At that time, due to a very low demand for new ships and a high supply capacity of the shipyards, the market was dominated by the ship-owners and the shipyards had to give in to the pressure from the owners to bring the ship's price down. By contrast, between 2003 and 2007, a supplier's market formed in which many owners rushed to have new ships built, whereas yards were unable to meet the explosive demands. In a supplier's market, the ship's price is set by the yards, since the owners have no alternative but to accept the prices quoted by the yards even if it is higher than normal. In a buyer's market shipyards often suffer heavy losses, whereas, in a supplier's market shipyards can benefit from high profits.

For example, the new build price of 170,000 DWT-class Capesize bulk carriers ranged between 40 and 50 million USD in 2002, but it soared up to 100 million USD in 2007 due to an explosive rise in speculative orders from a number of owners. A newly built ship which was immediately available once fetched 120 million USD, but as soon as it became apparent that there was an over-supply of carrying capacity in 2008, its price plummeted down to 50 million USD.

Well-managed large shipyards are usually not seriously affected by the rapidly fluctuating market in the short term, because they usually have enough ships on their order books. However, small cash-strapped shipyards with not much on the order book will inevitably suffer when the market turns for the worse. In desperation, they are forced to bring the ship's price down just to receive new orders, thereby hastening the collapse of the market. Once cut-price ships become available, the large shipyards have no alternative but to follow suit, thus starting a 'price war' and dragging the whole market into a slump.

4.4.2 Competitors' Activities

What concerns shipyard sales team the most in dealing with a shipbuilding project initiated by an owner is the intention of the competitors and their seriousness in their effort to win the order. When the market shows signs of recovery, yards are more likely to make concessions, assist owners to conceive new projects, and even share workload with the competitors. In a depressed market, on the other hand, the yards are driven to fight tooth and nail to secure orders. A bitter tug-of-war is fought in such conditions by the competitors in a bid to win the price game, exploiting others' weaknesses while making the best use of one's own strengths.

Until the moment of clinching the contract, yards continue to do everything possible to persuade the owners by offering the most favorable delivery or by proposing payment terms difficult for them to refuse. In the past, some government agencies also intervened in the fight of shipyards to win orders as a last resort, directly offering the owners unprecedented financing packages and trading opportunities. Sometimes they exaggerate the competitor's infringement of international regulations and they even bring diplomatic pressure to bear on the competitors.

4.4.3 Relationship with Owners

It is important for a shipyard to cultivate the goodwill of owners and give favorable treatment to those who have been loyal and cooperative over a long period. In good times, it is difficult to maintain good partner relationship with owners who find faults with the shipyard and squeeze the last drop of blood from it in difficult times when the yard has no alternative but to accept their orders. If, on the other hand, the shipyard is prepared to work only with the owners who are prepared to pay good prices in good times, many existing clients may turn away in difficult times.

5. Contract Negotiations

A new-building contract is finalized step by step through a process of long and painful negotiations. The principals of the negotiations are, of course, the owner and the yard. A number of other interested parties may also take part in the negotiations to iron out and clarify such complicated issues as subtle movement of ship's price, legal reviews of the contract and financing arrangements.

5.1 Conducting Negotiations

5.1.1 Methods of Negotiations

As shipbuilding contract involves a huge amount of investment from both parties, contract negotiations call for all available expertise and experience from both parties. In order to gain more favorable grounds in the negotiations, the parties may hire shipbrokers, lawyers and technical consultants, and sometimes ship investment & management companies (SIMC) are also brought in. The negotiations usually take place at a location most convenient for all participants, and this can be the owner's office, office of the owner's lawyer, or the shipbroker's office. If it is difficult for the technical and sales teams of the shipyard to travel, the owner may come to the yard for negotiations. In cases of several yards participating in a competitive bidding process, the owner may summon all parties in one place. Even though negotiations are conducted separately with each shipyard by turns, it is quite possible for the competing shipbuilders to come across each other. If this continues for a few days, a war of nerves can break out which can be highly stressful for the yard representatives.

In the Middle East one may experience a negotiation of 'horse-trade' style. At a horse trade a horse buyer summons several horse dealers at one place and gets them to compete against each other. The horse buyer receives the horse price from the first dealer, showing it to the second dealer and asks if he can better it. With the price from the second dealer, the buyer goes to the third dealer for another price cut. With the price from the third dealer, the buyer comes back to the first dealer and so on for endless price-chopping. Merciless repetition of this process usually drives the dealer who needs the cash the most to sell his horse often at a most miserable price. To resist being sucked into such an extreme price-squeezing practice the shipyard may resort to such tactics as discontinuing the price negotiations and bringing about re-tendering by proposing adjustments to the specifications. Another way out may be for the negotiating team to earn time by giving an excuse of having to consult the head office.

5.1.2 Ship Brokers

Ship brokering is one of the professions that naval architects can pursue, and is sometimes regarded as an enviable profession because of the glamor of large-scale business deals involved. A ship broker, of course, needs to know ships well, and must have good understanding of ship operations, chartering practices and management too. Brokers have to meet many people of diverse background, and so it is important for them to have good interpersonal and communication skills.

As addressed in Section 3.1.4 'Owner's Choices' earlier in this chapter, a ship broker is usually appointed by the ship-owner. However, they need to keep a good relationship with both the owner and the yard, since they have to act as facilitators of dialog between the two parties.

5.1.3 Lawyers

A new-building contract is a minefield with legal traps all over it. On signing the contract, some owners are reputed to file away the normal documents as well as starting another file in preparation for future legal disputes. This is because any documentary records of the shipyard's mistakes or weaknesses can be used by the owner as the necessary evidence for demanding compensation or even terminating/rescinding the contract. In a brisk market very few disputes arise, but in a sluggish market every clause, every word on the contract can become a bone of contention. It is not uncommon for the two parties to have long nerve-racking haggling even on relative minor points, such as which word to use, 'will' or 'shall,' and whether a definite article should be used or an indefinite article should be used.

Consequently, lawyers lead the contract negotiations on behalf of, and representing, the owner and the yard.

5.1.4 Technical Consultants

Negotiating on technical specifications is very arduous, because the specifications are voluminous and so wide in scope that dealing with them is very time-consuming and requires expert knowledge. It is not very surprising that the technical negotiations are usually delegated by the owner to external consultants. During this process the owner is thus represented either by engineers of a specialist consultant firm or by engineers of the chosen classification society. The scope of the consultants' involvement may later get extended to site supervision of the ship's construction at the shipyard. If the owner has sufficient number of technical personnel with appropriate expertise in-house, he/she may dispense with the services of the external consultants.

5.1.5 Financial Institutions

Both the owner and the yard may have huge budgets to run their business, but it is very rare for them to be able to afford the entire ship's construction cost from their cash reserve outright. Usually the owners meet anything between 10% and 20% of the ship's price from his/her own equity, but the remaining 80% to 90% is borrowed from financing organizations. To help ease the owner's cash flow, the shipyard may arrange 'deferred-payment' on behalf of the owner through providers of ship export finance or commercial banks. When the market cash flow is sluggish, financial support from banks is essential. When this happens, the financiers also become a party to the contract, and participate in the negotiations on such issues as performance guarantee, refund guarantee and payment method.

5.1.6 Ship Investment and Management Company (SIMC)

If the owner's balance sheet makes it burdensome to bring in loan capital, an SIMC may get involved. Such a company, when involved, takes over the contract with its terms and conditions including the ship's price. The vessel, when delivered, is chartered bareboat to the original owner for an agreed period of time at an agreed lease or charter rate. This can serve as a workable alternative when the market has collapsed, or a financial crisis renders banks unable or unwilling to provide ship's financing.

Under this scheme, the SIMC concerned usually raises funds from both individual small investors and institutional investors. Small investors can receive tax benefits for the amount of money they invest. The SIMC receives charter fees from the charterer (original owner), calculated from the ship's price with interest at an appropriate rate added, and the profit thus secured is distributed to the investors. The SIMC has the right to sell the ship, if the ship price rises during the charter period, and the profit from the sale is distributed to the owner and the investors by agreement. This alternative form of ship financing is similar to Germany's Kommandit Gesellschaft (KG) and Scandinavia's Kommandit Selskap (KS) systems, and many investors have shown interest in it.

5.2 Letter of Intent

On completion of negotiations with multiple bidders, the owner selects a preferred shipyard. Then, when the two parties come to an agreement on basic terms and conditions, the owner signs a letter of intent (LOI) before starting to negotiate for the final shipbuilding contract. The LOI reflects the agreement between the owner and the yard on the basic framework of the contract produced in the process of moving from 'open bid' to 'sole bid' status, and forms the basis of the subsequent negotiations. During the final contract negotiations slight changes due to minor technical and commercial modifications are unavoidable. However, the main framework must be adhered to, failing which legal disputes may arise later. The letter of intent usually contains ship type, number of ships, price, outline technical specifications, payment terms, delivery date, 'subject to' clauses and validity. The 'subject to' clauses set out the conditions on which the LOI can become 'null and void.' It is the usual practice to set the validity of LOI as short as possible.

6. Contract Documents

The so-called 'napkin contracts' once served as contracts before the end of World War II. The owner and the shipyard used to meet informally, for breakfast for example, and discuss the new ship to be built. They then wrote down whatever they agreed on a piece of napkin as a kind of gentlemen's agreement. Ships of an owner used to be built by the same yard for years and therefore there was no need for complex documents. A personal agreement between the yard and the owner was quite sufficient. For example, a napkin contract contained only the following information:

Contract

- Ship type: 40,000 DWT bulk carrier
- Units: two
- Price: one million dollars per ship
- Delivery: 31st January 1944
- Specs: Same as the previous ship, but with crane capacity changed to 30 tons and with speed improved to 12 knots

After World War II, the world seaborne trade expanded rapidly, interchangeable machines and components became widely available all over the world, and ships were often built in foreign countries, necessitating contracts with complicated clauses. In the process of catering for each complex clause, the contracts of present day have grown to nearly 100 pages supported by specifications of nearly 1000 pages. This, of course, has helped to provide much work for lawyers and has played a key role in specialization of technical personnel and component industry.

6.1 Contract

Even when a number of ships with identical specifications are built, a separate and independent contract has to be produced for each ship. The contract is like a constitutional law governing the whole processes of construction. Should any items in the specifications differ from the contents of the contract, the contract has priority. The contract is written in English and contains the following items:

- Description of the ship & class

  Describing the basic dimensions, principal particulars and the nationality of the ship; regulations of intended classification, national and international maritime regulations to be applied.
- Contract price & payment terms

  Payment is usually made in US dollars. Other currencies can also be introduced by mutual agreement. Depending upon the movement of currency exchange rate, euro and Korean won can also be the currency of choice. Payment methods (see Section 4.3), due dates and payment guarantee are all specified.

- Adjustment of the contract price

  A contract has provisions on what to do if major contingency events occur important enough to require adjustments to the contract price. The main provisions include delayed delivery, and shortfalls in performance from specifications in the contract, such as speed, fuel consumption and deadweight.
- Inspection and approval

  Methods and procedures of drawing approval and on-site inspections during the construction of the ship are specified.
- Modifications, changes and extras

  Provisions are given for the events of 'unavoidable' changes to the contract specifications by mutual agreement. The method of adjusting the ship price due to the modifications in the specifications is also specified.
- Trials and completion

  Individual trial runs of all equipment requiring verification of performance and the sea trials are comprehensively included in the contract. It specifies notification of test schedule and method as well as the attendance of the owner's representatives. The conditions for the owner's acceptance or rejection of test results are also given.
- Delivery

  The contract makes it clear what to do for any discrepancy between the contractual delivery date and actual delivery date. It includes causes of delay acceptable by both parties, procedures of delivery and acceptance, and accompanying delivery documents.
- Delays and extensions (force majeure)

  This clause describes all major unavoidable events ('force majeure') causing delays in the ship's construction, eventually leading to a delay in delivery date. A delay in the ship's delivery date equal to the duration of force majeure events is allowed. Therefore, the yard's claim for an extension in the ship's delivery date must be supported by evidences of force majeure events. The validity of these events as force majeure can be disputed by the owner, and therefore they must be clearly identified. Force majeure events include: war; acts of God or natural disasters such as earthquakes, floods and typhoons; riot, strike and fire; and other events normally beyond the reasonable control of the shipyard. If the total accumulated time of delays aggregates beyond the limit of agreeable delays, the owner may, at his/her discretion, cancel the contract.
- Warranty of quality

  Details of quality assurance post-delivery are specified, including (but not limited to) notification and repair of defects, method of settling the repair costs, etc. At the owner's request, the yard may appoint a guarantee engineer to attend onboard. Provisions for such eventuality are given by specifying the engineer's status on board and allotment of his costs.

- Termination/cancellation/rescission due to the shipyard's default

  Procedures for serving notices to rescind the contract, refunding of pre-payments and the issues regarding ship's total loss are specified.

- Termination/cancellation/rescission due to the ship-owner's default

  Owner's defaults which can lead to the contract termination are identified, and the procedure for terminating the contract and the method of compensation are also specified.

- Arbitration

  If disputes arise, legal judgement may be required. Against such eventuality procedures are specified for appointing the arbitration organization and the method of executing the decisions made by the arbitrator.

- Successors and assigns

  The contract may be assigned or transferred to another company ("Assignee" or "Successor") to be designated by the owner ("Transferor" or "Assignor") subject to the prior written approval of the yard. Defined in the clause are: relevant procedures of this contract assignment (or transfer); how the owner's rights, obligations, liabilities and responsibilities are transferred to and succeeded by the "Assignee" (or "Successor") in preparation for the event of a written request for such a contract transfer (or assignment) being made by the owner to the yard.

- Taxes and intellectual property rights

  The guidelines covered in the clause include how to handle various issues such as taxes, customs duties, patents, trademarks and copyrights as well as owner's supplies, insurance, language for warning/caution signs, validity of the contract, interpretation and governing law.

6.1.1 Documents Appended to the Contract

Some documents dealing with details of a contract are usually appended to the contract, and the typical examples are a refund guarantee from the shipyard and a performance guarantee from the owner. A refund guarantee is issued by the relevant bank and given to the ship-owner by the shipyard in exchange for the first payment from the ship-owner. A specimen refund guarantee is appended to the contract and constitutes an integral part of the contract. If the owner is a paper company without a clear physical presence or when it is judged that there is some doubt about the ability of the owner to discharge the duties stated in the contract, the shipyard can demand a performance guarantee from the owner in the name of a reliable parent company of the owner company or a bank.

6.2 Specifications

The specifications and the plans together with the maker's list are produced in English, containing all technical information necessary for constructing the ship. The specifications clearly state the expected ship's performance, and the capacity, installation method and test procedure of all on-board equipment. Should there be any inconsistencies or contradictions between the specifications and the contract, the contract prevails. Should there be any inconsistencies or contradictions between the specifications and the plans and other appended documents, the specifications prevail. The contents of specifications are as follows:

- General

  Covering the basic dimensions and principal particulars of the ship, the ship's purpose, the classification rules to be applied, the laws and regulations applicable, ship's performance, cargo capacity, cargo loading conditions, capacity of all tanks, model tests, details of trials (performance tests), allowable ship's vibration and noise level, drawing approval procedures, guidelines on drawing submissions, spare parts and the owner's supplies.
- Hull part

  Setting out generalities relating to materials for hull, structural design loads, methods of reinforcement for various parts of hull, reinforcement for installing equipment/machinery, scantlings of structural members, welding methods, steering gear-related parts, and other information relating to location/shapes/mounting methods for all steel structures mounted on the ship's hull.
- Prevention/protection against corrosion

  This is one of the most important parts, because ships are exposed to harsh corrosive environment due to their operation at sea and using sea water for ballast. Anti-corrosion measures are specified here, setting forth types of paints, painting methods, and surface preparation of steel plates for each and every part, and sacrificial anode system or impressed current cathode protection (ICCP) system. In the case of crude oil tankers, product carriers or chemical carriers, suitable paint for cargo tanks, having sufficient durability and capable of withstanding the characteristics of the cargoes the owner intends to carry, is selected and specified here.

- Hull outfitting

  This section contains the detailed specifications of major outfitting items needed for the ship's operation and cargo handling, including steering gear, anchoring and mooring systems, on-deck machinery & hydraulic systems, on-deck outfitting, cranes, hatch covers, ventilation systems, lighting systems and lifesaving appliances.

- Accommodation

  It sets out the facilities related to the crew's onboard living quarters: heating, soundproofing, ventilation systems, amenities and common rooms. The arrangement, size, structure and furnishings of cabins are determined in accordance with the ranks of the crew members.

- Hull piping

  It describes general specifications of piping: pipe materials, sizes, valves and pipe fittings; and specific requirements of overall piping systems fitted on board, including fuel oil system, ballast piping system, bilge water system, ventilation and air-conditioning system, and fire-fighting system.

- Cargo handling system

  This part specifies requirements of equipment and control systems suitable for transport and handling of cargo. For tankers they include pumping system, cargo piping system and cargo control system. For dry cargo carriers overall cargo handling systems are treated here, such as hatch cover system, loading/unloading system, cargo protection system and ventilation system. For liquefied gas carriers it specifies systems for loading/unloading of liquefied gas, re-liquefaction system, insulation system, inert gas system, and cargo control and monitoring system.

- Machinery systems

  This section sets forth required performance of major propulsion equipment such as the main engine, shafting and propellers; and the specifications of auxiliary machinery such as engine room boilers, generators, various pumps, fuel supply system, piping, air compressors, fuel oil purifiers and heat exchangers; size of various tanks; and methods of testing machinery performance.

- Automatic control systems

  This section specifies remote and automatic control and protection system of the main and generator engines and the related equipment. Also dealt with are all engine room instrumentation/gauges and warning lights installed in the engine control room and the wheel house and used for monitoring the running conditions of boilers and engine room auxiliary machinery. Also covered is the automation system in the engine control room and the wheel house, enabling the ship's operation unmanned or with a minimum manning level while sailing at night or on a route with no foreseeable obstacles.

- Electrical devices/electrical systems

  In this section power supply systems, electric motor systems, spare parts for onboard control systems, lighting systems, communication equipment, navigational equipment and electric cabling & wiring are specified. This includes the specifications of generators and power distribution, starting procedures of various electrical equipment and emergency procedures. It also describes the specifications and quantities of navigational equipment, onboard and external communication equipment, navigational lights and lighting fixtures for the crew's working and living spaces.

6.2.1 Annexes to the Specifications

In order to clarify some contents of the specifications general arrangement (GA) drawings are appended to the specifications. A GA drawing shows: a rough outline of the ship's overall shape; shape, location and number of cargo holds; arrangement and size of engine room and accommodation; major equipment for ship operation; and the positions of the decks. GA drawings also supplement the Specifications for items which are difficult to explain in words.

Another important annex is the list of equipment suppliers who supply the equipment/machinery/components as called for by the Specifications. Every shipyard maintains a standard maker's list for each ship type, with multiple entries for major equipment/machinery/components in order to encourage competition in quality and price. At times, the maker's list becomes a subject of heated discussion between the owner and the yard during the course of contract negotiations. The owner can focus only on quality, after-sales service and their relationship with the makers, while the yard needs to keep a stern eye on the prices as well.

7. Contract Management

With the signing of the contract, all relevant documents become legally effective, and therefore all the documents are handed over to the contract management department (CMD). The main remit of this department includes obtaining owner's approval of all design and construction plans and drawings, maintaining good co-operative relations with the owner's resident supervisory office, and tracking the major stages of the ship construction as stated in the contract. If a good relationship with the owner is maintained throughout the process and the owner is happy, he/she may become a return client. Therefore, the task of contract management takes on a greater significance than a simple project administration.

Invoices for installment payments are served upon the owner pursuant to the relevant provisions of the contract. Such invoices are issued when certain construction stages as specified by the contract are completed and when so certified by the owner's chief resident representative or by the chief classification society surveyor on site. As specified by the contract, the owner pays the first installment to the yard, three to four days after receiving a refund guarantee from the yard. Each of the subsequent installment payments becomes due on completion of each of the major construction stages, such as work commencement (or the first steel-cutting), keel-laying, launching and delivery. The first steel-cutting usually occurs about six months after contract signing.

Smooth and efficient construction of ships require timely approval of plans/drawings and inspection on the owner's part. Approval of plans/drawings by both the owner and the classification society should start at least several months prior to the first steel-cutting. Thereafter, the work of modifications/alterations and approval of plans/drawings is carried out continuously by the owner's resident supervisory office and the classification society until the ship is delivered. During the plan approval process, the owner must ensure that the plans/drawings reflect all the contents of the contract and specifications. During this process, the owner may request alterations to the specification, in which case the additional expenses and building time required have to be discussed and agreed to mutual satisfaction before amending the drawings. Inspections by the owner's site office normally start with the first steel-cutting phase. The yard's quality control department establishes the overall inspection schedule in close discussion with the owner's site representatives, revising the details as and when necessary in consultation with the representative. The yard's quality control department also keeps a complete record of all inspection results in writing.

1) Handling of force majeure events

Bearing in mind that force majeure events are directly linked with delays in the ship's delivery, the shipyard needs to receive the owner's confirmation of any such event in line with the procedures set forth in the contract. If rejected by the owner, it may lead to heavy penalty or in extreme cases become a ground for terminating the contract. Special care must be given to handling these events, since this is such a sensitive matter liable to give rise to disputes with the owner.

2) Non-fulfilment of the contract by either party

Failure to fulfill contractual obligations by either the owner or the shipyard can constitute a ground for terminating/cancelling/rescinding the contract. Therefore, utmost care and attention must be given to timely discharge of contractual obligations.

3) Preparing for arbitration

Even a minor or trivial issue can develop into a legal dispute. Therefore, whenever such an issue does arise, the yard must communicate with the owner in writing, clarifying the cause, progress and the outcome of such issues. The yard must carefully maintain all relevant records and documents, as they may serve as evidences in any possible future arbitration.

4) Adjustment of the ship's price

Amendments to the specifications and additional demand of the owner will affect the ship's price, and therefore it is important for the shipyard to keep a careful record of the details and obtain the confirmation from the owner's site office and agreement from the owner.

5) Organizing commemorative events

Various milestones during the construction of the ship are marked by commemorative events, including work commencement, keel-laying, launching and delivery. These events take on a greater significance than simple commemoration, since they are directly related to instalment payments. The naming ceremony is the most significant event in shipbuilding for both the owner and the yard. This celebratory event takes place just before delivery of the ship and is enjoyed by the owner and his/her guests, the yard's executives, and specially the yard's workforce who worked so hard over the ship. Gifts are exchanged between the ship-owner and the shipyard and utmost hospitality is usually extended by the shipyard to all guests many of whom have travelled a long distance to attend the event. Delivery takes place when the ship is completed, and this marks the end of the ship's construction. All delivery documents are handed over to the owner, and the final instalment is paid to the yard.

6) Miscellaneous

It is essential for the yard to ensure timely delivery of all equipment and component to be supplied by the owner (owner's supplies). Assisting the owner's representatives with their immigration formalities is also an important business of the contract management.

8. After-Delivery Service

After-delivery service refers to the service offered from the time of delivery to the end of the warranty period. Should the shipyard fail to render good service even after delivering a high quality ship, the owner may consider the shipbuilder unprofessional, and such experience by the owner will seriously jeopardize the prospect of obtaining future orders. All post-delivery matters are the responsibility of the after-sales service department.

1) Warranty of quality

The scope and duration of the warranty of quality is often one of the subjects of the most heated discussion during the contract negotiations. The defects that come to light after delivery but during the warranty period must be remedied or otherwise made good by the yard. However, owners often demand the yard to rectify even the defects discovered after the expiration of the ship's warranty period free of charge.

2) Warranty period

The normal contractual warranty period is one year from the ship's delivery date. However, disputes may arise about the warranty period of machinery/components that were repaired or replaced during the ship's warranty period. Most owners demand to receive an extended warranty for one year beginning from the date of such replacements. Such being the case, the yard provides one year warranty from the ship's delivery date, while the additional warranty for the replaced equipment is left to the manufacturer of the machinery/components to provide. Paints are sometimes guaranteed for four to five years, but usually the paint manufacturers provide the warranty to the owner directly and carry out the after-sales service through their global sales network.

3) Guarantee engineer

In the past a guarantee engineer was appointed by the yard to stay on board for up to one year or until such time as the crew becomes familiar with the ship's systems. However, due to possible disputes about the remuneration of the engineer and the demarcation of the responsibility, the engineer's stay onboard has often been shortened or his/her service is dispensed with entirely these days.

References

1. Khil, J. & Suh, S.W., 2010. Risk management lessons from 'Knock-in Knock-out' option disaster. *Asian Pacific Journal of Financial Studies*, 39(1), pp.28-52.

2. Korea Offshore and Shipbuilding Association (KOSHIPA), 1996~2012. *Shipbuilding yearbook* (in Korean). Seoul: Korea Offshore and Shipbuilding Association.

3. UNCTAD(United Nations Conference on Trade and Development), 1996~2012. *Review of Maritimetransport*, Geneva: UNCTAD.

CHAPTER 3

Ship Design

1. Definition and Characteristics of Ship Design
2. Principal Dimensions as a Design Consideration
3. Design Constraints
4. Design Process and Steps
5. Computer-Aided Ship Design and Application of Information Technology

1. Definition and Characteristics of Ship Design

1.1 Definition of Ship Design

Ship design is a series of decision-making process to realize a product satisfying the ship-owner's requirements. In this process the designers have to define the ship in a producible form, subject to a set of given design criteria. It is an optimization process of achieving a balance between various conflicting requirements while tempering the owner's demands with the scientific knowledge and technical experience of the designer. Nevertheless, the designers must continuously strive to accommodate requirements or even the aesthetic preferences of the owners and exercise his practicable creativity to its fullest extent. Ship design is far more than simple drafting or a set of calculations and will inevitably reflect the designer's abilities and creative intentions.

A ship is the largest engineering structure at sea which must be able to move as its operator wishes and, therefore, must be equipped to facilitate the commands of the operator. Ship design is, therefore, a long and complicated process by which conflicting elements are effectively expressed, modelled and coordinated so that all the requirements are met as much as is possible.

Since scientific principles must be adhered to in a ship design process, the designers must fully comprehend the nature of these principles and how they are applied. It is plain that they must have sufficient insight into the operator's needs so that they can find technical means of satisfying them. In addition, it must be remembered that merchant ships are usually engaged in international trades and they, therefore, must conform to all relevant international laws and regulations.

In recent times, green issues and their technical solutions have been receiving a great deal of attention from the designers, and environmentally friendly shipping has become the clarion call of all those concerned with shipping. As the regulations regarding environmental protection and safety have become ever more stringent and will be toughened further still, it is essential that ship design evolves to meet these new requirements.

Notwithstanding what has been said above, the main goal of any new merchant ship design is that the ship is competitive in the economic circumstances and climate of the contemporary shipping market.

Designing of a ship marks the first step of its construction as a result of an investment decision made by the owner pursuing profit based on a close scrutiny of the global market trends. The detailed design process includes reviewing, optimizing, adjusting and confirming the ship's hull form, structure and arrangement to meet the flotation and stability requirements and all other performance requirements specified in the contract. The most widely used approach to ship design is to use an existing ship, whose performance is widely recognized as good, as the basis ship and attempting to improve the performance of the new design incrementally by adjusting the design parameters. About 80% of improvement and innovation is implemented during the design process, since it is practically impossible to make radical changes during the construction or manufacturing phase.

Ship design is essentially an optimization process that involves reasoning based on multi-disciplinary theories, of which naval architecture and marine engineering are pre-eminent, and the designer's experience. During this process the designer not only strives to improve the product's performance and productivity, but also tries to reduce environmental impacts of the ship. It can be said, therefore, that ship design requires considerable amount of creativity and balanced decision-making on the designer's part, augmented by his/her engineering sensibilities.

1.2 Characteristics of Ship Design

Most stationary structures on land are supported by peripheral utilities to maintain their functionality, but, since ships operate independent of any external support while at sea, they have to carry all facilities and requisite supplies to last several days at a time. Unlike land-based stationary structures, ships and other floating structures must be designed to withstand the dynamic loads inherent in their movement and the ocean environment.

Depending on the shipping route, the cruising range of oil tankers often exceeds 6,500 nautical miles (1NM = 1,852 meters), and the sea conditions vary continually along the route. Unless ships are designed and constructed so that their safe and effective operation in such environmental conditions can be guaranteed, the ship operator cannot produce a sensible business plan. Furthermore, unless it's structural reliability or stability in waves can be guaranteed while operating in seaways, there is no point in building the ship in the first place.

The designer is also faced with conflicting requirements. For example, the ship's weight needs to be reduced for operational efficiency, whereas its structural safety calls for the ship's weight to be increased. Incidentally, it is crucially important to minimize the structural weight of ships, as it adversely affects resistance and cargo carrying capacity, while there is much less stringent requirement to reduce the weight of land based stationary structures.

Let us compare a 300,000-ton class oil tanker, which is considered technically simple in terms of modern ships, with a modern passenger car which is believed to require advanced technologies to produce.

Since oil tankers normally carry large quantities of cargo over long distances, they require continuous operation for several consecutive weeks, whereas cars are often used intermittently for short periods in a day. While a car has a 10-year service life, an average oil tanker is deemed to have a service life of 20 or 25 years despite operating in even harsher circumstances. Whereas the ship operates on a constantly moving sea surface and maneuver in fluid, the car moves on a solid pavement using four wheels, giving it good stability and maneuverability.

It is unanimously agreed that ship structures need to be made lighter as long as it is within the structural constraints, primarily for improving their economic efficiency and agility. However, often it is not realized that an oil tanker's shell plate thickness is less than 1/10,000th of its length in contrast to an egg's shell which is about 1/100th of its maximum dimension. The pressure on ship structural designers, therefore, is quite considerable.

A person riding in a car can normally feel the impact of a pool of water on the road only 5 cm deep. Consider, then, the impact a ship receives at sea. If an oil tanker undergoes pitch motion with a maximum angular velocity of one degree per second, the bow must go in and out of water at the speed of about 3 meters per second. Furthermore, this relative speed becomes even greater when the ship moves forward in waves. Even a layman would understand that if the ship receives impact of this sort from the wave surface repeatedly, the ship will suffer damage or even some sort of failure.

The oil tanker that we have discussed above costs about 100 million dollars (as of April 2011), which is equivalent to the price of 2,500 cars. This is a huge investment by any standard, and it is not surprising that a meticulous and thorough feasibility study and research is conducted to produce a set of requirements before ordering a ship. The shipyard then designs the hull form, structure and operating system to meet these requirements. Since the ship is more than likely to enter service in international waters upon completion, its performance in service becomes immediately known worldwide. A few good ships can build up a solid reputation for the designer and/or shipyard. It is pleasing to note that Korean shipyards are internationally recognized for their excellent design capabilities.

When designing a car, the car manufacturer predicts the changing tastes of target consumer groups and tries to reflect their financial abilities so that there is a healthy effective demand for the car. Therefore, the car maker pays close attention to the car's aesthetic appearance, while trying to appeal to the consumers' sensibilities for performance with appropriate technical specifications. Furthermore, since cars are designed for mass production, a new car model is not released to market until a long-term durability test is performed on a prototype and it is concluded that quality cars can be consistently supplied. All in all, cars are produced after demand forecasting; planning based on mass production; design; and thorough quality assurance.

On the other hand, the design of a ship begins by identifying the owner's requirements and the ship should be built to give good performance over 20 years or more without the benefit of durability testing of a pilot ship, the whole process is significantly different from the car design from the beginning. Most cars can be used nearly everywhere subject to some legal restrictions in certain regions. Ships, however, must not only comply with international standards like IMO, ILO and ISO, but they must also meet the classification rules and provide the level of performance that meets the major industrial standards.

2. Principal Dimensions as a Design Consideration

2.1 Principal Dimensions

Once the designer has determined the principal particulars of a ship, the routes that the ship will run are examined to identify restricted passages, such as canals and bridges, to ensure that the ship is not too large for them. The designer must also determine the minimum water depths of not only the routes but also the ports it will be calling at. Attention should also be given to cargo handling facilities and accessibility to piers at which the ship can berth and use shore cargo handling equipment. The Safety of Life at Sea (SOLAS) and the classification rules for ships define the principal dimensions of a ship as shown in **Figures 3-1** and **3-2**.

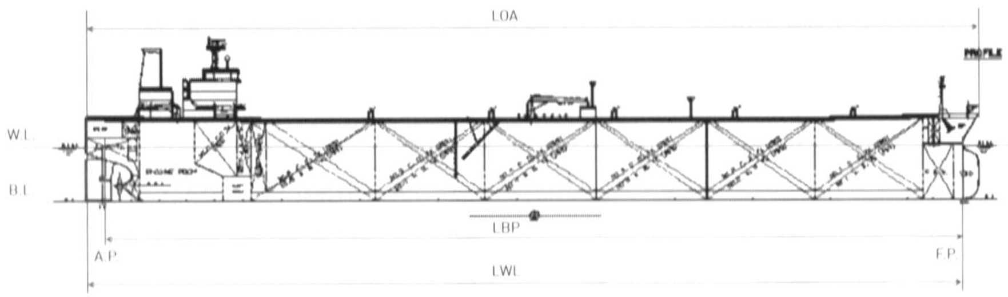

Figure 3-1 Principal longitudinal dimensions of a ship

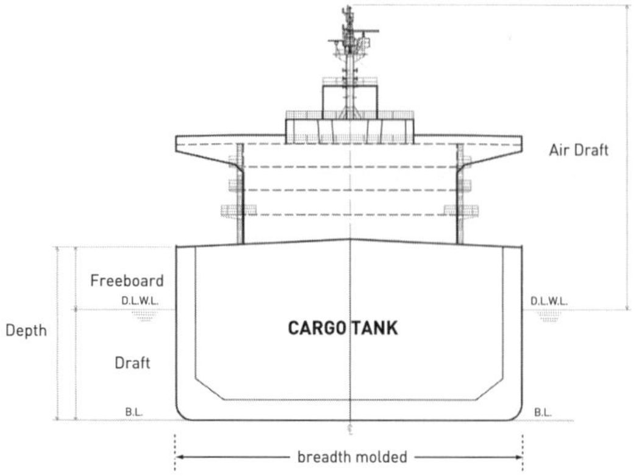

Figure 3-2 Principal transverse and vertical dimensions

- Length Overall (LOA): The horizontal distance between the foremost and the rearmost ends, including all fixed structures on board, and is used to find the required length when berthing at a pier or docking.
- Length between Perpendiculars (LBP): The horizontal distance between the forward perpendicular (FP) and the after perpendicular (AP); the forward perpendicular is an intersection between the front side of the stem and the design waterline, and the after perpendicular is the center of the rudder stock in a modern ship. LBP is used to represent the length of the submerged part of a hull.
- Load Waterline length (LWL): The length from the forward perpendicular to the end of the stern on the design draft is used to design a hull form. This length is also used as a reference length for estimating the ship hydrodynamic performance such as resistance and propulsion.
- Freeboard Length (L_f): 96% of the waterline length at 85% of the depth or the LBP whichever is greater; used as a reference length to calculate freeboard.
- Rule Length (L): The length between perpendiculars on the fully loaded waterline (LBP at maximum draft Ts); it must be at least 96 or 97% of the overall length and is used for calculations according to the classification rules.
- Molded Breadth (B): The breadth of a ship excluding the shell thickness. 'Molded' surface defines the outer edges of the frames which are then covered with a shell. The molded surface of a ship is perfectly smooth, while the outer surface of a steel ship is slightly uneven due to differing thicknesses of shell plating used at various different points of the shell. Consequently the designer usually uses molded dimensions for specifying the size and shape of steel ships and the shell thickness is excluded from these.
- Molded Depth (D): The vertical height from the molded baseline to the bottom of deck plate at side amidships.
- Design Draft (T_d): The draft of a ship when it is fully loaded with cargo and supplies including fuel; used as a reference for design.
- Scantling Draft (T_s): The greatest draft allowable by the structural strength, at which a ship can be actually operated.

- Freeboard (F): The reference freeboard set forth by the International Convention on Load Line (ICLL) is the distance from the load waterline to the upper surface of deck plate at side amidships.
- Air Draft: The emerged height above water, as opposed to the draft which refers to the immersed depth; the height from the load waterline to the top of the ship; or for cargo ships, the height from the load waterline to the top of the hatch coaming excluding the hatch cover. The designer must heck which definition is appropriate for the ship.

2.2 Shipping Routes

Since ships operate internationally, they must be designed in accordance with the regulations applicable in the waters where they are intending to call. For example, ships operating in the North Atlantic are subject to the most stringent regulations. Furthermore, some waters in the cold regions of the globe, such as the North Sea in Canada, the Baltic Sea and Sakhalin, freeze during the winter season and ships passing through these waters need structural reinforcement. For the Polar Regions like the Arctic Sea Route in Russia, ships must be designed to have robust hulls, protection against cold weather and de-icing equipment.

Since the primary aim of a ship is to carry passengers or cargo, a voyage must begin at a port, pass through certain regions and end at a port. Since all the waters and ports around the world impose substantial restrictions on the ships operating there, the designer must incorporate them into his/her designs.

2.2.1 Canals

1) Panama Canal

The Panama Canal connects the Atlantic and the Pacific Ocean and includes several lakes and two locks in its 46 nautical miles-long course. The existing two-lane lock has a passage width of up to 32.32 meters and, when a ship enters the lock, it is towed by the locomotives on both banks as it passes through the canal. Ships passing through this canal should be equipped with Panama chocks for this purpose. The maximum allowable draft is 12.04 meters at the passage way. For a ship to pass the Panama Canal, the ship's passage equipment and its drawing must be approved beforehand by the Panama Canal Authority.

Besides the existing two-lane lock, a new lock is currently under construction for the expansion of the waterway to allow passage for very large container ships. Once opened in

2015, it will be able to accommodate ships with a breadth of 49 meters and a draft of up to 15.2 meters as shown in **Figure 3-3**.

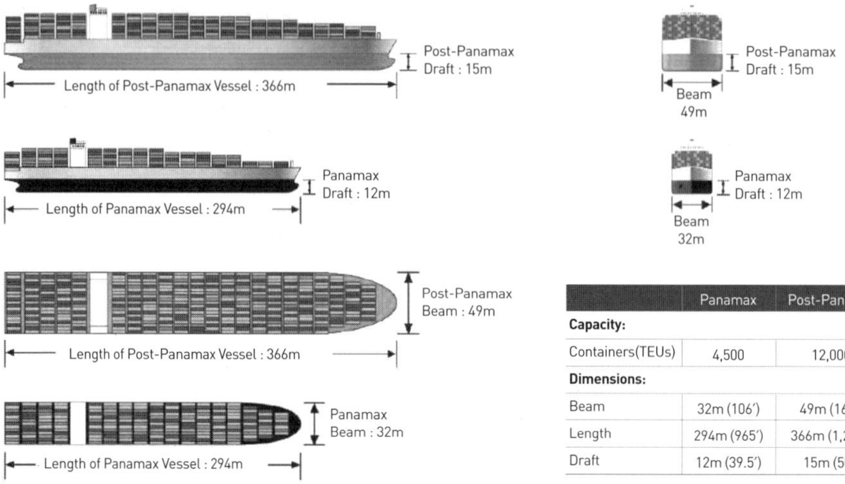

	Panamax	Post-Panamax
Capacity:		
Containers(TEUs)	4,500	12,000
Dimensions:		
Beam	32m (106′)	49m (160′)
Length	294m (965′)	366m (1,200′)
Draft	12m (39.5′)	15m (50′)

Figure 3-3 Comparison between Panamax and Post-Panamax container vessels

2) Suez Canal

The Suez Canal is about 88 nautical miles long connecting the Mediterranean Sea and the Red Sea. Unlike the Panama Canal, the Suez Canal is a sea-level waterway without the need for locks. Therefore, there is no limit in the length of ships that can pass through. However the breadth is limited to 64 meters and the draft to 18.9 meters only when the breadth is less than 49.98 meters. If the breadth is greater, the draft is subject to further restrictions determined by the canal authority, and the limit is usually inversely proportional to the breadth. Ships passing through this canal are required to pay high transit fees. Since the fees are determined based on Suez tonnage, and not the type of cargo, the ship with smaller enclosed space will pay less. Furthermore, ships wishing to pass through the canal are required to furnish a Suez searchlight on the bow.

3) Saint Lawrence Seaway

The Saint Lawrence Seaway, extending more than 4,000 km connecting several canals, from the Port of Montreal to the Great Lakes, lies between the northeastern part of the U.S. and Canada. The seaway includes 14 locks and 3 canals, where the maximum length of ships that can be accommodated is limited to 222.5 meters and the maximum breadth of 23.2 meters. After inspection of the drawings of the ship specially prepared for this purpose, it may be possible to extend the length limit to 225.6 meters and breadth to 23.8 meters. The air draft is limited to 35.5 meters and the maximum allowable draft is 7.89 meters.

4) Kiel Canal

The Kiel Canal is a fresh water canal and runs for 53 miles from Brunsbuettel in the North Sea to Kiel-Hotenau. Up to about 280 miles is saved by using the canal instead of going around the Jutland Peninsula. The maximum length of ship allowed is 235 meters, breadth 32.5 meters, draft 9.5 meters and air draft 40 meters.

2.2.2 Ports

As shown in **Table 3-1**, there can be several constraints imposed by each port. However, it is practically impossible to design a ship while taking into account all of the conditions of all ports. Usually ships are designed to be within the limits of the ports that they are likely to call in addition to the limits of a few widely-known ports. Therefore, while any special requirements can be incorporated into a ship's design, additional design and construction costs arising out of such special requirements are normally covered by the ship-owner.

Table 3-1 Ship constraints of main port for iron ore exports and imports (unit: m)

Port	LOA	B	T_s (Scantling Draft)	Air Draft
Newcastle (Australia)	290	47	15.3	18.5
Ponta Do Ubu (Brazil)	313	54	15.5	18.5
Beilungang (China)	290	45	17.1	20.0
Dunkirk Solac (France)	283	45	14.2	13.7
Pohang (Korea, Republic of)	310	45	14.5	15.0

2.3 Required Power and Propulsion Systems

Presently the power to propel a ship comes from the main engines of two main types: internal combustion engines such as low to medium speed diesel engines; and external combustion engines such as steam and gas turbines. The latest environmental regulations have introduced tougher emissions control, which makes it necessary to use high efficiency engines. Low to medium speed dual fuel engines, which allow alternative use of diesel fuel and natural gas, have been developed and are already used or scheduled to be used. Many research works have also been done on the deployment of fuel-cell-powered ships and nuclear-powered ships.

3. Design Constraints

3.1 Ship-owner's Requirements

A ship-owner, planning to purchase a new ship, determines a set of particulars and minimum achievable performance level for this ship so that its operational profit can be maximized. When the order is placed for this ship, the owner will specify the condition that the ship delivers certain level of performance, when built, as a part of shipbuilding agreement. These constitute the key owner's requirements, and they are crucial in all stages of construction of the ship from design to delivery.

Since most shipping companies around the world, large or small, include specific requirements in their orders, the ship designer should try to meet these within the constraints of classification rules and national and international regulations, taking into account the practicality of constructing the ship with the given shipyard resources.

3.1.1 Ship-owner's Key Requirements

Ship-owners present a variety of requirements in diverse formats. Some companies produce concise requirements to be delivered in a single page by fax, but others may request a detailed draft specification containing hundreds of pages. Some designers receive and grant approval of design specifications from reliable shipyards; others require minute details of a shipyard's health, safety, and environment aspects.

The following are the basic requirements of ship design:

- ship type and principal dimensions;
- type of cargo, cargo handling performance, cargo capacity and deadweight;
- main engines and the minimum operational speed;
- number of passengers, and the size and arrangement of the cabins;
- flag of registration and classification society;
- applicable international conventions, international standards and regulations, and the regulations of authorities concerned; and
- testing of shipyard equipment, and the performance requirements to be confirmed before delivery.

3.1.2 Incorporating Owner's Requirements into Design

The designer must not only meet the owner's requirements but also check the following list to design a ship that best meets the criteria, even if it is not specified:

- safety of the ship and its crew;
- prevention of marine pollution and use of environmentally friendly equipment;
- suitability to operate in designated shipping routes in calling ports; and
- acceptable shipbuilding cost and optimum operating costs.

3.2 International Standards

Since ships are used in international trade, they must be designed and built in accordance with the conventions of international organizations, regulations of countries of call, requirements of ports and canals, classification rules and other international standards.

3.2.1 International Maritime Organization (IMO)

The International Maritime Organization (IMO) was formed as the UN's agency for settling international disputes by establishing unified rules and regulations for ships engaged in international trade. The IMO prepares and recommends international treaties on marine safety and pollution prevention, which must be taken into account when designing, building, and operating ships, and is also responsible for stipulating related rules.

3.2.2 Major IMO Agreements that Affect Ship Design

- International Convention for the Safety of Life at Sea: SOLAS, 1974
- International Convention for the Prevention of Pollution from Ships: MARPOL, 1973
- International Convention on Load Line: ICLL, 1966
- Convention on the International Regulations for Preventing Collisions at Sea: COLREG, 1972
- International Convention on Tonnage Measurement of Ships: ITC, 1969
- International Convention on the Control of Harmful Anti-Fouling Systems on Ships: AFS
- International Convention for the Control and Management of a Ship's Ballast Water and Sediments: BWMS

3.2.3 Major National Authorities

- Australian Maritime Safety Authority (AMSA)
- Central Committee for Safety (CCS), France

- Belgian Maritime Inspectorate (BMI)
- Hong Kong Marine Department (HKMD)
- Indian Merchant Shipping (IMS)
- Ministry of Land, Transport and Maritime Affair (MLTM), Korea
- Maritime and Coastguard Agency (MCA), UK
- Norwegian Maritime Directorate (NMD)
- See-Berufsgenossenschaft (SBG), Germany
- United States Coast Guard (USCG)

3.2.4 International Organizations

- International Labor Organization (ILO)
- Baltic and International Maritime Council (BIMCO)
- International Association of Classification Societies (IACS)
- International Chamber of Shipping (ICS)
- International Electro-technical Commission (IEC)
- International Association of Independent Tanker Owners (INTERTANKO)
- International Shipping Federation (ISF)
- International Organization for Standardization (ISO)
- Oil Companies International Marine Forum (OCIMF)
- Society of International Gas Tanker & Terminal Operators (SIGTTO)

3.2.5 Other Standards that Affect Design

Other design constraints include the restrictions and requirements of ports and the restrictions of constrained seaways that the ship is designed to pass through, including the Suez and Panama Canals, and Malacca Strait. Some major companies have their own standards and regulations, including Chevron, BP, Shell, Exxon Mobil and Statoil.

3.3 Classification Standards

3.3.1 Classification Society

Some government departments engaged in maritime activities and a classification society in each country constitute the major national agencies responsible for interpreting the rules of the International Maritime Organization (IMO). They ensure the compliance to the rules and provide technical advice, as well as approving ship designs and supervising ship construction. The classification societies provide their own classification rules based on their experience and technology and ensure that the ships in their register comply with these rules.

Classification societies formed the International Association of Classification Societies (IACS) as a non-profit private organization to offer uniform classification rules and provide technical assistance on shipbuilding and marine engineering. The Korean Register of Shipping (KR) joined the IACS in 1975 and, in 1998 became an official member, which helped develop the Korean shipbuilding industry and elevate its international standing.

3.3.2 Fulfillment of Classification Requirements

Since a merchant ship has to be registered with at least one classification society for marine insurance, the shipyard must perform a variety of calculations and tests to obtain approval of the drawings from the relevant classification society and then build it in accordance with its rules. Moreover, the shipyard should then receive all kinds of proof from the classification society and keep them on board along with other drawings and documents in compliance with the rules. Even while the ship is in service, the ship must be surveyed by the classification society on a regular basis in order to ensure the safety of lives and properties at sea.

3.4 National Standards

All ships, like people, must have a nationality, and must be registered in a country which is then known as its flag state. A ship in operation is obliged to show the flag of the country of destination on the bridge and the flag of the country of registration at the stern. To acquire a certain nationality, a ship must meet the requirements set forth by the country covering all aspects from design to construction. For example, a ship under the Greek flag must feature a cofferdam between its ballast and drinking water tanks and meet the Greek stability requirements to load bulk cargo in the cargo hold. Furthermore, the flag state has the responsibility to inspect the ship regularly and issue necessary certificates. A ship can be registered under a certain nationality if the ship-owner is a citizen of that country, at least a certain percentage of the ship's crew members are citizens of that country, or the ship is built in that country.

A ship can acquire the flag of convenience as long as its owner is a citizen of the country. However, ships owned by a corporation registered in that country are also eligible for the flag of convenience, since the corporation is considered as a citizen. Therefore, in any country where corporations can easily register, the nationality of the ship-owner is often obscure.

Ship owners tend to prefer flags of convenience, such as Panama, the Republic of Liberia and Bahamas, mainly because the states rarely impose regulations more stringent than the minimum international standards and they do not impose any restrictions on nationality of the crew. The main benefit of such practice of open registry is economic, such as minimum tax and non-disclosure of financial details.

3.5 Environmental Standards

The Exxon Valdez oil spill occurred in Alaska on March 24, 1989 caused one of the most devastating man-made environmental disaster which was worsened by a series of storms. As a direct consequence of this, the requirement for double hull tankers (as illustrated in **Figure 3-4**) was introduced unilaterally by the USA to begin with, but since then has been adopted internationally.

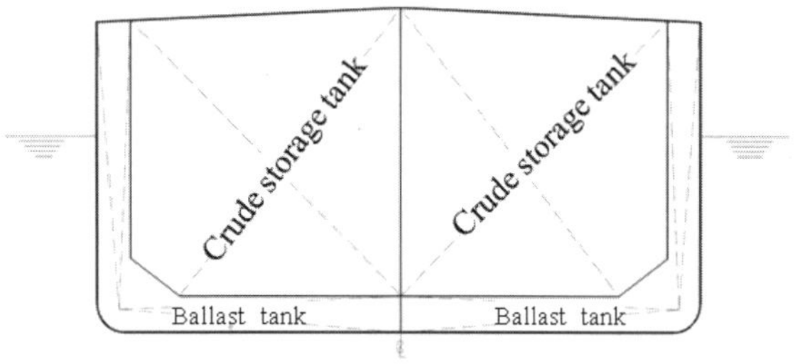

Figure 3-4 Midship section of crude oil tanker with double hull

Since then environmental regulations have also become more stringent, including the latest tightening of international emissions regulations. Ships carry at least 90% of cargo around the world, but as shown in **Figure 3-5**, they are much greener compared to other means of transport in terms of carbon emissions, although the actual quantity of emission depends on the ship type.

There is not much doubt that marine transportation is more eco-friendly and energy-efficient than other modes of transport. Nevertheless, the regulatory authorities want to reduce the harmful by-products of transport by tackling the mode which carries the most volume of cargo. In addition to the Energy Efficiency Design Index (EEDI) associated with carbon dioxide emissions, the environmental regulations also extend to emissions of sulfur oxides (SOx) and nitrogen oxides (NOx).

The emission standards for SOx and NOx are planned to be applied in stages. To do so, shipyards are required to fine-tune ship engines, install dust collectors to reduce emissions, or use low-sulfur fuel. The EEDI is an indicator of carbon dioxide emissions per tonnage and the formula for its calculation includes the output of main and auxiliary engines, fuel consumption, additional equipment that affects fuel efficiency, cargo being carried and the ship speed.

To prevent microorganisms and other pollutants being transmitted through ballast water and adversely affecting marine ecosystems of the world, ships are required to use appropriate ballast water treatment systems. There is also a variety of environmental regulations currently being adopted or under discussion on the important issue of ship recycling at the end of their lives in an environmentally friendly way.

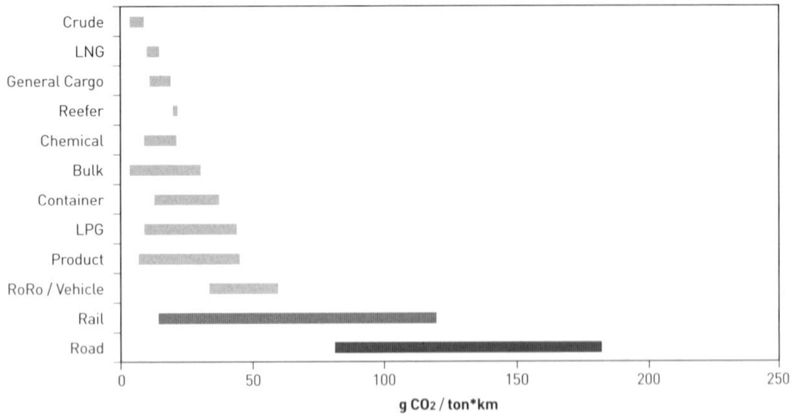

Figure 3-5 Carbon dioxide emissions by ship types compared to other means of transportation

4. Design Process and Steps

4.1 Commencement of Design

The ship-owner requests the shipyard to design a ship by specifying the hull form, speed, type of cargo, cruising range, type of main engines, effective load capacity, applicable laws and nationality. During the basic design stage, the designer determines the hull form, principal dimensions and speed. He/she then reviews relevant international standards, and develops a set of specifications and a basic drawing based on the design requirements. During this process the designer may consider a variety of hull forms and equipment, as long as they meet the specified requirements.

The overall economic evaluation of each option then provides the basis for selecting an optimum design. In this process existing ships of similar characteristics are also used for reference purposes. If the shipyard already has a standard hull form that is close to meeting the design requirements, it may be possible to negotiate with the ship-owner for adopting it either as it is or modified in some way.

4.2 Design Process

4.2.1 Determining Principal Dimensions

Since the principal dimensions of a ship have the following effects on its performance, they have to be carefully determined.

1) Length (L)

As the length of a ship has a significant influence on its propulsion performance and hull strength, changes in length directly affect the weight of the propulsion system and the hull. An increase in ship length enhances propulsion performance as well as directional stability during operation. On the other hand, a greater length will probably necessitate increased structural scantlings to cope with higher longitudinal loads and the consequent increase in ship price. Freeboard will also have to be increased with a consequent loss in loading capacity. Furthermore, higher directional stability or course-keeping ability usually implies lower course-changing ability or steering performance.

2) Breadth (B)

The breadth of a ship directly affects the ship's stability, rolling frequency and lateral strength. Given a constant ship length, a broader ship has higher transverse stability and loading capacity. However, this leads to an increase in roll natural frequency which is usually detrimental to crew and passenger comfort. A greater breadth is also associated with increased resistance and hence higher engine output. For a given draft, lowering L/B ratio usually leads to lower course keeping ability.

3) Depth (D)

The depth of a ship affects hull strength, and consequently structural scantlings and hull weight. Given a constant length, the greater the depth (smaller the L/D ratio), the higher the longitudinal strength, reducing hull weight as well as the ship's price.

4) Draft (T)

The draft of a ship affects the ship's resistance and displacement. Limited water depth of the seaways and ports that the ship intends to visit will limit the ship's allowable draft.

4.2.2 Hull Form and Power Requirement

The rapidly rising oil prices have encouraged researchers to study hull form optimization for reducing resistance and saving fuel. Much effort has been made in making ships more environment-friendly by improving energy efficiency and reducing gas emissions. Therefore, the ship designer's main task at this stage is to create a hull with less resistance at the speed required by the specification and to equip it with a propulsion system of high efficiency.

Usually the designer selects a hull form with proven track record, and uses his/her own experience and knowledge to modify it to suit the current needs. Often a physical scale model of the hull form is then built and a resistance test is performed in a towing tank. If the hull fails to meet the design speed, adjustments are made before more tests are performed. This process is repeated until the resistance characteristics of the hull form are satisfactory.

Recently computational fluid dynamics has been used instead of physical performance testing in some cases, but its results are not sufficiently accurate or reliable for practical design purposes. Therefore, model testing is still the preferred method of verifying vessel performance.

4.2.3 Overview of Compartments and General Arrangement

Since the general arrangement of a ship is closely related to the ship's functionality, it is important to develop a reasonable layout by taking into account the purpose of each compartment and the relations between them. For instance, fuel tanks, engine spaces, cargo and ballast tanks should be roughly grouped together. Their relative locations and size can then be determined. After that, the arrangements of the stern and the bow must be determined to complete preliminary general arrangement.

The general arrangement of a tanker is illustrated in **Figure 3-6**, and the main items of this arrangement are described starting from the aft end of the ship.

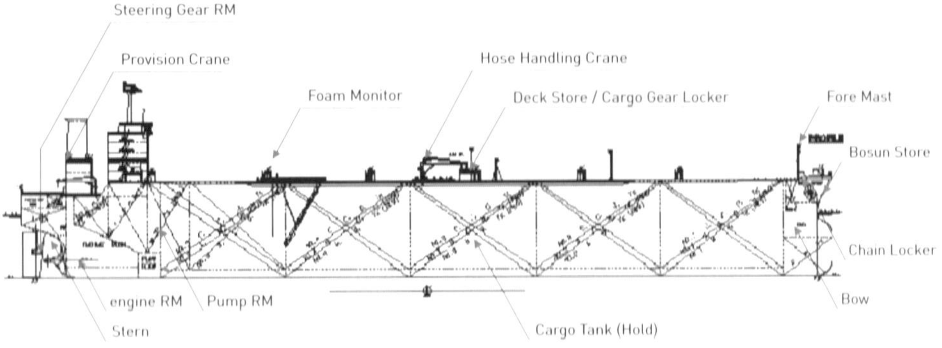

Figure 3-6 General arrangement of a tanker

1) Stern

The rearmost compartment from the aft end of a ship to the aft bulkhead of the engine room is called the stern. This compartment houses the ship's steering gear which controls the ship's rudder for maneuvering. Therefore, the size and arrangement of the steering gear and the rudder determine the length of the stern compartment. Since the stern is adjacent to the living quarters, allocation of the fresh water tank is also taken into account when determining the length of this compartment.

2) Engine Room

In general cargo ships, bulk carriers, oil tankers and gas carriers, the engine room is usually situated next to the stern; for large container ships, however, it can be arranged further forward to maximize the amount of containers that can be carried on deck without compromising the forward visibility requirement. The main items in the engine room compartment, of course, are the main engine(s) providing the power to propel the ship, auxiliary engines, generators, and boilers to supply steam for heating and power for some equipment to operate. Main fuel and lubricating oil storage tanks are often located in the double bottom space under the engine room. The length of the engine room is usually determined based on the size of the main engine for propulsion. The engine room should also have sufficient headroom to allow dismantling of the main engine for maintenance.

3) Accommodation Quarters

The crew's accommodation block is generally located above the engine room. Every effort should be made to reduce vibration and noise from the engine room and create healthy and pleasant living spaces for the crew.

4) Cargo Holds

Cargo holds are compartments where the cargo is carried. They have a double bottom to limit water ingress in case the ship's hull is broached, and the space in the double bottom is used for ballast tanks. In oil tankers, the whole cargo space is divided by two or three longitudinal bulkheads and 5 to 9 transverse bulkheads in order to prevent the cargo holds from getting too large.

For container ships and bulk carriers only transverse bulkheads are used. The number and length of the holds depend upon the ship size and the type of cargo carried. In some cases cofferdams may have to be installed between the holds too to insure that the cargo is not cross-contaminated.

5) Bow

The bow compartment refers to the space from the forward bulkhead of the foremost cargo hold to the stem of the ship. Often a ballast tank (fore peak tank) and chain lockers are located there with anchors and mooring equipment mounted on the forecastle deck.

4.2.4 Structural Design and Strength Analysis

Once all the compartments and cargo holds are arranged, the ship's structure is designed to have the requisite strength. The midship section structure is determined first as representing the structural features of the whole ship. A shell expansion plan is then drawn which illustrates the arrangement of the shell plates. Based on the compartment arrangement, midship section and shell expansion plan, the designer can now develop structural drawings by compartment or by location. At this stage the designer must determine the dimensions of the members so that the structure meets the classification and design requirements.

The longitudinal strength of the ship is the primary consideration during structural design. When a ship floats in equilibrium in calm water, the total buoyancy force equals the total gravitational force which is made up of lightship weight, cargo weight and weights of all other items on board. However, the uneven longitudinal distribution of these forces gives rise to a shear force and bending moment distribution. The arrangement of the cargo holds, fuel tanks, and ballast tanks directly affects the distribution of loads. Ballast tanks, in particular, cause a considerable load difference, depending on the operating condition, it is crucial to optimize their locations for various operating conditions.

If the weight of cargo or an external force is concentrated within a limited area, it can give rise to high structural loads and some measures must be taken to counter them. Good examples of this include cargo with a low stowage factor, main engine and propeller which present an additional problem of noise and vibration.

Shipyards generally use the hull structural analysis software developed by a classification society to evaluate hull structural designs. They include Safehull developed by American Bureau of Shipping (ABS), Nauticus by Det Norske Veritas Germanisher Lloyd (DNV GL), and ShipRight by Lloyd's Register of Shipping. Using such software, the designer assesses the global structural strength of the ship and attempts to optimize the scantlings. The programs can also be used to determine the scantlings of local structural members as required by the classification rules. Liquid cargo tanks, in particular, can be subjected to considerable impact loads created by the sloshing of the liquid. In this regard, the cargo tanks of LNG and LPG carriers must be carefully strengthened to cope with such sloshing loads.

4.2.5 Arrangement of Major Outfitting

The major outfitting items are as follows:

1) Steering Control System

The steering gear and the rudder are designed to control the ship's heading and direction of travel, and are generally situated at the stern. Sometimes side thrusters are installed at the bow and/or stern to assist steering in confined narrow waters. The installation of such thrusters will alter the arrangement of bow or stern compartments, since thruster rooms have to be provided together with spaces for extra electrical equipment associated with them.

2) Mooring System

A ship needs to be moored to piers or other structures usually during loading/unloading operations and occasionally it needs to be anchored, although very rarely in open sea. Mooring equipment on board a ship consists of anchors, anchor chains/cables, bollards, winches, windlasses and fairleads. The chains are stored in a chain locker usually situated in the bow, while cables are stored around a drum. Anchors are often carried lashed to the outer end of hawser pipes, but on some offshore work vessels they are carried at the stern or side of the vessel as well. The rest of the equipment is mounted wherever they are needed, most often on forecastle deck and poop deck.

3) Lifesaving Equipment

The quantity and types of lifesaving appliances to be carried on board depend largely on the number of people carried and is set forth by SOLAS. Lifeboats have to be carried near the mustering points for crew and passengers in such a way that they can be boarded and launched readily and safely. This necessitates them to be carried at the side of the deck or stern. Life rafts are carried in similar locations, although they are boarded after being launched. Other lifesaving equipment includes rescue boats, life jackets and life buoys.

4.3 Design Steps

Ship design is a process by which a ship is created satisfying various design requirements in a most advantageous way. It is customary to divide the ship design process into the following stages: conceptual design, basic design, detailed design, and production design.

4.3.1 Conceptual Design

Conceptual design is defined as a step during which the essential features of the ship are determined. The design is conducted during the ordering stage and should include all the elements that meet the owner's requirements. During this step, the designer determines initial specifications of the ship and clarifies its features which are distinct from existing ship designs. A general arrangement drawing is produced, and the basic properties and overall performance of the ship are calculated and predicted.

The designer then estimates the quantity of material and the amount of labor required to build the ship. This will enable him to make an estimate of the total cost of construction. The shipyard uses this information to quote a price to the client and proposes a building program.

4.3.2 Basic Design

Basic design is a phase during which a more detailed design is developed based on the concept design. As more design details are determined, it is possible to carry out more specific calculations and assessment. Some minor modifications of the outline specifications of the ship determined at the concept design stage may be made in consultation with the ship-owner.

The arrangement of major compartments is determined, and some basic calculations are performed, including stability, longitudinal strength and freeboard. The main engines are selected, propellers are designed and model tests are conducted. The basics of the structure and arrangement of major compartments, such as the engine room and accommodation quarters, are determined. The outcome of basic design, therefore, includes initial general arrangement, initial load estimate, midship section view of the hull, shell expansion plan, deck arrangement, and bulkhead layout, and estimated power, steam and electricity consumption.

4.3.3 Detailed Design

Detailed design is a stage where design guidance and design drawings are produced, based on the specifications and the basic design produced above, in such a manner that the owner's requirements are fulfilled while various functions do not contradict or interfere with each other. During this step the designer further develops the basic design and confirms the cargo compartments, ballast tanks, and fuel storage tanks, arrangement of auxiliary machinery in the engine room, areas for the auxiliary machinery, arrangement of cargo handling and ballast systems and piping, and steel outfitting. Care should be taken at this stage to prevent interference or discrepancies between the ship's functions from creeping in. Documents such as design guidelines can be useful for this purpose.

As for the structural design, the detailed structure of each compartment is determined based on the midship section view, shell expansion plan, and deck arrangement of the ship completed during the basic design stage. Detailed design will, therefore, produce the structural drawings of compartments such as the engine room, stern, bow and cargo holds, piping layout, outfitting layout, accommodation layout, electrical and navigation equipment layout, related design guidelines and basic naval architectural calculations.

4.3.4 Production Design

Production design is a step in which the designer defines the hull form suitable for construction down to its single component level and determines an efficient method and procedure to implement the drawings and design guidelines obtained in earlier design stages. Detailed part and assembly drawings are produced in accordance with the layouts and structural drawings, installation drawings are developed based on the facility available at the shipyard, and working drawings are created in accordance with the working standards, criteria and tools of the shipyard.

5. Computer-Aided Ship Design and Application of Information Technology

5.1 The Evolution of Information Technology in Ship Design

From the viewpoint of information technology (IT), ship design is a process of generating and expressing information on a product called ship. As shown in **Figure 3-7**, it was not until the late 1970s that shipbuilding information systems came into service in Korea, starting with the automated manufacturing systems such as the Management Information System (MIS) and the numerically controlled (NC) cutting machines. In the mid-1980s, the introduction of the computer-aided design (CAD) system triggered more wide use of IT in design.

The initial version of CAD was used as a tool for creating 2-D drawings and, in the 1990s, was integrated into 3-D CAD model-based design and production management. Since 2000, with the global spread of the internet, shipyards began to work with outside organizations, such as ship designers, classification societies and equipment manufacturers, on digital e-commerce projects. The evolution of information technology in the shipbuilding industry is illustrated as shown in **Figure 3-7**.

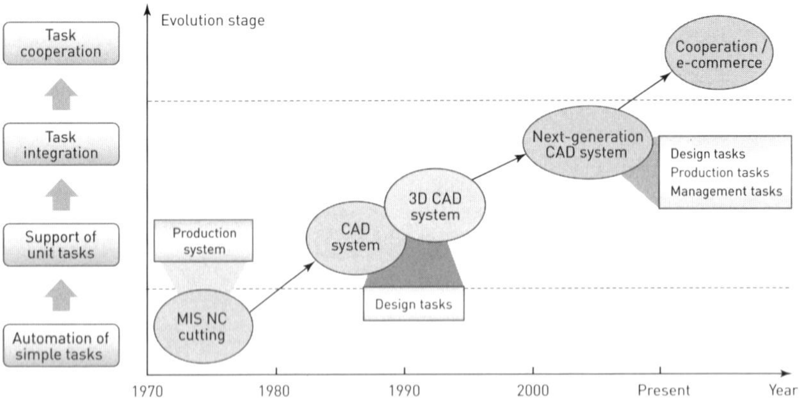

Figure 3-7 Evolution of IT application in the shipbuilding industry

5.2 Information Technology in Design

5.2.1 CAD

In ship design, the 2-D-based generic CAD systems, developed in Europe specifically for shipbuilding, had been widely used because they could be easily adopted and they reflected the shipbuilding knowledge and production processes prior to the introduction of such systems very well. Since 2000 large shipyards in Korea have worked with American and European CAD suppliers to develop proprietary CAD software using state-of-the-art technology. **Figure 3-8** is a conceptual illustration of the 3-D CAD system for shipbuilding with integrated hull and outfitting design currently under development.

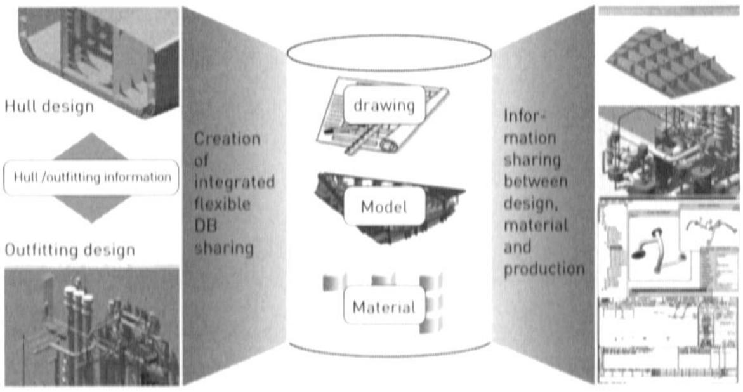

Figure 3-8 Concept of the 3-D CAD integration system for shipbuilding

5.2.2 CAE

Since a ship is a large and complex structure, it is practically impossible to test or verify every aspect of its performance before it is built, with the exception of some model experiments conducted to investigate certain critical elements of the performance. Consequently, computer-aided engineering (CAE) software is often used to analyze and assess the ship's performance in ship design.

As shown in **Figure 3-9**, the development of production design of a single vessel requires an overall evaluation and analysis of structural strength, fluid dynamics, vibration characteristics, noise level and production methods, and many CAE systems have been developed to support this process.

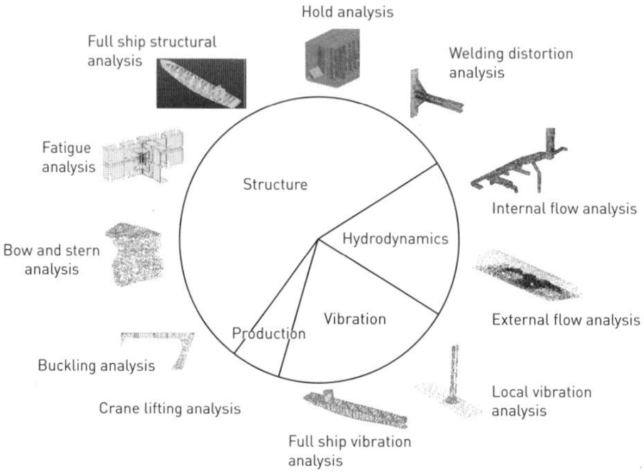

Figure 3-9 Application of CAE in ship design and construction

5.2.3 Computer-Aided Process Planning

In order to plan ship production, the construction methods appropriate for the building schedule and delivery have to be identified and communicated to the production department. Information on the production method includes block division, assembly, erection, transportation, and supports of blocks, as well as scaffolding and other work platforms. In the past such information was created manually, but shipyards now use computer-

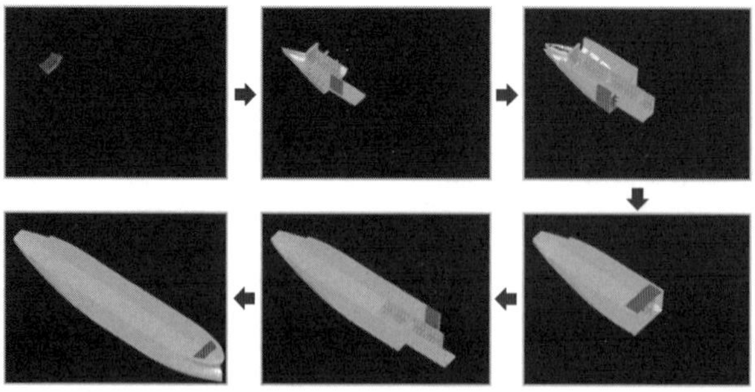

Figure 3-10 Example of the block erection produced by CAPP

aided process planning (CAPP) systems based on a 3-D computer model. **Figure 3-10** is an example of the block erection process using CAPP.

5.2.4 Enterprise Resource Planning

Since 2000 Korean shipyards have competitively adopted the enterprise resource planning (ERP) system to allow the integrated management of manpower, capital and equipment, and optimize procurement and production processes. Using the ERP system, the shipyards aim to revolutionize the way they work and make the flow of information, materials and capital 'customer-oriented' to cope more effectively with the changing market and gain a competitive advantage.

SD: Sales and distribution
MM: Material management
PP: Production planning
QM: Quality management
PM: Plant maintenance
HR: Human resources

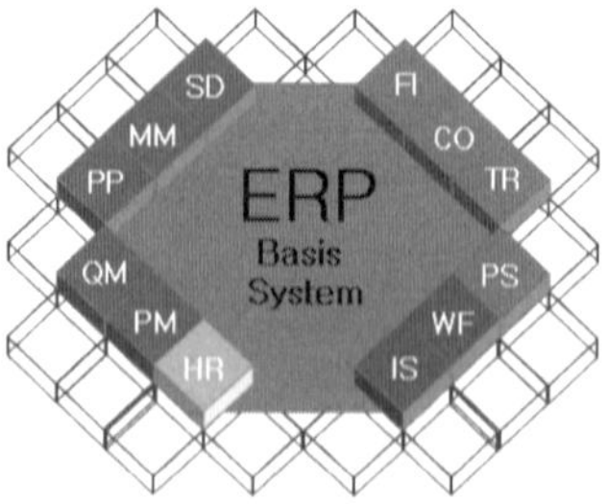

FI: Financial accounting
CO: Controlling
TR: Fund management
PS: Project system
WF: Work flow
IS: Information sharing

Figure 3-11 Typical modules of the shipbuilding ERP system applied in shipyards

5.2.5 Procurement Management System

Materials account for 60 to 70% of the entire shipbuilding cost. It is quite understandable, therefore, that shipyards try to minimize the costs of procuring the materials required for constructing a given ship. In the past procurement management systems were used to reduce procurement costs by minimizing the inventory level while trying to ensure stable supply.

Recently, however, information sharing and security systems have been developed to enable collaboration necessary in incorporating overseas materials and human resources into the company's strategic outsourcing resources. The scope of this practice has been extended to supply chain management (SCM) to include planning, implementation, accounting and making payments required in working with major suppliers on production, design, and logistics. In other words, the system is being further developed to strengthen such functions as logistics and inventory forecasting, enhanced management in response to trends in material and resource prices, and improved supplier management in keeping with the diversification of distribution channels. It is also intended to be used for analyzing procurement costs for the whole shipyard by re-classifying major resources, and for supporting strategic procurement decision-making. This requires the integration of manufacturing monitoring and the ERP system, and **Figure 3-12** shows the strategy for SCM after the deployment of ERP.

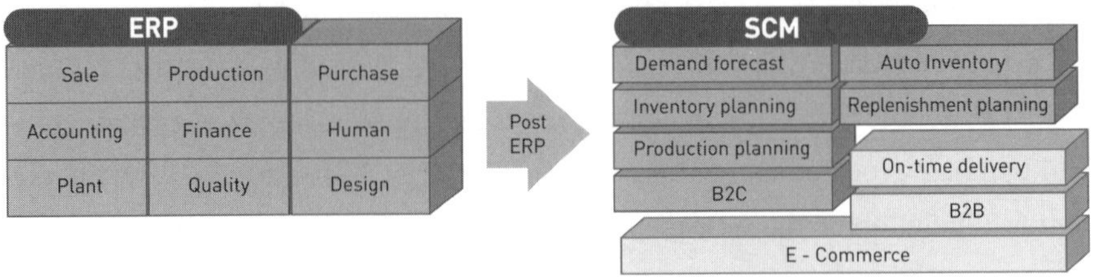

Figure 3-12 SCM strategy connected with ERP

5.2.6 Management System for Technical Knowledge

The growing need for experts and expert knowledge in shipbuilding has led to projects to apply knowledge management system (KMS) which systemize the process of sharing and transferring such knowledge. A few large shipyards have already deployed the KMS. Though at present insufficient amount of knowledge has been stored in the system, once sufficient information necessary, modeled on the optimum work patterns, is accumulated, the best practices and information templates can be utilized. This process also allows rapid learning and therefore can produce many experts in a short period of time.

5.2.7 Simulation-Based Design and Manufacturing

Since the adoption of 3-D CAD in the shipbuilding industry, there has been increasing efforts to verify design documents and construction processes before the actual ship construction. In particular, the growing demand for high-valued ships and novel concept designs has led to the development of simulation techniques aimed at optimizing functions, verifying feasibility, reducing manufacturing costs and identifying and eliminating risks during manufacturing. Therefore, there is an increasing necessity to create an environment for simulation-based design and manufacturing, which can process large quantities of design and manufacturing information and handle irregular changes in the working environment.

In this way, efforts are being made to develop a comprehensive system by integrating the traditional concept of computer-aided design and production with the computer-aided analysis of hydrodynamic performance and structural behavior and virtual realization of engineering technologies such as novel construction methods. Application of simulation-based design and manufacturing (SBD & M) using these technologies to parallel design, manufacture, assembly, installation, testing and maintenance to ship construction is being actively pursued. Such a development will effectively revolutionize the existing design concepts and workflow in the future.

Figure 3-13 presents a screen shot of an actual dynamic simulation of the production and erection of an offshore structure.

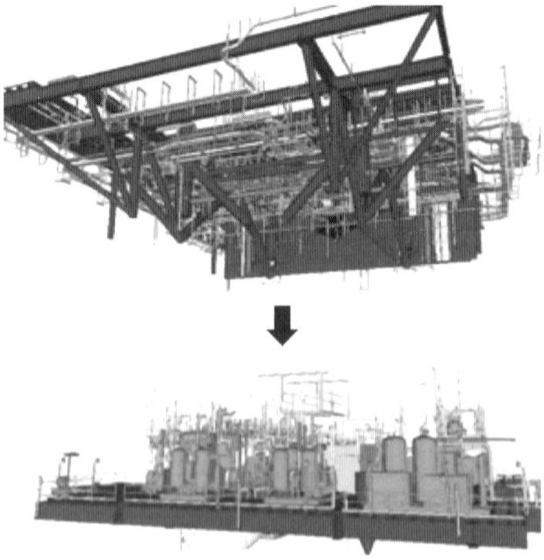

Figure 3-13 Erection simulation for offshore structure

5.2.8 Asset Management System and Computerized Maintenance Management System

Prices for offshore structures that are currently ordered range from a few hundreds of million to billions of dollars. Unlike moving ships, these offshore structures have to be installed at specific locations and operate for over 20 years. The system for repair and maintenance is the Asset Management System (AMS) in shipyards and the Computerized Maintenance Management System (CMMS) for the owners.

These systems are aimed at providing computerized management of information on the necessary equipment and parts, including specifications, makers, shapes, installation locations, and performance, as well as maintenance history so that the offshore structure can operate for as long as possible.

Such a system is indispensable, since improper repair and maintenance of equipment or parts can result in a huge loss through downtime. The maintenance system is an exemplary risk management tool aimed at preventing future losses, a topic which has been receiving much attention recently.

Figure 3-14 shows the Information Management System (IMS) to allow collaboration in the shipbuilding and ocean engineering industry. The IMS is internally linked to CAD or ERP software to create an asset management system while sharing information with external organizations. When connected to the maintenance management system of the owner, it can provide engineering information for repair and maintenance purposes.

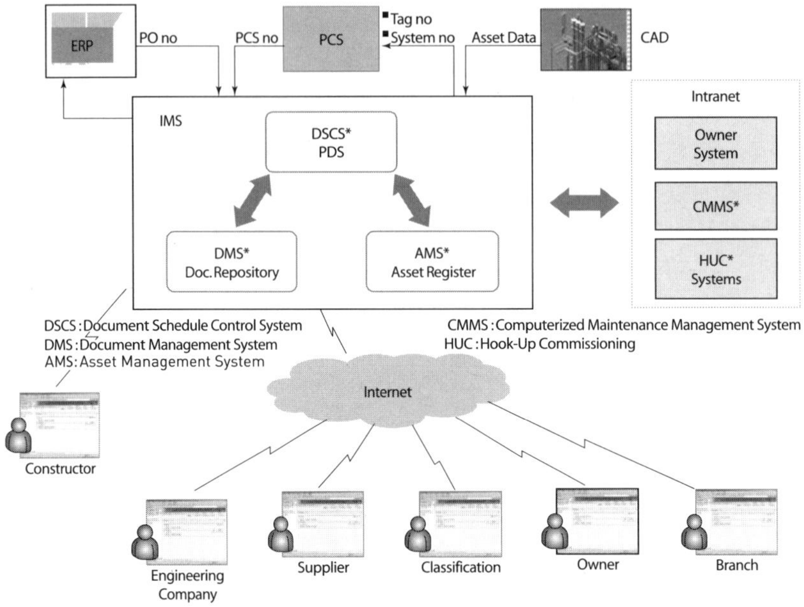

Figure 3-14 Information Management System (IMS) for shipbuilding and ocean engineering project

5.2.9 Information Technology in Manufacturing

Information technologies in shipbuilding are largely divided into production management IT and automated production IT. Production management encompasses design, materials management, process planning and control, scheduling, logistics management, quality management and facilities management. Typical production management systems are as follows:

- integrated resource management system;
- integrated logistics management system;
- integrated manufacturing planning, control and analysis system;
- digital manufacturing system by workplace;
- real-time manufacturing information system; and
- manufacturing information management using PDA.

Automated production covers all the equipment and software used by a shipyard to build ships. Such equipment includes tools/jigs, robots, machinery, cranes and surface plates, and software includes application software for controlling the equipment, knowledge and know-how. The major examples of automated manufacturing systems developed for shipbuilding tend to combine CAD and computer-aided manufacturing (CAM), and include 'welding robots for block assembly, mid-assembly and sub assembly,' 'robotic CAD interface and CAM systems including direct numerical control (DNC)' and the 'large-block 3-D measurement systems.'

References

1. CAD & Graphic ed., 2005. *PLM guide book* (in Korean). Seoul: BB Media.

2. Erleben, K. et al., 2005. *Physics-based animation*. Rockland, MA: Charles River Media, Inc.

3. Kwon, Y.J., 2006. *Theory of ship design* (in Korean). Seoul: Dongmyungsa.

4. Lee, C.E., 1984. *Ship Design* (in Korean). Seoul: Daehan Textbooks Co., Ltd.

5. Park, M.G. & Kwon, Y.J., 1995. *Basic ship design* (in Korean). Seoul: Hankookigonghaksa.

6. Shin, J.G., 2006. *CAD, digital manufacturing and PLM* (in Korean). Seoul: Sigmapress.

7. The Society of Naval Architects and Marine Engineers, 1971. *Principles of Naval Architecture*. Translated to Korean by Yim, S.J., 1991. Seoul: Daehan Textbooks Co., Ltd.

8. The Society of Naval Architects of Korea, 1993. *Introduction to Naval Architecture and Ocean Engineering* (in Korean). Seoul: Dongmyungsa.

9. The Society of Naval Architects of Korea ed., 2011. *Introduction to Naval Architecture and Ocean Engineering* (in Korean). Seoul: GS Intervision.

CHAPTER 4

Hull Form Design

KRISO 3,600teu Containership(F1+A2)

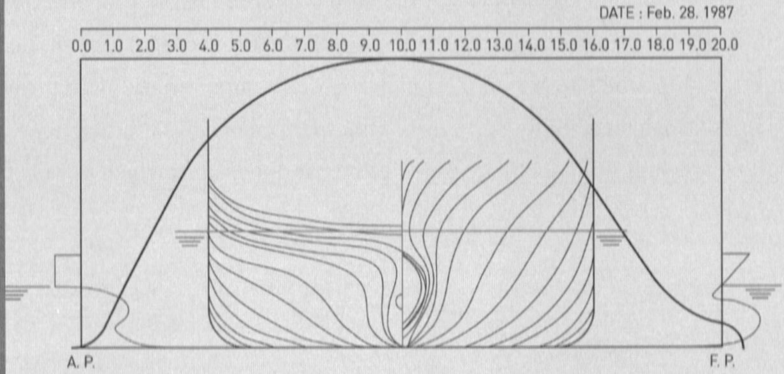

1. Ship's Principal Particulars and Hull Form Coefficients

1.1 Ship's Principal Particulars

For ship-owners, the primary considerations in determining the economic efficiency of a ship are its deadweight, speed and required engine power, and they will be anxious to minimize the structural weight and reduce the power requirement and shipbuilding costs. Ship designers, on the other hand, have to determine the principal dimensions, such as length, beam and draft, since their job is to design a ship that can fulfill the deadweight and speed requirements within the allowable limits of engine power in such a way as to minimize the ship's weight and building costs. Therefore, these six parameters (length, breadth, draft, deadweight, speed and power) become the principal particulars in the preliminary design of a ship and are treated as key data along with other additional requirements by the ship-owner and the shipyard.

Once the initial set of principal particulars is selected, the designer can outline the general arrangement of the ship's compartments and roughly assess its basic influence on trim, longitudinal strength and stability. The principal particulars are then adjusted according to the result of this assessment and the process is repeated. The appropriate principal particulars are decided, when, after a few iterations, the designer is satisfied with the assessment.

Only after the principal particulars are confirmed can the designer conceive the ship's hull form and render it as a lines plan corresponding to those particulars. To express the hull form as a lines plan, the designer must conduct a more detailed investigation upon the influence of the principal particulars on hydrodynamic performances. This investigative process is a phase in which a more effective hull form that meets the given principal particulars is determined. Since, upon deriving the hull form, the ship's hydrodynamic properties are also determined, the phase is the most important in the entire workflow of hull form design. When determining the ship's hull form, the designer can draw on a variety of empirical formulae for ship speed, maneuverability and sea-keeping performance as the functions of ships principal particulars, hull form coefficients and principal dimensions which have been derived and compiled for various ship types.

1.2 Hull Form Coefficients

A few hull form coefficients are used as measures of the fullness of the submerged hull form and the water plane area. These hull form coefficients are closely related to the ship's hydrodynamic performances such as resistance and stability. Therefore, finding the ship's hull form coefficients is an essential part of ensuring good hydrodynamic properties of the ship in the hull form design process.

1.2.1 Block Coefficient

Consider a rectangular block, with the same dimensions as the ship, being carved into the required hull form. The block coefficient (C_B) is then defined as the ratio of the displacement volume of the ship at the given draft to the volume of the rectangular block. This relationship is illustrated in **Figure 4-1**. It can be readily seen that for normal hull forms this coefficient cannot be greater than 1. A low C_B implies a slim hull, and the hull will become fuller and get closer to the rectangular block as it approaches 1.

C_B ranges from 0.5 to 0.6 for high-speed ships, such as passenger or container ships; 0.65 to 0.75 for medium-speed cargo ships; and 0.78 to 0.85 for low-speed, large oil tankers or bulk carriers. The displacement volume of a ship is usually represented by the symbol ∇, and the block coefficient of a ship can be expressed $C_B = \nabla / LBT$ where L is the length, B is the beam and T is the draft.

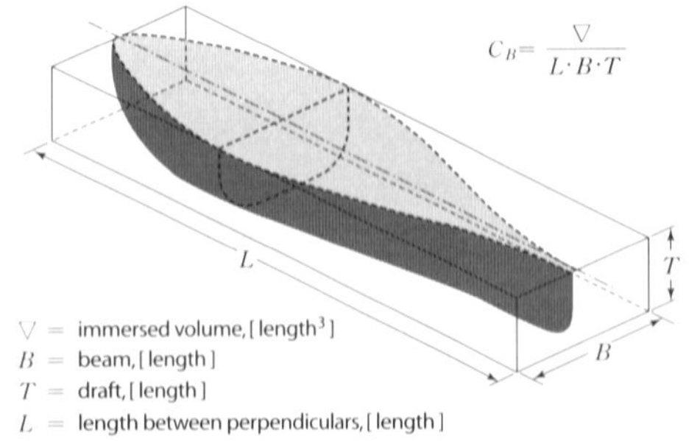

Figure 4-1 Block coefficient

1.2.2 Maximum Section Coefficient

Maximum section coefficient (C_X) is the ratio of the maximum cross sectional area A_X of the submerged hull to the area of rectangular cross section of the block. Thus the maximum section coefficient can be derived from the relation: $C_X = A_X / BT$ where B is the beam and T is the draft of the ship. As shown in **Figure 4-2**, a small C_X indicates the maximum cross section with a narrow bottom width, and as it gets larger to approach 1, the shape of the maximum section becomes closer to a rectangle.

The coefficient ranges from 0.85 to 0.97 for high-speed ships; 0.98 to 0.99 for medium speed ships; and about 0.995 for low-speed ships. The ship's midship section coefficient C_m is a similar dimensionless coefficient,

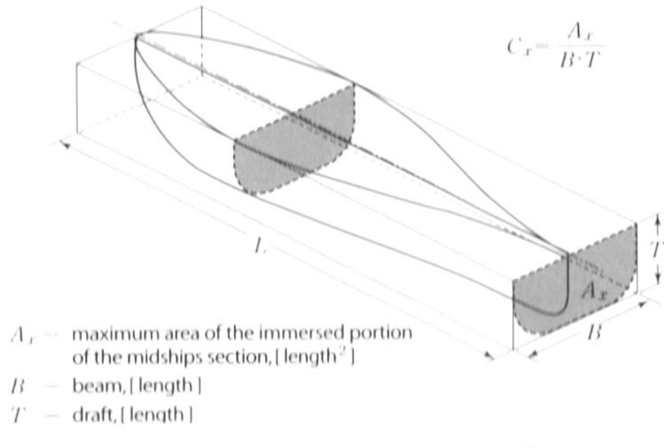

Figure 4-2 Maximum section area coefficient

and is determined in the same way as C_x except that the area of the midship section A_m is used in this case instead of the maximum cross section. For the majority of ships these two coefficients are equal.

1.2.3 Prismatic Coefficient

Prismatic coefficient (C_P) is the ratio of the ship's displacement volume to the volume of a prismatic body with the midship section as its uniform cross section as illustrated in **Figure 4-3**. It can be seen that a small C_P indicates a hull form with fine bow and stern, and as it gets closer to 1, the volume is distributed more evenly along the length of the ship. C_P can be derived by dividing the block coefficient with the maximum-section coefficient. For high-speed ships, C_P ranges from 0.56 to 0.62; for medium speed ships, 0.66 to 0.76; and for low-speed ships, 0.78 to 0.86.

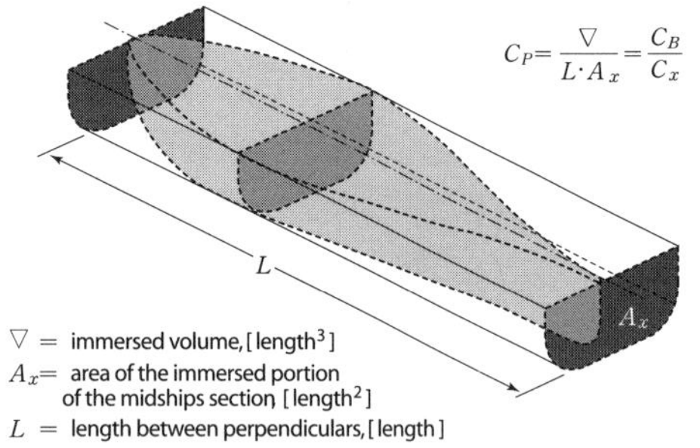

Figure 4-3 Prismatic coefficient

A sectional area curve can be drawn by plotting the sectional areas upon the horizontal axis of the longitudinal locations. Such a curve shows the longitudinal distribution of the ship's cross sectional area, and the area under this curve is equal to the displacement volume of the ship. This curve is closely related to the ship's resistance, since it indicates the distribution of displacement volume per unit length.

1.2.4 Waterplane Area Coefficient

As shown in **Figure 4-4**, the waterplane area coefficient (C_{WP}) is the ratio of the waterplane area (A_{WP}) to the area of a rectangle of the same length (L) and breadth (B), which is expressed as $C_{WP} = A_{WP} / LB$ and represents the fullness of the waterplane. For high-speed ships, C_{WP} ranges from 0.68 to 0.78; for medium-speed ships, 0.8 to 0.85; and for low-speed ships, 0.86 to 0.92. The area and shape of the waterplane is one of the factors that influence the transverse stability of the ship. Even for high-speed ships, therefore, there are occasions when this coefficient should be set higher than normal: for example, small passenger ships that require higher level of stability.

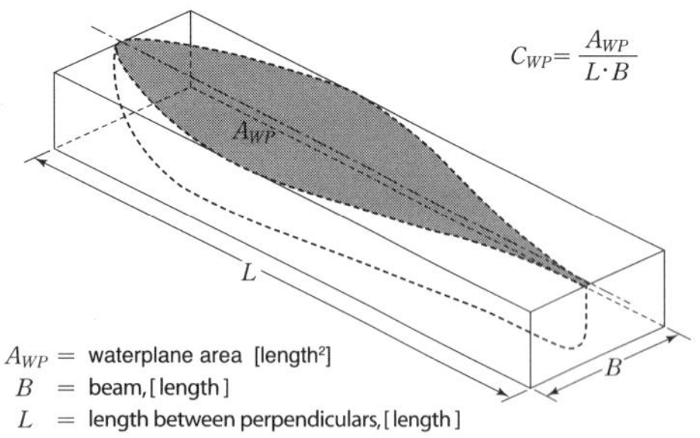

Figure 4-4 Waterplane area coefficient

1.3 Evaluation and Determination of Principal Particulars

1.3.1 Length, Beam and Draft

An increase in the ship's length (L) causes an increase in the construction costs – more than the variation of beam or draft does. It influences the ship's resistance a great deal: given the constant displacement, the ship's wave resistance gets smaller as its length increases, whereas its frictional resistance increases as its submerged surface area increases. Therefore, for ships traveling at high speed a longer ship has less total resistance, while for low-speed ships the opposite is true. Length is often constrained by the length of docks and canal locks which it is likely to use.

An increase in draft (T) generally leads to a reduction in resistance, but the change in construction costs is relatively small compared to an increase in length or beam. However, since draft is subject to the restrictions imposed by the water depth of ports, canals, rivers and docks, it is usually considered a design requirement along with displacement and speed rather than a variable design parameter.

Beam (B) is one of the dominant factors that have an impact on the stability of a ship. If the beam is increased to achieve the desired stability, the ship's resistance also increases. To counterbalance such an increase in resistance, the designer must reduce the block coefficient or change the hull form in some way. In the case of most low-speed ships, a reduction in length and an increase in beam to compensate for the subsequent reduction in displacement can not only improve the ship's stability, but also reduce its submerged surface area, leading to only a small net increase in resistance. This strategy is commonly adopted when attempting to reduce the cost of a ship in case of large crude oil tankers.

1.3.2 Length-beam Ratio

Along with the block coefficient, the length-to-beam ratio (L/B) is most closely related to the resistance and propulsion of a ship. **Figure 4-5** shows scatter plots of L/B ratios against C_B and Fn of some recent ships. Using these graphs, the designer can select an initial L/B ratio for the given values of C_B and Fn of a design hull form.

For a hull form with high C_B and low Fn, a low L/B will be preferred, because the proportion of the wave-making resistance to the total resistance is relatively small. Hull form optimization in this case will be focused on the reduction of the wetted surface area to minimize the frictional resistance. Hull forms with low C_B and high Fn, on the other hand, represent high-speed ships, such as container and passenger ships. For such ships the wave-making resistance is the dominant components in the total resistance and, consequently, a higher L/B is preferred.

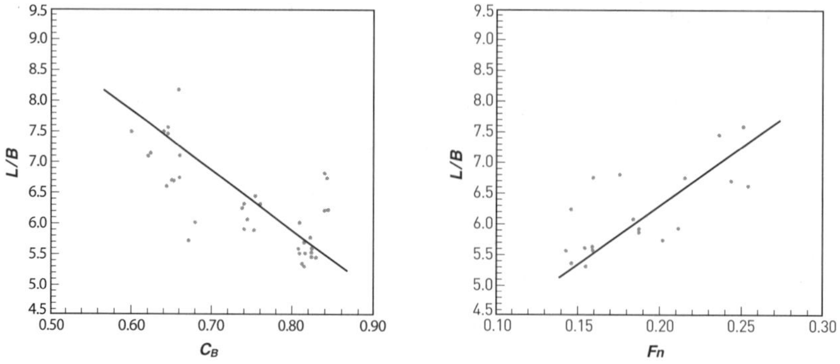

Figure 4-5 *L/B* vs C_B (left) and *L/B* vs *Fn* (right)

The *L/B* ratio selected based on C_B should be reasonably close to the *L/B* selected based on *Fn*. However, if the difference between them is too high, it probably means the hull form is unconventional. In that case careful efforts may have to be made to reduce wave resistance even for a low-speed ship and to reduce frictional resistance even for a high-speed ship.

Once displacement and draft have been determined, an increase in the *L/B* ratio leads to the reduction of resistance at all speed range. **Figure 4-6** compares the power required to maintain given speeds as functions of *L/B* for a container ship in case of a constant displacement. It shows that, as the *L/B* ratio gets smaller, the resistance increases sharply, and this tendency becomes more pronounced as the speed increases. When designing a high-speed ship, therefore, the designer must take care while attempting to reduce the *L/B* ratio.

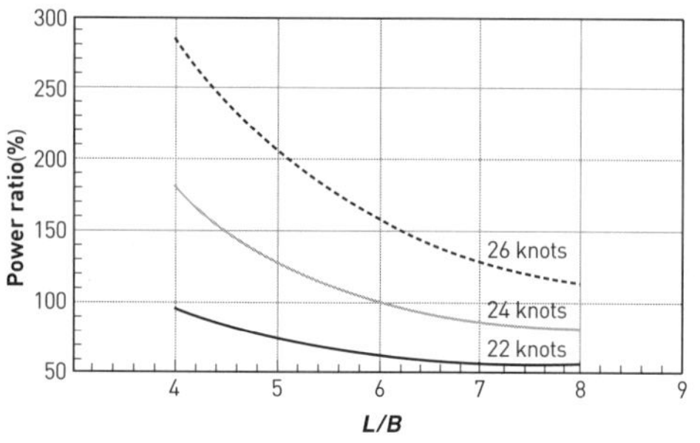

Figure 4-6 Power ratio vs *L/B* for various speeds

Figure 4-7 is a chart that is commonly used in designing container ship hull forms, when design speed, displacement and draft are prescribed by the owner's requirements. This chart shows contours of constant power ratios (thick solid lines) with *L/B* on the horizontal axis and C_B on the vertical axis.

Since displacement and draft are generally given from the design conditions, ∇/T can be expressed as a constant c^2. Since $C_B = \nabla/(LBT)$, $C_B = c^2/(L{\cdot}B)$.

If we multiply the right hand side with *L/L* and rearrange for *L*, we obtain $L = c\sqrt{(L/B)/C_B}$. Similarly if we multiply the right hand side with *B/B* and rearrange for *B*, we obtain $B = c\sqrt{(L/B)/C_B}$. These two expressions can be added to the constant power contour chart as iso-length lines (dotted lines) and iso-breadth lines (thin solid lines), forming a network of lines as shown. This chart can now be used to reduce the power requirements by varying the principal dimensions for any given condition.

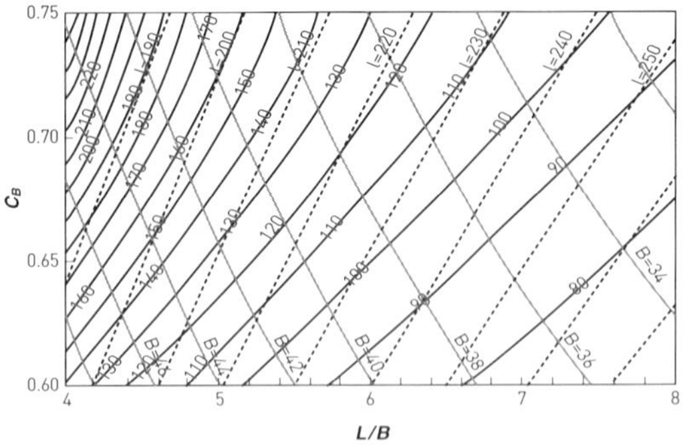

Figure 4-7 Iso-power lines for L/B and C_B variations

For instance, let us suppose that the designer has set initial values of 6.0 and 0.65 for L/B and C_B, and that the required horsepower for these values is 100%. By consulting the chart in **Figure 4-7** we can see that there are a few hull forms that require about 10% less power as summarized in **Table 4-1**. If the maneuverability and sea-keeping performance are examined in this process, the optimal principal particulars can be selected. L/B can be used as a general measure of maneuverability, i.e. a higher L/B ratio usually means greater course-keeping ability, while a smaller L/B ratio means greater course-changing ability.

Table 4-1 Examples of hull form design using the iso-power curves

	L/B	L	B	C_B	Power ratio
Initial selection	6.0	231	38.5	0.65	100%
Same L/B	6.0	237	39.6	0.615	90%
Same B	6.2	239	38.5	0.625	90%
Same C_B	6.6	242	36.8	0.65	90%
Optimum reduction	6.4	241	37.7	0.635	90%

1.3.3 Length-draft Ratio

The data compiled from existing ships reveal that a low-speed ship with high C_B has a smaller length-draft ratio (L/T) while a high-speed ship with high Fn has a greater L/T. **Figure 4-8** presents scatter diagrams of L/T against C_B and Fn together with trend lines obtained by linear regression analysis. The graphs show that, for container or passenger ships with low C_B and high Fn, the L/T is quite high, whereas low-speed ships with high C_B and low Fn have low L/T.

If a container ship has near-rectangular cross sections for maximum loading capacity, the effect of L/T becomes more significant. The fluid around a barge-like hull mainly flows along the buttock lines. A low L/T ratio makes the buttock lines steeper, and the fluid cannot follow the lines easily, causing an increase in resistance.

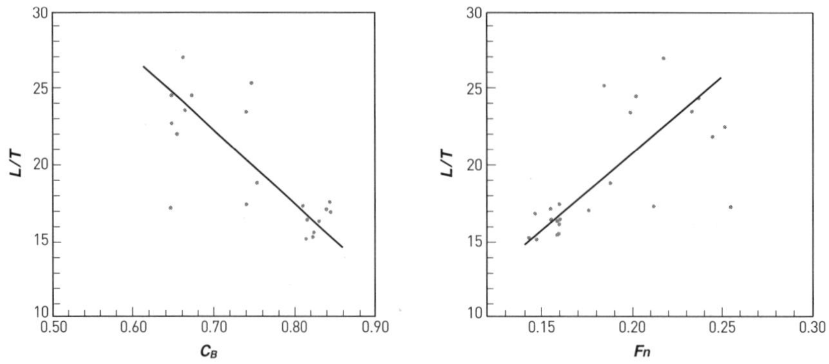

Figure 4-8 L/T vs C_B (left) and L/T vs Fn (right)

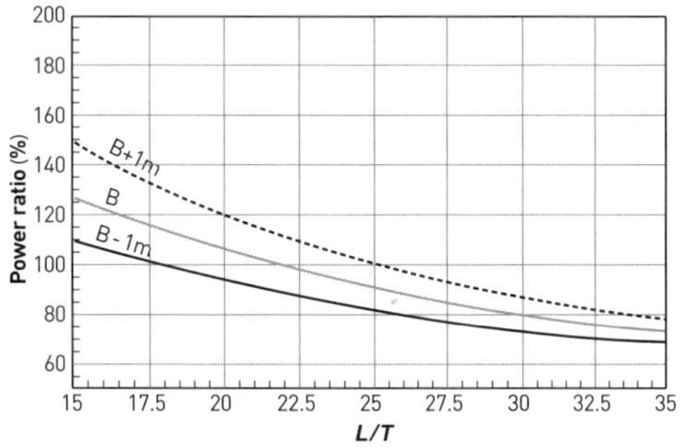

Figure 4-9 Power ratio vs L/T for various beams

Figure 4-9 shows the variation of the required power of passenger ships as functions of L/T for a given displacement and speed. Given the constant displacement and beam, any increase of L/T leads to lower resistance and, consequently, a reduction in required power. Since this ratio also influences the pitching motion, the designer must keep the risk of damage caused by bow slamming in rough seas in mind when deciding on the length to draft ratio.

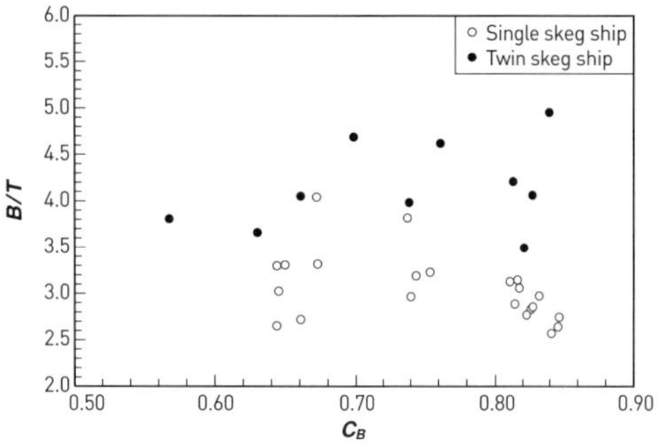

Figure 4-10 B/T vs C_B

1.3.4 Beam-draft Ratio

The data of the principal parameters of existing ships do not appear to exhibit any clear trend of beam to draft ratio (B/T) against C_B, unlike L/B and L/T ratios. As shown in **Figure 4-10**, however, the data of single screw ships, in general, appear to have lower B/T ratios than twin-screw ships.

There are a variety of reasons why twin screws are preferred to single screw for some cargo and passenger ships. Firstly, it is difficult for single screw ships with high B/T ratio to have good maneuverability, but the ships constrained in draft must be made beamy to meet the displacement requirement. Secondly, there is also a limit to the diameter of the propeller in this case and it may not be possible to obtain necessary propulsive power with just one such limited propeller. Thirdly, twin screw ships, when one of the propellers fails, can rely on the other, i.e. they have propeller redundancy.

B/T is also related to the ship's resistance. **Figure 4-11** represents the relationships between the power ratio (%) and *B/T* ratio for various L_{BP} and C_B for low-speed, full-form single screw ships. It can be seen from the graph that the L_{BP} and the overall resistance increase with *B/T* along the constant C_B curves. However, the constant L_{BP} curves indicate that the power ratio as well as C_B decreases as *B/T* increases, if the length is kept constant.

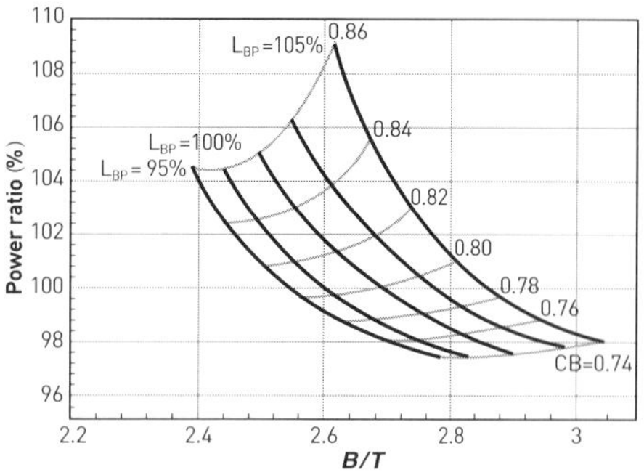

Figure 4-11 Power ratio vs *B/T* for various L_{BP} and C_B values

1.3.5 Longitudinal Center of Buoyancy

The most important factor after the hull form coefficients and principal dimensions that has an impact on the ship's resistance performance is the longitudinal center of buoyancy (*LCB*). **Figure 4-12** presents the data compiled from existing high-performance ships of *LCB* (as % of ship length *L*, positive forward from the midship) plotted against *Fn* with a solid trend curve obtained by least squares regression analysis. When the *LCB* is determined through the hull form design process, it should be compared against this figure. If it deviates too much from the norm presented in this figure, the cross section and waterplane shapes may have to be further optimized so that the ship's resistance is acceptable.

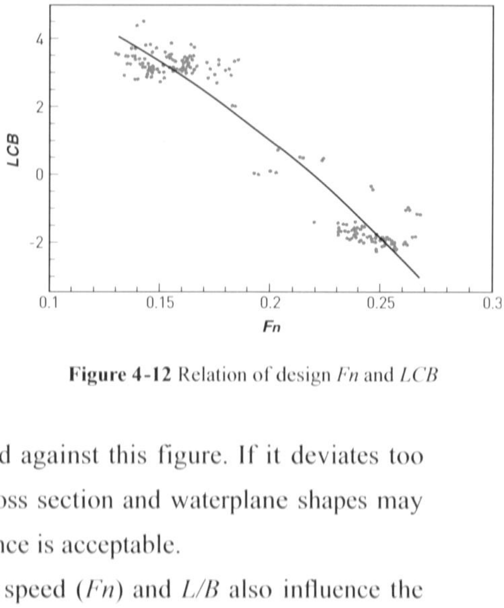

Figure 4-12 Relation of design *Fn* and *LCB*

Aside from the block coefficient (C_B), the design speed (*Fn*) and *L/B* also influence the resistance performance. Therefore, these factors must be considered simultaneously in determining the appropriate position of *LCB*. For example, the wave resistance of a ship with a high design speed is likely to be high. Therefore, the bow should be slender and more pointed and consequently *LCB* is likely to be further aft than usual.

On the other hand, for a low-speed ship, the *LCB* should be shifted forward to reduce form resistance which, as one of the viscous resistance components, is affected the most by the stern part. If the *L/B* of a low-speed vessel becomes smaller, the run angle of the stern will increase with accompanied increase of form resistance. In this case it is necessary to improve the flow by reducing the run angle through moving *LCB* further forward than normal. This increases wave resistance, but, for low-speed ships, the beneficial effect of reducing viscous resistance is greater than the increase in wave resistance.

2. Basics of Hull Form Design

2.1 General Scope of Hull Form Design

In general, hull form design is the first step in ship design. No further design work can be undertaken until the hull form is determined. While the designer has to satisfy the owner's requirements, the shipyard may ask the designer to develop a simple, easy-to-build design so that the ship production process is simplified as much as possible. This also plays a critical role in enhancing the shipyard's competitiveness.

The shipyard first selects the principal particulars that provide minimum light displacement to reduce the required power, and ultimately reduce shipbuilding and operating costs, as long as they meet the owner's requirements. An outline general arrangement is developed within the given initial principal particulars, and then the ship's basic properties, such as trim, longitudinal strength and stability, are examined. If these are satisfactory, it can be concluded that the principal particulars are reasonable.

The principal particulars and the arrangement of compartments are closely related to the hull form, and consequently the whole process of arriving at an optimum compromise of the conflicting factors so that the economic efficiency is maximized while satisfying all the design requirements can be comprehensively referred to as the hull form design. However, the principal particulars and an outline general arrangement have to be determined before signing the contract to build the ship, and the process in which the designer studies and determines various factors which are likely to influence the ship's performance and then produces a lines plan of the hull geometry which satisfies these factors is known as hull form design in practice.

For most ships, once the ship type and size are selected, the range of feasible principal particulars and hull form coefficients is determined. Once all the hull form factors are determined, about 50 to 60% of the ship's hydrodynamic performance, and building and operating costs are determined, and the rest is determined when the hull form is finalized. Therefore, it is very important to optimize the hull form through a systematic investigation to improve the hydrodynamic performance and develop a more efficient general arrangement.

2.1.1 Classification of Hull Forms

Normally, ships are classified by size, form, type and speed into full-form low-speed ships, medium-speed ships and high-speed slender ships. However, as shown in **Figure 4-13**, hull forms can also be categorized by the nature of the lift force they rely on to float, such as hydrostatic force, aerodynamic pressure and hydrodynamic lift. Ships supported by the hydrostatic force include conventional displacement ships, and special displacement ships, such as catamarans, trimarans and SWATH ships, belong to this category as well; ships supported by the hydrodynamic force include hydrofoil and planing craft; and lastly, ships supported by the aerodynamic pressure include hovercraft and surface effect ships, which feature an air cushion at the bottom of the ship and propelled by a separate engine. More recently, new types of ships have been developed using a combination of two or more of the three lifting forces. High-speed ships can also be divided into hovercraft, hydrofoils, mono-hull and multi-hull vessels.

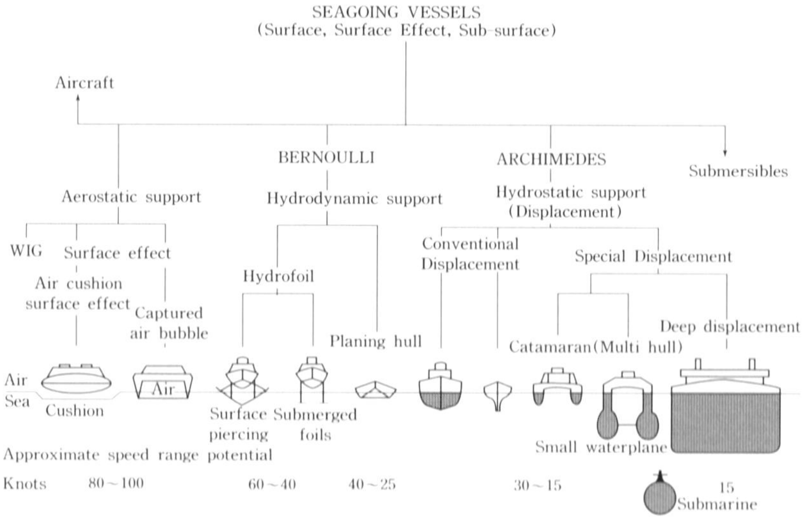

Figure 4-13 Hull forms classified by hull-supporting methods
(Courtesy of Resistance and Propulsion of Ships, SNAK, 2009)

2.1.2 Division of Hull Form

When designing a hull form, it is often more convenient to divide the hull into two bodies (fore-body and after-body) or into three bodies (bow, parallel middle body and stern). An important division used in hull form design is related to the behavior of flow around the hull. According to this division the part extending from the fore perpendicular to the point where the maximum cross section begins is called the 'entrance.' The entire body portion of the maximum cross section is the 'parallel middle body.' The body further aft of the parallel middle body is called the 'run.' For high-speed ships, the entrance and run take up most part of the ship's length with hardly any parallel middle body. For low-speed vessels, however, the entrance part is relatively short while the parallel body and the run part are longer.

2.2 Concept of Hull Form

Designers spend much time searching for a hull form that not only provides the ship with excellent hydrodynamic performance but also allows the efficient arrangement of compartments. Their preoccupation at this stage of the design process is with the question of "Which part should be shaved off, and which part should be filled up and how?" Put in another way, the task here is to determine the longitudinal distribution of buoyancy and the location of the center of buoyancy to minimize wave generation or wave breaking resistance while ensuring the most efficient arrangement of cargo holds. At the same time, the designer should ask what shape of stern has low viscous resistance, is appropriate for the engine arrangement, and provides high propulsive efficiency.

The most important factors that influence a ship's hydrodynamic performance are LCB, sectional area curve and the shape of cross section. Other important factors include the shape of bulbous bow and the profiles of the bow and stern. These factors vary depending on the vessel type, size, and operating conditions, as well as the block coefficient and the arrangement of compartments. To come up with the best hull form possible, the designer needs to understand fully their relationships and sensitivity. For example, the hull forms of a 300K DWT tanker and a 45K DWT bulk carrier are very different even though they have similar principal dimension ratios, block coefficient and design speed. This is because the characteristics of resistance and propulsion vary with the Froude number Fn, which can be regarded as the non-dimensional speed, based on the length of ship.

2.2.1 Sectional Area Curve

Sectional area curve (C_P curve) shows the longitudinal variation of the cross sectional area along the length of ship, or the longitudinal distribution of the ship's volume of displacement. The sectional area curve also indicates how the incident flow to the bow should split, flow aft past the maximum cross section and converge at the stern. Not only does this curve have a crucial effect on the hydrodynamic property, but it also shows the longitudinal variation of ship's interior space. It is, therefore, a critical design variable that

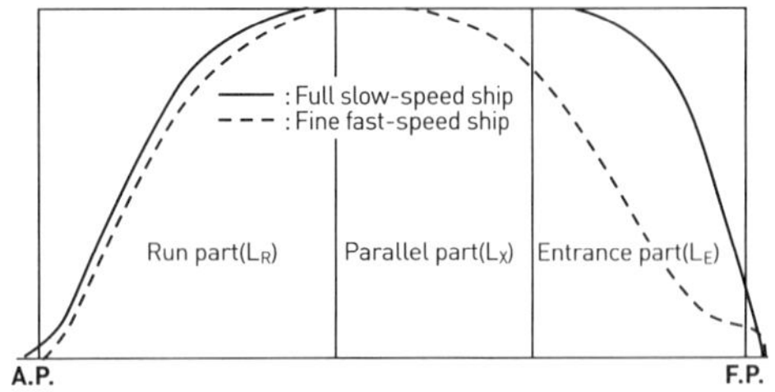

Figure 4-14 Typical sectional area curve (C_p curve)

significantly affects the arrangement of cargo holds and main engines.

The shape of this curve depends on the *L/B* ratio, block coefficient, *LCB* and *Fn* at the design speed. **Figure 4-14** represents examples of the sectional area curve of a high-speed ship (dotted line) and a low-speed vessel (solid line). The high-speed ship has a fine bow and consequently the *LCB* is located toward the rear while the slow ship has a blunt bow and, as a result, the *LCB* is situated much further forward.

A hull form designer often starts by reviewing the likely existing ships and selecting a parent hull. Its sectional area curve is then modified to a desired one based on the designer's experience. The methods of deforming the parent sectional area curve include 1-C_P method, Lackenby's method and the swinging method. Since the sectional area curve is sensitive to *LCB*, *Fn*, C_B and *L/B*, it is advisable to generate as many sectional area curves as possible from existing similar ships. The final sectional area curve can then be determined after hydrodynamic numerical analysis and/or model tests. At the same time, local geometry, such as the fore-shoulder and the cross sectional area of the bulb must be determined after carefully studying their effect on resistance characteristics.

The sectional area curve of the stern may not be as important as that of the bow. However, it should be designed to have gentle curvature and a smooth shoulder wherever possible in order to reduce viscous resistance and enhance propulsive efficiency. If the sectional area curve in the stern region is to be designed so that the stern transom can be submerged without causing vortices or vibration, then the curve has to be gentler near the stern, improving hydrodynamic performance. This is true especially for slow full-form ships or super-fast ships.

2.2.2 Shapes of Cross Section

Of the factors that determine the bow shape, the cross sectional area curve is the most important, followed by the shape of the cross section at the bow. As illustrated in **Figure 4-15**, the cross section shape of the bow is often divided into a U- and a V-shape; the U-shaped bow is used for low-speed ships, while the bow with a V-shaped cross section is selected for high-speed ships. The U-shape is good for wave resistance, arrangement of compartments and ease of construction. The V-shape can be effective in reducing wave-breaking resistance when used in a ship with a bulbous bow. Furthermore, the UV-shaped cross section, which combines the U and V forms, is used in small, medium-speed ships.

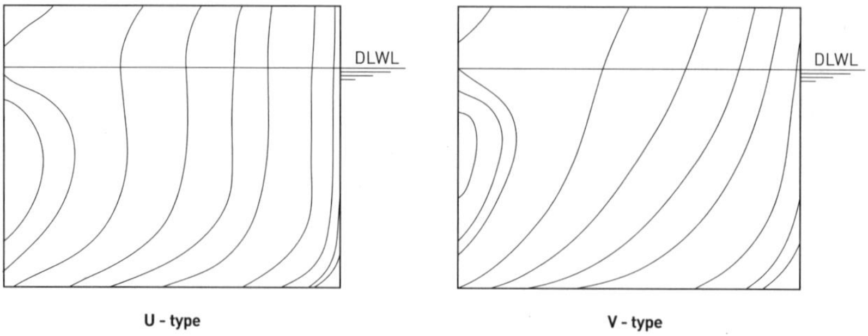

Figure 4-15 Bow shape of U- and V- types

The cross section shape of the stern has significant effects on viscous resistance, form resistance and propulsive efficiency. Since the stern shape is much more complicated than the bow, the reference longitudinal position of the stern cross section has to be determined first. The reference point is usually taken at 5-percent L_{BP} forward of the aft perpendicular (A.P.), namely, the first station forward of the A.P. when 10 equally spaced stations are used.

As illustrated in **Figure 4-16**, a tangent line is drawn to the cross section of the longitudinal reference point at the height of the shaft center. The section shape is categorized into U-type, bulb-type and V-type as illustrated in **Figure 4-17** depending upon the gradient of the tangential line. For the U-shaped stern, the skeg surrounding the shaft is relatively bulky, but the main hull is rather slender. For the bulbous stern, the water planes around the shaft are relatively narrow but the main hull is wide. So the U-shaped and bulbous sterns are usually used for low speed ships. The V-shaped stern, on the other hand, is usually used for high-speed slender ships.

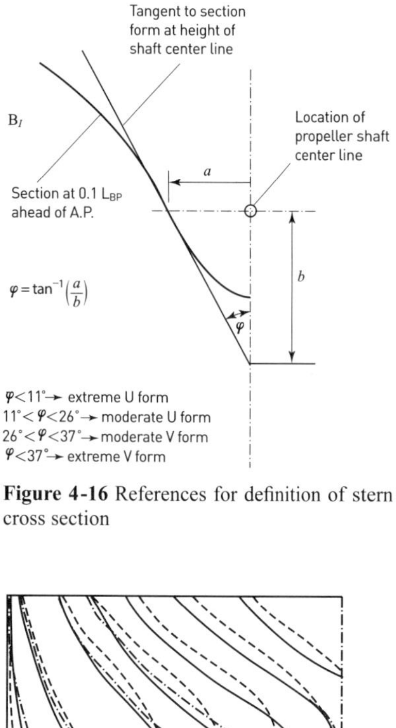

Figure 4-16 References for definition of stern cross section

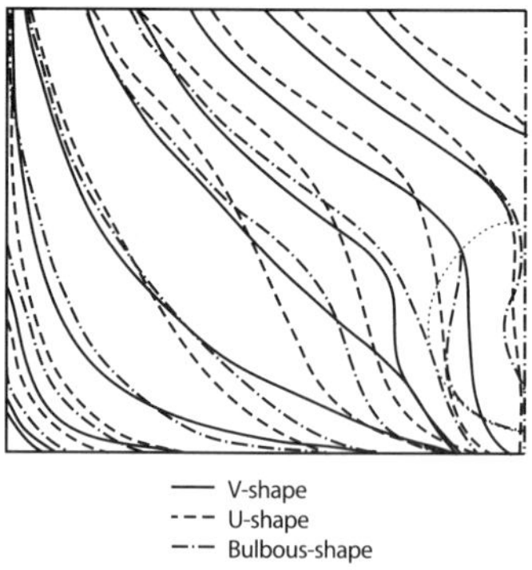

Figure 4-17 Stern shape of U-, V- and bulb-types

2.2.3 Bow and Stern Profiles

1) Bow profile

The bulbous bow is attached to the hull for reducing the wave resistance through the interference of the waves generated by the hull with the bulb waves. The height and length of the waves produced by the bulb is related to the bulb size, and the position of the bulb plays an important role in the interference effect. The main parameters used to express the bulb form when designing the bulb are illustrated in **Figure 4-18**. The shaded portion on the right is the side-projected profile of the bulb, and the one on the left is the front-view profile.

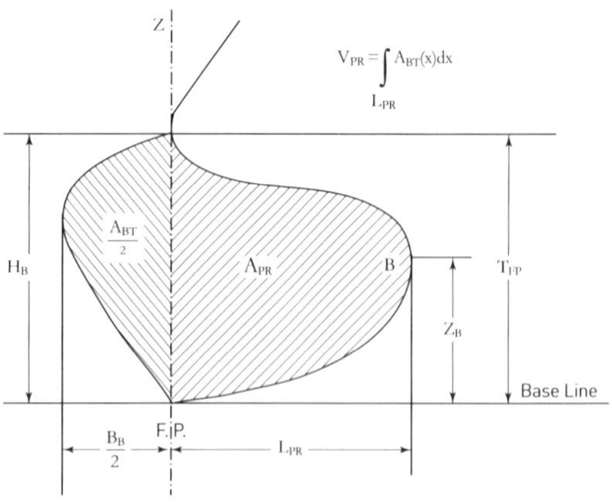

Figure 4-18 Definition of bulbous bow form factors

Among the principal particulars of the bulbous bow, the bulb width (B_B) and cross section area (A_{BT}) are used to represent the bulb size and the bulb height (Z_B) and the length of protrusion of the bulb (L_{PB}) is related to wave interference., There has been a trend recently to make the bulb longer to increase the volume (V_{PR}) of the bulb. The sectional area of the bulb is determined by the ship's speed; for low-speed ships, the appropriate cross section area is 12 to 18% of the midship section and for high-speed ships, 6 to 10%. The profile of the bulbous bow can be divided as shown in **Figure 4-19**, and its form is determined by taking account of the ship's loading conditions.

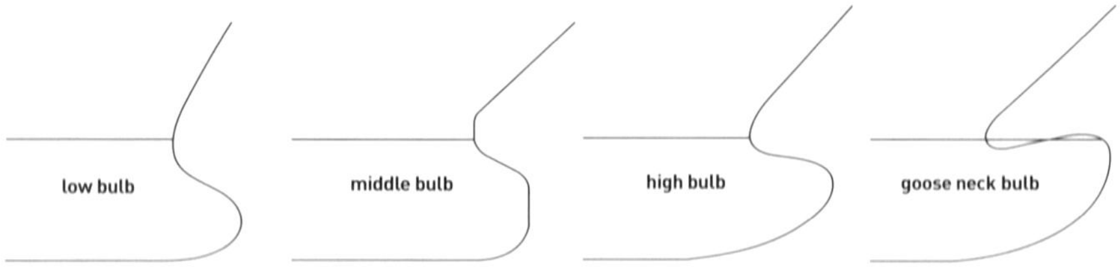

Figure 4-19 Various side profile shapes of bulbous bow

2) Stern profile

A typical stern profile is shown in **Figure 4-20**, and it is obvious that its shape will have a significant influence on the design and operation of the propulsion (propeller and shaft) and directional control systems (rudder). The stern profile should allow effective arrangement of these devices while ensuring low viscous and form resistance and excellent overall propulsive efficiency. Furthermore, the stern form should be determined so as to minimize the undesirable propeller cavitation and shaft vibration.

In general, the larger the diameter of the propeller is, its efficiency is higher. Moreover, the flow near the propeller should have axi-symmetrical distribution near the shaft for better

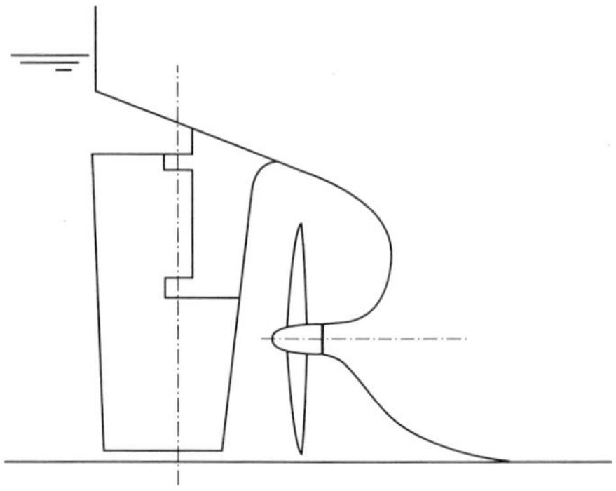

Figure 4-20 Stern profile

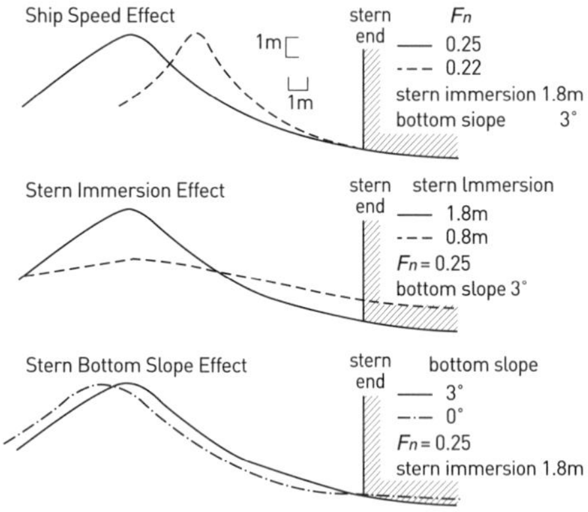

Figure 4-21 Flow around various stern profiles

propulsive efficiency and vibration properties. There must also be sufficient space for replacing the propeller as and when required and securing the clearance between the hull and the rudder to minimize vibration. As illustrated in **Figure 4-21**, the stern profile can have a significant effect on the stern wake, depending on the speed of the ship, submergence of transom stern and the longitudinal slope of transom. Therefore, care should be taken when determining the clearance between the hull and the propeller, the height of the transom and longitudinal inclination of bottom at stern.

For low-speed vessels, the submergence of the transom may be allowed up to 10% of the design draft to enhance the ship's resistance, propulsion and steering performance simultaneously. On the other hand, if the transom of a high-speed ship is deeply submerged, it may improve the ship's stability and the distribution of wake, but it reduces the clearance between the hull and the propeller, making it difficult to prevent vibration. Furthermore, it may also increase the resistance due to stern wake, and consequently the stern transom is usually designed to be above the design load waterline.

2.2.4 Water plane Shape at Design Draft

The horizontal surface obtained from the intersection line of the still water surface and the fully loaded hull surface is called the water plane of the ship. The entrance angle is defined by the angle between the centerline of the water plane and the tangential line at the stem of the water plane on either side. If the entrance angle increases, the flare angle around the fore body also increases concurrently. Therefore, for high-speed ships, high entrance angle is likely to induce wave breaking at the bow and large amplitude pitch motion will result in bow flare slamming. It follows, therefore, that a smaller flare angle will also bring about narrower waterline and smaller amplitude pitch motion resulting in lower wave–breaking resistance and bow flare slamming. On the other hand, the reduction of the breadth along the water plane in the fore body will adversely influence the cargo hold capacity. Therefore, the section shape of the fore body should be optimized with these factors in mind.

Similarly, in cases of low-speed vessels, the entrance angle of the water plane can be reduced to improve the hydrodynamic performances near the free surface. In this process, some of its area in the upper part of the cross section should be transferred to the lower part near the bilge making the section shape closer to U. In this way a reduction in the entrance angle can mitigate the wave–breaking resistance at the bow.

2.2.5 Location of Longitudinal Center of Buoyancy

The longitudinal center of buoyancy is the center of a ship's displacement and can also be said to be the centroid of the area under the sectional area curve. Since any change in the location of *LCB* is directly related to changes in the shape of the sectional area curve, and also to the changes in cross section shape, it has a significant effect on the ship's hydrodynamic performances as well as on the arrangement of compartments and the loading capacity.

When the required full load displacement is satisfied, the location of *LCB* is determined while maintaining appropriate block coefficient C_B and *L/B* to get the design speed. For a low-speed, full-form ships with a very high C_B, *LCB* is better placed forward of midship to reduce viscous resistance and improve propulsive efficiency. For high-speed ships, on the other hand, *LCB* is placed aft of midship to prevent high resistance occurring due to bow waves.

Maritime Research Institute Netherlands (MARIN) developed the design chart for single-screw ships without bulbous bow presented in **Figure 4-22** and it shows the optimal *LCB* location corresponding to the design speed for various C_B. This chart can be used to determine the location of *LCB* in the preliminary design stage as a first approximation, and then finer adjustments can be made manually to determine the definite location after the bulb form is selected.

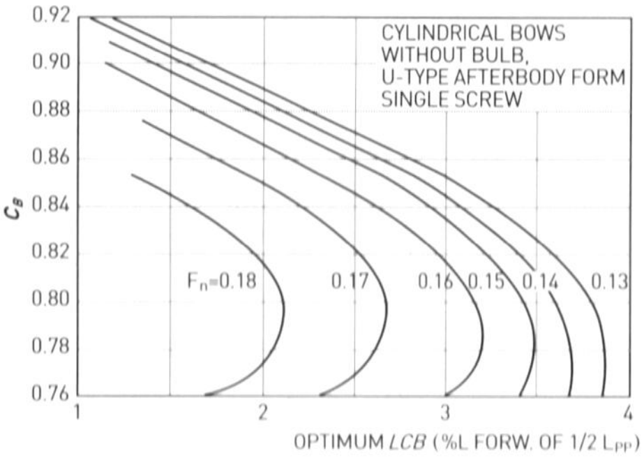

Figure 4-22 Optimum *LCB* for given C_B and *Fn*

2.3 Hull Form Design Methods

As stated earlier, the fore-body shape is closely related to wave resistance. Therefore, the designer should pay close attention to such principal particulars as the sectional area curve, cross section shapes, bulb shape and the entrance angle of the waterplane in connection with wave resistance. V-form bow cross section (C_B<0.7) is commonly used for high-speed ships; U-form bow (C_B>0.80) for low-speed full-form ships; and UV-form bow (0.7<C_B<0.8) for medium-speed ships.

On the other hand, the after-body is closely related to viscous resistance and propulsive efficiency. The designer should determine the profile of the stern before generating the sectional area curve and cross section shapes. Close attention should be paid to the disturbed flow caused by viscosity and the complex 3-dimensional geometry of the after-body, as well as the effects of the propeller, rudder, shaft and transom. It is clear that the use of an appropriate propeller effectively arranged at the stern can ensure favorable resistance and propulsion performance and minimize hydrodynamically induced noise and vibration. Some of the parameters important in this context include the stern profile and the vertical and horizontal clearance between the tip of the propeller and the hull. As shown in **Figure 4-21**, the protruding part of the stern or the shape of the transom can affect the stern wave, the ship's resistance and propulsion and cavitation characteristics. Moreover, it can limit the rudder area, thus directly influencing the ship's maneuvering performance.

2.3.1 Bow Design

Since the bow shape is directly related to the generation of waves, emphasis in its design should be laid on reducing wave resistance. Wave resistance can increase very sharply with an increase in speed at high speed ranges. Therefore, the bow shape should be designed for a specific design speed. The particulars that need to be determined as the first step in the bow form design include: the cross section and profile of the bulbous bow; the distribution of displacement at the fore-body which is defined by the sectional area curve on the bow; the water plane shape; and the cross section shape of the fore-body.

Bulbous bows can reduce wave-making resistance through beneficial interference of the bulb-generated waves which cancel waves generated by the hull. It has been shown that a bulbous bow reduces the total resistance even for low-speed ships, as it allows a smaller entrance angle preventing breaking waves at the bow and has lower form resistance. This is due to the gentler curvature of streamlines around the hull as a result of the more effective distribution of the displacement of fore-body. Furthermore, it gives an extra advantage of longer waterlines in ballast condition, which is also effective in reducing wave-making resistance and improving sea-keeping performance.

The cross sectional shape of the fore-body must be determined by considering the effect of water flow around the hull surface on wave resistance and viscous resistance. In a macroscopic view point, the sectional area curve plays a governing role on flow around the hull. However, to develop the desired fore-body hull form it should be derived from the cross section shape, based on the load waterline in a full load condition. If the volume of the lower part is increased without changing the sectional area curve, the section shape will become closer to a U-shape; if the volume near the load waterline is increased, the cross section will move towards a V-shape. Though the U-shape may cause an increase of form resistance due to a decrease in the bilge radius, it simplifies flow around the hull in other regions. Therefore, the U-shape is usually adopted except in high-speed craft, unless the bilge radius becomes excessively small.

At high speeds the incident flow should be able to separate easily into port and starboard sides, and form a smooth flow thereafter for good resistance characteristics. In this regard, a U-shaped cross section at the bow disturbs the incident flow too much due to the abrupt geometrical variations it introduces, making it difficult to have good hydrodynamic performance. Therefore, the cross section of the fore-body near the bow should be provided with a V-shape suitable for 3-dimensional flow around the bow. **Figure 4-23** compares the body plans of ships with an equal beam but having different design speeds. It can be seen that, as the speed increases, the cross section becomes more V-shaped.

It is to be noted that container ships and passenger ferries (Ro-Pax, Ro-Ro) that require high speed must be given a greater beam to ensure adequate transverse stability by maximizing the metacentric height.

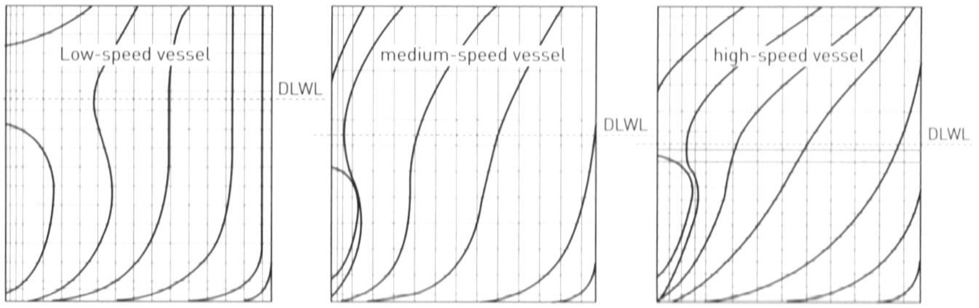

Figure 4-23 Characteristics of sectional form of bow for low-, medium- and high-speed vessels

2.3.2 Stern Design

Since the stern has a considerable effect on viscous resistance, it is one of the most decisive factors that determine the resistance characteristics. The after-body shape influences not only the propeller installation but also the propulsive performance. It is also directly related to propeller cavitation and pressure fluctuation on the hull surface. To maximize the ship's propulsive efficiency and reduce resistance, and to allow the propeller to operate in designed conditions, a variety of design elements must be considered, such as homogenization of flow field and minimization of wake peak. The ship's maneuverability is sensitive to the shape of the cross section and the profile of the stern, and therefore the design of the stern form of a full-form ship requires careful attention to detail on the part of the designer.

The following are the major design considerations and hydrodynamic properties which have an influence on the stern form:

- Minimization of viscous resistance and form resistance, homogenization of wake field, and minimization of wake peak
- Maximizing self-propulsion factors, optimization of propeller design, minimization of cavitation and pressure fluctuation
- Rudder design and consideration for optimum maneuverability
- Ergonomic general arrangement design for good working environment and safety.

Figure 4-24 illustrates the effects of the cross section shape of an after-body. The V-shaped stern, with a small wedge shape forming below the shaft centerline, is a low-resistance form used for a long time. The wedge shape must be designed to allow smooth three-dimensional flow. However, it has a disadvantage of non-uniform flow into the upper part of the propeller and much higher peak wake compared to the conventional U-shaped stern (bulb type), since the breadth has to be increased to secure sufficient area required for stern structure and the engine room layout.

On the other hand, the U-shaped stern shown in **Figure 4-24** generates a higher form resistance compared to the V-shaped stern. However, a partial U-form stern allows a hull form with higher propulsive efficiency despite the increase in resistance due to increased

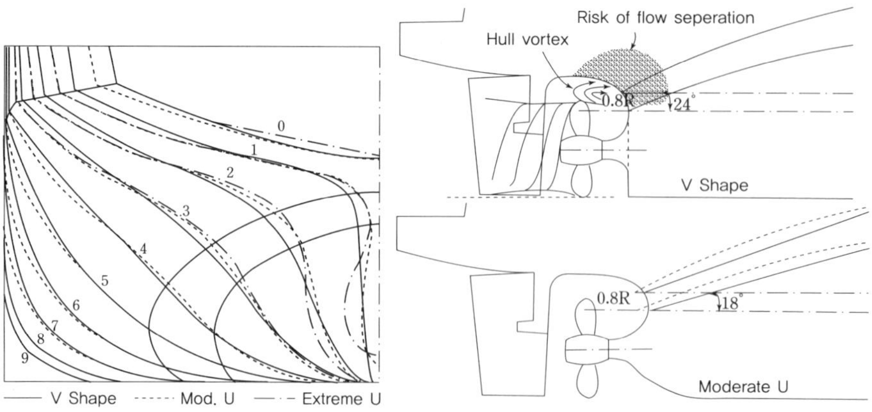

Figure 4-24 Examples of V- and U-type stern shapes

wake. Generally, the modern concept of the stern form encourages designers to opt for a hybrid form combining the advantages of U- and V-shaped sterns. In this case the V-shape with its lower resistance is often used for the upper part above the shaft center line while using the bulbous U-shape with its more uniform distribution of wake and higher propulsive efficiency for the lower part below the shaft line.

Figure 4-25 illustrates typical wake distributions on the propeller plane for various stern shapes. It is to be noted that the more uniform the flow into the propeller plane is, the higher the propulsive efficiency.

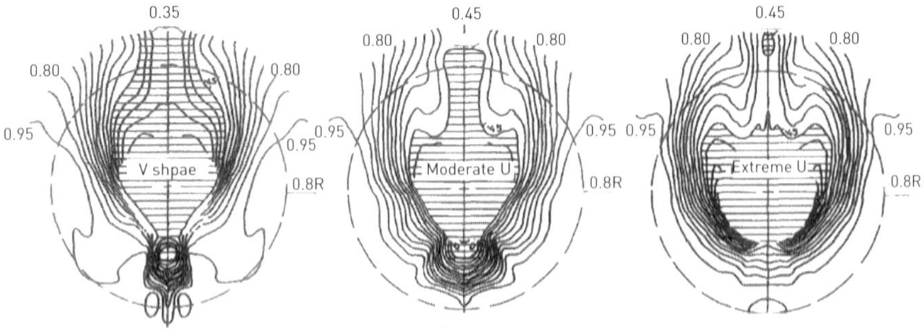

Figure 4-25 Wake distribution for different stern shapes

The stern profile must also be properly designed for satisfactory propulsive efficiency and maneuverability. The designer should first determine the area of the rudder to obtain sufficient lift force required for adequate maneuverability and then decide on the shape of the stern profile that allows an appropriate clearance between the hull and the tip of the propeller blade by considering the effect of pressure fluctuation, which occurs every time the tip of a blade passes by the ship's hull above the propeller.

When it is difficult to secure the area of rudder or to ensure the desired maneuverability due to the limitations of the ship's specifications, it is advisable to increase the area of the lower portion of stern profile as much as possible as shown in **Figure 4-26**.

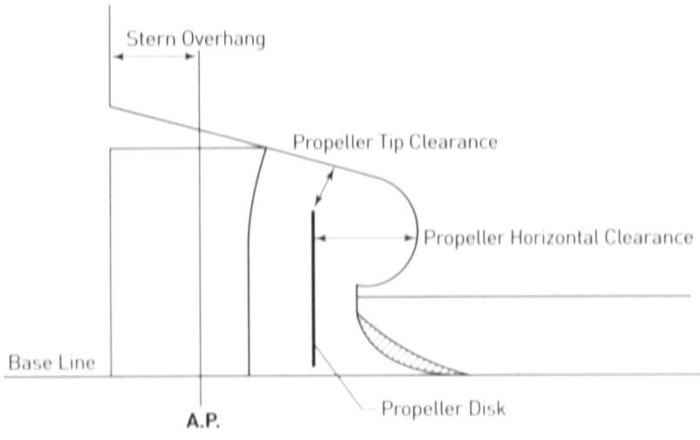

Figure 4-26 Design of stern profile shape

2.4 Hull Form Modification

A common way of obtaining an initial hull form is to select a parent ship and then adjust the principal particulars and hull form coefficients so that they are close to those of the new design. There are three methods of hull modification, viz. variation of principal dimensions, variation of the sectional area curve and variation of the frame shape.

The simplest method of hull form modification is to apply the ratios of the principal dimensions of the new design to the parent ship to obtain the new offset data. In the sectional area curve variation, a new curve is obtained by shifting the ordinates of the sectional area curve of the parent hull longitudinally by a certain amount determined by the form coefficients. The methods of sectional area curve variation include 1-C_p, Lackenby and swing methods.

2.4.1 Variation of Principal Dimensions

The dimensions of a parent ship can be adjusted to those of the new design. The principal dimensions, such as length, beam and draft, are modified along with bilge radius and the length of the bulbous bow. To obtain the desired hull form in this way, a new set of offset data must be obtained by applying the length, beam and draft ratios of the new design to those of the parent ship. If the offset data are made dimensionless, an appropriate hull form can be found without additional conversions.

When different dimensional ratios have to be applied lengthwise, beam-wise and draft-wise, the profile of the ship should be first determined by using the draft and length ratios. The profile of the bow is modified, based on the length and draft ratios, and brought into alignment with the location of each station in the body plan. Since the bulb length or transom length of the designed ship is determined based on the parent ship, they should be modified as and when necessary.

To convert the radius of the bilge the shape of the midship section should be found using the changed dimensions and modified along with the length of the parallel middle body. Therefore, when varying the principal dimensions, a body plan and the bow and stern profiles should be created identical to the parent ship in terms of dimensionless coefficients, excluding the parallel middle body which contains the bilge radius. Then sectional area curve and frame should be automatically modified.

2.4.2 Variation of Sectional Area Curve

As discussed already, the block coefficient represents the fullness of the submerged volume and can be calculated by dividing the displacement volume, which is equivalent to the area under the sectional area curve, by $L \times B \times T$. As shown in **Figure 4-27**, the sectional area curve can be divided into the entrance (L_e), the parallel middle body (L_x) and the run (L_r). Let the block coefficient of the entrance and run parts are C_{Be} and C_{Br} respectively, and the midship section coefficient be C_m, then the block coefficient of the entire hull can be expressed as **Equation 4-1**.

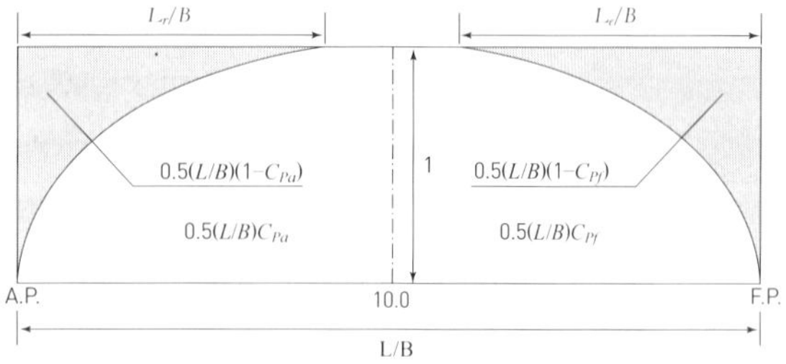

Figure 4-27 Sectional area curve (C_p curve)

$$C_B = C_{Be}\left(\frac{L_e}{L_{BP}}\right) + C_m\left(\frac{L_X}{L_{BP}}\right) + C_{Br}\left(\frac{L_r}{L_{BP}}\right) \tag{4-1}$$

The shape of the sectional area curve can be varied by moving the LCB longitudinally while keeping the block coefficient constant, and this can be achieved by moving the stations of the sectional area ordinates in longitudinal direction. As mentioned above, there are three methods of adjusting the LCB, viz., 1-C_p, Swing and Lackenby. The 1-C_p and Swing methods assume that the block coefficients of the entrance and the run are constant, and then the sectional area curve of a new modified hull can be found by adjusting the length ratios $\left(\frac{L_e}{L_{BP}}\right)$, $\left(\frac{L_X}{L_{BP}}\right)$ and $\left(\frac{L_r}{L_{BP}}\right)$. The Lackenby method determines the length of the parallel middle body first, and then the sectional area curve is varied by adjusting the block coefficient and the length ratios of entrance, and run. This method also allows hull variation for high-speed ships which do not have a parallel middle body.

1) 1-C_p variation method

The 1-C_p method is the most fundamental of all the methods used for sectional area curve modification. In this method the hull form is modified by adjusting $\frac{L_e}{L_{BP}}$ and $\frac{L_r}{L_{BP}}$ to obtain desired LCB while keeping the block coefficients of the entrance (C_{Be}) and the run (C_{Br}) unchanged. For the block coefficient of the whole ship to remain constant, the condition of **Equation 4-1** should be met. With the condition $L_{BP} = L_e + L_X + L_r$ the form of the new sectional area curve

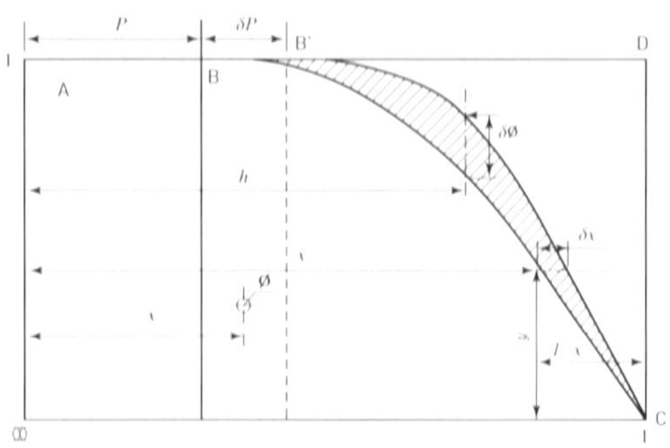

Figure 4-28 1-C_p variation method

can be determined.

The area of BDC outside the SAC in **Figure 4-28** is known as 1-C_p. Using the Guldhammer chart makes the calculation easier as well. However, this method has a certain limitation on the degree of hull variation for ships without the parallel middle body, such as high-speed ships.

2) Lackenby's variation method

Lackenby suggested two methods for improving on the 1-C_p method. The first one is for ships with no parallel middle body. In this case the sectional area curve can be expressed as a quadratic equation as shown in **Figure 4-29**.

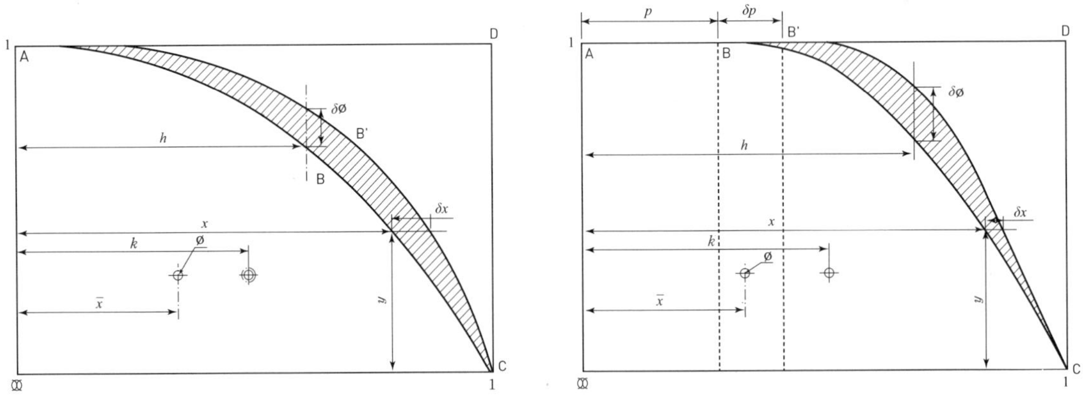

Figure 4-29 Lackenby hull form variation

Both ends of the sectional area curve remain unchanged while any change in the middle can be reflected in the quadratic equations, creating a new sectional area curve and relocating the stations. The second method treats the block coefficients of the entrance and run separately. In this case, like the 1-C_p method, the cross sectional area from the F.P. to the starting point of parallel middle body is expressed as a quadratic function for the entry and another quadratic function is used to express the sectional area curve from the A.P. to the end of the parallel middle body for the run. From this the required change in the length of the parallel middle body can be calculated.

The Lackenby's method allows direct hull form modifications from the input of the block coefficient of the entire hull form and the location of *LCB*. Furthermore, since the parallel middle body is intentionally adjusted and the sectional area curve is moved from each station according to the quadratic functions, the sectional area curve and the block coefficients of the entrance and the run are concurrently changed. The Lackenby's method is highly adaptable and is, consequently, the most widely used. For high-speed ships without the parallel middle body, in particular, this method is useful in modifying the block coefficient and moving the *LCB*. For more details on this method, see the appropriate references.

3) Swing variation method

The 1-C_P method varies the sectional area curve by keeping the block coefficients of the entrance and the run parts constant, while applying the length ratio of the entrance and the run to move the stations.

The swing method, on the other hand, divides the entrance and the run parts into a number of stations, and the sectional area ordinates at each station is moved to the stations inclined at a given angle α as illustrated in **Figure 4-30**. In this way a new sectional area curve with a new block coefficient and *LCB* location can be generated.

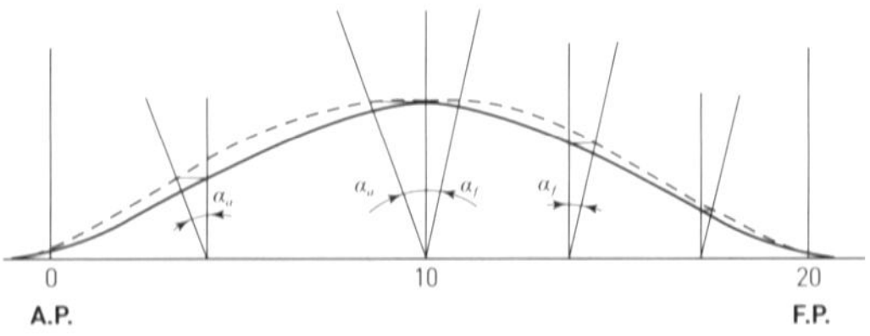

Figure 4-30 Swinging-C_p variation

By generating a number of sectional area curves through the variation of the angle α, and calculating their block coefficients, it is possible to get the relationship between the angle of inclination of the stations and the shape of the entrance and run. Then, by using interpolation on this data, the required angle of inclination can be determined for the desired block coefficient. By studying the relationship between the block coefficients of the entrance and the run and the inclination angle, new entrance and run curves are found through interpolation. A new sectional area curves and block coefficients can then be found by combining these curves.

Once a sectional area curve with the design block coefficient has been created, the block coefficient is kept constant and the curve is varied once again to move the *LCB* to the required position as shown in **Figure 4-31**. In this process the new sectional area curve is swung, both ends in the same direction, at a number of angles and a table is created showing the *LCB* positions for the angles. In exactly the same manner as the curve modification for block coefficient, these data are interpolated to find the appropriate inclination angle. Using this swing angle, the new curve with desired *LCB* and block coefficient can be obtained.

Since the swing method allows hull variation including the adjustment of the length of the parallel middle body, it is somewhat similar to the 1-C_P method. However, it can be used with the inputs of the block coefficient and *LCB* directly, and therefore is the easiest way of hull modification for the given initial data. In general, the swing method is used to perform

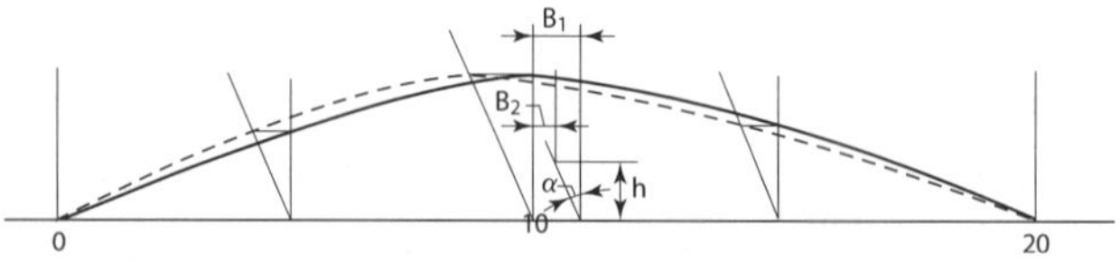

B_1 = longitudinal position of the center of gravity of the curve of sectional areas
B_2 = longitudinal position of the center of buoyancy
h = vertical position of the center of gravity of the curve of sectional areas
The transformation angle α is :
$\alpha = \arctan\left(\frac{B_2 - B_1}{h}\right)$

Figure 4-31 Swinging-*LCB* variation

primary hull variations and the 1-C_P and Lackenby methods are used to carry out further modifications according to the block coefficients of entrance and run.

3. Hull Forms and Power Requirements

3.1 Power Requirements and Selection of Main Engine

3.1.1 Definition of Power Requirements

Horsepower (HP) is a unit of work rate established by James Watt of Glasgow to express the power of the steam engine that he invented and is equivalent to the power required to lift a 330-pound coal basket at a speed of 100 feet per minute. In the metric system, metric horsepower (PS) is defined as a unit of work rate required to move a person of 75kg at 1 meter per second speed (75kgf·m/sec). In the SI system (International System of Units), a joule (J) is defined as the work done for moving a distance of 1 meter with a force of 1 Newton, and a watt (W) is the work rate required to do 1 J per second. The relationships between these units are as follows:

$$1\ \text{HP} = 330\ \text{lbf} \times 100\ \text{ft/min} = 550\ \text{lbf·ft/sec} = 746\ \text{W} = 0.746\ \text{kW}$$
$$1\ \text{PS} = 75\ \text{kgf·m/sec} = 735\ \text{W} = 0.735\ \text{kW}$$
$$1\ \text{W} = 1\ \text{Joule/sec} = 1\ \text{N·m/sec}$$

The required power to move a ship forward appears differently, depending on the point of measurement. The most common measures of power are as follows:

- Indicated Power (P_I): the power of an engine derived from the enclosed area of an indicated pressure diagram, which is obtained through a continuous measurement of pressure variation in a cylinder for a complete engine cycle.
- Brake Power (P_B): the power of an engine measured by attaching a dynamometer to the engine's output shaft, and is the indicated power minus the frictional losses during engine operation and the power required for driving a cam shaft.
- Shaft Power (P_S): the power delivered to a propeller through a shaft and is measured at the propeller shaft immediately in front of the stern tube. The shaft power is equivalent to the brake power minus power losses in reduction gear and bearings. In other words, shaft power = (brake power – power lost in reduction gear system and bearings)
- Delivered Power (P_D): the power actually delivered to the propeller, and is the same as the shaft power minus the power losses in the stern tube bearings, strut bearings and the sealing used to ensure water tightness. In other words, delivered horsepower = (shaft power – power lost in the shaft system)
- Thrust Power (P_T): the power exerted by a rotating propeller, and it can be estimated by the multiplying the propeller torque with the propeller RPM. The propeller torque can be measured from self-propulsion model tests in a towing tank and scaled up to full scale according to the law of similitude. In other words, thrust power = delivered power – power losses in the propeller
- Effective Power (P_E): the power required to overcome the resistance acting on the ship when it is moving at a constant speed in calm water. This can be obtained by multiplying the bare hull resistance with the advancement of the ship. The bare hull resistance of the ship can be estimated by measuring the resistance of a bare hull model in a towing tank at a number of speeds and then extrapolating the results to the full scale in accordance with the Froude's law.

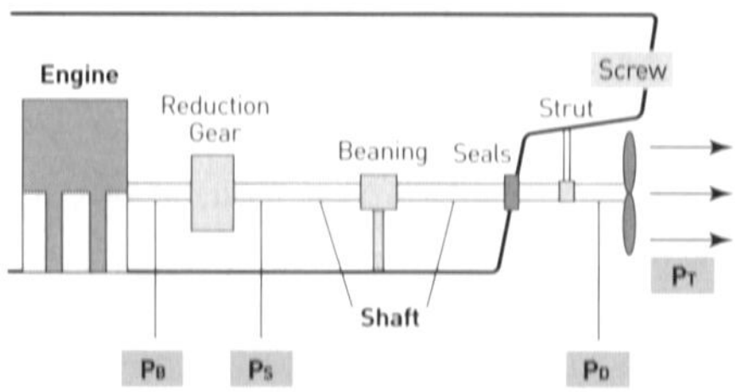

Figure 4-32 Measures of power

3.1.2 Selection of the Main Engine

Selection of the main engine is the process of finding the most appropriate engine, based on the ship's principal particulars and required speed. This is a very important process, since an inappropriate engine may produce superfluous power or, more critically, insufficient power leading to the ship failing to reach the design speed. The main engine also has a decisive impact on the overall weight, price and hull form of the ship. Therefore, the selection of the main engine is one of the most fundamental and essential steps in ship design, and require careful examination of the specifications provided by the engine manufacturers.

1) Classification of main engine output

Unlike automobile engines, marine engines should be able to operate continuously in rough seas for several months at a time. The output of the main engine is expressed in three different ways based on the characteristics of continuous operation as follows:

- Nominal Maximum Continuous Rating (NMCR): the maximum output that can be obtained during continuous operation of the engine, guaranteed by the engine supplier. It is a standard for estimating the engine size, weight, volume and price.
- Maximum Continuous Rating (MCR): the maximum continuous output that the ship designer expects the engine to produce and can be any value within the limits of NMCR. The ship designer, however, determines the operating conditions of the engine by considering changes in the required power due to changes in the propeller's RPM, fuel consumption and operational range of the propeller and the information regarding the engine output for various engine RPM. Clearly, when the designer sets the engine's operating conditions, the engine output and RPM must be set lower than the NMCR. Once the MCR has been determined, then the dimensions of the appropriate shaft system and auxiliary machinery can be determined. MCR is the standard for maximum continuous output and revolutions per minute during ship operation.
- Nominal Continuous Rating (NCR): refers to engine output during nominal ship operation. Most ships experience additional power requirements due to wind and adverse sea conditions. An allowance should also be provided for the decrease in ship speed due to the age-related increase in resistance and degradation of the engine through the lifespan of the ship. This is often known as 'sea margin.' The designer generally prefers to select a main engine that can operate continuously at up to 90% of MCR to provide such a sea margin, in which case the engine power margin is 10%.

2) Relationship between hull resistance and main engine output

The power required to operate a ship can be estimated by multiplying the total resistance of the hull by the corresponding speed, and is called effective power. Chapter 5 discusses the calculation of effective horsepower in detail.

Part of the power generated from a ship's engine is consumed in the power transmission system, and the power delivered to the propeller is transformed to thrust power through the revolution of the propeller. When thrust power is in equilibrium with effective power, the ship can move forward at the design speed.

The relationship between the effective and the delivered power can be explained with quasi-propulsive efficiency. In hull form design the influence of a propeller is not considered and in propeller design the influence of the hull on the propeller is also discounted. Therefore, since the ship speed is not equal to the speed of flow passing through the propeller, the quasi-propulsive efficiency cannot be easily estimated. See Chapter 5 for details on the estimation of quasi-propulsive efficiency.

We can estimate the engine brake power required to deliver the necessary power to the propeller by considering the power loss in the power transmission system. The brake power is estimated for ideal conditions. Therefore, in many cases, the ship fails to achieve the design speed due to the actual sea, weather and operating conditions.

The ship-owner naturally wants to have some margin for the engine output based on the ship's operating conditions and experience. The sea margin usually demanded is about 15%, but in some special cases a margin of up to 35% may be requested. The designer, then, estimates the brake power and calculates the continuous output of the engine using the sea margin.

3) General selection process for main engine

A proper hull form can be designed only when the information on the size of the main engine and propeller characteristics is available. On the other hand, the propeller can be designed only when the hull form and engine RPM have been specified. Furthermore, the engine supplier can set the engine design requirements only when the data on the ship's resistance and propulsion and sea margin are clearly provided.

Therefore, the designer often determines the nominal maximum continuous rating by an iterative process of determining resistance and propulsion, principal propeller dimensions, and selecting the main engine as shown in **Figure 4-33**.

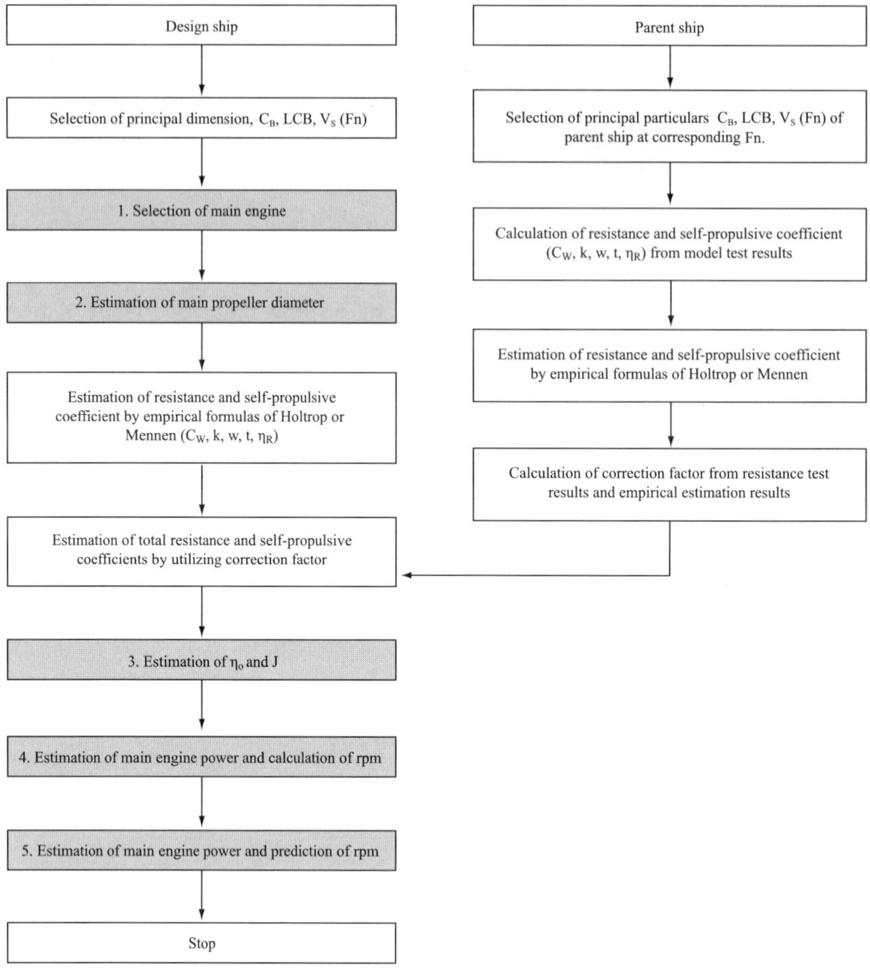

Figure 4-33 Procedure for selecting main engine

3.2 Estimation of Power Requirements

3.2.1 Systematic Series Data

During the preliminary design stage the power requirement can be estimated using the statistical data obtained from analyzing series test results. One of the drawbacks of this method is that a vast range of model tests must be performed beforehand. However, the results of a large number of model tests on hull forms and propellers conducted by numerous research institutions over the years can be used as a basis for estimating power requirements.

1) Representative series data

- The Taylor Gertler Standard Series (Gertler, 1954)
- DTMB Series 60 (Todd, 1963)
- The SSPA Cargo Liner Series (Williams, 1969)
- MARIN Mathematical Model (Holtrop & Mennen Method)
- MARAD Series ($C_B > 0.8$)

The typical hull form tests listed above have been updated several times through additional testing and re-interpretation. MARIN's mathematical model, for instance, was first developed by the institute's Holtrop and Mennen in 1977. In the following year an improved model was published which made allowance for bulbous bows and an improvement in estimating resistance of large waterplane area coefficient ships.

In 1982 the model was extended in order to improve the power prediction of ships with high block coefficients and low length-breadth ratios. The data used for developing the 1982 model contained the results of 191 model tests and sea trials of real ships which increased to 334 in 1984. For slow ships, MARAD (US Maritime Administration) developed a series of hull forms with $C_B > 0.8$, which serve as a good parent hull forms for full-form ships, such as tankers and ore carriers.

Even this vast scope of data, however, cannot cover all the design domains reflecting the recent design trend. Since the hull and propeller design technology are continuously evolving, the existing data is becoming increasingly unreliable, and consequently each towing tank continues to strive to update the data. **Table 4-2** represents the scope of application of the estimation method developed by MARIN.

Table 4-2 Range of the valid application of MARIN prediction method

Ship Type	*Fn*	C_p		*L/B*		*B/T*	
	max.	min	max.	min	max.	min	max.
Tankers, bulk carriers (ocean)	0.24	0.73	0.85	5.1	7.1	2.4	3.2
Trawlers, coasters, tugs	0.38	0.55	0.65	3.9	6.3	2.1	3.0
Container ships, destroyer types	0.45	0.55	0.67	6.0	9.5	3.0	4.0
Cargo liners	0.30	0.56	0.75	5.3	8.0	2.4	4.0
Ro-Ro's, car ferries	0.35	0.55	0.67	5.3	8.0	3.2	4.0

2) Representative propeller series data

- Wageningen B-Screw Propeller Series (van Lammeren et al, 1969)

- Gawn-Burrill Propeller Series (Gawn & Burrill, 1958)
- MAU Series (Yokoo, 1973)

Maritime Research Institute Netherlands (MARIN, former NSMB) developed the Wageningen B-Series propellers in 1936, after discovering that the A-Series blade sections developed for airplane wings were not appropriate for marine applications due to the risk of cavitation. The B-Series was based on the regression analyses and some corrections for Reynolds number effects on a series of tests on a total of 21 propellers by systematically varying the number of blades, expanded area ratio and pitch ratio. The B-Series propeller blade has a circular section near the tip and airfoil sections near the root. **Table 4-3** shows the scope of the tests performed.

Table 4-3 Range of valid application of B-Series propellers

	Number of blades					
	2	3	4	5	6	7
EXPANDED AREA RATIO (A_P/A_O)	0.30 0.38	0.35 0.50 0.65 0.80	0.40 0.55 0.70 0.85 1.00	0.45 0.60 0.75 1.05	0.50 0.65 0.80	0.55 0.70 0.85
Pitch ratio(P/D)	0.5–1.4	0.5–1.4	0.5–1.4	0.5–1.4	0.6–1.4	0.6–1.4
Blade thickness ratio(t/D)	0.055	0.050	0.045	0.040	0.035	0.035
Boss diameter ratio(d/D)	0.180	0.180	0.167	0.167	0.167	0.180
Pitch reduction at blade root	0	0	20%	0	0	0
Blade rake angle(φ)	15°	15°	15°	15°	15°	5°

Gawn-Burrill Propeller Series is a series of model tests performed on the 16-inch 4-blade propellers in the cavitation tunnel of King's College in Newcastle (present University of Newcastle-upon-Tyne), England. The test parameters were the pitch ratio from 0.6 to 2.0 and the propeller area ratio from 0.5 to 1.1. The test data is useful in judging suitability of particular design parameters during preliminary design stage.

In Japan another modern propeller series was developed called the MAU Series. The range of variables, such as the number of blades, the expanded area ratio and pitch ratios, are similar to those of the B-Series. The MAU Series was developed to have improved open-water efficiency compared to the B-Series propellers.

The series test results of hull forms and propeller geometry are expressed in mathematical

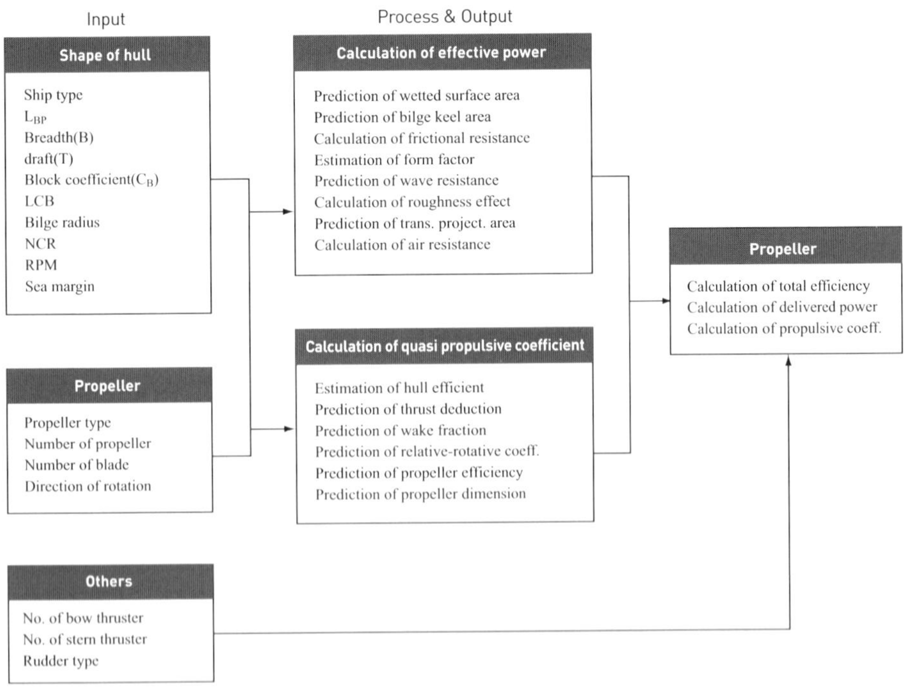

Figure 4-34 Procedure for the prediction of required power

relations, and the procedure of predicting the required power using these relations is shown in **Figure 4-34**. The prediction procedure might be different by model basins and institutions in detail, but the key aspects of the main procedures are all similar.

In addition to the normal predictions, hull appendages produce extra resistance and **Table 4-4** shows the increase of power requirement due to some appendages. Of course, the predicted power will be slightly different depending on the shape of appendages and the location of attachments on the hull surfaces.

Table 4- 4 Example of power increment for various appendages

Appendages	Power Increment	Remark
Bow / Stern Thruster	1.0 / 0.5%	
Schilling / Beck Rudder	6.0 / 1.0%	
CPP(Controllable Pitch Propeller)	1.5%	

3.2.2 Model Tests and Numerical Calculations of a Parent Hull Form

1) Power prediction using model tests of a parent hull form

It is necessary to estimate power requirements even before a specific hull form is determined. To do so, three methods are commonly used:

The first method uses series data alone and the second uses model test results on similar ships to verify the accuracy of the estimation and adjust it if necessary. The second method is more reliable than the first as it does not rely on historical data exclusively. The third method is used for ships with dimensions well outside the conventional ships and novel ship types. In this case the designer must prepare an initial lines plan and use computational methods to predict resistance and propulsive efficiency numerically. **Figure 4-35** is a conceptual diagram showing the first two methods of power prediction.

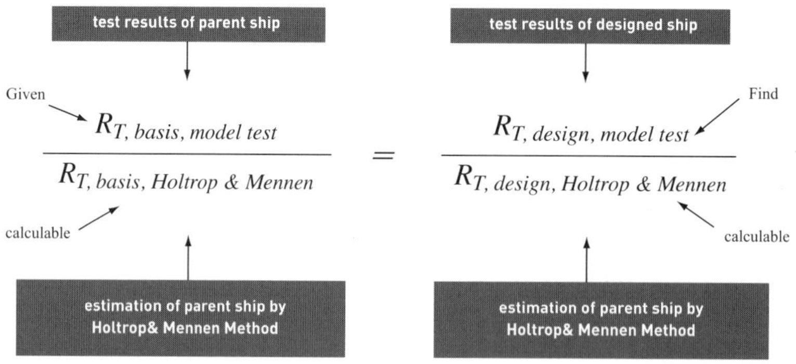

Figure 4-35 Required power prediction using statistical method and the model test data of parent hull

Since power requirements prediction must be made at the time of finalizing the contract prior to hull form design and subsequent model experiment, the accuracy of the prediction is extremely important. This is because the predicted power requirement determines the ship's speed, which is one of the most important owner's requirements to be fulfilled by the shipyard. The accuracy of power estimation, therefore, is one of the decisive factors that determine the successful conclusion of a ship construction contract.

2) Prediction of power requirements using numerical calculations

Since Kelvin's theoretical studies on the wave formation by a moving point in the 1890s there has been much theoretical research on wave resistance. In the late 1950s T. Inui proposed a theoretical approach to hull forms with a bulbous bow and succeeded in reducing resistance through applying simplified approximations to actual ships. This has brought about more theoretical research on wave resistance, but without much notable success due to the limitation of computing power at the time. Since the 1970s, when Kwang June Bai proposed a numerical method in hydrodynamic analysis based on finite element method, these numerical methods began to be adopted in the study of viscous flow around ships.

Meanwhile, the meteoric rise of computer technology has heightened the expectation that the numerical hydrodynamics can one day replace the expensive and time-consuming model tests. As a consequence, an international conference was held in 1979 regarding the computer-aided calculation of wave resistance. In the following year another international meeting took place to discuss the calculation of viscous resistance, which expanded the scope of computer applications in marine hydrodynamics.

The recent advancements in parallel computing and computer-aided hydrodynamics have enabled a number of powerful commercial software directly solving the RANS (Reynolds Averaged Navier-Stokes) equations to be applied to the evaluation of flow around the hull as well as estimation of resistance and required power. These software packages include FLUENT, STAR CCM, and WAVIS.

Such numerical power prediction not only allows quick verification of initial estimation based on empirical formulae and experience but can also save much time and cost compare to model experiments. It is thus a useful tool for hull and propeller optimization. There are two methods of power prediction.

The first method is to use a direct numerical analysis on the ship. This requires numerical analyses of the hull and propeller at a high Reynolds number region. However, the high Reynolds number makes it difficult to estimate the power requirement accurately due to the difficulties in generating the grids for numerical analysis and the limitations of turbulence models. The second method is similar to model experiments. Numerical simulations are applied on the prediction of resistance, propeller open water characteristics and self-

Figure 4-36 CFD application for a wind turbine installation vessel

propulsion performance. From these numerical results power requirement of a real ship can be predicted by the ITTC-1978 method as used in the analysis of model test results. Numerical analysis can be useful when estimating the power requirements of special ships, such as wind turbine installation ships, the statistical data for which is scarce or non-existent. **Figure 4-36** is an example of the numerical analysis of the waveform near the hull and flow at the stern of a wind turbine installation ship and its hydrodynamic performance during operation.

The ITTC's earlier efforts to increase the accuracy of model tests were extended to the development of numerical analysis techniques based on theoretical hydrodynamics resulting in active research work in the improvement of the CFD techniques and their applications. This laid the foundation for the rapid evolution of numerical or computer-aided hydrodynamics. In 1978, ITTC proposed a new resistance estimation method and undertook international joint research projects on the computer-aided numerical analysis of wave and viscous resistance of ships. Subsequently, to secure the validity of the theoretical calculation, 22 leading member organizations of ITTC participated in an international joint project on the same ship model used in the theoretical research. In 1987 the test results presented in Kobe showed that Korea provided the most reliable results, despite being the last country to equip itself with test facilities.

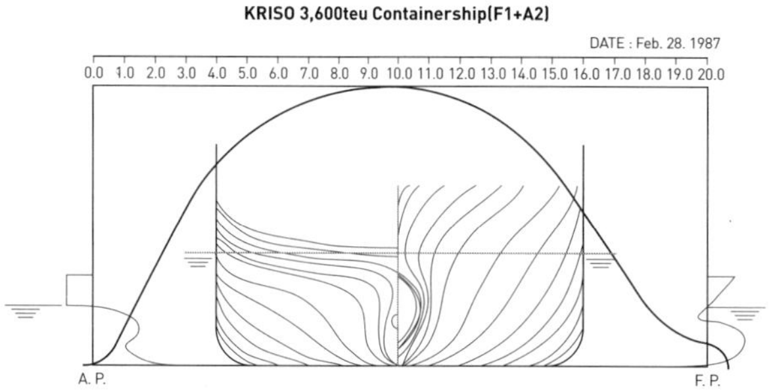

Figure 4-37 KCS 3,600TEU containership

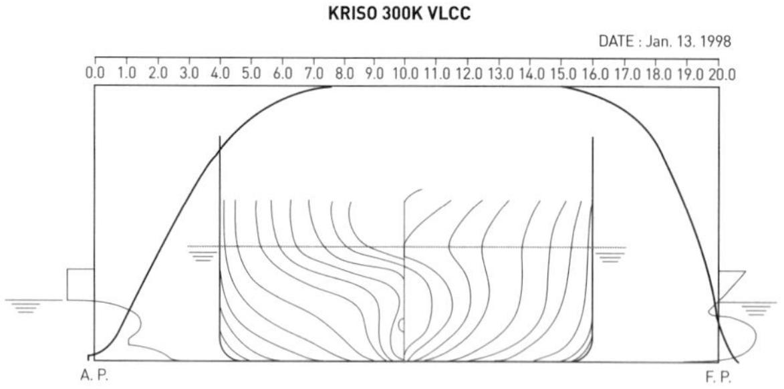

Figure 4-38 KVLCC 300K DWT oil tanker

Further international joint research laid the foundation for the global standardization of the test methods. The ITTC have selected two hull forms developed in Korea as the international standard hull forms for the validation of numerical research. The first one was the hull form of a 300K DWT oil tanker named KVLCC and the other one was the hull form of a high speed container ship named KCS. These two ships are illustrated in **Figures 4-37** and **4-38**.

4. Ship Stability and Righting Moments

4.1 Ship Stability

4.1.1 Equilibrium of Floating Bodies

Floating bodies like ships are subject to the buoyancy of water, which is equivalent to the weight of the water displaced by the body. This force is exactly offset by the gravitational force acting on the body, i.e., its body weight. A body tends to rise if the buoyancy, or buoyant force, exceeds the weight, and sink in the opposite situation. The magnitude of buoyancy is the vertical component of the force obtained by integrating the hydrostatic pressure acting on the wetted surface of the object.

The buoyant force acts vertically upward through the center of buoyancy or the centroid of the displaced volume, and the weight force acts vertically downward through its center of gravity. For a motionless floating object the weight and buoyancy should act along a same vertical line, but in opposite direction. When an external moment is applied to the object, it will experience a small, angular displacement to move the lines of action apart from each other. The object will be stable, if this misalignment generates a moment which tries to restore the initial equilibrium condition.

To better understand the rotational stability of a floating body, let us examine the equilibrium of objects set on a flat surface. Consider the case of the objects resting still in equilibrium on the surface, when a small angular displacement is induced by an external disturbance. Depending on what happens then, the equilibrium can be classified into three states: stable, neutral and unstable.

These are illustrated in **Figure 4-39**. The initial positions of three rigid bodies are shown in dotted lines and those after the disturbance in solid lines. In the figure, case (a) shows

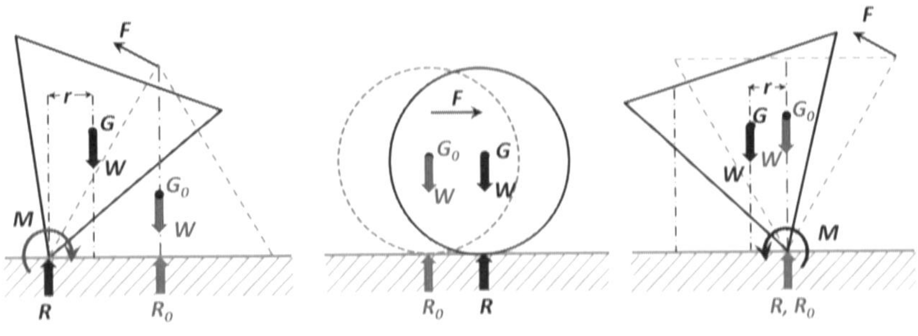

Figure 4-39 Three states of equilibrium

stable equilibrium. It can readily be seen that a small angular displacement in anticlockwise direction brings about a clockwise moment which tries to restore the object to its original position. Case (b) shows the state of neutral equilibrium in which the forces act on the same vertical line regardless of any angular displacement. Case (c) illustrates unstable equilibrium, where a small angular displacement results in a moment in the direction of increasing the displacement.

4.1.2 Response of Ships to External Forces

The stability of a floating body can be explained in a similar manner as above. The moment that makes a floating body incline is called *heeling moment* and the moment that tends to return the body to its original position is called *righting moment*. A ship must be designed to be able to recover its upright attitude when heeled due to waves, wind, movement of mass on board and its turning maneuver.

A floating object experiences motions in all six degrees of freedom induced by external forces and moments: three rectilinear components of surge, sway and heave motions along the x, y and z axes, respectively; and rotational motions of roll, pitch and yaw around the x, y and z axes, respectively. In a right-handed Cartesian coordinates system, o-xyz, where the origin o is lying at the center of gravity of the ship and x is defined to be positive forward. Heave, roll and pitch motions alter the parts of hull immersed and thereby generate a force or moment, which normally tends to return the body to its original position. The response of a ship in waves is typically treated in ship motion analyses and further details will be discussed in Chapter 6.

4.1.3 Transverse Metacenter

When a ship is floating in an upright position in calm water, the weight of the hull W acts vertically downwards through the center of gravity G and the buoyancy force, $g\,\Delta$ acts vertically upwards through the center of buoyancy B, which is also the centroid of the submerged volume, where g is the gravitational acceleration, and Δ represents the mass of displacement. Since the ship is in equilibrium, $W = g\Delta$ and G and B are co-linear, usually on the centerline.

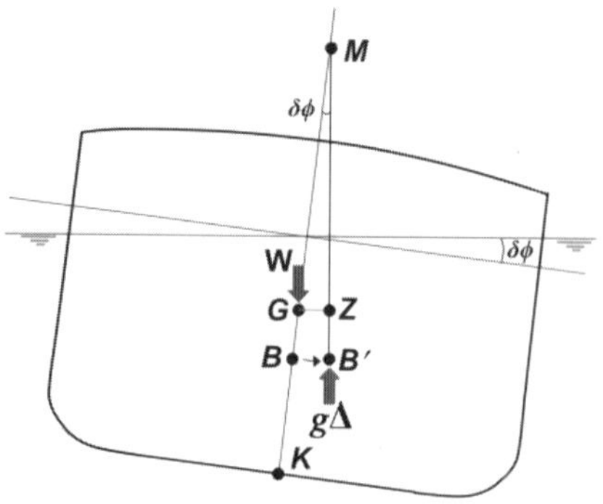

Figure 4-40 Metacenter and initial righting moment

If the ship is heeled at an infinitesimally small angle $\delta\phi$ by an external force, as shown in **Figure 4-40**, the center of buoyancy of the submerged volume moves off the centerline to B'. The vertical line through the new center of buoyancy B' intersects the ship's centerline at point M and this point is called the transverse metacenter.

Since G and B' are not on the same vertical line, the weight and buoyancy force create a righting moment that tends to restore the ship to its original position, and this moment is equal in magnitude and opposite in direction to the heeling moment of the external force, which inclined the ship in the first place. It can be seen from the illustration that the righting moment is equal to the buoyancy force multiplied by the lever $\overline{GZ}$, which is equal to $\overline{GM}$ $\sin(\delta\phi)$.

At a small angle of heel, therefore, the metacentric height $\overline{GM}$ is a sensible measure of initial stability. However, if the angle exceeds 10 degrees or more, the line of action of the buoyancy does not pass through M any longer because the original and new waterlines do not intersect at the ship's centerline. In such cases, the metacentric height $\overline{GM}$ becomes meaningless and indeed it can be misleading.

Figure 4-40 shows the height of the metacenter above the keel $\overline{KM}$ can be obtained from $\overline{KM} = \overline{KB} + \overline{BM}$. Therefore, $\overline{KM}$ can be found if the metacentric radius $\overline{BM}$ is known, since $\overline{KB}$, the height of the center of buoyancy above the keel, can be readily calculated for a given underwater hull form. $\overline{GM}$ can then be calculated from $\overline{GM} = \overline{KM} + \overline{KG}$.

According to the theorem derived by the French hydrographer Pierre Bouger, the metacentric radius $\overline{BM}$ is given by

$$\overline{BM} = \frac{I}{\nabla} \tag{4-2}$$

where I represents the moment of inertia of the entire water plane about the ship's centerline and ∇ represents the volume of displacement. Supposing that a ship is a hexahedron with the length L, beam B, and draft T, then the initial transverse metacentric radius $\overline{BM}$ and the longitudinal metacentric radius $\overline{BM_L}$ can be found in the **Equations** below **4-3**:

$$\overline{BM} = \frac{B^2}{12T}, \quad \overline{BM_L} = \frac{L^2}{12T} \tag{4-3}$$

Therefore, considering that the length-to-beam ratio of a ship is generally 5 to 7, the ratio of metacentric radii $\overline{BM_L} / \overline{BM}$ will be 25 to 49. In addition, it is apparent that the transverse metacentric radius is proportional to the square of the beam.

4.2 Static Righting Moment

4.2.1 Initial Righting Moment

As previously stated, the location of the metacenter M is determined for an infinitesimally small angle of heel and is known as initial metacenter. As stated previously,

$\overline{GZ}(\phi) = \overline{GM}\, sin\, \phi$, and the righting moment is

$W \cdot \overline{GZ}(\phi) = W \cdot \overline{GM}\, sin\, \phi$ as shown in **Figure 4-40**.

Table 4-5 Roll Periods of Various ships

Type of Ship	Tϕ (sec)
Cruise ship	20.0 ~ 25.0
Cargo-Passenger ship	10.5 ~ 14.5
Cargo ship	9.0 ~ 13.0
Tanker	9.0 ~ 10.0
Fishing vessel	5.5 ~ 7.0
Whaler	9.0 ~ 11.5
Battle ship	14.5 ~ 17.0
Cruiser	12.0 ~ 13.0
Destroyer	9.0 ~ 9.5
Torpedo boat	7.0 ~ 7.5

This is called *initial static righting moment.*

Therefore, the transverse metacentric height $\overline{GM}$ is an index of initial transverse stability at small angles of heel:

- $\overline{GM} > 0$: stable
- $\overline{GM} = 0$: neutral
- $\overline{GM} < 0$: unstable

The safety of a ship at sea depends upon many things, but any ship with insufficient stability can be exposed to the risk of capsizing even at a small angle of inclination. If the transverse righting moment of the ship is too large, however, roll frequency of the ship increases leading to jerky rolling motion and consequent passenger discomfort.

If roll damping, normally very small for round-bilge ships, is ignored, the roll natural period T_ϕ of a ship in still water can be found by an approximate expression given in **Equation 4-4**:

$$T\phi = \frac{2\pi k_x}{\sqrt{g \cdot \overline{GM}}} \qquad (4\text{-}4)$$

Where k_x is the radius of gyration of the ship's mass about the longitudinal axis through its center of gravity. The relation indicates that a ship with a small $\overline{GM}$ has a small initial transverse righting moment and a long roll period and is often known as a 'soft' ship, whereas a ship with a large $\overline{GM}$ has a short roll period and is referred to a 'stiff' ship. In other words, the larger the transverse metacentric height $\overline{GM}$ is, the larger the righting moment and the higher the roll frequency; thus, the ship becomes more stable, but less comfortable. By contrast, a small $\overline{GM}$ usually implies low stability but also low roll natural frequency. Roll motion is a very lightly damped system and is, therefore, quite susceptible to resonance phenomenon. Consequently, it is very important to ensure its natural frequency is as far removed as possible from the normal high energy wave frequency band without impairing the stability.

These conflicting conditions should be balanced out when designing the hull form of a ship. **Table 4-5** shows typical roll natural periods by ship types.

4.2.2 Righting Moment at Large Angles of Heel

When a ship inclines to a small angle, the original and the new waterlines can be assumed to intersect at the centerline of the ship, as shown in **Figure 4-40**. It should be noted that, for conventional ships, the two waterlines will not intersect at the centerline if heel angle grows beyond, say, about 10 degrees. It follows that the transverse metacenter of a ship at a large angle of inclination moves away from the initial metacenter. To begin with, it moves up and away from the centreline in the opposite direction to the heel, as the angle of inclination is increased. At very large angles, however, it levels out and may even move down or cross the centreline, depending on the particular hull geometry. These metacenters at large angles are known as pro-metacenters. As pointed out earlier, therefore, the metacentric height $\overline{GM}$ is an index of initial stability and should not be used to measure the stability of a ship at large angles of heel.

Since the righting moment is the product of the ship's weight and the righting lever $\overline{GZ}$, the transverse stability can be expressed with a $\overline{GZ}$ curve. One example of such a curve is shown in **Figure 4-42**.

4.2.3 Longitudinal Metacenter and Trim

When a heavy weight is loaded onto a ship floating in still water, and the weight is moved along the lengthwise direction as shown in **Figure 4-41**, the ship experiences a change in longitudinal attitude called *trim* by the resulting moment.

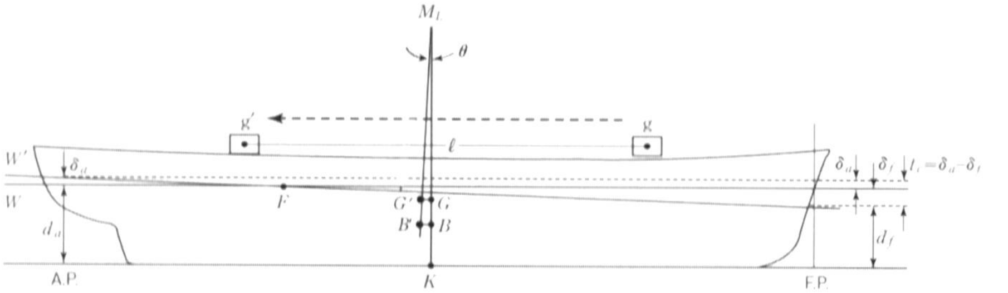

Figure 4-41 Longitudinal metacenter and trim

If the weight w is moved from g to g' by a distance l, the ship's weight W remains unchanged but the center of gravity G shifts to a new location G' parallel to the line $\overline{gg'}$ due to the moment $w \times l$ arising from the weight movement. The movement of the center of gravity can be obtained by the relationship

$$W \times \overline{GG'} = w \times l$$

In order to restore the equilibrium the center of buoyancy has to move to a point B' vertically in line with the new center of gravity. This is the mechanism of trim as the trim causes the center of buoyancy to move longitudinally. If the trim angle θ is small, the longitudinal metacenter M_L is the intersection of a vertical line through the center of buoyancy B at even keel and a vertical line through B'. From the geometry we can observe that

$$\overline{GG'} = \overline{GM_L} \tan \theta$$

Trim T_c is defined as the difference between the draft forward T_f measured at the FP and the draft aft T_a measured at the AP, i.e.

$$T_c = T_a - T_f \tag{4-5}$$

When the draft forward is greater than the draft aft, it is known as trim by the bow and is usually defined as positive trim; and the opposite case is called trim by the stern. The trimming moment that causes the ship to change trim by one centimeter is termed MTC and is calculated from the shapes of waterplane. Therefore, the change of trim ΔT_c due to the longitudinal heeling moment $w \times l$ induced by the weight movement can be calculated as:

$$\Delta T_c = \frac{w \times l}{MTC} \tag{4-6}$$

Once ΔT_c has been found, T_a and T_f can be also calculated using the characteristics of the waterplane.

4.3 Dynamic Stability

When a heeling moment is applied on a ship, the ship begins to incline until it reaches equilibrium at an angle of heel where the righting moment becomes equal to the heeling moment. An example of static stability curve, or $\overline{GZ}$ curve, drawn against heel angle ϕ is shown in **Figure 4-42**. Then, the area under the curve up to the angle A is the amount of work W_A required to incline the ship to that angle. This can also be viewed as the potential energy E_A acquired by the ship by heeling to the angle A, and can be expressed in the form:

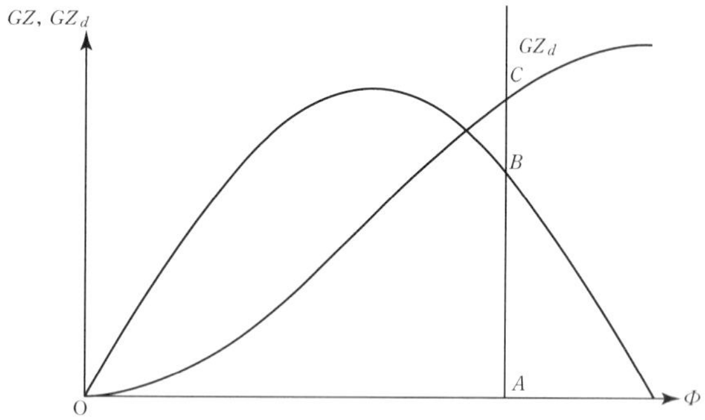

Figure 4-42 The static and dynamic stability curve

$$W_A = \int_0^A Md\phi = \int_0^A W \cdot \overline{GZ}(\phi)\, d\phi = W \cdot \int_0^A W \cdot \overline{GZ}(\phi)\, d\phi = W \cdot GZ_d = E_A \tag{4-7}$$

If the external moment M is then removed, the righting moment will try to restore the ship to the original upright position. In the process the potential energy will be transformed into kinetic energy to produce an angular velocity in the direction of the movement; when the ship rolls back to the upright position, the kinetic energy will reach its maximum value equal to E_A less the energy lost in the movement, and so will the angular velocity reach a maximum. Therefore, the ship will continue to roll in the opposite direction of the initial heel until the kinetic energy – less energy loss – is completely transformed into potential energy. The process will be repeated until the initial potential energy E_A is completely dissipated. The term dynamic stability comes from such behavior.

4.4 Hull Form and Stability

4.4.1 Effects of Center of Gravity

As mentioned above, the righting moment of an intact ship is dependent on its hull shape and weight distribution. The position of the center of gravity directly affects the stability and it is determined by the weight distribution even if the total weight of the hull remains constant. If the center of gravity rises from G to G' as shown in **Figure 4-43**, the new righting lever $\overline{G'Z'}$ can be calculated from the original lever $\overline{GZ}$ as follows:

$$\overline{G'Z'} = \overline{GZ} - \overline{GG'}\, sin\, \phi \tag{4-8}$$

If the center of gravity is shifted downward, the value of $\overline{GG'}$ in **Equation 4-8** is negative, making the new righting lever $\overline{G'Z'}$ larger than the original lever.

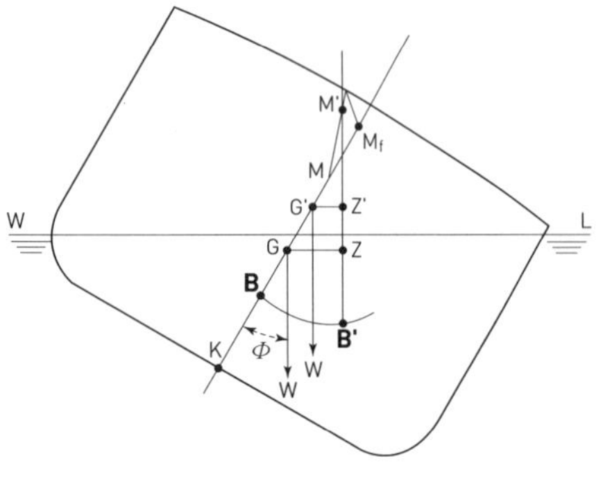

Figure 4-43 Righting arm for different height of the center of gravity

Through **Equation 4-8**, the righting arm for any value of $\overline{KG}$ can be easily found for all heel angles, once the static stability curve is determined for an arbitrary value of $\overline{KG}$. It is plain that as the center of gravity moves up, the righting arm reduces, causing the ship to heel more easily. If the center of gravity moves up too much, the metacentric height can become zero or negative, making the ship highly vulnerable to capsize. A ship with a negative $\overline{GM}$ will have a static angle of loll if not capsized.

If a ship is believed to have insufficient stability, a number of actions can be taken to rectify the situation. For example, some heavy items on board can be moved downward to reduce the height of the center of gravity. Alternatively fixed ballasts such as steel plates or concrete blocks can be placed at the hull bottom or temporarily fill the ballast tanks normally used for the adjustment of trim. It should be noted, however, that the use of water ballast can introduce free surface effects which are detrimental to stability.

4.4.2 Effects of Principal Dimensions

Needless to say, a ship's stability is influenced a great deal by its principal dimensions, and their effects are examined in this section. Note that the length, beam, depth, draft, displacement volume, moment of inertia of the waterplane and the metacentric radius of the reference ship are denoted as L_o, B_o, D_o, T_o, ∇_0, I_x, and BM_0, respectively.

1) Effects of length

Changing only the length of a ship may not be practical and in most cases other principal particulars must also be adjusted appropriately to maintain the ship's performance. Let us assume, however, that the length is increased in proportion to an increase in the displacement, while keeping the cross sectional shapes, draft and location of the center of gravity constant.

If the length of the ship is increased λ times, the waterplane area, moment of inertia and volume also change according to this ratio. However, the transverse metacentric radius $\overline{BM}$ of the elongated ship remains unchanged, as shown below

$$\overline{BM} = \frac{I}{\nabla} = \frac{\lambda I_x}{\lambda \nabla_0} = \overline{BM_0} \tag{4-9}$$

Consequently the metacentric height does not change. Therefore, the righting mass moment of the modified hull shape only changes to:

$$M_x = \rho\nabla \cdot \overline{GM}\, sin\,\phi = \rho\lambda\nabla_0 \cdot \overline{GM_0}\, sin\,\phi \tag{4-10}$$

2) Effects of beam

If only the beam is increased by the ratio of λ, the metacentric radius $\overline{BM}$ is increased due to an increase in moment of inertia as shown in the equation below

$$\overline{BM} = \frac{I}{\nabla} = \frac{\lambda^3 I_x}{\lambda \nabla_0} = \lambda^2\, \overline{BM_0} \tag{4-11}$$

and the righting mass moment of modified hull shape changes in proportion to λ^3 as follows:

$$\rho\nabla \cdot \overline{(BM + KB - KG)}\, sin\,\phi = \rho\lambda^3\nabla_0 \cdot (\overline{BM_0} + KB - KG)\, sin\,\phi \tag{4-12}$$

3) Effects of draft

If the draft increases by the ratio of λ with the length and beam unchanged, $\overline{BM}$ is reduced in the inverse proportion of λ as the equation below:

$$\overline{BM} = \frac{I}{\nabla} = \frac{I_x}{\lambda \nabla_0} = \frac{1}{\lambda}\overline{BM_0} \tag{4-13}$$

The initial mass moment undergoes no change and is

$$\rho\nabla \cdot \overline{(BM + KB - KG)}\, sin\,\phi = \rho\nabla_0 \cdot (\overline{BM_0} + KB - KG)\, sin\,\phi \tag{4-14}$$

It should be cautioned here, however, that the increase in draft will accompany a change in the VCB and an increase in weight not only below but also above the waterplane, resulting in a change in KG as well. All this makes the effect of draft change to stability difficult to summarize neatly.

4) Effects of simultaneous changes in length, beam and draft

If the length, beam and draft simultaneously increase all by the uniform ratio of λ, $\overline{BM}$ will be changed to:

$$\overline{BM} = \frac{I}{\nabla} = \frac{\lambda^4 I_x}{\lambda^3 \nabla_0} = \lambda \overline{BM_0} \tag{4-15}$$

In other words, $\overline{BM}$ increases by a ratio of λ.

4.4.3 Effects of Hull Form

1) Effects of cross sectional shapes of bow and stern

Given an equal displacement, $\overline{BM}$ of a ship is proportional to I_x, the moment of inertia of the waterplane about its longitudinal axis. Comparing two equivalent ships the cross sections of which are the same in the parallel middle body but different near the bow and stern with one having a U- and the other having a V-shape, the ship with the U-shaped bow and stern has less waterplane area and so a smaller I_x and $\overline{BM}$ than the V-shaped hull form. Therefore, the hull form with the V-shaped cross sections at the bow and stern has better stability than the counterpart with the U-shaped cross sections.

Even when a ship operates in waves, its transverse righting moment in beam waves can be easily estimated from the static stability curve, if wavelengths are sufficiently long. However, in head or following seas, as the waves pass by, the area of the waterplane can vary greatly in time, particularly if the wavelength is approximately the same as the ship's length. In such cases a dangerous situation may result due to the great changes in breadth near the bow and stern if the ship is poised on the crest of the wave (hog). Unexpected roll motions in head or quartering waves can occur in this situation and they are known as parametric roll.

2) Effects of hull form coefficients

In general, the range of stability of a ship increases as its block coefficient C_B increases, as long as B/T, f/T, and $\overline{KG}$ remains constant, where T and f are the draft and freeboard of the ship, respectively. Given a constant block coefficient, the range of stability also increases as the waterplane coefficient C_w increases, because the cross section becomes V-shaped so that the moment of inertia of the waterplane increases.

3) Effects of tumblehome and flare

Narrowing and widening hull forms above the waterline are called 'tumblehome' and 'flare,' respectively. Tumblehome is commonly found at the stern of sailing ships and useful when rigging the ship with a mast. Flare is often used to prevent deck wetness due to waves or spray.

Compared with the wall-sided vessels, flare tends to increase the righting arm while tumblehome decreases it. The effects are similar to a gradual change in the beam and become more pronounced as the angle of heel increases.

4) Effects of appendages

Appendages attached to the exterior of the hull provide their own buoyancy forces, affecting the ship's stability. **Figure 4-44** shows bulges installed on a hull. In case (a) the bulges are completely submerged in water and thus they will lower the center of buoyancy and consequently lower $\overline{GM}$. This type of bulges was employed on warships in the past as a passive defense against torpedo attacks and was known as anti-torpedo bulges or blisters. In case (b) the bulges are installed near the waterline to increase the initial righting moment. In case (c) the bulges are installed above the waterline to increase the stability at large angles of heel to compensate for the loss of stability caused by mounting heavy anti-shell armor plates on deck. If a ship is found to have insufficient stability upon completion, appendages may simply be added to improve its stability without changing the principal structure which can be costly. Even a newly built ship may require some modification to the hull form to ensure stability if there are special reasons such as a change of use.

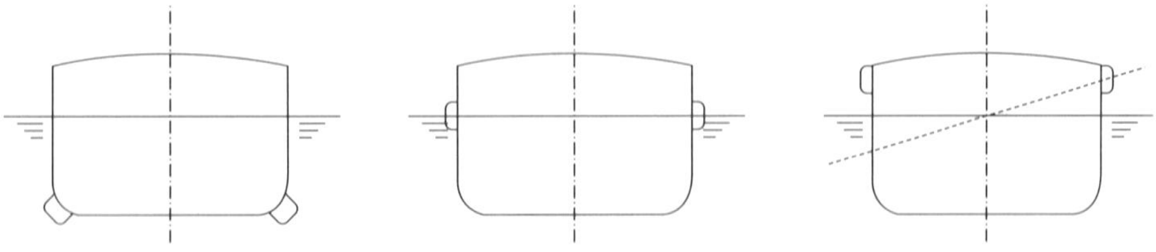
(a) Low bulge (b) Mid-bulge (c) High bulge

Figure 4-44 Locations of bulge

4.4.4 Effects of Other Changes in Form

1) Effects of freeboard

Higher freeboard tends to make the weight near the upper deck heavier and thus the center of gravity rises somewhat. However, the increase in weight is small compared to the ship's weight. Higher freeboard increases the range of positive stability and the magnitude of the maximum righting arm and the angle at which it occurs. If the freeboard is reduced, the range of positive stability decreases and this deteriorates the dynamic stability, even if it has sufficient beam and $\overline{GM}$. In summary, therefore, an increase in freeboard results in improvement of stability.

2) Effects of sheer

The sheer is a measure of longitudinal curvature of the main deck of a ship to raise its exposed deck at the bow and stern. The sheer is effective in preventing deck wetness and submergence in waves and is also effective in improving the profile aesthetics. The sheer also increases freeboard at the bow and stern regions and the overall righting moment of the ship.

3) Effects of superstructure

The superstructure on deck will increase the ship's heeling moment due to wind pressure, but it is considered favorable to the static stability, since it can provide reserve buoyancy at large angles of heel when the deck begins to be submerged.

4) Effects of trim

During the operation a ship may have a trim that is different from the design trim. Such a variation in trim, in turn, makes the ship's righting moment differ from the one estimated for the design trim. Therefore, the effect of trim on stability is dependent on the hull form.

5. Characteristics of Hull Forms by Ship Types

5.1 Cargo Ships

A cargo ship should be designed with plain accommodation quarters and large cargo holds. The design should be focused on cargo handling equipment so that large amounts of cargo can be loaded and unloaded safely and fast. Cargo ships are often classified according to the types of cargo they carry, but from the viewpoint of hull form design, the design concept and strategy depend on the speed of the ship. Therefore, the characteristics of cargo ships are examined below under the headings of low-, medium- and high-speed ships.

5.1.1 Low-speed Ships

Low-speed ships include oil tankers and bulk carriers, and **Figure 4-45** represents the general arrangement (GA) of an oil tanker. The parallel middle body accounts for most of the ship's length; wheel house, living quarters and engine room are all located near the stern

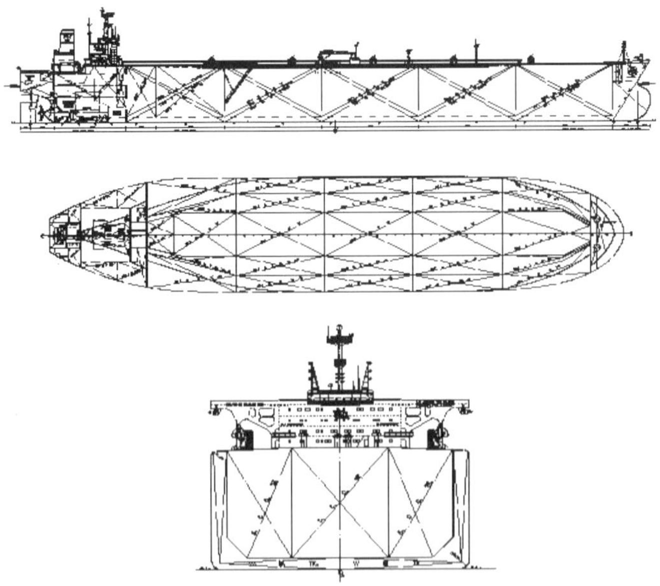

Figure 4-45 General arrangement of a low-speed oil tanker

of the ship and cargo tanks are placed forward of them. Since the ship usually experiences trim by the bow due to cargo being loaded relatively forward, the *LCB* should also be placed reasonably forward to maintain approximately even keel.

A low-speed ship usually has a relatively high block coefficient C_B in the range of 0.75 to 0.86 and viscous component of the resistance is important for this type of ships. When designing a low-speed ship, therefore, a relatively slender stern should be selected to place *LCB* as forward as possible and reduce viscous resistance. This strategy of placing *LCB* forward also helps more efficient loading of cargo as explained above. A slender stern delays flow separation and makes the flow going into the propeller plane more uniform. Therefore it improves propulsive efficiency and has a direct, decisive influence on propeller pressure fluctuation and maneuverability.

Ordinarily, wave resistance of a low-speed ship accounts for a small portion of the total resistance. However, if the bluntness of the bow exceeds a certain limit determined by the Froude number, the wave resistance increases rapidly. In this context care must be taken not to

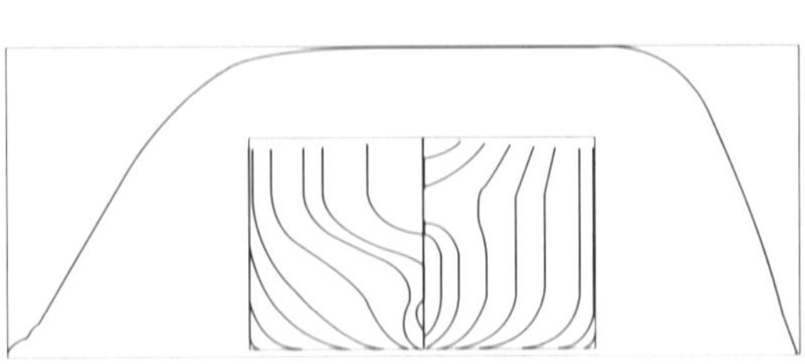

Figure 4-46 A typical sectional area curve and body plan of a low-speed oil tanker

increase the wave resistance by enlarging the foremost cargo hold too much in an effort to maximize the cargo capacity for a given ship's length.

Figure 4-46 represents a typical sectional area curve and body plan of a slow ship. As discussed earlier, as the *LCB* is located forward of the midship, the overall distribution of displacement is slanted towards the bow. The cross section of the bow is nearly U-shaped and the stern is also U-shaped to reduce resistance, increase efficiency and ensure the uniformity of flow into the propeller plane. In the past the stern of a slow ship used to be often designed to have V-form cross sections to reduce resistance, but most modern slow ships are now designed to have U-form stern cross section to increase the propulsive efficiency.

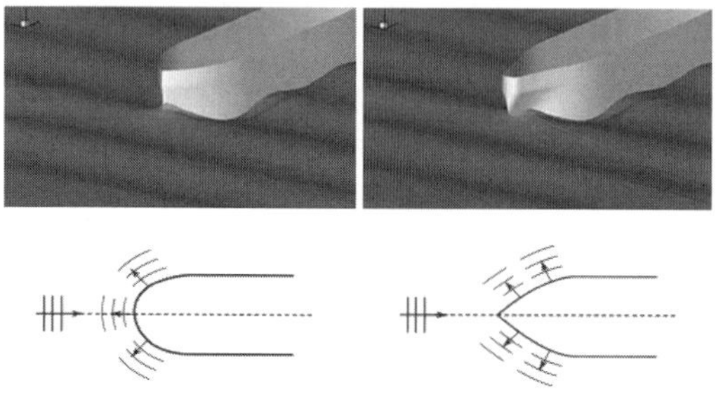

Figure 4-47 Reflection of head waves against the bow

Ships are subject to an additional resistance component caused by waves. For example, incoming waves are reflected at the bow and this increases the resistance. As shown in **Figure 4-47**, a sharper bow reduces the wave reflections and, thus, the wave-induced resistance.

5.1.2 Medium-speed Ships

LNG (liquefied natural gas) carriers are medium-speed ships, and so are LPG (liquefied petroleum gas) carriers. Both carriers carry liquefied gases which are high volume low density cargoes and require relatively spacious cargo holds. Therefore, the C_B ranges between 0.65 and 0.75 and the ship's speed is required to be relatively high at 19 to 20 knots, and this calls for caution in designing the bow and stern.

The stern of medium-speed ships is designed in fundamentally the same way as that of a low-speed ship. But, they have higher speed so the flows are prone to separate near the stern. Therefore, the run angle of the stern should be reduced to diminish viscous resistance. Due to its relatively high design speed, the ship is often placed at a critical point where wave resistance begins to grow rapidly. Therefore, the entrance angle of the bow should be made as small as possible to reduce the resistance and prevent wave breaking. The bulb should be designed to maximize wave cancellation.

Figure 4-48 shows a typical sectional area curve and cross sectional shape of a medium-speed ship. When designing a low-speed vessel, the fore-shoulder is made smooth to reduce shoulder wave, even if it makes the angle of entrance at the bow slightly larger. However, when designing a medium-speed ship, the entrance angle of the bow needs to be reduced to avoid wave breaking and resistance. That is because the waves produced at the entrance of the bow are more influential to the total resistance than the waves generated at the fore shoulder at this speed range.

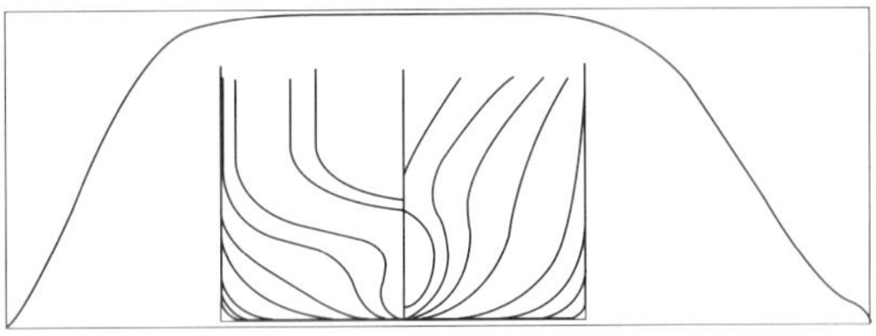

Figure 4-48 A typical sectional area curve and the body plan of a medium-speed ship

LNG carriers can be divided by the shapes of containment system into Moss and membrane types. As shown in **Figure 4-49**, the Moss type LNG carriers are loaded with self-supporting spherical containment systems while the membrane types have prismatic containment system to allow a better use of the interior space of the ship.

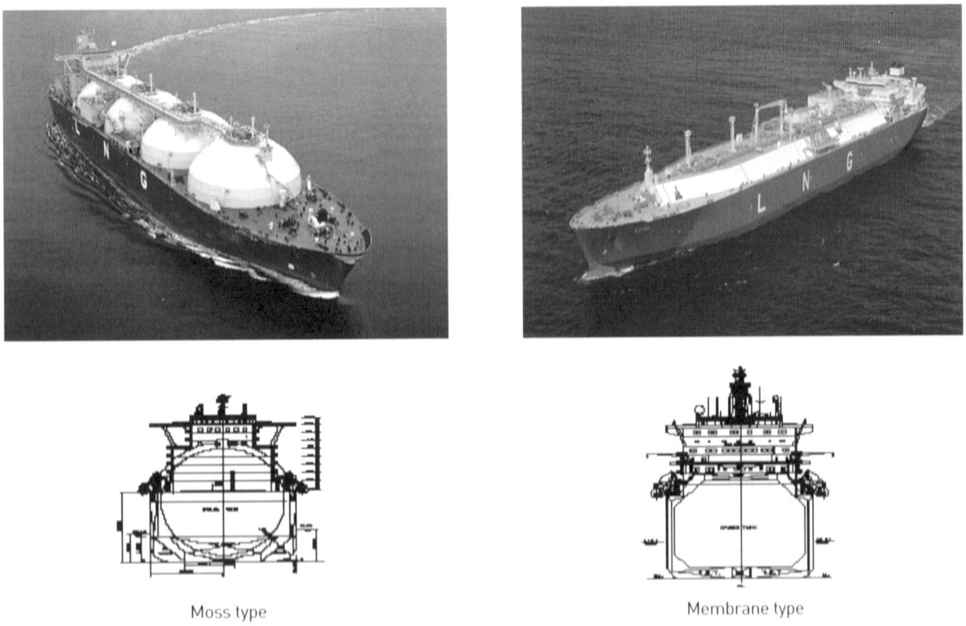

Figure 4-49 LNG carriers and their midship sections

The midship section of the Moss type design has a large bilge radius to accommodate spherical containment systems while the membrane design has a smaller bilge radius suitable for prismatic containment systems. The Moss type LNG carriers, thus, have smaller C_m, greater beam, and lower block coefficient than the membrane type of an equal loading capacity. The Moss type is subjected to higher resistance due to the increased breadth, but its spherical cargo tanks are more robust and resistant to sloshing force and it has much potential for offshore storage, such as in F-LNG which frequently has to contend with bad weather in normal operations.

Since LPG is heavier than LNG, the *LCB* of LPG carriers has to be placed further forward than the equivalent LNG carriers to maintain appropriate trim and, consequently, the designer should carefully distribute the increased displacement towards the bow.

5.1.3 High-speed Ships

Container ships transport cargoes contained in standardized containers for intermodal transport and are categorized as high-speed ships and the general arrangement of a typical container ship is shown in **Figure 4-50**.

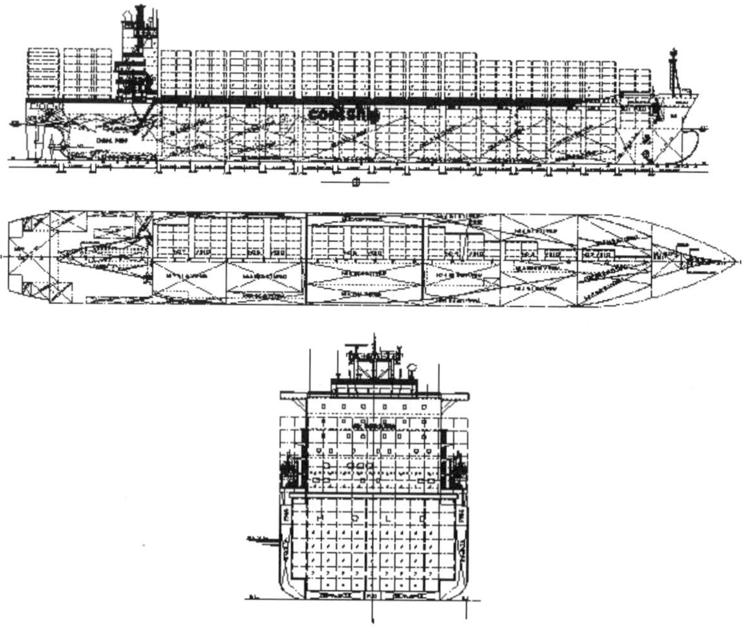

Figure 4-50 General arrangement of a container ship

Unlike other cargo ships, the container ships carry cargo stacked both in cargo holds and on deck. Therefore, they require a large deck area to maximize its loading capacity. Furthermore, if they are to move fast, high wave resistance at such speeds must be avoided. Therefore, care should be taken to optimize the form factors related to the bow shape, since they are closely related to wave resistance.

Figure 4-51 represents a typical sectional area curve and body plan of a container ship. The entrance of the bow is designed to be as sharp as possible to avoid excessive bow waves and wave breaking. In addition, the V-shaped cross section is used to increase the waterplane area and ensure smooth flows below the waterline. The stern is designed as a barge form to simplify the flow field in the region and reduce resistance, and a small skeg is installed near the propeller plane to enhance propulsive performance.

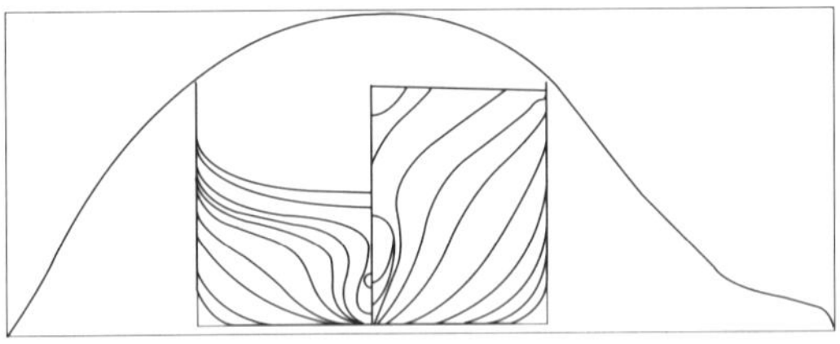

Figure 4-51 A typical sectional area curve and the body plan of a container ship

Figure 4-52 compares the bow shapes of a container ship (on the right) and an oil tanker (on the left), which highlight different hull forms designed to serve different purposes. The container ship has a more protruding and higher bulb so that the waves generated by it can cancel the waves generated by the stem and shoulders, thus reducing the overall resistance. The oil tanker's bulb, on the other hand, is used to reduce the bow entrance angle of the waterplane.

Figure 4-52 Difference in bulb shapes of low-speed and high-speed ships

5.2 Passenger Ships

Passenger ships are used for transporting people and typical ones are passenger liners or ferries which operate on given routes according to published timetables and cruise ships which cruise around tourist spots. Since the tall superstructures of passenger ships make their center of gravity quite high, $\overline{KM}$ should be made large to meet the high stability requirements. In determining the $\overline{KM}$, designers should strike a happy balance between the stability of the ship and the comfort of the passengers.

- In general, the block coefficient of a passenger ship is not large, but its speed is relatively high at Froude number of over 0.25. Therefore, the design should take wave resistance into serious consideration. A typical sectional area curve and body plan of a cruise ship is given in **Figure 4-53**. If the ship is designed to have waterplane with moderate curvatures in order to reduce wave resistance, as shown in the body plan, the waterplane area can become too small to make it difficult to secure the required $\overline{KM}$. It is critical that the bow of passenger ships should be designed with these two conflicting requirements in mind.

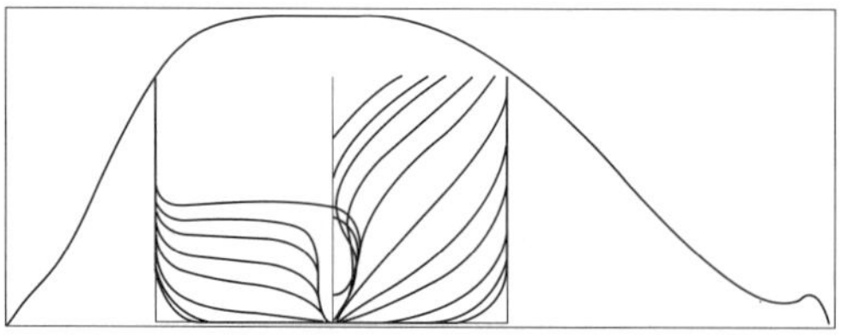

Figure 4-53 A typical sectional area curve and the body plan of a cruise ship

The stern shape of passenger ships is a barge form to simplify the flow field and reduce resistance. Since the direction of the main streamline is determined largely by the buttock lines, they should be designed to induce smooth flows. Furthermore, as the design speed increases, the resistance due to the transom becomes more significant. The transom resistance is related to the slope of the buttock line at the end of the stern. As the slope of the buttock line becomes smaller and the length from after perpendicular to the stern end larger, transom resistance becomes smaller. In addition, various appendages are developed and used to reduce resistance, such as a wedge, ducktail and flap, as illustrated in **Figure 4-54**.

Passenger ships require considerable engine power to reach a high speed, but the required propeller diameter can become too large for the design draft. In such cases a twin-propeller system is often used. Furthermore, to enable a maximum propeller diameter within the limited draft, the designer often resorts to modifying the stern to have a tunnel-shape. Sometimes inclined propeller shafts or podded propulsion system are used to maximize uniformity in flow going into the propeller plane.

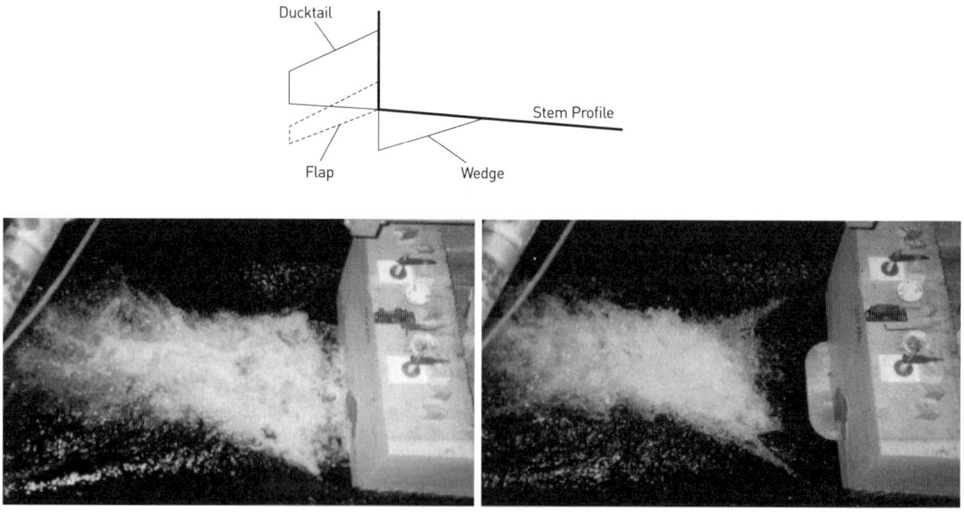

Figure 4-54 Various transom appendages for the reduction of resistance. Photographs show the effect of a stern flap

5.3 Icebreaking Cargo Ships

There has been a growing interest in icebreaking cargo ships since the official opening of the Arctic Northern Sea Route by Russia. Icebreaking ships are double acting ships (DAS) which can carry cargo while breaking thin ice when necessary. Therefore, they can operate in both open waters and thin ice independent of ice-breakers. To do that an icebreaking cargo ship must be capable of delivering its desired performance in both frozen and open waters and consequently its hull form must be designed for its major operational routes and scenarios.

The bow of an icebreaking cargo ship must be designed to enhance the ship's icebreaking performance and effectively push broken ice sideways. **Figure 4-55** shows the design parameters of the bow in relation to icebreaking ability. In general, the smaller the entrance and flare angles, the better the ship gets at breaking ice; the wider the waterplane near the bow, the better. Aside from the entrance angle, the buttock line angle is also related to the ship's icebreaking performance: the smaller the angle, the smaller the ship's resistance to ice.

Since each of the parameters must be adjusted to the ice conditions of waters where the ship operates, the bow form is designed as shown in **Figure 4-56**. The designer of this ship chose this kind of bow form because the design parameters that permit higher ice-breaking

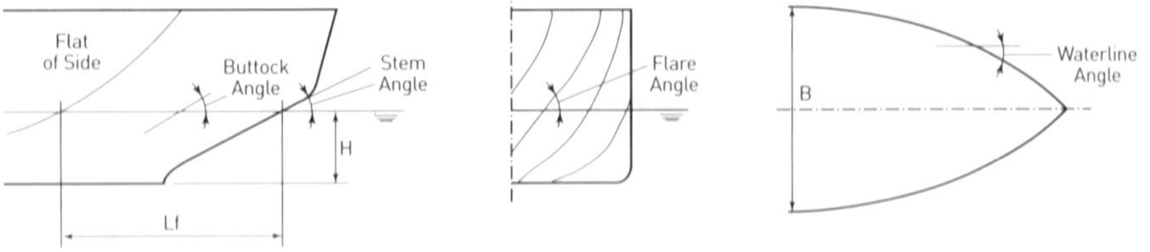

Figure 4-55 Major parameters for designing the bow shape of an icebreaking cargo ship

Figure 4-56 Photographs for the bow shape of an icebreaking cargo ship

performance adversely affect the performance of the ship in open waters.

The stern is designed to provide the waterplane as wide as at the bow to ensure the ice breaking performance of the stern area, and both the buttock lines and cross sections are flatter compared to ordinary merchant ships.

Since an ice-breaking ship breaks through ice to move forward, it requires more engine power than ordinary merchant ships of similar size. In addition, the propeller torque may increase rapidly as the propeller interferes with ice, and it is important to reserve a sufficient torque margin. The propeller of an icebreaking cargo ship requires more power to move astern than ordinary ships, as it may require backing out from a thick ice area from time to time. Therefore, the pod propellers or controllable pitch propellers (CPP) are sometimes preferred to provide astern power. The ship should also be able to operate under different load conditions of icy and open waters.

References

1. Chun, H.-H. et al., *1993. Resistance and propulsion of small waterplane catamarans.In: theSociety of Naval Architects of Korea, Summer Seminar.* Busan, 5~7 July, 1993. The Society of Naval Architects of Korea: Seoul

2. Gertler, M., DTMB (David W. Taylor Model Basin), 1954. *A reanalysis original test data for the Taylor standard series*-DTMB Report 806, Washington DC: U.S. Navy Department.

3. Kang, K.J. et al., 2000. The resistance and propulsion of 2,500-ton trimarans (in Korean). In: the Society of Naval Architects of Korea, *Spring Meeting.* Kwangju, 20~21 April, 2000. the Society of Naval Architects of Korea: Seoul.

4. Kim, H. et al, 2006. *Advancedships in Korea* (in Korean). Seoul: Jisungsa.

5. Koo, J.-D., 2000. *Introduction to naval ship* (in Korean). Seoul: Dongmyungsa.

6. Kracht, A.M., 1978. Design of bulbous bows, *SNAME Transactions*, 86, pp.1-15.

7. Lackenby, H., 1950. On the systematic geometrical variation of the ship forms. *RINA Transactions*, 92, pp.289-316.

8. Lee, C.J., 2003, *Astudy on the systematization of semi-planing hulls and on the estimation of resistance coefficient.* Ph. D.Daejeon: Chungnam National University.

9. Schneekluth, H. & Bertam, V., 1998. *Ship design for efficiency and economy*, 2nd ed. Oxford: ButterworthHeinemann.

10. Shin, M.S., 1993, Resistance and propulsion of high-speed ships (in Korean), In: the Society of Naval Architects of Korea, *Summer Seminar.* Busan, 5~7 July, 1993. The Society of Naval Architects of Korea: Seoul.

11. The Society of Naval Architects of Korea, 2009, *Ship resistance and propulsion* (in Korean). Seoul: Jisungsa.

12. Versluis, A., 1977. *Computer aided design of shipform by affine transformation.* Rotterdam: International Periodical Press.

13. Ward, G., 1982. The Application of current vibration technology to routine ship design work. In: RINA, *Spring Meeting.* London, ??~?? 1982, RINA: London.

14. Yokoo, S. et al., 1973. *Propeller Design Method and Reference Charts* (in Japanese). Tokyo: Sanseido Publishing Co., Japan.

CHAPTER 5

Resistance and Propulsion

1. Introduction

Energy efficiency is one of the most important goals to be achieved in designing a ship and for that reason naval architects try to reduce resistance and improve propulsive efficiency from the outset. For this purpose, designers must have various theoretical and experimental tools at their disposal for predicting the resistance and propulsive efficiency of a ship.

It was in the 19th century that William Froude suggested a practical method to extrapolate test results of a small scale model to the full scale as a means of estimating the resistance of a ship, the central idea of the procedure has remained the same; and Froude's method has been the most frequently used one up to now. However, the method contains inherent uncertainties arising from the inability to meet all the similarity requirements between the flow fields around an actual ship and the model, despite subsequent improvements achieved so far.

Owing to the rapid growth of computer and numerical techniques in recent years, numerical methods (computational fluid dynamics or CFD) can now be applied to predict the resistance and the flow field around a ship and these methods yield results qualitatively similar to those of the experiments. However, more progress will be necessary before the full potential of CFD is realized in simulating full scale turbulent flow fields and the shape of the free surface around a real ship with sufficient quantitative accuracy.

The resistance of a ship may be estimated from sets of series data obtained through tests with a series of models constructed by varying their principal characteristics in a systematic manner, but the method is rarely utilized these days mainly because of the rapid growth in CFD.

1.1 Decomposition of Resistance

A surface ship moves on and near the air-water boundary (hereafter 'the free surface') and has to overcome the resistance of both fluids against its motion. The total resistance of a ship is too complicated to comprehend simply, because it is composed of diverse components arising from different physical origins and cross coupled with each other. Therefore, it is normal to explain the total resistance of a ship moving on the calm water surface as a linear combination of the major components as shown in **Figure 5-1** after discounting the contribution from the air resistance.

William Froude divided the total resistance of a ship into two components: namely,

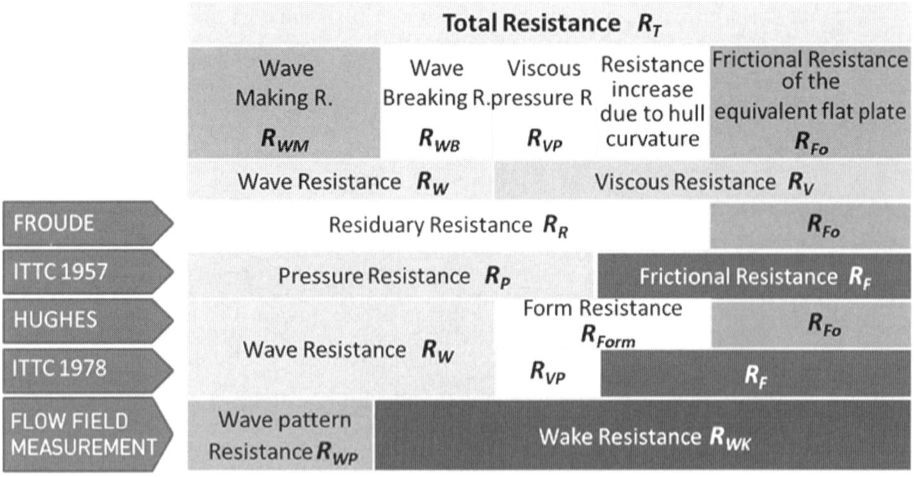

Figure 5-1 Decomposition of the total resistance of a ship moving on calm water

frictional and residuary resistance. Furthermore, he claimed that the frictional resistance of a ship can be approximated as the drag of a flat plate having an identical length and wetted surface area with the ship (hereafter 'the equivalent flat plate') and moving at the same speed. He also proposed the law of comparison, which states that the residuary resistance of geometrically similar ships is in the ratio of the cube of their linear dimensions, if their speeds are in the ratio of the square roots of their linear dimensions.

Subsequently, it has become more customary to split the ship resistance into wave resistance owing to the energy spent on generating a wave system around the ship, and viscous resistance due to the energy loss by the viscosity of water.

The wave resistance of a ship can be subdivided into wave pattern resistance(or \-making resistance), which can be calculated from the wave height measured sufficiently far from the ship, and wave-breaking resistance due to the loss of energy occurring when a wave is broken near the hull surface, typically close to the bow. In many cases, it is not unusual to neglect the wave-breaking resistance and treat the wave-making resistance as the wave resistance.

The viscous resistance can be subdivided into the frictional resistance of an equivalent flat plate, form resistance due to the hull surface curvature, and viscous pressure resistance due to the presence of a turbulent boundary layer and wake behind the ship. Alternatively, the viscous resistance of a ship can be divided into the frictional resistance of an equivalent flat plate and a form resistance.

In summary, a ship sailing on calm water experiences a resistance from the water which can be subdivided into five components, i.e., wave-making resistance, wave-breaking resistance, viscous pressure resistance, frictional resistance of an equivalent flat plate, and resistance increase due to the curvature of the hull surface. Some procedures for the model-ship extrapolation can be devised by combining the components in different ways as shown in **Figure 5-1**.

A ship also experiences some additional resistance components. All the appendages attached to a ship give rise to extra resistance, i.e., the appendage resistance. The above-water part of the hull moving through the air is subject to air resistance, the rudder produces steering resistance, and water spray occurring at the bow or the intersection of a strut and free surface causes spray resistance for a high-speed ship. The motion of a ship in waves increases the resistance (added resistance) and likewise for a ship in restricted or shallow waters.

In general, the frictional and the wave-making resistance occupy the most of ship's resistance. Moreover, the frictional resistance dominates in the low speed range but the wave- making resistance increases rapidly as the speed increases and exceeds the other eventually. For example, **Figure 5-2** compares the resistance components for different types of ships, where the horizontal axis indicates the percentages of the components in the total resistance coefficient. The approximate magnitude of the total resistance coefficient $C_T = R_T/(\frac{1}{2}\rho SV^2)$ is given under each bar where R_T denotes the total resistance, ρ is the density of water, V is the speed of the ship, and S is the wetted surface area of the ship.

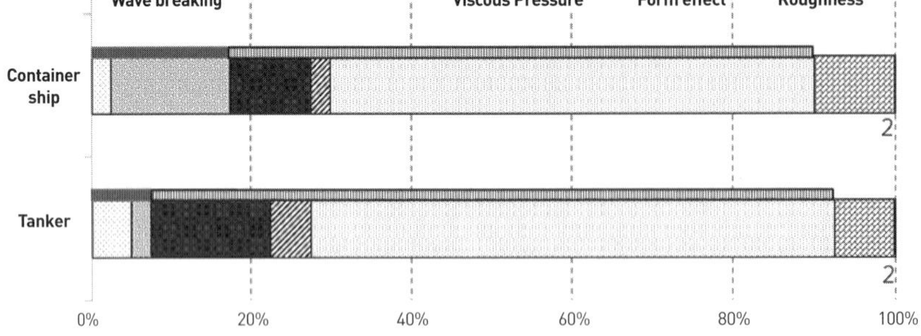

Figure 5-2 Relative composition of resistance for different types of ship
(Based on data published in Larsson & Raven, 2010)

1.2 Dimensional Analysis in Ship Resistance

Since it is impractical to measure directly the resistance of a ship by towing a full scale ship both in terms of time and cost, it is usual to obtain the quantitative data of resistance through towing tests conducted on a scaled model. The use of a scaled model, however, requires the detailed knowledge of the relationship between the model test conditions and the full scale conditions. The relationship is known as the law of similitude and this requires careful analysis. The procedure normally involves investigation on geometric, kinematic and dynamic similarities, and dimensional analysis is one of the most powerful tools to confirm the relations.

There are seven primary dimensions in classical physics and among them, length L, mass M, and time T are the most important ones in this subject. The rest are temperature, electric current, amount of light and amount of matter, and other physical quantities can be formed by some combination of the primary dimensions. The dimensional analysis relies upon the law of dimensional homogeneity, stating the basic principle that every additive term in an equation must have the same dimensions.

To apply dimensional analysis to the ship resistance problem, consider a ship moving on a still water surface as shown in **Figure 5-3**. The ship disturbs the water and the air around the hull and eventually generates a free surface wave system at the interface under the influence of gravity. We can neglect the air resistance, for simplicity, since the density and viscosity are small compared to those of the water. The influence of the atmospheric pressure P_a is also excluded, as it is almost constant near the free surface.

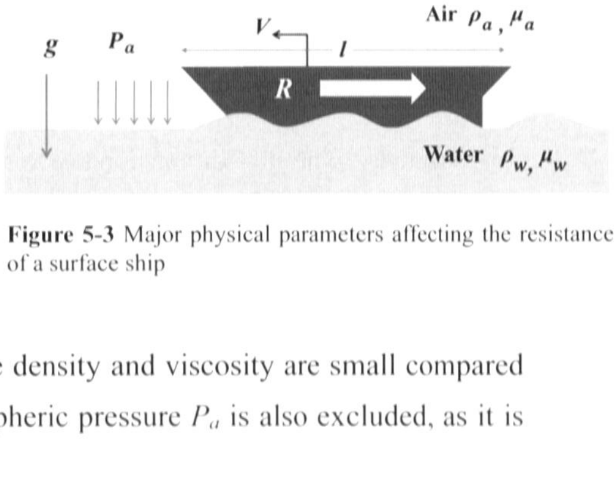

Figure 5-3 Major physical parameters affecting the resistance of a surface ship

Then, the resistance R of a ship moving at a constant speed V can now be written as an unknown function of the parameters and forcing functions, or in terms of unknown powers of them

$$R = f_0(\rho, V, l, \mu, g) = \rho^a V^b l^c \mu^d g^e \tag{5-1}$$

where l is the characteristic length of the ship, g is the gravitational acceleration, and ρ and μ are the density and viscosity of water respectively.

Since R is a force, its dimensions are MLT^{-2}. Similarly the dimensions of the density are ML^{-3}, ship speed LT^{-1}, the viscosity $MT^{-1}T^{-1}$, and the gravitational acceleration LT^{-2}.

Introducing these dimensional quantities into **Equation 5-1**, we have

$$[MLT^{-2}] = [ML^{-3}]^a [LT^{-1}]^b [L]^c [ML^{-1}T^{-1}]^d [LT^{-2}]^e \tag{5-2}$$

From the law of dimensional homogeneity, three simultaneous equations are generated, since both sides of the equation should have the same dimension and only three primary dimensions are involved in the problem. Eliminating *a, b* and *c* with the equations, we have:

$$\frac{R}{\rho V^2 L^2} = \left(\frac{\mu}{\rho V l}\right)^d \left(\frac{gl}{V^2}\right)^e = f_1\left(\frac{\mu}{\rho V l}, \frac{gl}{V^2}\right) \tag{5-3}$$

All the terms in the above equation are now non-dimensional and therefore there is no dimensional restriction on the exponents *d* and *e*. Therefore, it is more customary to write the above equation in the form of:

$$C_T = \frac{R}{\frac{1}{2}\rho V^2 L^2} = f\left(\frac{Vl}{\nu}, \frac{V}{\sqrt{gl}}\right) = f(Rn, Fn) \tag{5-4}$$

where C_T is the non-dimensional total resistance coefficient and $\nu = \mu/\rho$ is the kinematic viscosity. The non-dimensional parameter *Fn* is called the Froude number and *Rn* is called the Reynolds number.

The Reynolds number denotes the ratio of inertial force to viscous force, whereas the Froude number stands for the ratio of inertial force to gravitational force. In special cases, in which the atmospheric pressure or surface tension must be considered, the similarity will be ensured when values of the Euler number $P_a/(\rho V^2)$, the cavitation number $(P_a - P_v)/(\frac{1}{2}\rho V^2)$ and the Weber number $\rho V^2 L/\gamma$ are also equal, where P_v is the vapor pressure of the water and γ is the coefficient of surface tension.

The results of the dimensional analysis can be summarized with the conclusion that the total resistance coefficients of geometrically similar but different sized bodies are the same, if the values of Reynolds and Froude numbers are made equal. Since both viscosity of the water μ and the acceleration of gravity *g* remain almost constant in normal situations, the only plausible way to match the values of *Fn* and *Rn* for a ship and its scale model is to change the speed of the model. For example, to make the Froude number of the model the same as that of the ship, the towing speed of the model has to be reduced by the ratio of the square root of the scale ratio $\lambda = L_S/L_M$ where L_S and L_M are the characteristic length of the ship and the model, respectively. On the contrary, however, the speed of the model must be increased by the ratio of λ if the Reynolds numbers are to be the same. Now, it is apparent that it is impossible for both Fn and Rn of the ship and its model to be the same simultaneously unless the lengths are the same.

Therefore, it is a usual practice in model tests to neglect the influence of the viscosity and adjust only the values of the Froude number to be the same. In such a case, however, ignoring the viscosity violates the law of similitude and the total resistance coefficient C_T of the ship and the model is no longer identical. Therefore, it is necessary to devise a robust procedure for the extrapolation of model results to full scale so as to estimate the resistance of a ship as accurately as possible.

2. Viscous Resistance

The viscosity of a fluid is a measure that represents the internal resistance of a fluid to gradual deformation by shear or tensile stress. In other words, viscosity is a property of a fluid that resists against relative motion and results from friction among neighboring fluid particles moving with different speeds. Since all fluid has viscosity, more or less, if a body moves in a fluid, friction causes the momentum being transferred into the fluid particles. Then, convection, diffusion and dissipation of the momentum delivered to the fluid characterize the flow field around the body.

Figure 5-4 shows the flow phenomena occurring around a ship-like body moving at a constant speed in still water. If the body moves, the water particles attached to the body

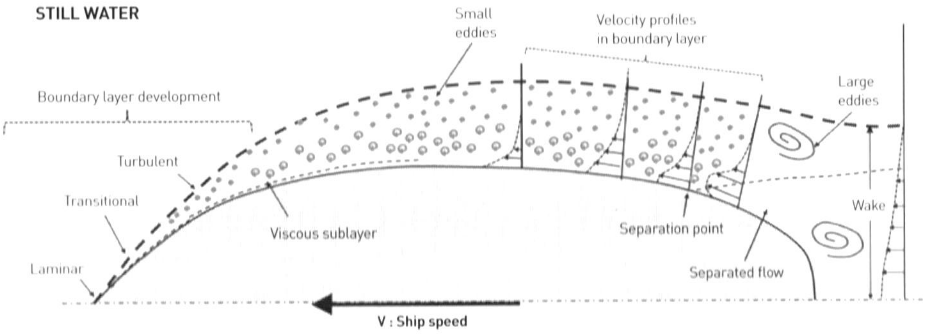

Figure 5-4 Characteristics of flow field around a ship-like body moving in still water

surface cannot leave the surface and have to travel with the ship due to the presence of friction. The moving particles drag adjacent ones in turn but at a relatively lower speed. This mechanism of momentum transfer persists to the flow field far from the body surface but gradually diminishes with distance. Therefore, there exists a flow region in the vicinity of the body surface where the influence of the viscosity is prevalent, and the region is called the boundary layer.

The boundary layer around a hull thickens as it gets closer to the stern and sometimes an adverse pressure gradient that typically exists at that region may cause the flow in the boundary layer near the surface to move faster than the surface itself to result in flow separation eventually. The boundary layer will be lifted off from the surface if the flow separates or passes through the stern to produce a lump of flow following behind the ship: this flow is called the wake. Therefore, the main source of the viscous resistance is the energy loss due to the boundary layer and wake around a moving ship and such loss should be compensated for by the ship in order to move at the same constant speed.

The boundary layer is a flow domain in which viscous force dominates the inertial counterpart and the Reynolds number Rn characterizes the flow. The flow in the boundary layer has low energy compared to the free stream and is more sensitive to changes in pressure. If a fluid starts to move along a body surface, viscosity tends to retard the flow speed so as to produce the boundary layer near the surface. In the leading part of the boundary layer, the flow maintains orderly behavior (laminar) but soon becomes unstable as it moves downstream (transitional) and finally develops into a fully turbulent boundary layer.

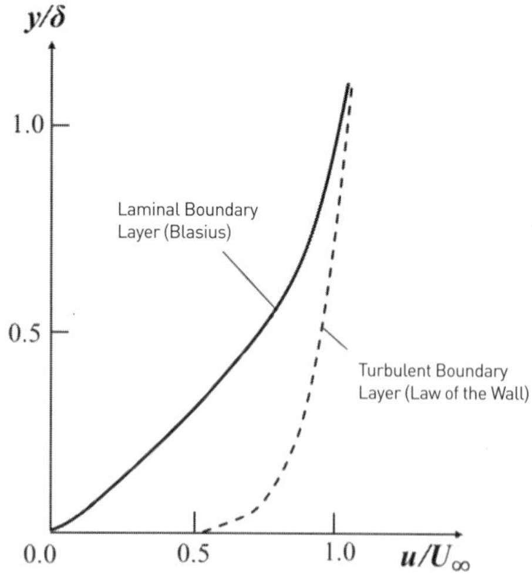

Figure 5-5 Comparison of laminar and turbulent flat plate boundary layer profiles normalized by boundary layer thickness

The same analogy applies to the flow field around a ship moving at a constant speed. The viscosity begins to influence streamlines as they pass through the edge of the boundary layer at the fore-body and progress downstream. The flow behavior within the boundary layer depends on the local Reynolds number Rn_x defined with the length of the streamline measured from the stem. The internal flow remains laminar and the stream-wise velocity changes uniformly as it moves away from the wall until Rn_x reaches some value above which the flow becomes turbulent. In this condition the stream-wise velocity is characterized by unsteady swirling flows inside the boundary layer. In other words, the turbulent boundary layer contains numerous tiny eddies of variable sizes and therefore the subsequent active momentum transfer results in the time-averaged velocity profile more uniform than the laminar profile, except in the narrow region close to the surface where abrupt changes in the velocity gradients are inevitable to satisfy the wall boundary condition as shown in **Figure 5-5**. The large velocity gradient brings an increase in surface shear stress and the frictional resistance as well. Normally, the laminar boundary layer occupies a small region near the bow and the turbulent boundary layer over the rest of the hull surface.

2.1 Decomposition of Viscous Resistance

The famous D'Alembert's paradox states that an unbounded, inviscid and incompressible flow produces no drag on an object of any shape moving in it at a constant speed. More specifically, in such a flow, neither the pressure drag nor the viscous drag is acting on the body, since the surface integral of the pressure vanishes and no friction exists either.

It is well understood now that the occurrence of the paradox is due to the negligence of the viscosity, and a ship moving in the water, a real fluid, necessarily meets with resistance against the motion. Despite the viscosity, the pressure distribution over the fore portion of a ship does not change much, but fast growth of the boundary layer over the aft portion and the presence of the wake result in insufficient pressure recovery near the stern.

Consequently, the integration of the pressure on entire wetted surfaces does not vanish but yields a net force against the ship motion and is called the viscous pressure resistance; the resistance is higher, especially if a three dimensional separation occurs near the stern. For example, under the influence of the bilge vortex, the wake region grows larger to increase the viscous pressure resistance considerably. **Figure 5-6** shows the pressure distributions along a two-dimensional body and illustrates the phenomena in a somewhat simplified manner.

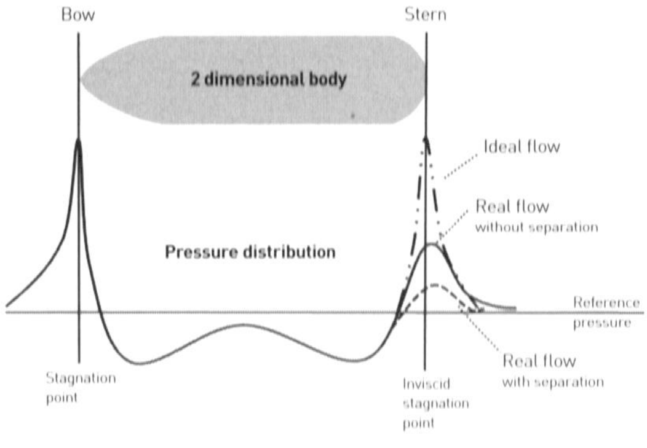

Figure 5-6 Comparison of pressure distributions along a two-dimensional moving body

The viscous resistance is the sum of the viscous pressure resistance and the frictional resistance. It can be obtained by integrating the pressure and shear stress acting on the hull surface. The viscous resistance occupies the majority of the total resistance in the lower speed range and a full inspection on the complicated flow field around the hull surface of complex geometry is needed for rigorous estimation of the magnitude. However, this may not be practically feasible.

William Froude is the first person who approximated the frictional resistance of a ship as the friction of an equivalent flat plate and many researchers have followed to advance his work. More details on the topic will be given in the following section but it is worthwhile at this time that a superficial difference between the viscous resistance of a ship with a three-dimensional curvature and the friction of an equivalent flat plate was found and it was named the form resistance. In a sense, the form resistance is the sum of the viscous pressure resistance and the increase in the frictional resistance due to three-dimensional curvature of a hull surface.

2.2 Frictional Resistance

The French hydrodynamicist Ferdinand Reech is known for his first formulation of the hydraulic model law of gravitational similitude necessitating the proportionality between the velocity of a ship and the square root of its length (more generally known as Froude's law of similarity). It is said to have been included in his lectures as early as 1831, first mentioned in his memoir of 1844 and then discussed in detail in his course notes of 1852 but he had not demonstrated how it could be applied to practical problems in ship resistance.

William Froude is generally recognized as the first to quantify ship resistance. He constructed a series of scale models and measured their resistances at various towing speeds and proposed in 1868 the law of comparison, stating that "the (residuary) resistance of geometrically similar ships is in the ratio of the cube of their linear dimensions if the speeds are in the ratio of the square roots of their linear dimensions." Here, the residuary resistance is the total resistance minus the frictional resistance of an equivalent flat plate as mentioned previously.

Froude also undertook a series of towing tests to find the frictional resistance of smooth planks in his 300 ft long basin containing fresh water at Torquay, England. He towed planks 19 in × 3/16 in with the length varying from 2 ft to 50 ft and coated with various substances at speeds up to 13.3 fps. He found that the frictional resistance of a plane surface is proportional to $S\,V^n$ where S is the wetted surface area, V the towing speed, and n depends on the length and nature of the surface. Later in 1888, Robert E. Froude supplemented his father's formula to show that the frictional resistance of a flat plate is proportional to $SV^{1.825}$.

The formula was adopted by ITTC (International Towing Tank Conference) in 1935 and was widely used by most towing tanks until the 1960s.

In 1932, Schoenherr collected some experimental data available then and compared it to the theoretical formula of von Karman and Prandtl to determine the constants to find that the following formula is a good fit to the experimental data.

$$\frac{0.242}{\sqrt{C_F}} = \log_{10}(RnC_F) \tag{5-5}$$

This Schoenherr curve was adopted by ATTC (American Towing Tank Conference) in 1947 and is sometimes known as the 'ATTC line.' ATTC recommended an allowance that is ordinarily 0.0004 to account for the effect of hull roughness but the formula has a tendency to underestimate the value of C_F if the Reynolds number is relatively low ($Rn < 10^7$). Furthermore, Schoenherr used experimental data from various sources and the planks were not geometrically similar. Therefore, the data was influenced by aspect ratio and edge effects and can hardly be regarded as the friction formula for a flat plate in two- dimensional flow.

Hughes then performed resistance tests on planks and pontoons of various lengths and aspect ratios. He extrapolated the resistance coefficients thus obtained to infinite aspect ratio and proposed in 1954 a curve of minimum turbulence resistance for plane, smooth surfaces in two-dimensional flow

$$C_{F_0} = \frac{0.066}{(\log_{10} Rn - 2.03)^2} \tag{5-6}$$

where C_{F0} denotes the frictional coefficient in two-dimensional flow. The Hughes line appears to be a rigorous formula on the friction of a flat plate in two-dimensional flow but for this reason it may not predict the friction of a real plate with finite width precisely.

Based on Hughes proposal, ITTC recommended the 'ITTC 1957 model-ship correlation line' with a minor change in his formula

$$C_F = \frac{0.075}{(\log_{10} Rn - 2.0)^2} \tag{5-7}$$

This gives C_F values almost the same as those of the Hughes line increased by 12% uniformly. The line is called a *model-ship correlation line*, since it is not meant to be used to predict the frictional resistance of a plate but to extrapolate the result of model tests to full scale.

Granville proposed a new formula considering the velocity distribution in the turbulent boundary layer in 1977.

$$C_{F_0} = \frac{0.0776}{(\log_{10} Rn - 1.88)^2} + \frac{60}{Rn} \tag{5-8}$$

The formula shows good agreement with the 1957 ITTC line for the values of Reynolds number less than 5×10^5. If Reynolds number exceeds 10^8, all the lines in **Figure 5-7** are in good agreement except the Hughes line.

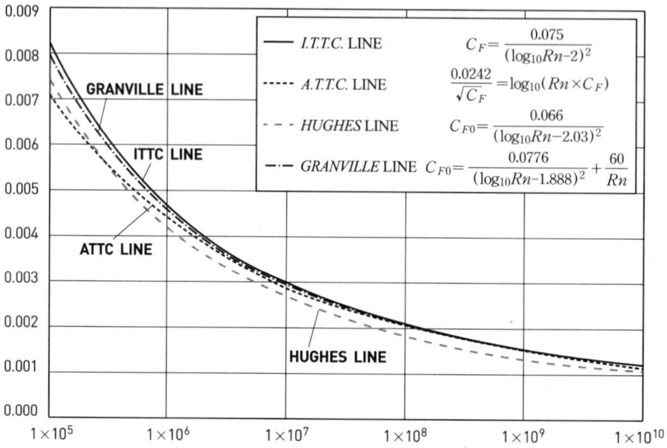

Figure 5-7 Comparison of friction lines

In summary, Froude's method defines the difference between the values of the total model resistance coefficient C_{TM} and the model frictional resistance coefficient C_{FM}, estimated by one of the above formulae, as the model residuary resistance coefficient C_{RM}. It is further assumed that the ship's residuary resistance coefficient C_{RS} has the same value as C_{RM} and we can now proceed to find the total ship resistance. The procedure is called 'two-dimensional extrapolation method'

$$C_{TS} = C_{RS} + C_{FS} = C_{RM} + C_{FS} = (C_{TM} - C_{FM}) + C_{FS} + C_A \tag{5-9}$$

where the second subscript '*S*' denotes values for a ship and '*M*' is for a model, and C_A is an allowance for roughness and ordinarily has a value of 4×10^{-4}.

2.3 Form Resistance

As previously discussed, the viscous resistance of a ship can be defined as a sum of the frictional resistance and the viscous pressure resistance or as the frictional resistance of an equivalent flat plate and the form resistance. Thus, the form resistance of a ship is the sum of viscous pressure resistance and an increase in the frictional resistance due to hull curvature.

Hughes assumed that the total model resistance coefficient C_{TM} can be divided into viscous resistance coefficient C_{VM} and the wave-making resistance coefficient C_{WM}. Among those, C_{WM} reduces rapidly at low Froude numbers since it is proportional to Fn^4 or even the higher order. At a point where C_{WM} is not substantial, the curve of C_{TM} will become almost parallel to the two-dimensional friction line C_{F0M} and Hughes called it the run-in point. At the run-in point, C_{TM} can be considered as a sum of C_{F0M} and the form resistance coefficient C_{FORM}. If we further assume that C_{FORM} is proportional to C_{F0M} and the form factor k can be defined by **Equation 5-10**;

$$C_{TM} = C_{WM} + C_{VM} \xrightarrow[Fn\,<\,0.1]{} C_{TM} \fallingdotseq C_{VM} = C_{F_0M} + C_{FORM} \fallingdotseq (1+k)\,C_{F_0M} \tag{5-10}$$

when the Froude number is sufficiently small and k is assumed to be invariant with the Reynolds number.

As shown in the above equation, Hughes initially used two-dimensional flat plate frictional coefficient C_{F0M} defined in **Equation 5-6** for full scale estimation and his choice stands to reason in some sense. But later in 1978, ITTC recommended to replace C_{F0M} with C_{FM} found by the ITTC 1957 model-ship correlation line as normally practiced presently, although C_{FM} already takes account of form effects.

Moreover, it may be challenging to find the form factor, since neither the measurement of the total resistance of a model in a low speed range nor the determination of the run-in point where wave almost vanishes is easy in practice. In 1978, ITTC recommended the use of Prohaska's method (1966) for determining the form factor. The method is based on the assumption that the wave-making coefficient is proportional to the nth power of the Froude number where $n = 4$ is the most common choice

$$C_{WM} \fallingdotseq cFn^n,\quad 4 \le n \le 6 \tag{5-11}$$

If we take $n = 4$, the total model resistance coefficient defined in **Equation 5-10** becomes

$$C_{TM} = C_{VM} + C_{WM} \fallingdotseq (1+k)\,C_{FM} + cFn^4 \tag{5-12}$$

where c is an unknown constant to be determined from the measurements. Division of both sides with C_{FM} yields

$$C_{TM}/C_{FM} = (1+k) + cFn^4/C_{FM} \tag{5-13}$$

Therefore, if Prohaska's assumption is correct, C_{TM}/C_{FM} is a simple function of Fn^4/C_{FM} and the measured points should fall on the straight line in the graph of C_{TM}/C_{FM} to Fn^4/C_{FM}. Such a tendency is observed ordinarily in the low Froude number range, and in that case, the intercept of the line on the vertical axis is $1+k$, and c the slope. The use of Prohaska's method, however, necessitates measurements of resistance at a very low speed range: typically as low as $Fn \approx 0.1$; and one may fall into the difficulty in measuring very small forces or being confronted by the laminar flow, especially when a small model is used.

The model wave-making resistance coefficient C_{WM} is defined as the difference between the total model resistance coefficient C_{TM} and the model viscous resistance coefficient $(1+k)$ C_{FM}. If the ship wave-making resistance coefficient C_{WS} is assumed to be identical to C_{WM}, the total ship resistance coefficient C_{TS} can be determined as

$$C_{TS} = C_{WS} + C_{VS} = C_{WM} + C_{VS} = \left[C_{TM} - (1+k)\,C_{FM}\right] + (1+k)C_{FS} \tag{5-14}$$

This is the three-dimensional extrapolation method recommended by ITTC in 1978. It is a common practice, however, to add the roughness allowance and the air resistance coefficient to the value of the total ship resistance coefficient found by the above equation.

2.4 Numerical Simulation of Viscous Flow Fields

Normally, the resistance of a ship has to be determined through a model test. However, at an early design stage, extensive model tests are not desirable since the tests are expensive and time consuming and repeated tests may have to be undertaken. Nevertheless, it is essential for ship designers to understand flow characteristics around the ship hull, and the propeller as well, if they are to develop an efficient hull form and propeller.

With rapid growth of numerical techniques and fast computers, CFD is more frequently used in the initial design stage of a ship. The Navier-Stokes equations can be directly solved numerically to find a turbulent flow field around a ship (DNS, Direct Numerical Simulation), but the process usually consumes prohibitive computing time and resources in order to simulate small eddies in the turbulent boundary layer. The Large Eddy Simulation (LES) method is devised to alleviate such an excessive burden by simulating mean flow and larger eddies, while smaller eddies are modeled. The Detached Eddy Simulation (DES) method employing the DNS methods in the flow field near a body surface and the LES in the far field is also used to avoid modeling of small eddies.

The Reynolds-Averaged Navier-Stokes (RANS) equations, which are essentially time-averaged Navier-Stokes equations, are solved in most cases by various numerical techniques, since the direct solution of the Navier-Stokes equations is yet to be realized. To solve a RANS equation, it is necessary to model the Reynolds stress introduced into the equation during the averaging process of turbulence fluctuation. Turbulence modeling, i.e., modeling of the Reynolds stress has been one of the most challenging topics in fluid mechanics and currently numerous models are being used which include zero-, one-, and two-equation models, and algebraic and Reynolds stress models. Details on this subject can be found easily in textbooks and elsewhere and it is sufficient at this time to indicate that all the models are not generally applicable, because they include empirical constants specified to certain class of flows.

The information on the characteristics of the turbulent flow field around a ship, and more

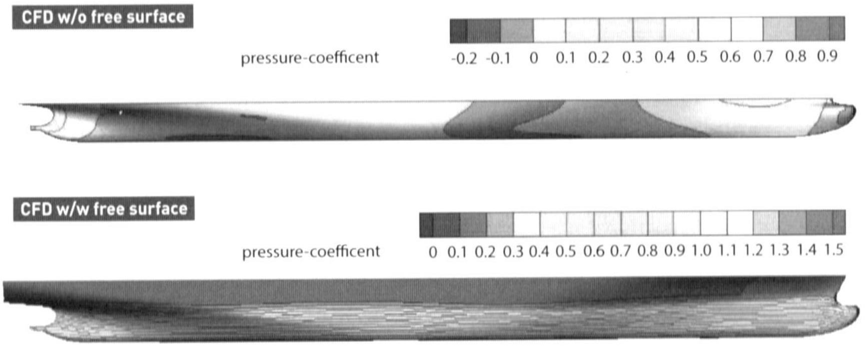

Figure 5-8 Pressure distributions on a hull surface (Courtesy of Hyundai Heavy Industries)

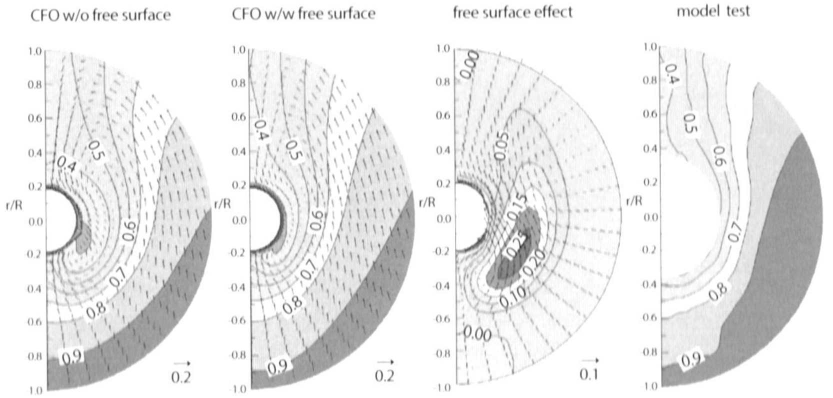

Figure 5-9 Wake distributions on a propeller plane (Courtesy of Hyundai Heavy Industries)

specifically the frictional resistance, the form resistance and the wake distribution on the propeller plane, among others, are normally the most sought after to develop excellent hull form, and various numerical packages have been developed and utilized. **Figures 5-8** and **5-9** exemplify a pressure distribution on a hull surface and a wake distribution on a propeller plane simulated with WAVIS program developed by KRISO, and is being used by most shipyards in Korea.

3. Wave-Making Resistance

If a stone is thrown into a pond, the impulse pressure generates circular waves of innumerable frequencies around the stone which spread out on the surface successively. In the same manner, if a ship moves on the free surface of the water, variation in the hull surface pressure generates a wave system around the ship hull. All the waves around a ship need energy to sustain and are continuously supplied with energy from the ship. In other words, the energy of the ship-generated waves is identical to the work done by the ship to the water if the viscosity of the water is assumed to be negligible. The work done by a ship will be used to overcome the pressure force against its motion.

Consequently, the work done to the water by a ship per unit time is the product of the wave-making resistance and the ship speed, and the wave-making resistance is the pressure force against the ship motion and can be obtained by integrating the pressure acting on the hull surface. The magnitude of the wave-making resistance is dependent on the characteristics of the ship-waves generated. As previously mentioned, the wave-making resistance is a function of the Froude number and the portion of the wave resistance in the total resistance increases with the increase of the Froude number, i.e., as the ship speed increases.

3.1 Ship Waves

Kelvin considered a single pressure point traveling at a constant speed on a straight line over the surface of the water and investigated the characteristic pattern of the waves generated by the movement. The wave pattern consists of a system of transverse waves, crest lines of which are almost perpendicular to the moving direction and follow behind the point, and divergent (or diverging) waves radiating from the point laterally as shown in **Figure 5-10**.

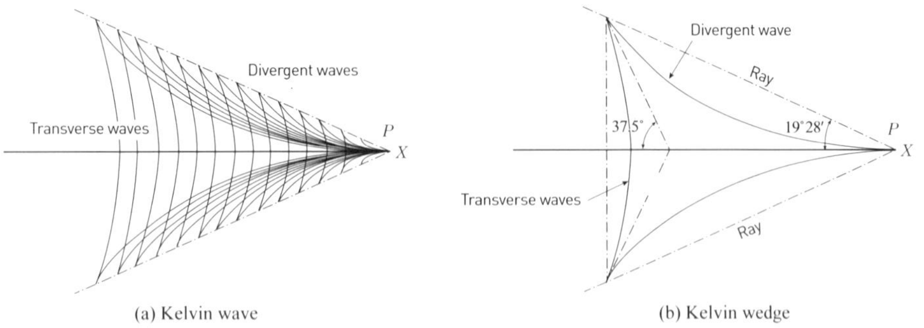

Figure 5-10 Kelvin wave pattern and wedge behind a moving pressure point

Furthermore, wave energy is confined to a triangular sector, two sides of which originate at the moving point with a half vertex angle of 19°28′ and the sector is called the Kelvin wedge after Lord Kelvin; and the sides are called the rays. Divergent waves are prevalent near the rays since much of the transverse wave system remains close to the centerline.

If a ship is observed from high enough for it to be considered as a point, the waves generated by the ship will look similar to the Kelvin pattern consisting of transverse and divergent wave systems. An actual ship is not a single pressure point but a sum of all points on the hull inside the water (the wetted surface) and may be simplified as a set of the most prominent pressure points existing near the bow, shoulders and stern of the ship. Then, each pressure point will generate a wave system contained in its own Kelvin wedge and these are then superposed to form the total ship wave system.

To simplify the problem further, let us consider only the wave profile along the hull or centerline which is dominated by the transverse waves. Wigley (1931) showed that the wave profile along the hull of a double-wedge-shaped body contains five terms, as shown in **Figure 5-11**.

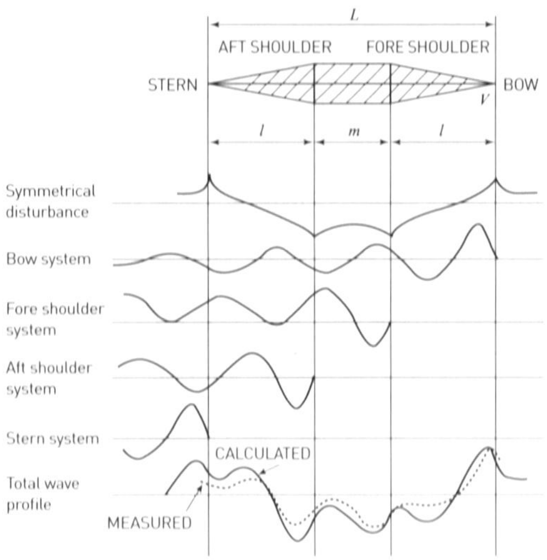

Figure 5-11 The wave systems around a wedge shaped hull (Wigley, 1931)

① A symmetric near-field disturbance of the free surface: it has a peak at bow and stern and troughs at the shoulders. It travels with the hull, dying out quickly ahead and astern of the hull thereby leaving no wave behind, and does not absorb any energy at constant speed because of its symmetry. The disturbance is also called the local disturbance or the Bernoulli wave.

② The bow wave system: beginning with a crest.

③ The fore shoulder wave system: starting with a trough.

④ The aft shoulder wave system: also starting with a trough.

⑤ The stern wave system: beginning with a crest.

The local disturbance has the fore-and-aft symmetry because of the body symmetry. The bow and stern wave begins with a peak, since the pressure at these locations is higher than the reference, while waves start with troughs at fore and aft shoulders where the pressure is lower. The calculated wave profile along the body is the sum of the above five terms and is in good agreement with the measured one in the qualitative sense.

The first crests or troughs of each wave system are fixed in position but the second ones shift downstream as the model speed increases. Therefore, the total wave profile will change in shape if the speed is altered. The wave-making resistance, which is proportional to the square of wave amplitude, will be high at which a crest meets a crest of each wave system but small if a crest meets a trough.

3.2 Theoretical Prediction of Wave-Making Resistance

In principle, the wave-making resistance of a ship or its model can be found if one measures pressure distributions on the wetted surface and integrates them in the direction of the ship motion. However, the method is far from practical and the wave-making resistance of a ship is normally calculated from the wave profile measured sufficiently far from the ship (the far field waves). This method is called the wave pattern analysis. In numerical methods, wave pattern analysis can also be used to estimate the wave-making resistance and in such cases, of course, simulated pressure distributions or wave profiles are used instead of measured ones.

It is customary to take viscous effects into account only when the viscous pressure resistance is concerned and it may be neglected in the present discussion. Once the flow around a ship is assumed inviscid and incompressible, one may be reminded of the D'Alembert's paradox stating that a body moving in an unbounded, inviscid and incompressible fluid does not experience a drag against its motion. However, a ship moving on or near the water surface experiences the wave-making resistance against its motion because the pressure distribution on the hull surface changes due to the ship wave, and the pressure integral does not vanish.

The wave-making resistance of a ship moving along the positive x axis can be defined as

$$R_W = -\iint_S (p\vec{n}_x)\, d\vec{S} \tag{5-15}$$

where S is the wetted surface area under the wave profile along the hull surface, $\vec{n}_x$ is an x-component of an outward unit normal vector on the hull surface, and p is total pressure acting on a differential area of the hull surface.

In 1898 J. Michell developed an approximate method of evaluating the wave-making resistance of a thin ship of narrow beam moving over the free surface of inviscid flow. Since the ship has a small beam / length ratio, the boundary conditions are imposed on the center-plane instead of the hull surface. The method is called the thin ship theory but produces inaccurate results if applied to a ship of practical hull form and usually shows exaggerated hollows and humps on the wave-making resistance coefficient curve at low Froude numbers. However, the theory finds its practical application in some areas: hull form optimization, where reduction in computing time is crucial, and the computation of far field wave patterns of a high speed craft.

If a ship overcomes the wave resistance defined in **Equation 5-15** and is to move at a constant velocity, the energy necessary to maintain the wave system should flow out of the ship. The energy flow can be calculated from the wave pattern generated by the ship measured in the far field. Havelock (1934) gave an expression in which the wave (or wave pattern) resistance of a ship moving in a straight line at the speed U is

$$R_W = -\frac{1}{2}\pi\rho U^2 \int_{-\pi/2}^{\pi/2} \left[A(\theta)^2 + B(\theta)^2 \right] \cos^3\theta d\theta \tag{5-16}$$

where θ is an angle of a wave crest line propagating off the bow, and $A(\theta)$ and $B(\theta)$ are amplitudes of the cosine and sine components, respectively, of a ship wave measured in the far field. It can be seen in the equation that the wave resistance of a ship is proportional to the square of wave amplitude and the weighting term $\cos^3\theta$. Also, it indicates that transverse wave systems contribute to the wave resistance much more than the diverging waves.

Revolutionary growth in computer technology and numerical techniques has enabled more accurate and plausible numerical prediction of the wave-making resistance. Numerical methods for solving the RANS equations satisfying the free surface boundary conditions are often used to find the wave-making resistance of a ship. However, panel methods proposed

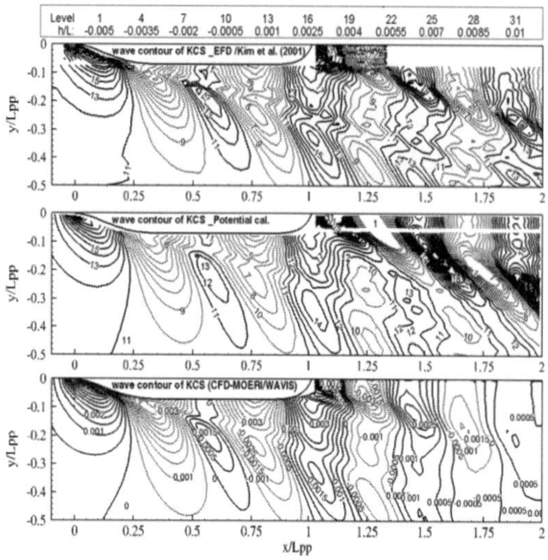

Figure 5-12 Wave pattern of KRISO container ship (KCS): measured (top); potential flow calculation (middle); viscous flow calculation (KRISO) (bottom)

by Dawson in 1977 and extended to nonlinear wave problems later are more widely used for this purpose.

In panel methods, boundary surfaces encompassing an inviscid and incompressible flow region around a ship are subdivided into small panels containing singularities of unknown strength which should satisfy the boundary conditions. Panel methods do not consider the influence of viscosity and cannot predict viscous behaviors such as the boundary layer or flow separation. Nevertheless, for most of the regions around a ship – except the vicinity of the stern – the method generally produces numerical results well enough for hull form design at an early design stage. **Figure 5-12** compares the wave pattern measured around a KRISO container ship (KCS) with computed ones to show good agreement in most regions upstream of the stern. Here, KCS along with KVLCC (KRISO very large crude oil carrier) and its variant KVLCC2 were originally designed at the Korea Research Institute of Ships and Ocean Engineering (KRISO) and have been extensively used worldwide for the validation and verification of numerical methods since plentiful and detailed flow data measured around the hull forms have been made public.

4. Determining the Ship Resistance

The speed of a ship is one of the major targets to be achieved in sea trials. To design a ship with good speed performance from limited engine power, a designer has to make constant effort to reduce the resistance of the ship, while trying to improve the propulsive efficiency of the propeller at the same time. As mentioned above, numerical methods for the prediction of ship resistance have grown to a remarkable level, but the application of the methods to full scale is yet to be realized. Therefore, estimating the resistance of a ship still relies on model tests in a towing tank at present time.

4.1 Model Testing and Towing Tank

Various tests are performed in a towing tank with a scale model of a ship or propeller towed at a constant speed to predict the hydrodynamic performance of a ship in calm water. Resistance tests, propeller open water tests and self-propulsion tests are the most common, but other tests including the measurement of wake distribution in a propeller plane or the flow visualization of limiting streamlines with paints, dyes or tufts are routinely carried out in the tanks as well.

In 1872, William Froude built a towing tank in Torquay, England with the support of British Admiralty. It was equipped with a mechanically driven towing carriage and Froude showed his law of comparison could be used for the practical prediction of ship resistance through model tests. Thereafter, many towing tanks have been constructed around the world and today there are well over 100. The dimensions and maximum towing speeds of some selected tanks are shown in **Table 5-1**.

Seven towing tanks exist in Korea presently as shown in **Table 5-2**. The first of these, a 27 m long tank using a falling weight to tow ship models, was constructed at Seoul National University in 1962. Later, a 75 m long more modern towing tank equipped with an electrically driven towing carriage was constructed at Inha University in 1971. The first commercial towing tank, 203 m long, shown in **Figure 5-13** was constructed at KRISO in

Table 5-1 Dimensions and towing speeds of the selected tanks

Institute	Tank dimensions (m) (Length × Breadth × Depth)	Maximum towing speed (m/sec)
Krylov SRI (Russia)	1,324.0 × 15.0 × 7.0	20.0
NSWCCD (previously DTMB, USA)	Deep: 575.0 × 15.0 × 5.0 × 6.7 Shallow: 363.3(271.0+10.0+82.3) × 15.5 × (6.7+3.0) High speed: 904.0(514.0+356.0+34.0) × 6.4 × (4.9+3.0)	Deep : 10.3 Shallow: 9.3 High speed: 6.5, 25.7, 30.9
CSSRC (China)	474.0 × 14.0 × 7.0	15.0
INSEAN (Italia)	470.0 × 13.5 × 6.5	15.0
NMRI (Japan)	400.0 × 18.0 × 8.0	15.0
CEHIPAR (Spain)	320.0 × 12.5 × 6.5	10.0
HSVA (Germany)	300.0 × 18.0 × 6.0	8.0
MHI (Japan)	270.0 × 13.0 × 5.0	5.0
SSPA (Sweden)	260.0 × 10.0 × 5.0	10.0
MARIN (Netherlands)	250.0 × 10.5 × 5.5	9.0

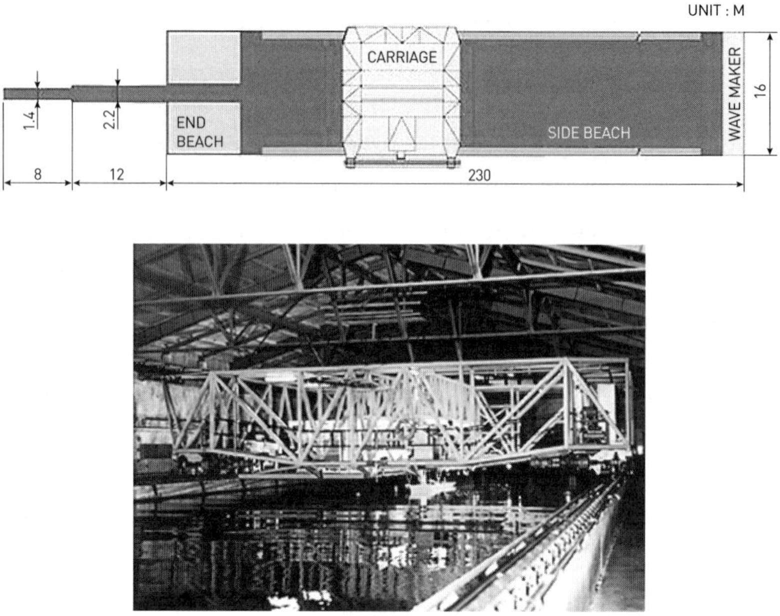

Figure 5-13 Schematic diagram and a photograph of a towing tank (KRISO)

Table 5-2 Dimensions and towing speeds of towing tanks in Korea

Institute	Tank Dimensions (m) (Length × Breadth × Depth)	Maximum towing speed (m/sec)	Year Built
Seoul National University	26.6 × 3.1 × 1.5	1.2	1962 (closed in 1980)
Inha University	75.0 × 5.0 × 2.7	3.0	1971 (closed in 2014)
Busan University	100.0 × 8.0 × 3.5	7.0	1974 (rebuilt in 1989)
KRISO	203.0 × 16.0 × 7.0	6.0	1978
Hyundai Heavy Industry (HMRI)	188.0 × 14.0 × 6.0	7.0	1984
Seoul National University	110.0 × 8.0 × 3.5	5.0	1984
Samsung Heavy Industry (SSMB)	400.0 × 14.0 × 7.0	5.0/18.0	1996
National Fishery Research & Development Institute	85.0 × 10.0 × 3.5	3.0	2001
Inha University	50.0 × 3.5 × 1.8	5.0	2015

1978 and this was later followed by tanks at Hyundai (HMRI) and Samsung (SSMB) Heavy Industries.

As previously stated, similitude between the ship and model must be achieved to predict ship resistance through model tests and therefore the Reynolds number as well as the Froude number of the ship and its model should be identical. However, it is customary in model tests to match the Froude numbers only since it is practically impossible to make both numbers equal simultaneously.

Since the Reynolds numbers of similar models with a same Froude number are proportional to 1.5th power of the length, the gap between ship and its model Reynolds numbers will be very large, especially since models are very much smaller than full scale ships. Therefore, ignoring the Reynolds number inevitably causes the scale effects and for this reason models must be built as large as the circumstances allow.

Moreover, a larger model can be built more accurately, produces a larger force to be measured, and consequently makes the tests more accurate. However, the larger the model, the more expensive it is to manufacture and test. There are more constraints besides the cost in increasing the model sizes. For example, if a model becomes large, the towing speed has to be increased for the Froude numbers to be identical and this reduces the time available for measuring and observing, even if it is possible for the carriage to achieve the required speed. Moreover, the flow characteristics around the model may be affected from the tank walls if the cross sectional area of the model exceeds a certain limit, say about 1/200 of that of the towing tank, producing what is known as blockage effect. The size of the model will also be restricted by the capacity of the model-making workshop, but it is normal practice for a commercial towing tank to use models 7~8 m long, even though the facilities are capable of handling models more than 10 m long.

The Reynolds number of a model usually lies in laminar or at best transitional flow regimes, while the flow around a full scale ship is turbulent. It is very important to reconcile this difference. A useful way of overcoming this problem is to trip the laminar flow around a model forcibly by placing turbulence stimulators at judicious points on the model hull. Studs, wires, or sand strips are often used for this purpose and they are placed around a hull at 5% aft of the stem (19th station) and leave the flow upstream undisturbed to compensate for the additional drag of the stimulators. If a ship is equipped with a bulb, additional stimulators have to be placed at one third or a half of the bulb length. If the ship is bulb-less, the bow stimulator is placed on both sides of the stem parallel to the bow rake.

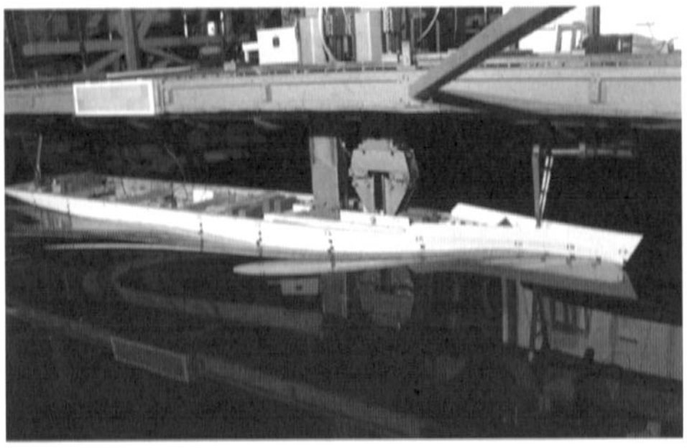

Figure 5-14 A ship model being towed in a tank (HMRI)

Figure 5-14 shows a model attached to a carriage and being towed at a constant speed. The carriage is equipped with a measurement system including a resistance dynamometer to measure the resistance of the model. It is also equipped with trim guides to prevent yawing motion of the model while allowing free trim and sinkage, and a cramp holding the model to protect sensitive measuring devices from excessive forces which may arise during acceleration and deceleration of the carriage.

4.2 Model Resistance Tests

As discussed above, ship model experiments for determining the resistance of a model and extrapolating the results to full scale has been practiced widely for a long time ever since Froude proposed the law of comparison.

There are basically two methods for extrapolating model results to full-scale, viz. two- and three-dimensional methods. In the two-dimensional method, the total ship resistance coefficient C_T is estimated by summing the frictional resistance coefficient C_F calculated by a model-ship correlation formula and the model residuary resistance coefficient C_R found by subtracting C_F of the model from the total resistance coefficient C_T of the model as described in **Equation 5-9**. On the other hand in the three-dimensional procedure, C_T is thought to be a sum of the wave resistance coefficient C_W and the viscous resistance coefficient C_V and it is further assumed that C_V is linearly proportional to the frictional resistance coefficient C_F. The form factor $(1+k)$ is then defined as the ratio of C_V and C_F and assumed independent of the Reynolds number. The relation was shown in **Equation 5-14**. Both methods may introduce the roughness allowance C_A and typical procedures are discussed in more detail in the following sections.

4.2.1 Two-dimensional Method

Froude assumed that the total resistance of a ship R_T is the sum of a frictional resistance R_F which is equal to the friction of an equivalent flat plate and a residuary resistance R_R accounting for the rest. In other words, he suggested the approximate relation, $C_T(Rn, Fn) = C_{F_0}(Rn) + C_R(Fn)$, in non-dimensional forms, and that C_R of the ship and a scale model is identical if the Froude number is the same. The ITTC recommended replacing C_{F0} with the frictional resistance coefficient C_F found from a new 1957 model-ship correlation line and thus the method is also called the ITTC-1957 method.

The process of extrapolating experimental results to the full scale resistance by the two-dimensional method is as follows:

(1) Manufacture a geometrically similar model of the ship with a scale ratio $\lambda = L_S/L_M$,

where L_S and L_M are length of the ship and the model respectively. Perform resistance tests at a number of corresponding speeds $V_M = V_S \lambda^{-0.5}$ to measure the total model resistance coefficient C_{TM}, where V_S is the speed of the ship.

(2) Compute the model Reynolds number Rn_M to find the frictional resistance coefficient C_{FM} of the model by the ITTC-1957 correlation line

$$C_{FM} = \frac{0.075}{(\log_{10} Rn_M - 2.0)^2} \quad \text{where } Rn_M = \frac{V_M L_M}{\nu} \tag{5-17}$$

(3) Compute the model residuary resistance coefficient C_{RM}

$$C_{RM} = C_{TM} - C_{FM} \tag{5-18}$$

(4) Assume the ship residuary resistance coefficient C_{RS} to be the same as C_{RM}

$$C_{RS} = C_{RM} \tag{5-19}$$

(5) Compute the ship frictional resistance coefficient C_{FS}

$$C_{FS} = \frac{0.075}{(\log_{10} Rn_S - 2.0)^2} \quad \text{where } Rn_S = \frac{V_S L_S}{\nu} \tag{5-20}$$

(6) Compute the total ship resistance coefficient C_{TS}

$$C_{TS} = C_{FS} + C_{RS} + C_A \tag{5-21}$$

where C_A is the roughness allowance coefficient and 4×10^{-4} is normally used for it.

(7) Lastly, estimate total resistance R_{TS} of the ship

$$R_{TS} = C_{TS} \times \frac{1}{2} \rho S V_S^2 \tag{5-22}$$

and the effective power P_E

$$P_E = R_{TS} \times V_S \tag{5-23}$$

The procedure is also called Froude's method if friction coefficients of equivalent flat plates are used instead of those by ITTC-1957 line.

4.2.2 Three-dimensional Method

In the three-dimensional method, the total resistance R_T of a ship is subdivided into viscous resistance R_V and wave resistance R_W. Then, it is assumed that the viscous resistance is the sum of a flat plate frictional resistance R_{F0} and a form resistance which is linearly proportional to the former at a ratio of k. Then, the total resistance has the relation of $C_T(Rn,Fn) = (1+\mathrm{k})C_{F0}(Rn) + C_W(Fn)$ in non-dimensional forms where k is called the form factor and assumed invariant of the Reynolds number. The ITTC recommended in 1978 to replace the friction coefficient of a flat plate C_{F0} with the frictional resistance coefficient C_F calculated by ITTC-1957 model-ship correlation line and the method is also called the ITTC-1978 method.

The process of extrapolating experimental results to the full scale resistance by the three-dimensional method is as follows:

(1) Manufacture a geometrically similar model of a ship with a scale ratio $\lambda = L_S/L_M$ and perform a resistance test at a number of corresponding speeds $V_M = V_S\,\lambda^{-0.5}$ to measure the total model resistance coefficient C_{TM}.

(2) Compute the model Reynolds number Rn_M to find the frictional resistance coefficient C_{FM} of the model by the ITTC-1957 model-ship correlation line

$$C_{FM} = \frac{0.075}{(\log_{10} Rn_M - 2.0)^2} \quad \text{where } Rn_M = \frac{V_M\, L_M}{\nu} \tag{5-24}$$

(3) Determine the form factor k by Prohaska's method in **Equation 5-13**.

(4) Compute the model wave resistance coefficient C_{WM}

$$C_{WM} = C_{TM} - (1+k) \times C_{FM} \tag{5-25}$$

(5) Assume the ship wave resistance coefficient C_{WS} to be the same as C_{WM}

$$C_{WS} = C_{WM} \tag{5-26}$$

(6) Compute the ship frictional resistance coefficient C_{FS}

$$C_{FS} = \frac{0.075}{(\log_{10} Rn_S - 2.0)^2} \quad \text{where } Rn_S = \frac{V_S\, L_S}{\nu} \tag{5-27}$$

(7) Compute the roughness allowance C_A and the air resistance coefficient C_{AA}

$$C_A = \left[105\left(\frac{k_S}{L}\right)^{\frac{1}{3}} - 0.64\right] \times 10^{-3},\ C_{AA} = 0.001\frac{A_T}{S} \qquad (5\text{-}28)$$

where A_T is the transverse projected area of the ship above the waterline and k_S is the surface roughness in microns (μ) measured by the mean apparent amplitude (MAA) method. Typically a straight line 750mm long is divided into 15 intervals of equal length and the normal distances from the highest peak to the lowest trench in each interval are measured. Then, k_S is chosen as the mean of the 15 values. The equation should be used only in conjunction with a form factor method and ITTC recommends 150μ as a typical value.

(8) Compute the total ship resistance coefficient C_{TS}

$$C_{TS} = (1+k)\, C_{FS} + C_{WS} + C_A + C_{AA} \qquad (5\text{-}29)$$

(9) The total resistance R_{TS} and effective power P_S of the ship are estimated as before

$$R_{TS} = C_{TS} \times \frac{1}{2}\rho S V_S^2 \quad \text{and} \qquad (5\text{-}30)$$

$$P_E = R_{TS} \times V_S \qquad (5\text{-}31)$$

The two- and three-dimensional procedures are also illustrated in **Figures 5-15** and **5-16**, respectively.

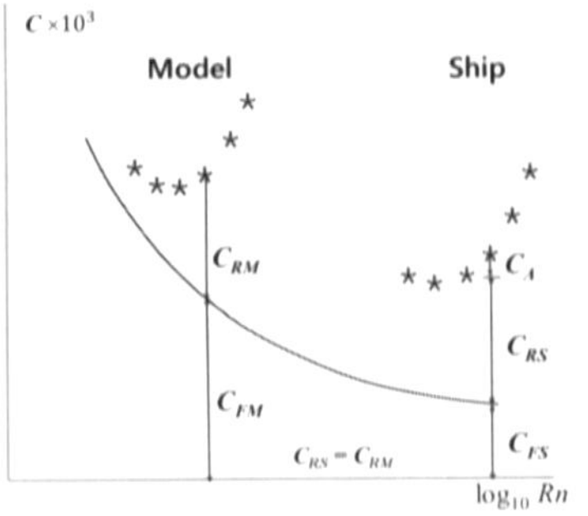

Figure 5-15 Two-dimensional method (ITTC-1957)

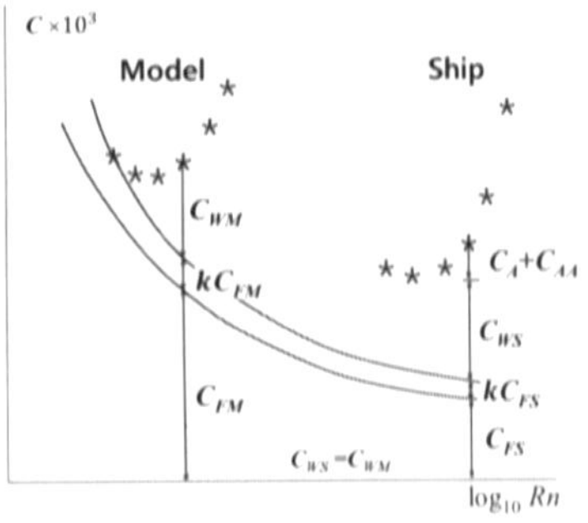

Figure 5-16 Three-dimensional method (ITTC-1978)

5. Local Flow Measurements

To achieve the target speed of the ship under design, model tests in towing tanks are needed. More specifically, propeller open water and self-propulsion tests as well as the resistance tests are required since performance aspects of the propeller, such as thrust, torque and rpm and the necessary engine power have to be known to estimate the ship speed.

In order to design an efficient hull form or a propeller it is very important to understand the local flow properties around them. For this purpose, quantitative measurements of flow characteristics, such as velocity vectors and local pressure distributions, and visualization of limiting streamlines, will be required.

Techniques of measuring flow velocities can be categorized into intrusive and non-intrusive methods. Intrusive methods are widely used, since they are comparatively simple, inexpensive and easy to practice, but sensors put directly into the flow field under investigation inevitably disturb the field and reduce the accuracies of measurements. A Pitot tube which measures the pressure difference between two different locations is one of the typical intrusive instruments.

A non-intrusive method most widely used in towing tanks is laser Doppler velocimetry (LDV), which uses a pair of split laser beams to measure the velocity in a tiny measurement volume. Another is particle image velocimetry (PIV), which uses thin laser sheet illuminating tracer particles in a measurement plane twice in a very short time interval to measure the velocity distributions tangential to the plane at once. These methods do not disturb flow fields, usually provide more accurate information and can be applied to measuring details of a flow such as turbulent fluctuations. However, many intricate and expensive instruments required and narrow measuring ranges are some of the obstacles for them to be more widely used in towing tanks. Having said that, the mechanisms are getting simpler and smaller, and they are encountered more often in towing tanks these days, since they possess a crucial merit of being able to provide substantial information on a local flow field if successfully employed.

5.1 Wake Measurement

Wake distribution, or the velocity field in a propeller plane behind a ship's stern, is very influential in determining the efficiency of the hull and the performance of the propeller. Measurement of wake distribution is usually carried out in a towing tank with Pitot tubes or a rake of Pitot tubes. The measuring system normally consists of Pitot tubes themselves, a transport system, pressure transducers, signal amplifiers, A/D converters, and a computer for signal processing. The wake measuring system is classified into rotational and rectangular types, according to the mode of Pitot tube movement and shown in the photos in **Figures 5-17** and **5-18**.

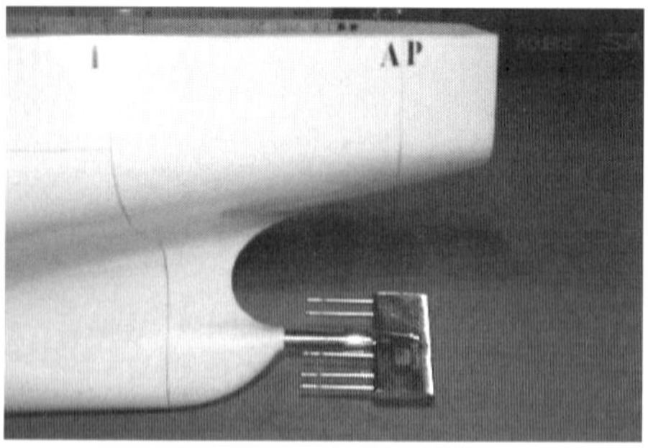

Figure 5-17 Rotational wake measurement with a rake of Pitot tubes

Figure 5-18 Rectangular wake measurement with a single Pitot tube

In rotational wake measurements, a rake – housing an array of Pitot tubes along the span – is rotated around a propeller shaft center to measure wake field efficiently at various angles and radii as shown in **Figure 5-17**, while a Pitot tube that can be moved up and down and side to side to measure the wake at a grid of points is used to cover a rectangular area and this is shown in **Figure 5-18**.

Five-hole Pitot tubes are used to measure all three velocity components at a measuring point and have a spherical or conical head with 5 holes on it, as shown in **Figure 5-19**. For example, a vertical velocity field can be determined from the pressure differences among three vertical holes and horizontal velocities by three horizontal holes.

Figure 5-20 compares the wake distributions in the propeller plane of KVLCC2 hull form measured by a rake with the numerical results.

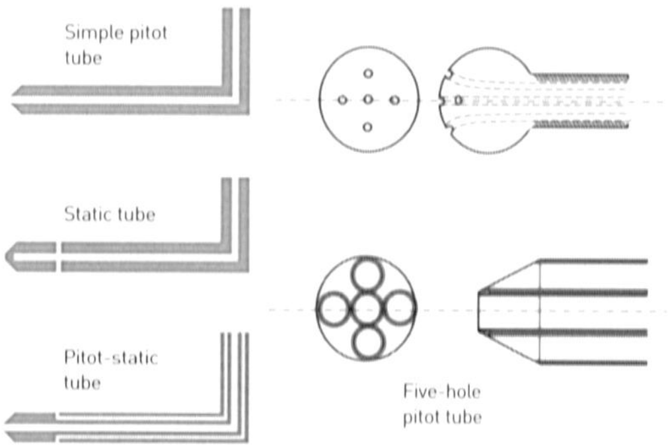

Figure 5-19 Various types of Pitot tubes

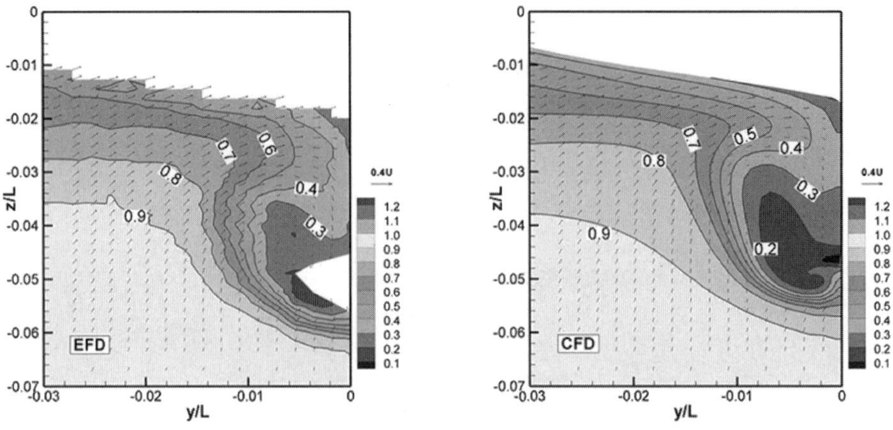

(a) measurement (b) numerical prediction

Figure 5-20 Wake distributions in the propeller plane of KVLCC2 hull form (KRISO)

5.2 Flow Visualization

Flow Visualization is a measurement technique to make a transparent flow visible, so as to obtain an insight into its qualitative behavior. The technique is frequently employed for determination of bow and stern shapes or for design of appendages including propellers, rudders, stabilizer fins and struts.

The techniques include wall tracing tests to outline limiting streamlines with paint, dye or oil; tuft tests to visualize unsteady local flow fields with tufts mounted on the surface; and hydrogen bubble tests utilizing hydrogen bubbles as flow tracers.

Recently PIV or PTV (particle tracking velocimetry) techniques using laser optics and digital image processing (see **Figure 5-21**) are becoming increasingly popular for flow

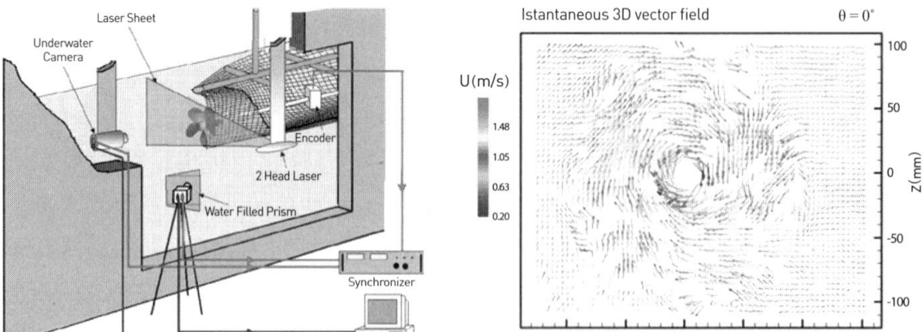

Figure 5-21 Visualization of wake distribution behind a propeller by PIV techniques
(Calcagno et al., 2002. Cited in 24th ITTC Rc Report 2005)

visualization work. However, paint and tuft tests are still more often used than PIV in towing tanks, primarily due to limitations in experimental environments.

In a paint test, special paint is applied in girth-wise strips along the hull surface before immersing in the water. It shows up limiting streamlines clearly as in the photograph of **Figure 5-22**. The paint is a judicious mixture of an oil color, enamel paint, polywax and diesel oil, or else produced according to a recipe of the tank based upon its own experience.

Figure 5-22 Paint test around a bow thruster opening

Tuft tests shown in **Figure 5-23** can visualize unsteady flow fields produced by vortices or flow separation near the stern, all of which are indispensable pieces of information for the improvement of propulsive efficiency.

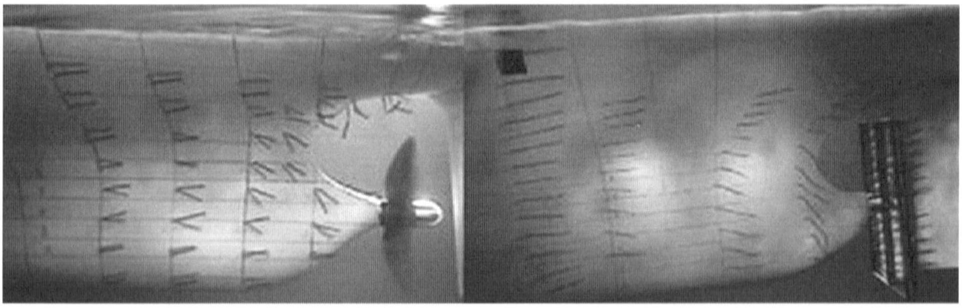

Figure 5-23 Tuft test on the aft-body of a ship model

Recently, numerical flow visualizations are also practiced as shown in **Figure 5-24**. The figure shows a pressure distribution on – and limiting streamlines around – the hull surface of a KVLCC2 model. The pressure distribution is not directly related to flow visualization, but shown below because it determines the direction of the cross flow inside the boundary layer and is helpful for understanding the behavior of the limiting streamlines.

Close inspection on the behaviors of limiting streamlines obtained either by flow visualization or numerical simulation facilitates understanding of the flow characteristics around a ship. For example, **Figure 5-24** shows that the streamlines on the fore body above the bulb gently curve down to become horizontal further downstream. On the other hand, the streamlines around the bulb are bent downward to reach the bottom and flow underneath

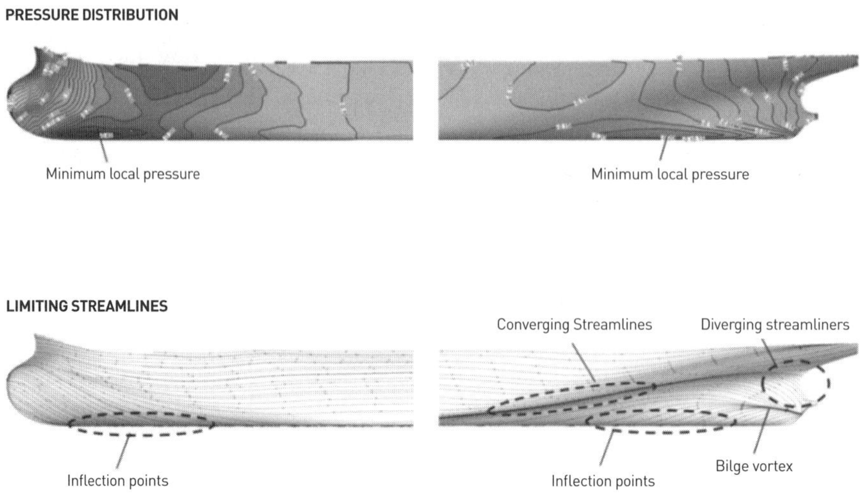

Figure 5-24 Surface pressure distribution and limiting streamlines for KVLCC2 (KRISO)

the flat of bottom. Thus, there exist inflection points in the streamlines at the bilge region near the bow. At an inflection point, the pressure gradient across a streamline changes sign, causing nearby streamlines to converge downstream as shown in the figure, and eventually producing a bilge vortex. The bilge vortex in the fore body is normally quite weak, unless the bilge radius is too small.

Similarly, the bottom streamlines of the after body curve upward to create inflection points near the bilge. Subsequent changes in the pressure gradient cause the nearby streamlines to converge and produce the stern bilge vortex which is much stronger than the one at the bow, since the boundary layer is fully developed and much thicker than at the bow. The stern bilge vortex develops just upstream of the propeller and significantly affects the flow into it. A little upstream, the streamlines flowing along the flat of bottom rise up to meet those flowing along the side of the hull. In this case, unlike at the bilge, the neighboring streamlines diverge due to the pressure gradient. Diverging streamlines usually indicate the risk of bubble separation, but it can be seen from **Figure 5-24** that the well-designed KVLCC2 hull does not have this problem.

6. Propeller

6.1 Basic Propeller Terminology

Among many propulsors used to propel ships, the screw propellers, or simply the propellers, are the most widely used for their ability to deliver high power with high propulsive efficiency. This section defines the basic terminology related to the shape and the performance of the propeller (see **Figure 5-25**).

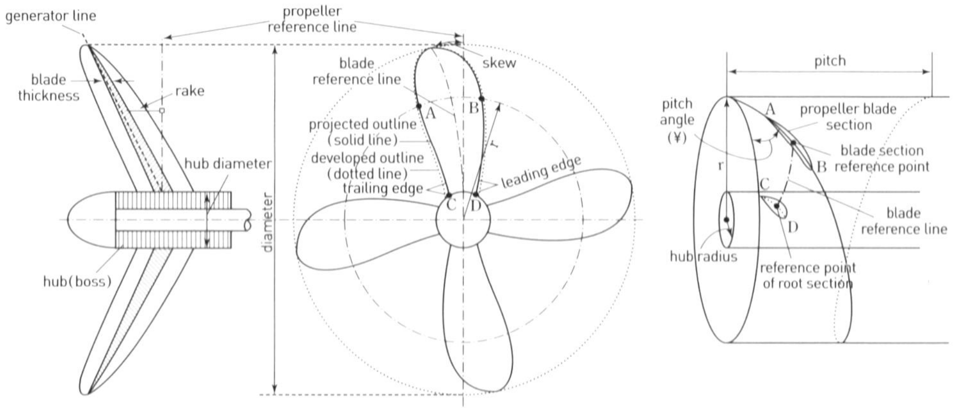

Figure 5-25 Basic geometric characteristics and various parts of the propeller

1) Diameter

As the propeller makes one revolution around the shaft axis, the maximum diameter of the circle drawn by the blade tip is called the diameter (D) of the propeller. This diameter has to be less than 2/3 of the stern draft TA to avoid drawing air into the blade surface due to the suction of the propeller and to avoid reducing propulsive efficiency when the blade tip moves close to the free-surface, especially in ballast conditions and/or during the pitching motion of the ship.

The clearance of the blade tips from the hull surface is normally greater than 20% of the diameter, and is determined to minimize the forces on the hull surface induced by the propeller loading and cavitation. The combination of the larger diameter and lower shaft rotational speed of the propeller is known to produce higher propulsive efficiency, and therefore the propellers of many merchant ships are made as large as possible and designed to operate at a low rotational speed range.

The propeller diameter is ultimately determined through the trade-off between the propeller-hull clearance and the diameter, while meeting the criterion to minimize the harmful effects of cavitation.

2) Pitch

The pitch (P) is the axial distance the propeller advances through the water when it is rotated once. Propellers with constant pitch values at all radii are called constant pitch propellers. In uniform flow, without the presence of a ship, the constant pitch propellers show the highest propulsive efficiency. In this condition, the water is pushed downstream with more or less uniform axial induced velocity. Propellers with non-constant pitch values in the radial direction are called variable pitch propellers. The pitch is often expressed by the dimensionless ratio P/D. When designing propellers for merchant ships, the pitch is determined to meet the speed-power requirement at a single design speed. Propellers with the fixed pitch values at all times are called fixed pitch propellers (FPP). For naval vessels and ships required to operate at multiple speeds, the propellers are often designed so that the pitch can be controlled. These propellers are called controllable pitch propellers (CPP). In general, the pitch ratio (P/D) for low-speed ships is about 0.7 and for high-speed ships, greater than 1.0.

3) Pitch angle

At a particular radius r, the slope angle defined by the ratio of the axial advance distance to the distance traversed in the circumferential direction by the blade section is called the geometric pitch angle (ϕ):

$$\phi = \tan^{-1}\frac{P}{2\pi r} \tag{5-32}$$

4) Propeller reference line

An arbitrary reference line is selected to define the propeller shape. This line, called the propeller reference line, is normally perpendicular to the propeller shaft centerline and the direction opposite to the direction of gravity has positive sign. The plane obtained by rotating this line about the propeller shaft is called the propeller plane.

5) Generator line

A line or a curve, drawn in the plane formed by the propeller reference line and the propeller shaft axis, is called the generator line (or generatrix) and is used as the reference line for the manufacturing of the propeller. The generator line is normally a straight line and is obtained by tilting the propeller reference line. The trajectory obtained by rotating the generator line around the propeller shaft while advancing in the axial direction becomes the helical surface.

6) Rake

At each radius, the axial distance from the propeller reference line to the generator line is called the rake. The positive rake points in the downstream direction. When the generator line is a straight line, the rake is linear. The angle between the propeller reference line and the generator line is called the rake angle. The rake is introduced to maintain the distance between the propeller and the stern hull to reduce the propeller-induced fluctuating pressures on the hull (which are a significant source of hull vibration) to an acceptable level.

For merchant ships the clearance between the stern frame and the propeller blade tip is used as a reference to determine the rake. For controllable pitch propellers, widely used in naval ships, the rake is determined to maintain the minimum torque required to turn the blades about the spindle axis.

7) Propeller blade section

Propeller blade sections refer to the cross-sections of the propeller blade with the cylindrical surface of radius r concentric with the propeller shaft. The blade section is aligned with the helix formed on the cylindrical surface by the pitch angle ϕ. The surface of the blade section further away from the hull is called the pressure-side (or face) of the blade. The pressure is positive when the propeller is generating forward thrust. The surface closer to the hull is called the suction-side (or back) of the blade. The speed of the water is higher here, and the pressure is lower, than on the pressure-side of the blade. Because of its ease in manufacturing, the ogival section, which has a relatively straight face and a quadratic-curved back, has been widely used. Recently, however, with the introduction of the numerically controlled machines, the airfoil sections have become more popular as the propeller blade section. Typical propeller sections include the Troost-section of the Netherlands, the MAU-section of Japan and the NACA airfoil series of the U.S.A.

8) Leading edge and trailing edge

When a propeller rotates, it generates the forward thrust for the ship. The front end of the blade section divides streamlines into the suction-side and pressure-side of a blade. This end is called the leading edge (L.E.), and the trailing end of the blade, where the streamlines meet again, is called the trailing edge (T.E.). The leading edge of the blade section has a relatively thick, round shape and the trailing edge is relatively sharp.

9) Blade outline

Blade outline (or blade contour) is formed by continuously connecting the leading and trailing edges at all radii to the blade tip. The most commonly used blade outline forms are the projected contour and the expanded contour. The projected outline is the view of the blade outline projected onto a plane perpendicular to the propeller shaft, and the expanded outline is the outline formed by connecting the leading and trailing edges of blade sections, which are then expanded and arranged at the corresponding radii of the sections. The expanded area ratio (E.A.R.) is defined as the ratio of the area of the expanded outline AE to the area of the propeller disk $A_0 = \pi D^2/4$, and is the important parameter governing the magnitude of thrust and drag forces acting on the surface of propeller blades.

10) Blade section reference point

The blade section reference point is a single point representing the position of the blade section. The linear distance between the leading and trailing edges, when the blade section is expanded, is called the chord (or chord-length), and the mid-chord point (half-way between the leading and trailing edges) normally becomes the blade section reference point. A blade section reference line is obtained by connecting the blade section reference points at all radii.

11) Skew

When the blade section reference line is projected onto the propeller plane, the angle measured from the propeller reference line to the blade section reference point at each radius is defined as the skew (θ_m) of the section at that radius. The skew angle is positive when the blade section reference point is downstream from the generator line. When the propeller rotates in the non-uniform ship wake, the blade section at each radius generates lift force, reacting to the sinusoidal (or combination of the sinusoidal) velocity components with some phase angles. The skew is an effective design parameter in minimizing the fluctuating loadings acting on the blade by suitably-varying the phase angle of the lift on blade sections at different radii. Currently there is a trend to increase the skew to reduce the excitation force levels, but the increase of the skew may deteriorate the structural strength of the blades and also may decrease the propulsive efficiency.

12) Hub

The hub (or boss) connects propeller blades to the propeller shaft. The maximum diameter of the hub at the propeller plane is called the hub diameter d and the hub ratio is defined as d/D. The propeller blades and the hub are usually cast at the same time in a single mold, called the solid-type. The blades for controllable pitch propellers are manufactured separately from – and integrated later with – the hub which houses the mechanical and hydraulic pitch-changing devices.

13) Blade thickness

The maximum thickness of a blade section at each radius is called the blade thickness t_0 at that radius. The blade thickness ratio is defined as t_0/D. The blade thickness increases as it gets closer to the hub. When the blade thickness is extended to the propeller shaft center, the hypothetical thickness t_{max} is defined at the shaft center. The blade thickness ratio at shaft center t_{max}/D is used frequently to represent the propeller blade thickness. For a given diameter of the propeller, thicker blades have lower propulsive efficiency. Therefore, it is desirable to have blade sections as thin as possible, as long as the blade strength requirement is met. Normal blade thickness ratio is approximately 0.03 ~ 0.07.

6.2 Principle of Propeller Action

The advancing ship experiences a resistance R_T against its motion. To overcome the resistance, the ship needs a thrust T, which is slightly greater than R_T. The device generating the thrust is called the propeller or the propulsor. The propeller receives the power delivered from the engine and converts it into the reactive mechanical power propelling the ship forward by pushing the water downstream. There are a number of theories which explain and model this physical phenomenon and the most important of these will be examined in this section.

1) Momentum theorem

The thrust $T(N)$ generated by the propeller advancing at a constant speed $U(m/s)$ may be expressed by the product of the increase of the axial velocity $\Delta u(m/s)$ from the far-upstream to the far-downstream points as shown in **Figure 5-26** and the mass flux passing per unit time $\dot{m}(kg/s)$ through the propeller disk of diameter D. If we denote the speed increase at the propeller disk as $u^*(m/s)$, the mass flow rate may be expressed as

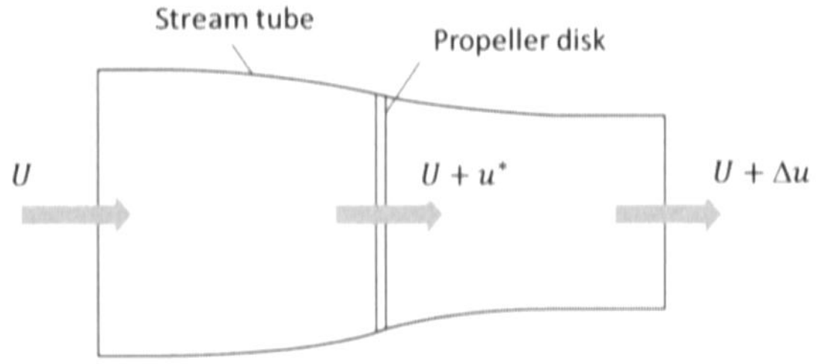

Figure 5-26 Process of the water being accelerated in the propeller slipstream

$$\dot{m} = \rho \frac{\pi D^2}{4} = (U + u^*) \tag{5-33}$$

Because the work done at the propeller disk by the thrust T in unit time is equal to the increase of the kinetic energy within the slipstream, we may derive the relation $u^* = \Delta u/2$ and the ideal efficiency $\eta_I = 1/(1 + u^*/U)$. The momentum theorem neatly expresses the basic principle of propulsion of the propeller, but cannot relate quantitatively the geometric shape of the propeller blade with the propulsive action of the propeller.

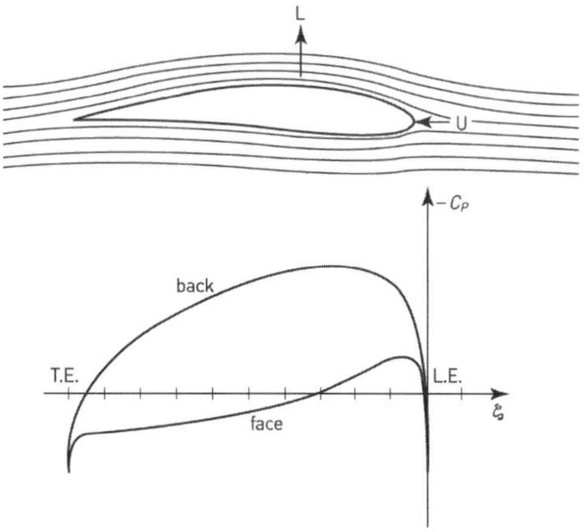

Figure 5-27 Pressure distribution on the 2-D blade section

2) Blade element theory

The propeller blade section at a radius r, when expanded into a plane, is shown at the top of **Figure 5-27**. This section, originally developed for airplanes, has typically a round leading edge and a curved back (upper) surface. The figure shows the streamlines around the blade placed in the uniform stream U flowing from the right to the left of the figure. We may observe that the streamlines above the blade are more tightly packed than those below.

The law of conservation of the mass of the water can only be satisfied if the flow speed on the upper side of the blade with a narrower width is greater than the flow speed on the lower side of the blade. Applying Bernoulli's equation, we see that the pressure on the upper surface of the blade is lower than the pressure on the lower surface. Integration of the pressure around the blade surface leads to the lift L pointing upward and perpendicular to the incident flow direction. In this case, the dimensionless pressure coefficient C_P is defined as

$$C_P = \frac{P - P_\infty}{\frac{1}{2}\rho U^2} \tag{5-34}$$

The bottom of **Figure 5-27** shows the pressure distribution on the blade with the negative pressure coefficient $-C_P$ or the vertical axis. Plotting in this way, the pressure on the upper (back) surface is made to appear on the upper side of the pressure plot and the pressure on the lower (face) surface on the lower side. The lift characteristics for the blade sections may be available from the experiments in a wind tunnel or from theoretical computations.

The blade element theory assumes that the three-dimensional blade can be divided into a set of blade elements of finite width, Δr, in the radial (span-wise) direction and that the interactions between the neighboring blade elements are negligible. We may then sum the piece-wise constant lift values acting on the blade elements in a manner akin to strip theory to get the total thrust and torque on the blades. This theory visualizes how the fluid flow around the blade sections are related to the propeller thrust by the fundamental principles of fluid mechanics. The theory, however, has an inherent drawback stemming from ignoring the interactions of blade elements at different radii and also interactions between the blades.

3) Circulation theory

Let us assume that a blade, advancing at a constant speed U in the fluid of density ρ, has the circulation Γ around the blade, as shown in **Figure 5-28**. The law of Kutta-Joukowsky says that the lift L per unit span of the blade is $\rho U\Gamma$ and is perpendicular to the direction of advance.

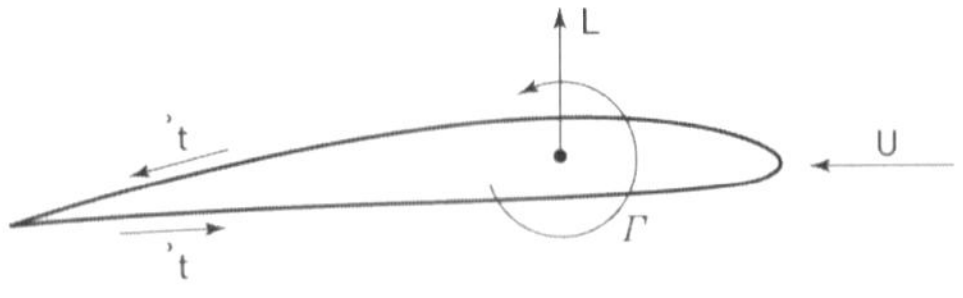

Figure 5-28 Circulation around a 2-D blade section

If we apply Kutta-Joukowski's law to the blade section at the radius r with the circulation Γ and the incident velocity of $V_R = \sqrt{V_A^2 + (\omega r)^2}$ as shown in **Figure 5-29**(a), we can get the lift δL per unit span of the blade element at that radius r. Applying this law to all section elements of the blade, as shown in **Figure 5-29** (b), we can get the total force and moment in vectorial form. The circulation theory can be extended to the lifting-surface theory, which distributes the vorticity continuously on the blade camber surface to solve the lifting problems more accurately. For numerical computation, the continuous distribution of the vorticity is replaced by a set of discrete vortex lattice, strengths of which are determined so as to satisfy the boundary condition on the blade camber surface. This method, called the discrete vortex method (DVM), is well-developed and is widely used around the world for the practical design and analysis of the propellers. The DVM considers the interaction of the blades including the induced velocities by all vortex segments representing the blade loadings and the trailing wakes.

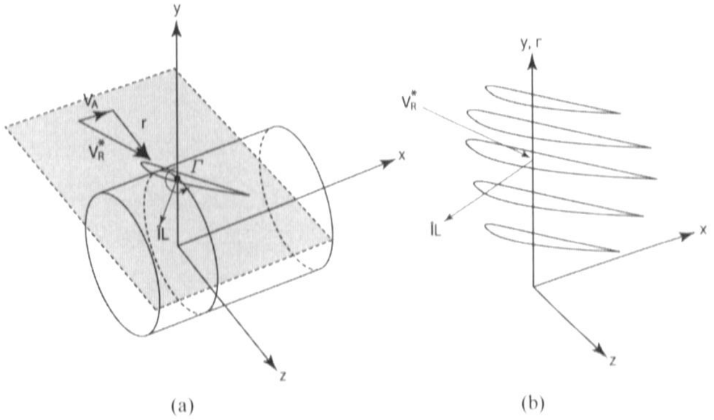

(a) (b)

Figure 5-29 Illustration of procedure building a 3-D blade through piling-up of lift-generating 2-D blade sections

6.3 Propeller Open-water Experiment

The propeller on an actual ship is operating in the non-uniform flow field behind the ship. But the propeller-only performance in the uniform inflow is usually tested in undisturbed water in a towing tank. This test is called the propeller open-water (POW) experiment. The POW performance data are essential in analyzing the propeller-hull interaction during the self-propulsion tests where the ship resistance is measured with the propeller working in place behind the ship.

The theoretical prediction of the POW performance is not yet fully reliable. Consequently the POW tests are performed in towing tanks, producing performance charts such as shown in **Figure 5-30**. During the test, the propeller is attached to the shaft end of the propeller dynamometer installed on the towing carriage. The advance speed $V_A(m/s)$ and the rotational speed of the shaft n $(1/s)$ are varied systematically and the thrust $T(N)$ and the torque $Q(Nm)$ are measured by the dynamometer. Test results are presented in the form of graphs shown in **Figure 5-30** or in numerical tabular form using the following dimensionless parameters:

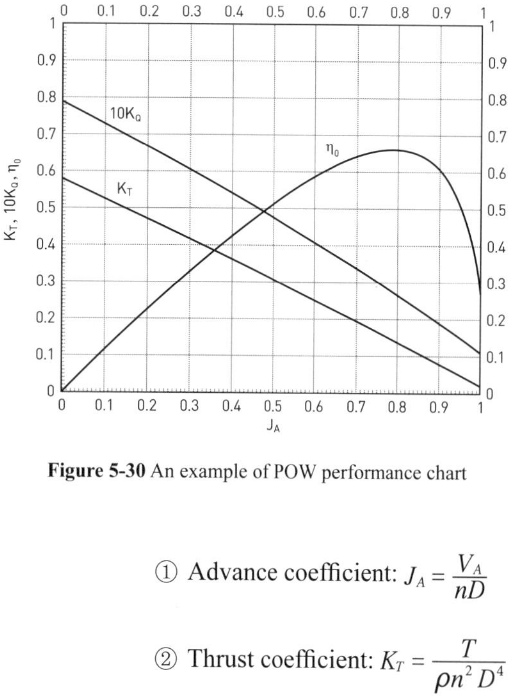

Figure 5-30 An example of POW performance chart

① Advance coefficient: $J_A = \dfrac{V_A}{nD}$

② Thrust coefficient: $K_T = \dfrac{T}{\rho n^2 D^4}$

③ Torque coefficient: $K_Q = \dfrac{Q}{\rho n^2 D^2}$

④ Propeller open-water efficiency: $\eta_0 = \dfrac{TV_A}{2\pi nQ} = \dfrac{J_A}{2\pi}\dfrac{K_T}{K_Q}$

7. Propulsion of Ships

To determine the propeller thrust required to overcome the resistance of the ship advancing forward at a given speed, it is essential to understand the hydrodynamic interaction between the hull and the propeller.

7.1 Propulsion Factors

Because the propeller operates in the ship's wake, the propeller-hull interaction starts with the ship's influence upon the thrust-generating propeller in the form of oncoming flow. At the same time, the propeller pulls the ship backward with the lower (vacuum) pressure on the suction-side of the blades in the form of an additional resistance. Meanwhile, it is traditional to separate all the involved propulsion factors following the recommendations of the 1978 ITTC Performance Committee.

1) Wake

When the ship advances forward, the water particles near the hull obtain the momentum following the ship's motion, mostly due to the viscosity of the water. If the speed of the ship is V and the speed of the water at some point (mostly at the propeller plane) in the water is V_A, then the difference of the speeds V-V_A is defined as the wake velocity V_W. Therefore, the value of the wake is positive when the water particles move in the same direction with the ship, and is negative otherwise. **Figure 5-31** shows the iso-velocity contours derived from the distribution of the velocity V_A measured at the propeller plane and non-dimensionalized by the ship's speed V.

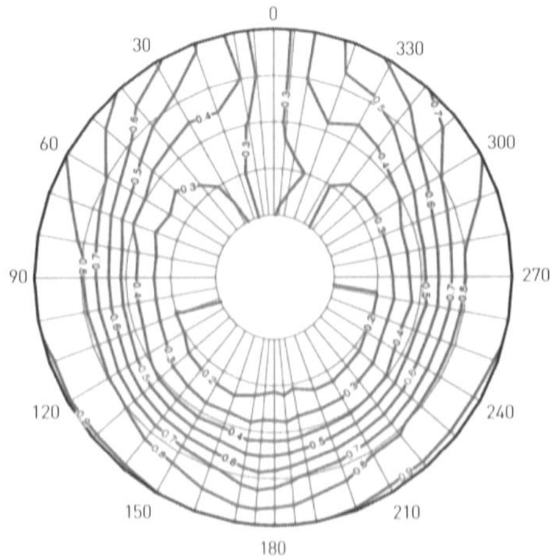

Figure 5-31 Iso-speed contours at propeller plane

The wake at the propeller plane consists of the following three components.

① Potential wake: Even a double-body model of a ship, consisting of the hull under water and its mirror image, which is introduced whenever the free surface effects need to be neglected, advancing in the inviscid, incompressible fluid of infinite extent, inevitably disturbs the flow field around it. The resulting difference of the flow velocity with the ship speed is called the potential wake, and has positive values near the stem and the stern and negative values near the mid-length of the ship.

② Frictional wake: Frictional wake refers to a wake caused by the viscosity of the water around the hull. The value of this frictional wake is influenced by the ship's length, wetted surface area, and hull roughness. The frictional wake becomes very large near the stern, that is, at the propeller plane.

③ Wave wake: Wave wake is caused by the motion of the water particles in the stern wave. The value of the wave wake varies slowly in the transverse direction, but rapidly in the depth-wise direction. For single-screw ships, the wave wake is very small compared to the other two components of the wake, but for twin-screw ships, the wave wake component is sometimes so large that the total wake values may become negative.

At the propeller plane, the three components of the wake appear at the same time and, in general, are inseparable. The axial velocity V_A measured at the propeller plane is used to compute the wake fraction (w) defined as follows:

$$V_A = V(1 - w) \Rightarrow w = \frac{V - V_A}{V} \tag{5-35}$$

The wake, measured without the presence of the propeller in place, is called the nominal wake. The wake, when the propeller is operating in place, is called the effective wake. The direction and magnitude of the wake in the propeller plane varies, depending upon the radial and circumferential location in the propeller plane. The distributed wake is normally integrated within the propeller disk to produce the representative mean wake. The non-uniform wake distribution causes the variation of the inflow angles to the blade sections, leading to the fluctuating thrust and torque on the blades, and ultimately becomes the source of ship vibration.

2) Thrust deduction

A propeller installed behind a ship moving at a constant speed takes in, accelerates and expels the water in the uneven velocity fields as shown in **Figure 5-31**. The resulting increase in flow velocity in the space between the propeller and the hull decreases the hull surface pressure and causes an increase in the ship's pressure resistance, along with an increase in the frictional resistance. If we denote the total increase in resistance as ΔR, the thrust generated by the propeller must be balanced with the resistance of the ship R_T by the relation

$T = R_T + \Delta R$ to maintain the ship speed. In reality, the hull and propeller interaction leads to the variation in both the propeller thrust and ship resistance, but we assume from an engineering point of view that the ship resistance R_T is invariant and the thrust is reduced by ΔR at the same speed.

In other words, the resistance of the ship R_T is balanced with the propeller thrust T reduced by a fraction t such that $\Delta R = tT$, leading to the following relation:

$$R_T = T(1-t) \tag{5-36}$$

where t is called the thrust deduction fraction. The **Equation 5-36** may be rewritten as follows:

$$t = \frac{T - R_T}{T} \tag{5-37}$$

The thrust deduction fraction t changes with hull shape and with the relative position of the hull and the propeller, and can be obtained through experiments. t and the wake fraction w are not independent values but are related to each other. The following equations are examples of their relationship obtained through statistical analysis:

$$t = \frac{2}{3}w + 0.01 \quad \text{(single-screw ships)} \tag{5-38}$$

$$t = \frac{2}{3}w + 0.03 \quad \text{(twin-screw ships)} \tag{5-39}$$

3) Propeller efficiency behind hull

The thrust power P_T which the propeller delivers to the water by pushing the water downstream with the thrust T at the propeller-disk-averaged speed V_A is defined as $P_T = TV_A$.

For the propeller operating in uniform flow, namely in an open-water condition, the power P_D delivered to the propeller through the shaft torque Q_o at the rotational speed n_o is defined as $P_D = 2\pi Q_o n_o$ and the open-water efficiency, the ratio of P_T to P_D, as follows:

$$\eta_0 = \frac{TV_A}{2\pi Q_0 n_0} \tag{5-40}$$

However, the propeller is working behind the ship where the flow is non-uniform and the shaft torque necessary to produce the same thrust T will become different at the same rotational speed n. Therefore, the propeller efficiency behind hull (η_B) defined as the ratio of the output power P_T to the input power P_D is defined as follows:

$$\eta_B = \frac{P_T}{P_D} = \frac{TV_A}{2\pi Qn} \tag{5-41}$$

where Q is the shaft torque at that time.

From **Equations 5-40** and **5-41** we note that, while producing the same thrust, the delivered power behind the hull is $2\pi Qn$ and differs from $2\pi Q_0 n_o$ of open-water conditions. The relative rotative efficiency (η_R) is defined as the ratio between the delivered power in open-water and behind-hull conditions as in

$$\eta_R = \frac{\eta_B}{\eta_0} \tag{5-42}$$

η_R is normally larger than, but very close to, 1.

4) Hull efficiency

Hull efficiency (η_H) is defined as the ratio of the effective power P_E, used for a ship advancing at the speed V to overcome the resistance R_T, to the power P_T supplied to the ship from the propeller by producing the thrust T at an advanced speed of the propeller V_A relative to the surrounding water. It represents a measure of the interaction of the hull and the propeller and can be expressed using the **Equations 5-35** and **5-36** as follows:

$$\eta_H = \frac{P_E}{P_T} = \frac{R_T V}{TV_A} = \frac{1-t}{1-w} \tag{5-43}$$

The single-screw ship has a relatively higher hull efficiency value, because the wake fraction is larger than the thrust deduction fraction as seen in **Equation 5-38**. For instance, if the block coefficient is 0.7, the hull efficiency is approximately 1.16. On the other hand, a twin-screw ship has a hull efficiency of around 1.04. The difference in efficiency shows that single-screw ships may use less fuel than twin-screw ships.

5) Quasi-propulsive efficiency

The power P_D delivered to the propeller from the main engine differs with the power P_E effectively used to move the ship forward after suffering from various forms of losses caused by the propeller-hull interaction. A measure of such losses is defined as the ratio of effective power to delivered power, or namely, quasi-propulsive efficiency (η_D):

$$\eta_D = \frac{P_E}{P_D} = \frac{P_E}{P_T} \cdot \frac{P_T}{P_D} = \eta_H \cdot \eta_B = \eta_H \cdot \eta_0 \cdot \eta_R \tag{5-44}$$

7.2 Self-propulsion Test

A self-propulsion test is conducted to determine the overall powering performance of a hull and a propeller. The test is arranged similarly to a resistance test. The difference is that a self-propulsion test requires the presence of a propeller. The thrust and torque acting on the propeller and the number of revolutions are measured. To compensate for the difference of frictional resistance coefficients between a prototype ship and a model ship, a towing force is applied to tow the model ship during a self-propulsion test.

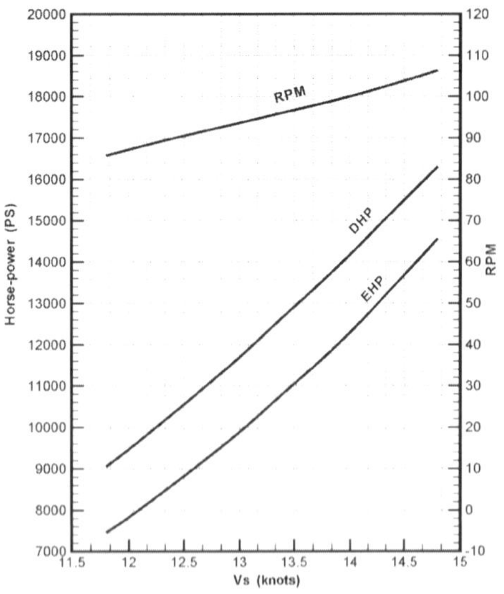

Figure 5-32 Chart showing results of self-propulsion test

According to the 1978 ITTC Performance Prediction Method, which is widely applied around the world, the propulsive factors, including wake fraction (w), thrust deduction fraction (t), relative rotative efficiency (η_R), and open-water efficiency (η_O) are determined during a model test.

These values obtained during the model test are extrapolated to full-scale based on empirical and statistical data. As shown in **Figure 5-32**, for a given draft, the final test results are expressed in a chart with the speed V_S (knots) on the x-axis and the delivered horsepower (DHP) P_D, the effective horsepower (EHP) P_E, and the number of revolutions N (RPM) on the y-axis.

8. Powering of Ships

8.1 Types of Power and Efficiency

For the power of a main engine to be delivered to a propeller, produce thrust, and be used as effective power for the propulsion of a ship, the power is transmitted through a number of stages. At every stage, there is always a power loss. This section discusses the types of power and the definition of efficiency at each stage.

1) Indicated power

Indicated power can be calculated from a chart illustrating the pressure-versus-volume relationship within the cylinder of a reciprocating internal combustion engine. Given the mean effective pressure p(Pa), the stroke length of a cylinder l(m), the circular cross-section area of a cylinder a(m^2) and the number of combustion cycles per unit time n(1/s), indicated power P_I (W) can be calculated using:

$$P_I = p \cdot l \cdot a \cdot n \times \text{(number of cylinders)} \tag{5-45}$$

2) Brake power

Brake power is measured at the end of a crankshaft outside an internal combustion engine. Brake power is the remainder of the indicated power delivered to the end of this crankshaft after subtracting the frictional power loss within the engine. The ratio of brake power to indicated power is called the engine's mechanical efficiency. In short, $\eta_M = P_B/P_I$. The η_M ranges between 80 and 85% for diesel engines, but varies even for the same engine, depending on the load conditions, and declines during partial load operation.

3) Shaft power

Indicated power P_I for rotary engines, such as steam turbines, cannot be measured mechanically. Instead, the torque delivered to the intermediate shaft is measured using a torsion meter. Given the torque Q_S(Nm) and the number of revolutions n(1/s), the shaft power P_S can be determined using:

$$P_S = 2\pi Q_S n \tag{5-46}$$

4) Delivered power

Delivered power is the power actually delivered to a propeller and used as a reference power when designing a propeller. It is the remainder of power produced by an engine after subtracting friction losses in a bearing supporting the shaft and in a stern tube and other power losses in the shaft system. The ratio of delivered power P_D to brake power P_B (or shaft power P_S) is called transmission efficiency η_T.

$$\eta_T = P_D/P_B \tag{5-47}$$

The η_T values vary depending on the engine types, installation location (midship or stern), the connection between the main engine and propeller shafts (directly or by a reduction gear), type of thrust bearing, number of bearings in between, precision in shaft assembly, and draft condition. It is extremely difficult to directly measure delivered power, and η_T is usually assumed in the range from 0.95 to 0.98.

5) Thrust power

Thrust power is the power used to propel a ship and determined by multiplying the thrust produced by the propeller T(N) by the speed of advance V_A (m/s) of the propeller against the surrounding water. In short, $P_T = TV_A$. The ratio of thrust power to delivered power is called the propeller efficiency behind hull η_B.

$$\eta_B = P_T / P_D \tag{5-48}$$

In other words, propeller efficiency behind hull can be found when the propeller is working at the ship's stern, and varies depending on the propeller type, hull, and hydrodynamic interaction between the propeller and hull-generated wake.

6) Effective power

Effective power is required to overcome the resistance R_T and move a ship forward at the speed of V_S.

$$P_E = R_T \cdot V_S \tag{5-49}$$

7) Relationship between the power components

Propulsive efficiency η_P from the indicated power P_I produced by an engine to the effective power P_E can be related as $P_E / P_I = (P_E / P_T) \cdot (P_T / P_D) \cdot (P_D / P_B) \cdot (P_B / P_I)$ based on the aforementioned definitions of various powers, and the propulsive efficiency η_P can be computed by multiplying the efficiency of various stages:

$$\eta_P = \eta_H \cdot \eta_B \cdot \eta_T \cdot \eta_M \tag{5-50}$$

8.2 Required Power of Engines

Once the power required to propel a ship at a given speed is identified in connection with the hull and the propeller, the data can be useful in designing a new ship. The required power of an engine can be estimated in one of the following methods:

① Method based on self-propulsion test in a towing tank: Using the delivered power P_D obtained during a self-propulsion test, the required power of an engine can be calculated after assuming transmission efficiency η_T and mechanical efficiency η_M.

② Method based on effective power and propulsion factors: Using the effective power P_E obtained during a resistance test or based on the resistance of a similar ship, the delivered power P_D or directly the indicated power P_I can be found after estimating the propulsion factors or propulsive efficiency.

③ Method using Admiralty coefficient: The required power of an engine to propel a ship at a given speed (indicated power, brake power or shaft power) is assumed to be proportional to the displacement Δ(ton) to the power of 2/3 and the ship's speed V_R (knots) cubed. The proportionality constant here is called the Admiralty coefficient. This coefficient can be calculated from the existing ship data using the **Equation 5-51**:

$$P_I\,(\text{or } P_B, P_S) = \frac{\Delta^{2/3} \cdot V_R^3}{C_{adm}} \tag{5-51}$$

where C_{adm} indicates the Admiralty coefficient. Among ships that have a similar hull shape, the Admiralty coefficient is also deemed to be equal for same Froude numbers.

④ Method based on the data of similar ships: If accurate navigation records, sea trial results or the results of self-propulsion tests in towing tanks of the ships similar to the ship in question in terms of hull form, engine, propeller, velocity and draft condition among others are available, then the data can be manipulated to predict the power requirements.

9. Cavitation

9.1 Causes of Cavitation

As demonstrated in the pressure-temperature (p-T) relationship chart in **Figure 5-33**, there are two types of processes by which water changes from a liquid phase to a gas phase. The first is the process of water turning into vapor as the temperature increases at a constant pressure (for instance, 1 bar pressure). This is called boiling (the A-B process in **Figure 5-33**). The second process of water changing phases from a liquid to a gas is when the pressure within water drops below the vapor pressure pv at a constant temperature (for instance, 15°C). This is called cavitation (the A-C process in **Figure 5-33**).

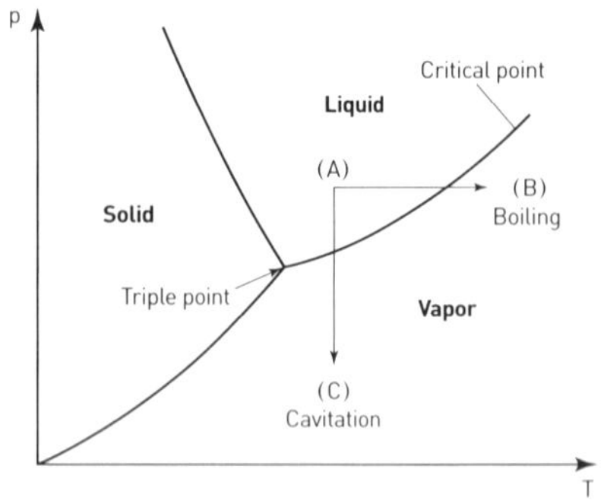

Figure 5-33 Pressure-temperature diagram of water in 3-phasest

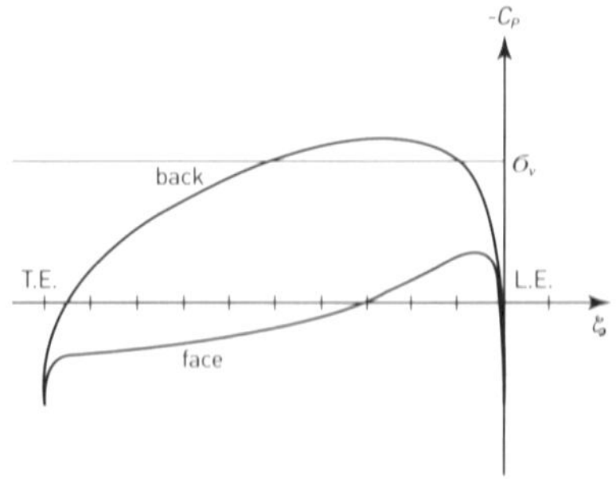

Figure 5-34 Pressure distribution around 2-D blade and the cavitation number

Literally, cavitation refers to the formation of cavities in water. It is called so because, considering the water liquid-to-vapor density ratio is about 800:1, the area inside a cavity is deemed relatively empty in terms of dynamics.

Along the streamlines near the hydrofoil moving forward with a constant speed U as shown in **Figure 5-27**, Bernoulli's equation is expressed as follows:

$$p+\frac{1}{2}\rho V^2=p_\infty+\frac{1}{2}U^2=\text{const.} \tag{5-52}$$

Therefore, on the hydrofoil surface, if the local speed V increases, the pressure p decreases. If the speed continues to increase, the pressure drops below the saturated vapor pressure, causing cavitation. Since the pressure along the streamline on water-cavity interface is equal to the vapor pressure p_v, the cavitation number σ_v , like the pressure coefficient, can be defined as follows:

$$\sigma_v=\frac{p_\infty-p_v}{\frac{1}{2}\rho U^2} \tag{5-53}$$

Though the area where $-C_p > \sigma_v$ in **Figure 5-34** becomes, basically, the extent of the cavitation, the length of cavitation actually extends downstream. While the saturated vapor pressure p_v changes slightly with temperature, it remains approximately 1.7 kPa at 15°C. Meanwhile, the air pressure is 101.3 kPa under the standard atmospheric conditions.

9.2 Cavitation on Two-dimensional Hydrofoils

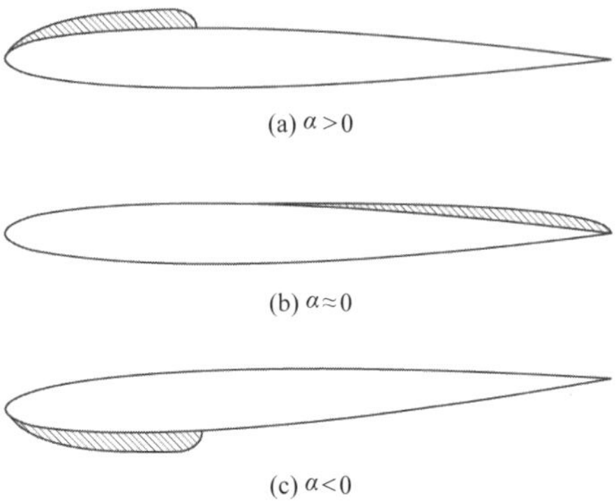

Figure 5-35 Cavity shape and location depending on α

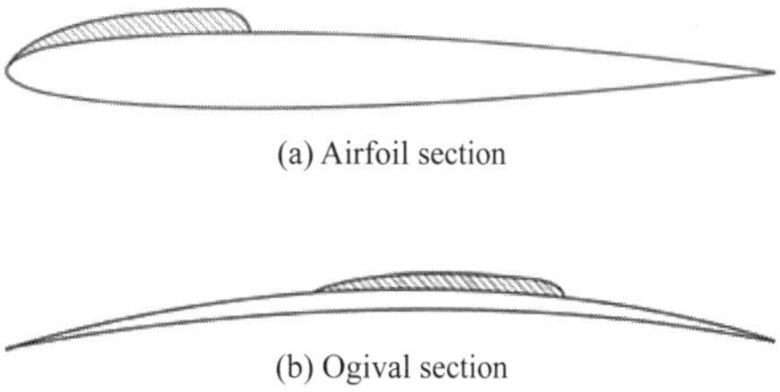

Figure 5-36 Cavity shape and location depending on the section type

Cavitation is closely related to the shape of a blade section. Its location and shape vary, depending on the magnitude of the angle of attack α and on whether it is positive or negative. As shown in **Figure 5-35**(a), if the angle of attack has a positive value and is relatively large, cavities start at the leading edge and extend downstream on the blade back, regardless of blade shape. If the angle of attack is close to zero, cavities initiate near the maximum thickness point and extend to the trailing edge, as shown in **Figure 5-35**(b). However, if the angle of attack has a negative value, cavities start at the leading edge and extend downstream on the blade face, as shown in **Figure 5-35**(c).

When comparing the sectional shapes of the blade sections, as shown in **Figure 5-36**(a), cavities are usually detected near the leading edge. However, for an ogival section, cavities easily occur at a small angle of attack near the mid-chord of the blade back as shown in **Figure 5-36**(b).

9.3 Cavitation on Propellers

As shown in **Figure 5-29**, a propeller blade can be considered to be formed by continuously piling up two-dimensional sections. Each blade section in a non-uniform stern wake is subject to the incoming fluid flow, whose angle of attack varies continuously as the blades rotate around the propeller shaft. Therefore, different types of cavitation occur, depending on the location of the blade section. **Figure 5-37** shows the inception, growth and decay of cavitation occurring on a propeller blade.

Since the influence of cavitation on propeller performance varies depending on the location, it has specific names in accordance with its location to distinguish one from another. **Figure 5-38** represents the most typical cavitation types, and the following discusses their behaviors and influences on propeller performance.

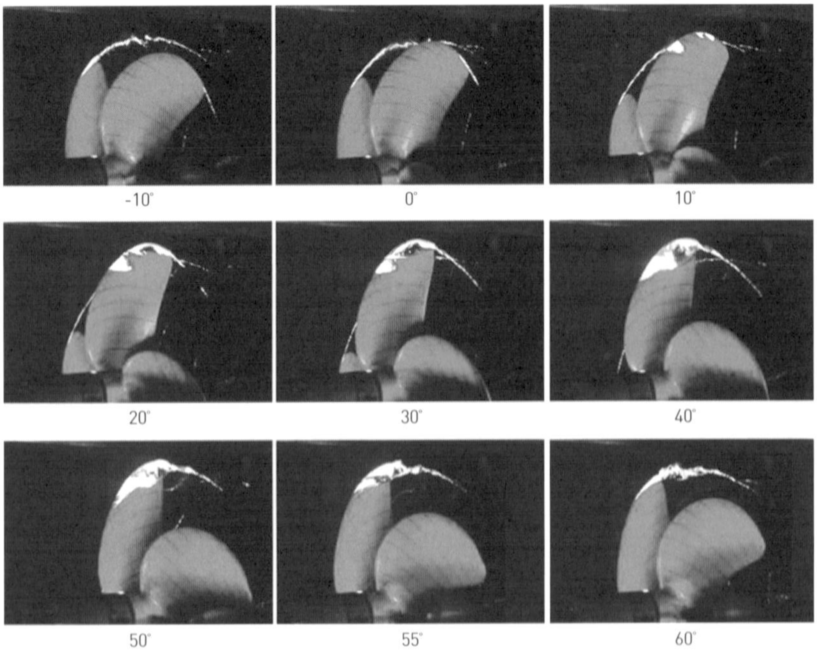

Figure 5-37 Variation of cavity extent on propeller blades
(Seen downstream direction from hull)

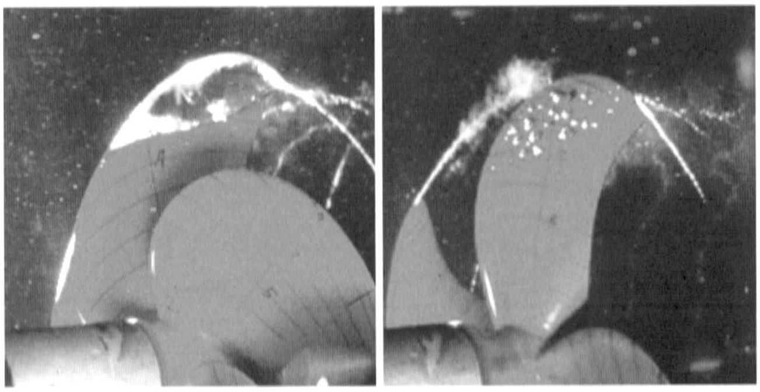

Figure 5-38 Sheet cavitation(left) and bubble cavitation(right)

1) Sheet cavitation

Sheet cavitation refers to cavities in the form of thin layers occurring on the leading edge of a propeller blade due to negative pressure on the back surface of the blade at a moderately large angle of attack. **Figure 5-38** (left) shows sheet cavitation commonly found on real propellers. This type of cavitation is in a relatively stable state, but fluctuations in cavity volume become a major source of ship hull vibration.

2) Bubble cavitation

As shown in **Figure 5-38** (right), if a propeller blade is relatively thick and has a relatively small angle of attack, bubble cavities occur near the maximum thickness point of the blade. Since this type of cavitation is in a very unstable state, it can result in the degradation of propeller performance and the erosion of the blade surface. Once bubbles are detected during cavitation tests, the propeller must be re-designed to avoid such cavitation.

3) Face cavitation

Face cavitation occurs near the leading edge of the blade face when the angle of attack is relatively small or has a negative value. This type of cavitation is frequently observed on the blade section near the propeller hub, or at the blade tip when the blade is in the 6 o'clock position. Since the cavitation of this type is in an extremely unstable state, it is a major cause of blade erosion.

9.4 Cavitation Tunnel

A cavitation tunnel is an experimental facility where a model propeller (behind a model ship or in simulated wake) is mounted and observed in flowing water for cavitation occurring on the propeller blade surface. The tunnel consists of closed loop ducts, as shown in **Figure 5-39**, inside which the internal pressure is controlled to satisfy the cavitation similarity law between the model and prototype propellers. The thrust and torque acting on the propeller and the fluctuating pressure on the ship's hull induced by a propeller are measured.

Figure 5-39 is a side profile of a cavitation tunnel (Large Cavitation Tunnel, LCT, run by KRISO; a lateral length of $60m$ and a height of $19.8m$). **Figure 5-40** shows a model ship about 8 meters long installed inside the measuring section of the tunnel (cross-section dimensions: $2.8m$ width $\times$ $1.8m$ height $\times$ $12.5m$ length; maximum flow speed of $16.5m/s$). **Figure 5-41** is an image of cavities occurring on the blade surface of a propeller working at the stern of a model ship in the large cavitation tunnel.

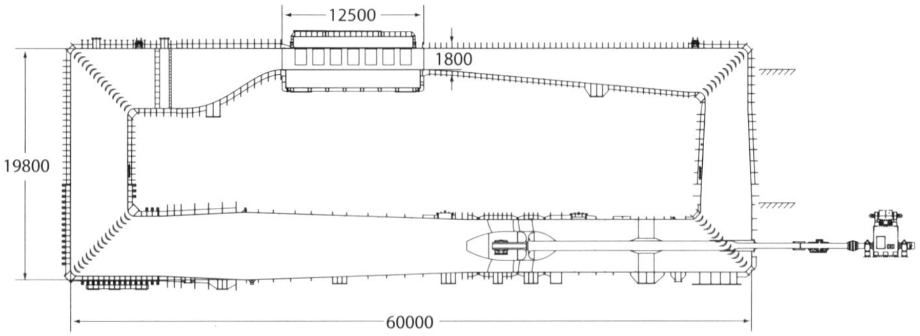

Figure 5-39 Drawing of a large cavitation tunnel (Courtesy of KRISO)

Figure 5-40 Measuring section of cavitation tunnel

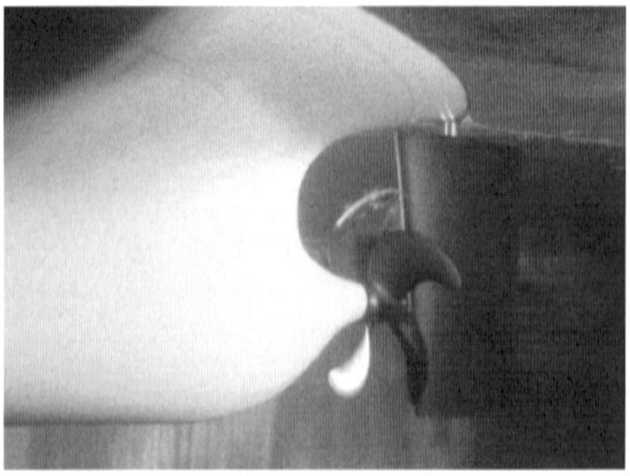

Figure 5-41 Observation of cavitation

10. Unconventional Propulsors

10.1 Water Jet Propulsors

An ordinary screw propeller experiences the dramatic degradation of propulsive performance as cavitation worsens with an increase in ship speed. For very-high-speed ships, an impeller installed inside the duct within a hull, as shown in **Figure 5-42**, is used in the same way as jet propulsion in an airplane. This propels the ship forward by a reaction force obtained by a jet of water drawn through an intake under the ship's bottom and expelled through a nozzle at the stern.

The theoretical maximum propulsive efficiency of water-jet propulsion is only 50% and

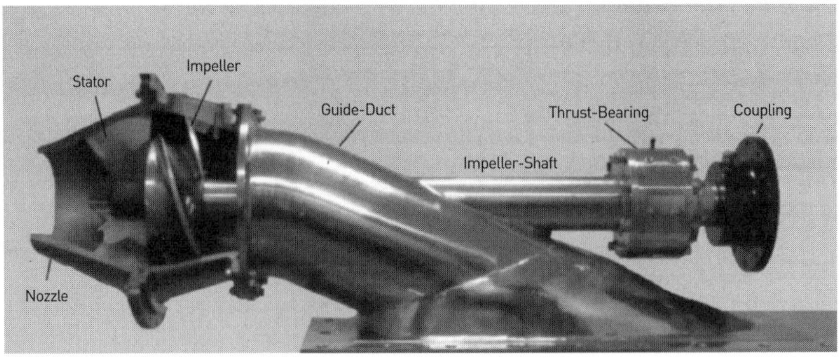

Figure 5-42 Water-jet propulsor (Courtesy of Daesung Marine Tech.)

the efficiency tends to decrease further due to a head loss caused by the height difference between intake and discharge nozzles, and due to the frictional loss in the internal duct piping. In general, therefore, it is more difficult to obtain efficient propulsion from a water-jet than an ordinary screw propeller. Nevertheless, in high-speed ships, water-jet propulsion is more widely used than screw propellers with a high risk of cavitation.

10.2 Contra-rotating Propellers

As demonstrated in **Figure 5-26**, for a propeller to produce thrust, it must accelerate water at the speed of u^* in the axial direction. An actual propeller accelerates a fluid at the same time in the direction of rotation. The fluid accelerated in the axial direction reacts to the propeller generating the forward thrust and hence is able to fulfill the ship's basic function of advancing with the payload onboard. However, acceleration in the direction of rotation is a pure loss with no relevance whatsoever to propulsion and leaves rotational kinetic energy in the wake of the propeller.

Figure 5-43 A CRP installed behind a model ship (Courtesy of HHI)

A Contra-Rotating Propeller (CRP) consists of two screw propellers, located fore and aft, but connected separately to coaxial shafts so that the two propellers rotate in opposite directions. As the accelerated flows in the different directions of rotation induced by two working propellers tend to cancel each other, the rotational kinetic energy shed downstream from CRP is reduced to improve the propulsive efficiency of the total CRP system.

An ordinary propeller develops torque on a shaft by

reacting to the rotation of a fluid. On the other hand, the CRP system is capable of balancing the effects of torque by rotating in two opposite directions. Since the CRP system requires a shaft rotating in opposite directions on top of the coaxial shaft, reliability in gear designing and manufacturing pose the biggest problem. Therefore, though it is difficult to use the CRP system for large ships that require great power, it has been widely used for torpedoes that require less power and good torque balance. **Figure 5-43** shows a CRP installed at the stern for testing.

10.3 Ducted Propellers

A ducted propeller is a propulsion system fitted with either an accelerating duct to produce high thrust through duct-propeller interaction and an increase in efficiency, or with a decelerating duct to improve cavitation behavior by increasing the pressure around the propeller. The accelerating duct, in particular, is commonly used on tugboats which require high thrust.

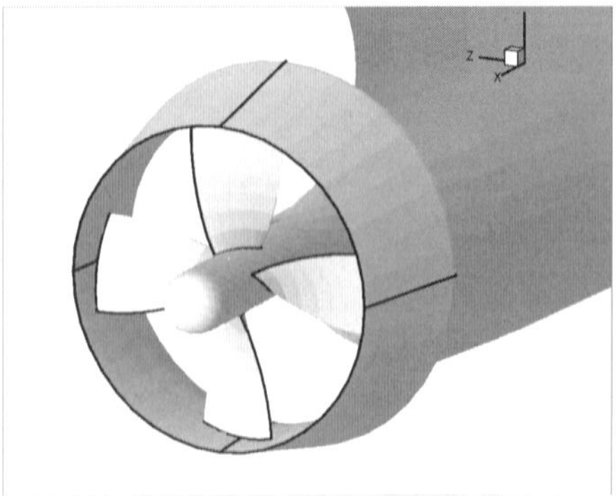

Figure 5-44 A ducted propeller installed behind a model ship

As shown in **Figure 5-44**, the propeller can achieve the maximum propulsive efficiency when placed in the middle of the duct chord. Keeping the gap between the blade tip and the duct as close as possible helps increase efficiency, but a consequent increase in load on the blade tip often results in cavitation erosion on the tip and on the inner surface of the duct.

References

1. Calcagno, G. et al., 2002. Propeller wake analysis behind a ship by stereo PIV. In: *24th Symposium on Naval Hydrodynamics*. Fukuoka, Japan 8~13 July 2002. National Academy Press, Washington, D.C.

2. Fluid Mechanics Research Group, the Society of Naval Architects of Korea, 2009. *Ship resistance and propulsion* (in Korean), Seoul: Jisungsa.

3. Harvald, Sv. Aa., 1983. *Resistance and propulsion of ships, Ocean Engineering: A Wiley Series*. New York: John Wiley & Sons.

4. ITTC, 2005. Resistance Committee Report, In: *24th International Towing Tank Conference*. Edinburgh, United Kingdom 4~10 September, 2005. ITTC: Edinburgh.

5. Kim, M.C. et al., 2010. Comparative study on the performance of Pod type waterjet by experiment and computation. *International Journal of Naval Architecture and Ocean Engineering*, 2(1), pp.1-13.

6. Kim, W.J. et al., 2001. Measurement of flows around modern commercial ship models. *Experiments in Fluids*, 31, 567-578.

7. Larsson, L .& Raven, C.H., 2010. *Ship resistance and flow, The Principle of Naval Architecture Series*. Jersey City: SNAME.

8. Lee, C.-S. et al., 2007. *Ship propulsion and propeller design* (in Korean). Seoul: Munundang.

CHAPTER 6

Ship Motions and Maneuverability

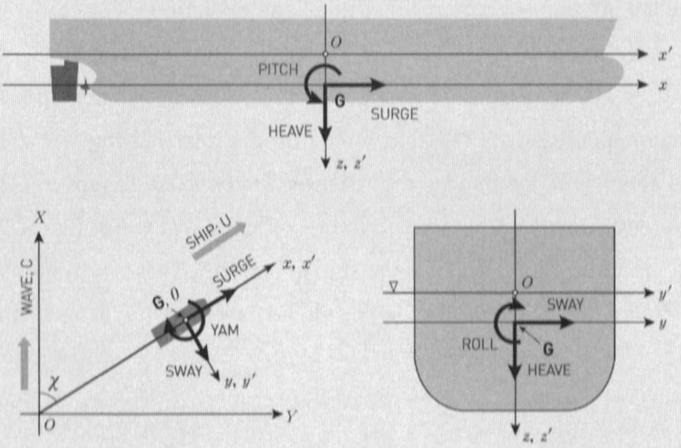

A ship advancing in waves experiences a sequence of complex motions, and in general they are undesirable in many ways. For example, they can cause seasickness to passengers and be detrimental to the operation of sensitive equipment and devices onboard the ship. Most notable of all, the relative motions between the ship and the water surface result in wave loads and added resistance. The added resistance leads to involuntary speed loss of the ship and the wave loads may be severe enough to cause permanent structural damage to the hull or even break up the ship.

In high waves, the relative motions may be so large that a considerable area of the ship's bottom can be exposed in the air. The subsequent reentry of the bottom may result in an undesirable impact, or a slam, as the bottom collides with the water surface. The vibration of the hull at its natural frequency following a slam is called whipping. Thereafter, severe deck wetness occurs mostly at the bow when slamming, since some of the water thrown up by the slamming and drawn up by the emergence of bow as well will be shipped onto the deck due to the forward speed of the ship. The dynamic behavior of a ship travelling in waves is very complex; the ship is subjected to various dangerous physical phenomena, especially in quartering/following seas, such as surf-riding and broaching, reduction of intact stability when riding on a wave crest, and synchronous or parametric roll motions, among others.

It used to be a common practice for ship designers to concentrate on meeting the owner's requirements for speed, displacement, stability, and structural strengths of a proposed ship before the behavior of the ship in waves was examined. However, recent demands for a higher quality ships has made the sea keeping ability one of the important performances to be assessed in an initial design stage and regulated by classification societies. In this chapter brief discussion will be given to various topics related to ship motions including basic properties of waves, mechanisms, estimations and consequences of the motions, and application of knowledge in these matters to ship design and operation.

One of the major issues to be discussed in this chapter is the maneuverability of a ship. Maneuvering of a ship is all the controlled changes in the direction of the ship's advance. A ship moving in water depends on the reaction force of the water to change its course, just as a wheeled vehicle changes its course by the reaction forces from the ground. Unlike a car, however, it is not easy for a ship to get the force sufficient to control its motions and consequently it takes a relatively long time for the ship to respond. That is why ships can run into accidents, even though the ship's master may be already aware of the dangers ahead.

1. Ocean Waves

Since motions come about in response to water waves, it is important to understand waves in order to investigate the motions. Ocean waves are very complex and usually approximated as a mixture of regular, small amplitude waves of various periods propagating in different directions. **Figure 6-1** shows the distribution of ocean wave energy where waves are classified by the periods (or wavelengths) into several groups ranging from capillary waves with periods of less than 0.1 seconds to the trans-tidal waves with periods of 24 hours or more. The primary forces responsible for wave-making and restoring for each group are also identified in the figure.

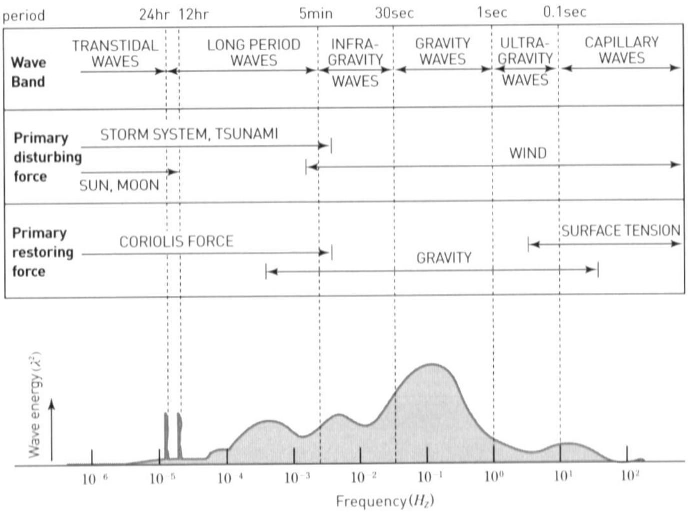

Figure 6-1 Approximate distribution of ocean surface wave energy

The wavelength of a water wave is proportional to the square of the period in deep water but proportional to the square root of the water depth in shallow water. Seismic waves are caused by volcanic eruptions, earthquakes or tectonic movements on the seabed, and they may travel across the Pacific Ocean within hours, if their periods are long enough. The seismic waves in the deep ocean are hardly recognizable since the heights of these waves are usually less than 50cm and the wavelengths may amount to a few hundred kilometers or more. However, once the waves reach a shallow coastal region, the wavelengths will be rapidly reduced since then it is proportional to the product of the period and the square root of the water depth. This leads to a drastic increase in the wave heights, and may develop into a powerful tsunami, since the wave energy remains almost the same throughout the process.

The longest waves in the ocean are the tides due to the gravitational attractions of the sun, moon and other celestial objects acting upon the rotating earth. The periods of the tidal waves are between a half (semi-diurnal) and a full lunar day (diurnal) where one lunar day is approximately 24.84 hours. In many cases, the tidal waves are not considered as waves but are currents.

The waves of most practical importance in engineering are the ones that arise in the ocean through the actions of the wind. If wind of a specific speed blows over a sufficient length of water (fetch) and lasts for long enough, wind-driven waves are generated in the area. Among those, the gravity waves of periods covering 1 to 30 seconds and lengths of about 1 to 1,400 m are the most important to naval architects and ocean engineers, since the lengths are comparable to the characteristic lengths of engineering structures operating in the ocean. Waves of heights more than 20m are not uncommon in the North Atlantic or North Pacific in winter.

Ocean waves are irregular and the motion of a ship in such a complex environment can only be treated with statistical methods based on the assumption that the real wave in the ocean is a linear superposition of a numerous regular waves having different lengths, heights and directions. Therefore, understanding ship motions in regular waves is an essential first step in coming to grips with ship motions in a realistic irregular wave environment.

1.1 Regular Waves

Figure 6-2 shows a plain progressive wave travelling with celerity c in positive x-direction in deep water. The motion of the wave is two-dimensional and sinusoidal in time with amplitude a and circular frequency ω. This type of wave is called a regular wave and appears to be steady to an observer moving with the wave train at the same speed c. The elevation of free surface must be of the general form shown in the figure, where the parameter $k = \omega/c$ is the wavenumber, the number of wavelengths per unit distance along the x-axis.

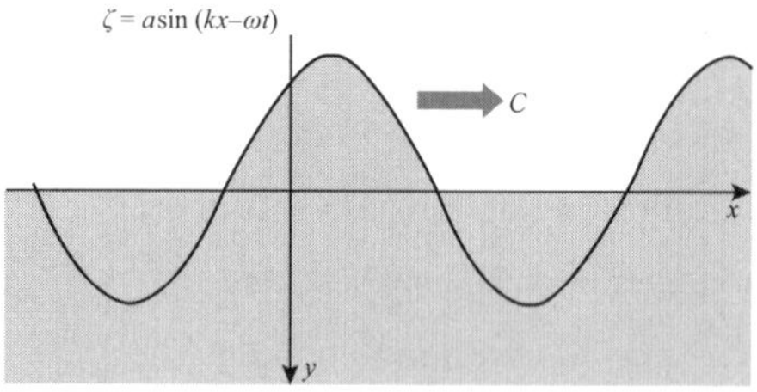

Figure 6-2 Regular waves

- Wavelength and period

  Since the sine function repeats over a length of 2π along the x-axis, it is clear that the wavelength λ is $2\pi/k$ and similarly the period $T = 2\pi/\omega$. An additional condition relating the wavenumber and the frequency must be introduced in the form $k = \omega^2/g$ in deep waters in order to satisfy the boundary conditions imposed on the free surface where g is the gravitational acceleration. This is called the dispersion relation, implying that the frequency and the wavenumber are mutually dependent parameters.

- Phase velocity (wave celerity)

The phase velocity or celerity of a regular wave can be easily determined from the relations discussed above in the form $c = \lambda/T = \omega/k = (g\lambda/2\pi)^{1/2}$, since the wave propagates exactly one wavelength during one period. It is clear from this that the longer waves propagate faster.

- Motion of fluid particles

While the wave advances with the celerity c, the fluid itself moves with much smaller velocity. If some non-linear effects (drift) are ignored, the mean position of the fluid particles remains the same, as the wave propagates with the celerity which is independent of the wave amplitude. The fluid particles in deep water move in circular orbits of radii proportional to the wave amplitude and exponentially decaying with increasing wavenumber (decreasing wavelength) and depth of submergence. In shallow water, the trajectories of fluid particles will be elliptical and the major axis is horizontal as shown in **Figure 6-3**.

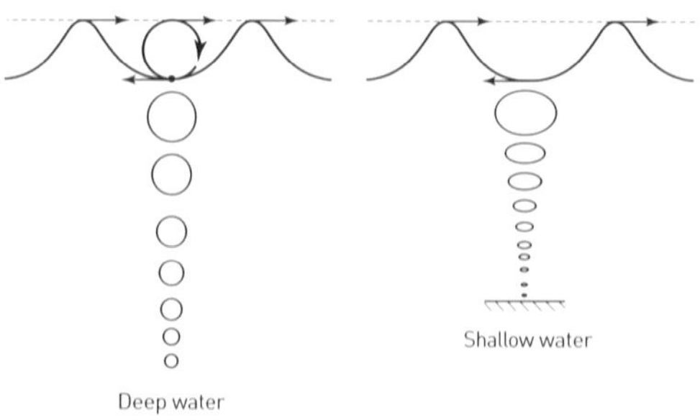

Figure 6-3 Particle trajectories in a regular wave

- Pressure distribution and the Froude-Krylov force

As mentioned earlier, the primary exciting force of a ship motion arises from waves and the force is called the wave or wave exciting force. Therefore, a basic knowledge of wave theories is indispensable to study ship motions excited by the wave force. Pressure is a normal force exerted by a fluid per unit area of the surface of an object in a fluid and the pressure force acting on the body can be found by integrating the pressure over the surface. **Figure 6-4** shows temporal

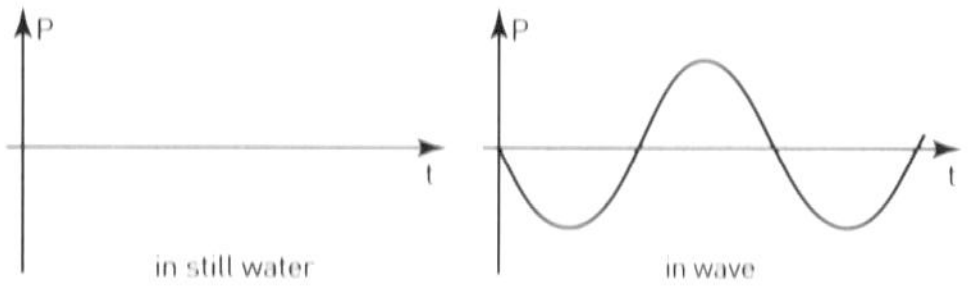

Figure 6-4 Temporal variations in pressure at a point beneath the free surface

variation of pressure at a fixed point underneath the free surface. In calm water, the pressure at the point remains constant since there is only the static pressure which is linearly proportional to the depth of submergence. Under a regular wave, the pressure varies with time to fluctuate around the static value by the wave motion and the difference is called the dynamic pressure. The amplitude of the dynamic pressure decays exponentially as the depth of submergence increases.

The simplest representation of pressure distributed on the surface of a floating body fixed in position in a regular wave is shown in **Figure 6-5**. In this diagram it is assumed that the pressure field is not affected by the presence of the body (the Froude-Krylov hypothesis) and in such a case the total pressure acting on the body surface is a linear combination of static and dynamic pressure. The force obtained by integrating the static pressure over the surface is called buoyancy force or buoyancy and cancels out the weight of the body. Therefore, the Froude-Krylov exciting force and moment acting on a floating body in waves are consequences of the dynamic pressure generated by the waves.

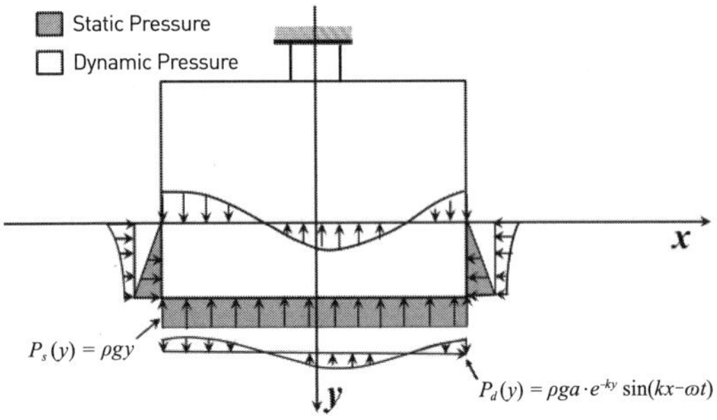

Figure 6-5 Static and dynamic pressure distributions on the surface of a 'transparent' body (the Froude-Krylov hypothesis)

Obviously, the real exciting force and moment acting on a ship moving in waves is more complex, since the presence of a body in the fluid results in diffraction of the incident wave (scattering effect). It is worth noting at this time that the Froude-Krylov exciting force and moment are the leading order contributions in the heave, pitch and roll motions of a thin or slender ship if the wavelength is large compared to the beam [Newman 1977].

1.2 Irregular Waves

1.2.1 Statistical Analysis of the Time History

Waves in the real ocean are irregular in time and space and somewhat random in nature. Therefore, one of the powerful ways for modeling the phenomenon mathematically is approximating them as a Gaussian random process. For this purpose, it is customary to measure wave elevation at a single fixed location in the ocean and to describe the general characteristics of the complete time history in terms of the following mean values of measured quantities:

$\bar{\zeta}_a$ mean wave amplitude
$\bar{H}_a$ mean wave height
$\bar{T}_p$ mean peak period
$\bar{T}_z$ mean zero-crossing period
$\bar{T}_1$ the average period (mean wave period)

Additional quantities may also be used:

$\bar{H}_{1/3}$ significant wave height – mean value of the highest third of all measurements of H_a in a time period
$\bar{\zeta}_{1/3}$ significant amplitude, $\zeta_{1/3} = ½H_{1/3}$
$\bar{H}_{1/10}$ mean value of the highest tenth of all measurements of H_a
σ standard deviation or root mean square wave height, $\bar{\zeta}_{1/3} = 2\sigma$

In order to analyze an irregular sea, it is necessary to take measurements of the surface elevation for duration sufficient to obtain proper spectral shapes and statistical values. This duration is represented by the total number of wave cycles N and records of $N = 200$ or above are considered to be excellent data [ITTC 1984]. In practice wave records of about 15~20 minutes duration are often used.

1.2.2 Wave Energy Spectrum

It is customary to represent the time history of irregular wave elevation by a Fourier series. In other words, an irregular sea can be reasonably approximated by a linear superposition of regular waves. A wave amplitude energy density spectrum – or a wave energy spectrum for short (or further abbreviated to wave spectrum) – is defined so that the area under the curve for a given frequency band is proportional to the total energy contained in the waves of that frequency band. In principle, a wave spectrum has to be derived from a wave record measured at a specific location and time in the ocean and is a unique property which cannot be reproduced at different locations or times. Usefulness of such wave spectra may be strictly limited for design of a ship moving across the ocean unless a series of measurement data is available along the route. Therefore, it is the usual practice to use idealized wave spectra instead. Two important spectra frequently found in the literature are Bretschneider spectrum for open sea and JONSWAP wave spectrum for fetch-limited coastal waters. The former is also called Modified Two-Parameter Pierson-Moskowitz wave spectrum, the ITTC spectrum, or ISSC spectrum. The latter spectral formulation is a product of an extensive wave measurement program known as the Joint North Sea Wave Project (JONSWAP) in

1968 and 1969. The spectra were recommended by the 17th ITTC in 1984.

The mathematical formulations of these normalized unidirectional wave energy spectra are based on two parameters: the significant wave height $H_{1/3}$ and average wave periods T defined by T_1, T_Z or T_p:

- Bretschneider (ITTC or ISSC) wave energy spectrum (for open sea)

$$S_{BS}(\omega) = \frac{A}{\omega^5} e^{-B/\omega^4}, \; A = 172.75\frac{\overline{H}_{1/3}^2}{\overline{T}_1^4}, \; B = \frac{691}{\overline{T}_1^4}$$

- JONSWAP wave spectrum(for limited fetch F)

$$S_{JS}(\omega) = \frac{Ag^2}{\omega^5}\exp\left[-\frac{5}{4}\frac{\omega_p^4}{\omega^4}\right](3.3)^r,$$

$$r = \exp\left[-\frac{(\omega-\omega_p)^2}{2\omega_p^2\sigma^2}\right], \; A = 0.076\left(\frac{U_{10}^2}{Fg}\right)^{0.22}, \; \omega_p = 22\left(\frac{g^2}{U_{10}F}\right)^{1/3}, \; \sigma = \begin{cases} 0.07 & \omega \le \omega_p \\ 0.09 & \omega > \omega_p \end{cases} \tag{6-1}$$

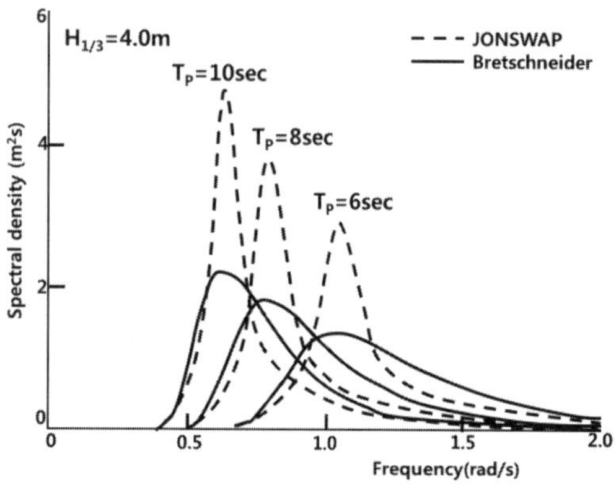

Figure 6-6 the wave energy spectra

where $U_{10}(m/s)$ is the wind speed at a height of 10 meters above the sea surface.

Figure 6-6 compares the Bretschneider and mean JONSWAP wave spectra for three sea states with a significant wave height $\overline{H}_{1/3}$ of 4 meters and peak periods $\overline{T}_P$ of 6, 8 and 10 seconds, respectively.

The area under the curve for a given frequency band is a wave energy per square meter contained in the band if multiplied by the specific weight ρg. The variance $m_0 = \sigma^2$ of the irregular wave time history is equal to the total area under the curve and some useful quantities can also be derived from m_0 and the first and second moments of the area about the vertical axis, m_1 and m_2. For example, for narrow band spectra:

$$\overline{H}_1 = 2.5\sqrt{m_0} \quad \overline{H}_{1/3} = 4.0\sqrt{m_0} \quad \overline{T}_1 = 2\pi\frac{m_0}{m_1} \quad \overline{T}_Z = 2\pi\sqrt{\frac{m_0}{m_2}}$$

In general, m_n ($n = 0, 1, 2 \ldots$) are called spectral moments.

Ocean waves closely connected with the design of ships and ocean structures are generated by the wind over a good fetch. Wind fetch is defined as the unobstructed distance that wind can travel over water in a constant direction; the faster the wind over a longer fetch, the larger the wind-generated waves. The duration of the wind is also influential to the development of winds. The Beaufort scale is an empirical measure that relates wind speed to observed conditions at sea or on land. It is based on observation rather than accurate measurement but is the most widely used system to measure wind speed today. **Table 6-1** shows the Beaufort scale at 12 degrees of sea conditions ranging from calm air to hurricane-force.

Table 6-1 Beaufort scales

Beaufort scale	Wind speed				Description	Wave height	
	km/h	mph	kts	m/s		m	ft
0	<1	<1	<1	<0.3	Calm	0	0
1	1-5	1-3	1-2	0.3-1.5	Light air	0.1	0.33
2	6-11	4-7	3-6	1.5-3.3	Light breeze	0.2	0.66
3	12-19	8-12	7-10	3.3-5.5	Gentle breeze	0.6	2
4	20-28	13-17	11-15	5.5-8.0	Moderate breeze	1	3.3
5	29-38	18-24	16-20	8.0-10.8	Fresh breeze	2	6.6
6	39-49	25-30	21-26	10.8-13.9	Strong breeze	3	9.9
7	50-61	31-38	27-33	13.9-17.2	High wind, moderate gale, near gale	4	13.1
8	62-74	39-46	34-40	17.2-20.7	Fresh gale	5.5	18
9	75-88	47-54	41-47	20.7-24.5	Strong gale	7	23
10	89-102	55-63	48-55	24.5-28.4	Whole gale/storm	9	29.5
11	103-117	64-72	56-63	28.4-32.6	Violent storm	11.5	37.7
12	≥118	≥73	≥64	≥32.6	Hurricane-force	≥14	≥46

2. Basic Equations of Ship Motions

Ships moving in ocean waves experience motions in six degrees of freedom, namely surge, heave, sway, roll, yaw, and pitch, as shown in **Figure 6-7**. More details on each component motion will be discussed later, but the rectilinear or translational components

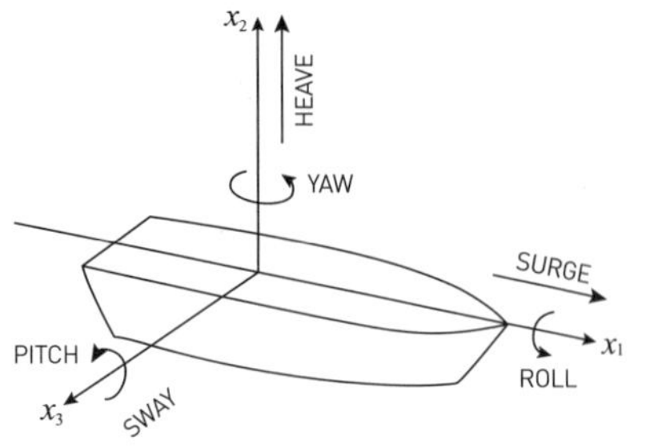

Figure 6-7 Ship motions in six degrees of freedom

will be examined in much of this section for simplicity, since the remaining three rotational components can be explained by analogy.

Small amplitude motions of a body can be modeled by a typical mass-spring-dashpot system whose behavior is generally described by

$$M\ddot{x} + b\dot{x} + cx = F(t) \tag{6-2}$$

where M is the system mass, b is the linear damping coefficient, c is the restoring force coefficient, and $F(t)$ is the force acting on the mass. If the body moves in a vacuum, no additional force is produced by its motion and the motion can simply be expressed by the equation

$$M\ddot{x} = F(t) \tag{6-3}$$

However, if the body moves in the water, the situation changes considerably and all the terms in **Equation 6-2** must be retained. In addition, a body accelerating or decelerating in fluid must accelerate/decelerate some volume of surrounding fluid with it. The inertia force required to accelerate/decelerate the volume of the fluid opposes the motion of the body and can be factored into the system as if the mass of the body is increased. This virtual increase of the mass is called the added mass or the virtual mass.

Therefore, for a ship moving near the water surface:

- The added or virtual mass a should be added to the mass M of the ship to take account of the inertia force required to accelerate the surrounding water.
- The force corresponding to the energy spent for the generation of waves around the hull is called the damping force or the wave damping force. In most cases, loss of energy due to the viscosity of the water is ignored as it is small compared to the wave damping. However, the viscous damping may play an important role in roll motions and for the motions of deeply submerged bodies.
- If the submerged volume of the ship and its shape undergoes a change due to a motion for example, the buoyancy and its distribution will change. This change results in a restoring force and moment to the ship motion. If the motion amplitude is small, the restoring force tends to, although not always, increase with the increase in the displacement from the equilibrium position.

In conclusion, a rectilinear motion of a ship in waves near the water surface can be described as the following equation

$$(M+a)\ddot{x} + b\dot{x} + cx = F(t) = F_{FK} + F_D \tag{6-4}$$

Note that the main exciting forces for the rectilinear motions are assumed to be linear sums of the Froude-Krylov forces, F_{FK}, and diffraction forces, F_D. The mass-spring-dashpot systems are the most frequently used model to describe the motion of a linear dynamic system and the solution can be found in various textbooks. The difficulties of the problem here are in the calculation of the coefficients a, b, and c and the exciting forces F_{FK} and F_D.

2.1 Wave Excitation Forces

As stated above, it is a common practice to approximate wave excitation forces as linear sums of the Froude-Krylov and diffraction forces. The Froude-Krylov hypothesis states that, if a body is assumed to be sufficiently small so that the presence does not affect the pressure field due to an incident wave, the diffraction effect can be ignored completely. Therefore, within the limitations of this simplification, the Froude-Krylov force can be easily estimated from the integral of the undisturbed pressure filed of the incident wave taken over the wetted surface of the ship.

However, the presence of the body in the fluid does affect the incident waves in real situations by adding a disturbance to them. The consequent fluctuation in the pressure field produces the diffraction force which can be calculated by the surface integral of the fluctuating pressure component determined by making the fluid in the wave regime satisfying an additional boundary condition on the motionless body surface. Therefore, the diffraction force is dependent upon the submerged shape of the body and the frequency and direction of the incident wave. Theoretical calculation of diffraction forces involves the solution of Laplace's equation subject to the boundary conditions on the free surface, bottom and body surface, but it is beyond the scope of this book.

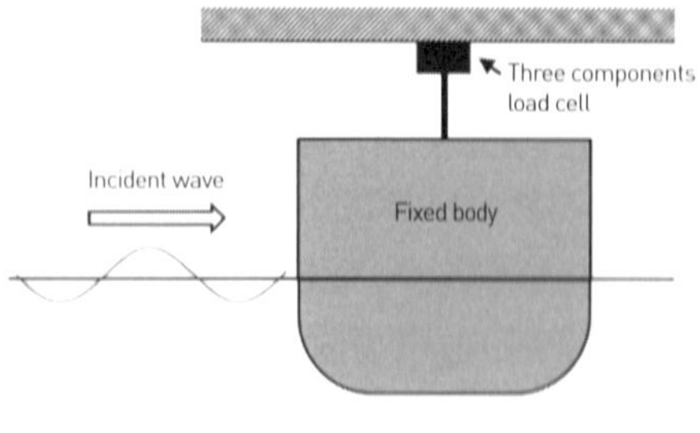

Figure 6-8 Schematic diagram of an experimental set-up for measuring wave excitation forces

An alternative to calculating wave excitation forces is measuring force components acting on a body fixed in waves of various lengths and incidence angles as shown in **Figure 6-8**. Clearly, the measured forces will be the sum of the Froude-Krylov and diffraction forces.

2.2 Added Mass and Damping – the Radiation Problem

Once wave excitation forces are known, it is necessary to determine the added mass coefficient a, damping coefficient b, and spring constant c to complete the motion equation. Of these three, the spring term corresponds to the hydrostatic restoring force, and this can be determined easily since hydrostatics is one of the most elementary topics of fluid mechanics. On the other hand, computation of added mass and damping force encompasses hydrodynamics and a deeper understanding of the physical phenomenon is required. When a body moves in an unbounded fluid, the fluid reacts to produce hydrodynamic forces acting on the body. These forces can be decomposed into a component which is in phase with the acceleration of the body and another in phase with the velocity. The former, when divided by the acceleration, is akin to the mass of the body and is consequently known as the added mass a. The latter is known as damping and it results from the waves generated by the body motion which radiate outward on the free surface. Since work must be done to oscillate the body and generate the radiating waves on the free surface, there is a direct relationship between the damping force and the amplitude of the generated waves. The quantity b is known as the damping coefficient.

There is also an experimental alternative to the theoretical approach. **Figure 6-9** shows a schematic diagram of an experimental setup for a body subjected to forced heave motion in quiescent water. The body is attached to a forced oscillator and the force produced by the motion is measured by a dynamometer.

Let us consider a body in a forced harmonic motion as shown in **Figure 6-9**. Then

$$z = Z_0 \sin \omega t \tag{6-5}$$

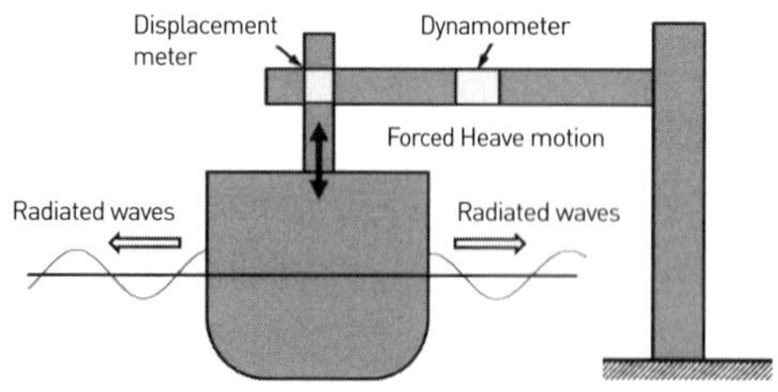

Figure 6-9 Schematic diagram of an experimental setup for measurement of wave excitation forces

where, Z_0 and ω are the amplitude and frequency of the motion, respectively, and given as input values. The force applied to produce the body motion in still water is measured by the dynamometer, in the form of

$$f = F_0 \sin (\omega t + \alpha) \tag{6-6}$$

where F_0 is the amplitude of the measured force and α is the phase difference with the motion.

From **Equation 6-4**, it follows that the equation of the motion is

$$(M+a)\ddot{z} + b\dot{z} + cz = F(t) = F_0 \sin(\omega t + \alpha) \tag{6-7}$$

In the equation, unknown quantities are the added mass a and damping coefficient b, and these values are to be found experimentally. For a ship in heave motion, M corresponds to the displacement of the ship and the hydrostatic or the restoring force coefficient c is the restoring force per unit change of draft, i.e. $\rho g A_w$, where ρg is the specific weight of the water and A_w is the waterplane area at the given draft.

Substituting **Equation 6-5** into **Equation 6-7** and expanding the sine function on the right side gives

$$Z_0[-\omega^2(M+a)\sin\omega t + \omega b\cos\omega t + c\sin\omega t] = F_0(\cos\alpha\sin\omega t + \sin\alpha\cos\omega t) \tag{6-8}$$

Equating the terms on both sides, the added mass and damping coefficient are found in the form

$$a = \frac{F_0\cos\alpha}{\omega^2 Z_0} - M, \qquad b = \frac{F_0\sin\alpha}{\omega Z_0}$$

Figure 6-10 shows added mass and damping coefficient for a family of two-dimensional rectangular cylinders heaving in deep water. It is clear in the figure that the added mass and damping coefficients depend on frequency ω and are greater than zero. The added mass coefficient is obtained by dividing the added mass by the body mass.

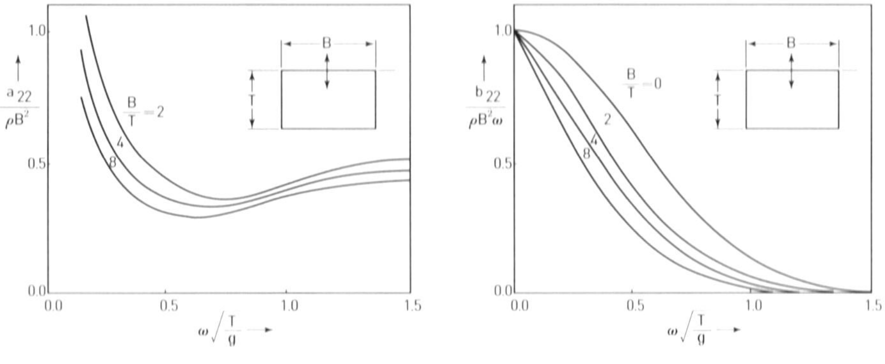

Figure 6-10 Added mass and damping coefficients for a family of two-dimensional rectangular cylinders, heaving in deep water: redrawn from Newman (1981)

Even when the body is submerged so deep as to generate no waves, the added mass does not vanish but has a positive value. On the other hand, the damping coefficient vanishes in this case, since it is the direct result of body motion-generated waves.

2.3 Motions of a Two-dimensional Body in Regular Waves

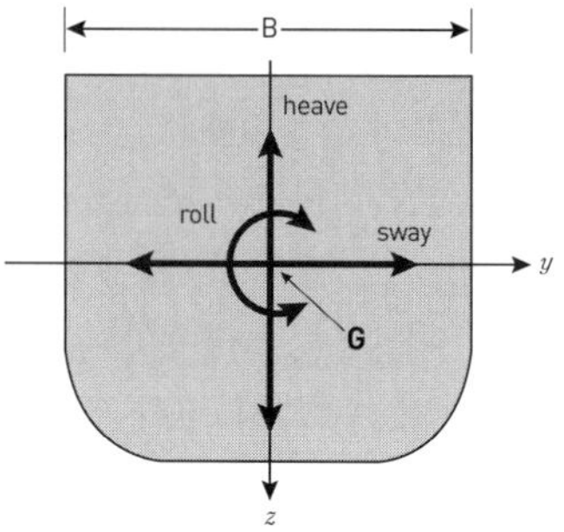

Figure 6-11 Motions of a two-dimensional body in regular waves

It has been shown that the coefficients in the equation of motion for a two-dimensional body moving in regular waves can be found experimentally. A two-dimensional body, infinitely long and having a constant cross section over its entire length, experiences three modes of motion, viz. heave, sway, and roll, in regular waves as shown in **Figure 6-11**. The procedures to find added mass, damping coefficient and wave exciting forces have already been discussed in the previous section and only the brief explanations on the characteristics of the motions are given below.

2.3.1 Heave

Heave is a linear vertical motion of a floating body. The vertical displacement z of the body is taken positive downward. If z is very small, the waterplane area A_w can be assumed not to change during the motion. Therefore, the restoring force coefficient c for the motion is $\rho g A_w$ as already mentioned in the previous section. For two-dimensional bodies waterplane area per unit length the beam (B) can be used instead of A_w.

2.3.2 Sway

Sway is a horizontal linear motion of a floating body. The restoring coefficient c vanishes since submerged part of the body does not change as it moves in this mode.

2.3.3 Roll

Roll is an angular rotational motion about the longitudinal axis perpendicular to the center of gravity G as shown in **Fig 6-11**. Therefore, roll motion is not governed by Newton's second law but by the law of conservation of angular momentum. It is also well known that angular momentum is conserved if linear momentum is conserved. The equation of roll motion of a two-dimensional body differs from those for translational motions in that each term in the roll equation is moment instead of force, and can be written in the form

$$(I+a)\ddot{\phi} + b\dot{\phi} + c\phi = M_0 \sin(\omega t + \beta) \tag{6-9}$$

where ϕ is an angle of rotation and I, a, b, c, and M_0 are the rotational counterparts of **Equation 6-9**, representing mass moment of inertia, added moment of inertia, damping moment coefficient, restoring moment coefficient and the amplitude of wave exciting moment, respectively.

The mass moment of inertia – or moment of inertia, I for short – of a two-dimensional body can be calculated by integrating the mass per unit length per unit cross sectional area times the square of the distance from the center of rotation over the section. If the radius of gyration k is known, the moment of inertia I equal mk^2 where m is the mass of the body per unit length. The radius of gyration is often assumed to be a quarter of the beam, i.e. $k = 0.25B$. The added inertia a and damping moment coefficient b can be determined by solving the radiation problem or performing experiments and wave exciting moment by solving the diffraction problem or experiments. Finally, the restoring moment coefficient c for the section in roll motion is $m \cdot g \cdot \overline{GM}$, i.e. weight of the body per unit length multiplied by the metacentric height.

From the above discussions it is clear that the most difficult task in the investigation of ship motions is finding the coefficients of the equations of motion. Measuring the coefficients or wave exciting forces/moments through experiments is impractical for routine calculations. Instead, theoretical methods are used to calculate the hydrodynamic coefficients and wave exciting forces, as described previously. The methods formulate boundary value problems in which Laplace's equation is solved under appropriate boundary conditions. A few numerical methods exist for the solution of the problems: multi-pole expansion and singularity distribution methods are among the most popular, whereas numerical methods such as finite element or boundary element methods are also frequently used.

3. Ship Motions in Regular Waves

3.1 Coordinates and Definitions

A ship experiences motions in six degrees of freedom in a seaway as shown in **Figure 6-12**: three linear translations, surge, sway and heave and three angular rotations, roll, pitch

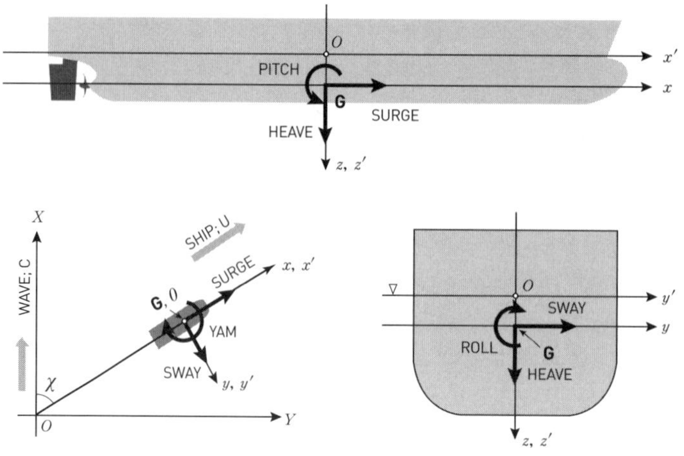

Figure 6-12 Coordinate systems for ship motions

and yaw. Of these, surge, sway and yaw do not cause any change in the submerged shape since they occur on the mean water surface, a plane perpendicular to z-axis.

Consider a ship moving with a speed U at an angle χ relative to the direction of a train of waves propagating with celerity c as shown in **Figure 6-12**. The speed of the wave relative to the ship is then c - $U\cos\chi$ and the corresponding encounter frequency ω_e is

$$\omega_e = \omega - kU\cos\chi \tag{6-10}$$

where ω and $k = \omega^2/g$ are the frequency and wavenumber, respectively. If the waves are incident forward of the beam ($90° < \chi < 270°$), the encounter frequency is higher than the wave frequency, and the two frequencies are equal for beam seas ($\chi = 90°, 270°$). For the remaining incident angles ($0° < \chi < 90°$, $270° < \chi < 360°$), $\cos\chi$ is positive and therefore the encounter frequency is lower than the wave frequency. The encounter frequency is zero if $c = U\cos\chi$ since $k = \omega/c$. It is interesting to note that the effect of the ship speed on the encounter frequency is identical to the Doppler shift in other fields of wave motion.

3.2 Strip Theory

In order to quantify the ship motions in waves direct measurements of models in wave tanks may be the most desirable whenever possible. However, time-domain simulations of ship motions through numerical solution of the Euler or Navier-Stokes equations

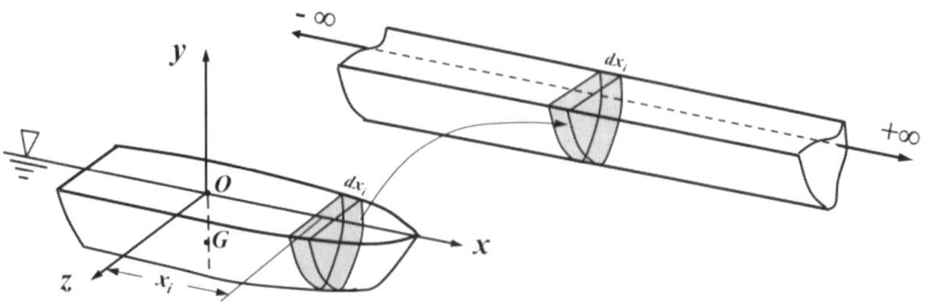

Figure 6-13 Two-dimensional section for strip theory

have become increasingly popular owing largely to the rapid progress in computational fluid mechanics, although these methods still call for a considerable expense of time and resources. Therefore, it is a common practice to solve the equations of motion with coefficients derived for two-dimensional sections, despite the fact that a ship is patently a three-dimensional body.

Let us assume now that a ship can be divided into a series of two-dimensional sections of small thickness, or strips, along its length and the total force and moment acting on the ship can be expressed by a linear summation of the force and moment acting on each strip. This approach is known as strip theory and is based on the assumption that a ship is slender, i.e. the draft and the beam are much smaller than the length and consequently mutual interference between the strips and, more broadly, all three-dimensional effects, including flow 'leaks' around the bow and stern, can be ignored. Despite this seemingly drastic assumption, the strip theory tends to yield reasonably good results, since ships are slender in most cases as shown in previous chapters. But the theory may yield unrealistic results if applied to offshore structures whose draft or beam is comparable to its length and high speed vessels with abrupt changes in cross sectional shapes. In these cases three-dimensional methods should be used.

To return to the strip theory, let us consider that a ship is sliced into n strips and the yaw and pitch motions are sufficiently small to treat the motions of each strip to take place in the plane perpendicular to the longitudinal axis. Then, the total force F_{ship} acting on the ship can be approximated with a sum of forces acting on each strip

$$F_{ship} = \sum_{i=1}^{n} f_i \cdot dx_i \tag{6-11}$$

where f_i are two-dimensional forces acting on i-th strip and dx_i the thickness of the strip.

There are various strip methods which are generally similar but differing in mathematical details: OSM (ordinary strip method), NSM (new strip method) and STF (Salvesen, Tuck and

Faltinsen) methods are the most well-known. In STF method, for example, wave diffraction force is found directly by solving the diffraction problem. However, the diffraction problem is quite difficult to solve and it is bypassed in OSM and NSM by employing the Haskind relation for reciprocity between the radiation and diffraction problems enabling the wave exciting force to be determined from the solution of the radiation problem. Results of the above methods are substantially the same.

Sway and yaw are weakly coupled to each other and they also influence surge motion, although surge does not affect the other modes at all due to the port and starboard symmetry of most ships. Heave and pitch motions are relatively strongly coupled to each other, but for small-amplitude motions, no coupling occurs between the motions and the other motions and (sway, yaw, roll) therefore they can be treated separately.

3.2.1 Heave and Pitch Motions in the Vertical Plane

Most strip theories ignore surge motion as an insignificant component, since effects of cross-flows in x-direction have been ignored along with other three-dimensional effects. The motion of equations for coupled heave and pitch can be derived by summing up the force and moment acting on each strip in the form

$$\text{Heave} \qquad \sum_{i=1}^{n}\left(f_{inertial}+f_{damping}+f_{restoring}-f_{FK}-f_{D}\right)_{z}\cdot dx_i=0$$

$$\text{Pitch} \qquad \sum_{i=1}^{n}\left(f_{inertial}+f_{damping}+f_{restoring}-f_{FK}-f_{D}\right)_{z}\cdot x_i dx_i=0 \qquad (6\text{-}12)$$

where subscript 'z' in the integrands denotes z-component of corresponding forces, and the moment arm x_i in the pitch equation is the distance from the i-th strip to the center of gravity of the ship.

Solving the two simultaneous equations for regular waves yields the heave and pitch response amplitude operators (RAOs) of the ship. Heave and pitch responses for a 175-meter long, 6500TEU class container ship is shown for various wave headings in **Figure 6-14**.

As seen in the figures, it is customary to non-dimensionalize the amplitudes of translational motions by the amplitude of incident wave A and rotational motions by the maximum wave slope k_A. The non-dimensional motion responses in frequency domain are called response amplitude operators (RAOs) and they are used for determining ship motions in irregular waves as well.

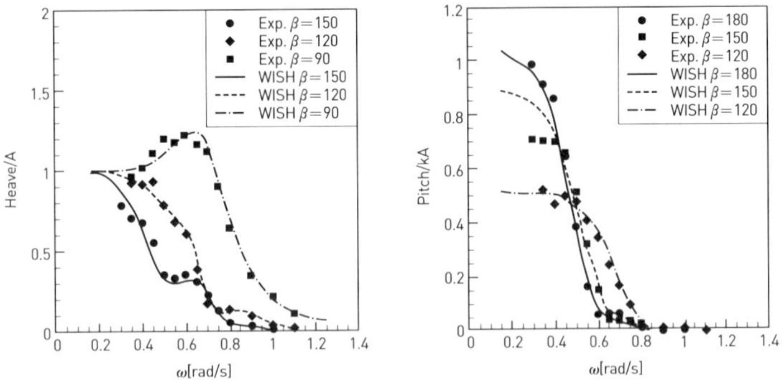

Figure 6-14 RAOs of heave and pitch motions at various wave headings (Song et. al., 2011)

3.2.2 Sway, Yaw and Roll Motions

In a similar manner as in the coupled heave and pitch, equations for coupled motions of sway, yaw and roll are given by

$$\text{Sway} \qquad \sum_{i=1}^{n}\left(f_{inertial}+f_{damping}+f_{restoring}-f_{FK}-f_{D}\right)_{y}\cdot dx_i=0$$

$$\text{Yaw} \qquad \sum_{i=1}^{n}\left(f_{inertial}+f_{damping}+f_{restoring}-f_{FK}-f_{D}\right)_{y}\cdot x_i dx_i=0$$

$$\text{Roll} \qquad \sum_{i=1}^{n}\left(M_{inertial}+M_{damping}+M_{restoring}-M_{FK}-M_{D}\right)_{x}\cdot dx_i=0 \qquad (6\text{-}13)$$

where subscript 'y' denotes y-components of the forces, and subscript 'x' denotes moments of forces about the x-axis. The solutions of the simultaneous equations for the 6500 TEU class containership are shown in **Figure 6-15** in non-dimensional forms.

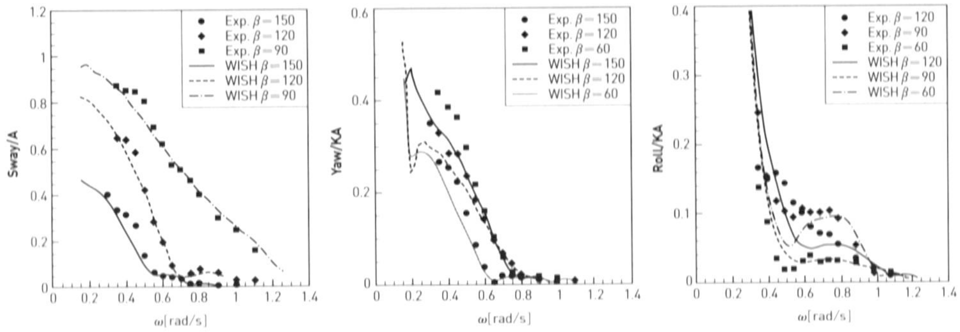

Figure 6-15 RAOs of sway, yaw and roll motions at various wave headings (Song et. al., 2011)

4. Ship Motions in Irregular Waves and Application to Design

Once the RAOs, the amplitude element of the transfer functions, are known, one can transform any wave energy spectrum to a corresponding motion energy spectrum by multiplying the square of RAOs. However, a ship sailing in irregular waves generally encounters waves of various amplitudes, frequencies and incidence angles. Therefore, the wave energy spectrum for a fixed point in the ocean needs to be transformed from a wave frequency domain to an encounter frequency domain first.

The motion energy spectrum obtained above is, therefore, a function of the encounter frequency ω_e and various relations derived in section 1.2.2 for the wave energy spectrum are still valid. For example, significant amplitude of heave motion is given by

$$\overline{Z}_{1/3} = 2.0\sqrt{m_0} \tag{6-14}$$

where the motion variance m_0 is the area under the motion energy spectrum curve. If $\overline{Z}_{1/3}$ exceeds twice the draft, there may be a good possibility of slamming occurring as the keel re-enters the water after emergence. If the significant heave amplitude exceeds the freeboard of the ship, deck-wetness is highly likely.

The motion energy spectrum in the encounter frequency domain may be transformed back into the wave frequency domain. The area under the spectral density curve in the wave frequency domain, and consequently the motion variance m_0, remains the same. However, the curve in the wave frequency domain is not a true motion energy spectrum and has no physical meaning.

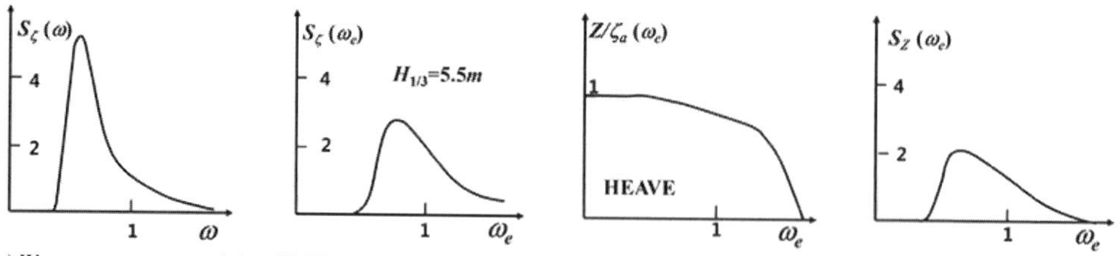

(a) Wave energy spectrum (ω) (b) Wave energy spectrum (ωe) (c) RAO (d) Motion energy spectrum

Figure 6-16 derivation of motion energy spectrum from wave energy spectrum: (d) = (c)^2 × (b)

4.1 Application to Design and Operation

Ships' hull forms must be designed to withstand the roughest waves expected in a planned service route. For this reason ship designers try to reduce the motion amplitudes of the ship, since the wave energy spectrum for a specific location in the ocean is already known. In other words, the designers attempt to find a hull form with the RAOs that produce the minimum area under the motion energy spectrum of the ship for given environmental conditions.

Once a ship is constructed, however, the hull form cannot be changed and RAO characteristics may be difficult to modify readily. Therefore, only the wave energy spectra determine the motion energy spectra of a given ship. The RAO information can be found in operation manuals prepared by the shipyard, usually stored in a computer. Thus, the ship's personnel or the shore staff can assess the risk the ship might be exposed to en route and adjust the speed and route accordingly, if they are provided with real time information on sea states or measured wave spectra.

5. Control of Ship Motions

Ships sailing in ocean waves inevitably experience motions. Therefore, ships and their systems are designed to tolerate the motions to a certain level. However, it is obvious that excessive motions are undesirable. This often calls for some effective methods for controlling ship motions, especially for passenger ships where passenger comfort is of prime importance and naval vessels which may have sensitive instruments and weapons systems. It was already shown at the beginning of this chapter that oscillatory motions of a ship in waves can be represented by the classical mass-dashpot-spring system model. Such systems often exhibit resonance phenomenon in which the response motion amplitudes become very high at or near the natural frequency if damping is small, but systems with very high damping have no resonant frequency and the maximum response amplitudes occur near zero frequency. Therefore, in principle, any ship motion can be reduced by simply increasing the damping, although an increase in damping of already highly damped systems will be ineffective.

Among the six degrees of freedom motions of a ship, roll is the only component which has low damping, because it alone does not involve large translational displacements. Therefore, wave-making damping of roll motions is very small for most practical hull forms and viscous effects are a more significant source of roll damping, especially at low speed. On the other hand, a ship with relatively sharp bilges or protruding appendages may experience extra damping in addition to skin friction.

Limited successes in controlling heave and pitch motions of small craft have been reported but this may not be possible for larger ships, since their inherent damping is so high and therefore unacceptably large forces are required to achieve the control.

5.1 Bilge Keels

Bilge keels have been used for a long time and are the most simple but very effective device for roll stabilization. As shown in **Figure 6-17**, bilge keels are long and narrow keels mounted at the turn of the bilges. Suppose, for instance, that a ship rolls clockwise in calm water. The bilge keels rotate in the same direction and meet the flow against their motions. Therefore, incident flows to the keels are in counter clockwise direction if observed in a reference frame moving with the keels. Now, the flow around the keel is analogous to the flow around a flat plate set perpendicular to the flow direction in that a drag force will be generated parallel to the incident flow due to the difference in pressures acting on the front and back surface of the plate. The drag force multiplied by the perpendicular distance from the point of action to the center of roll (usually taken to be the ship's center of gravity G) yields a counter-clockwise moment of force resisting the roll motion. Though using bilge keels offers a number of advantages, one disadvantage is that it increases the frictional and viscous pressure resistance of the ship, but the latter can be minimized by carefully aligning the keels with the streamlines around them.

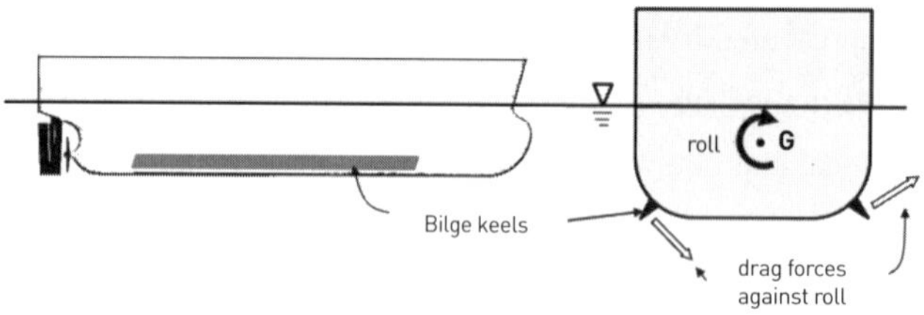

Figure 6-17 Sketch of bilge keels and damping against roll

5.2 Anti-rolling Tanks

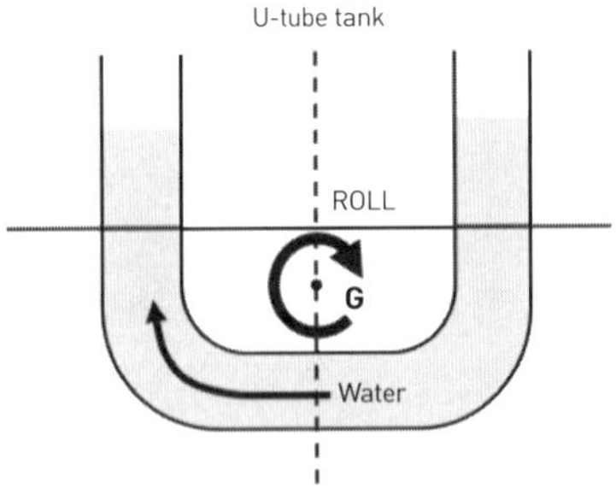

Figure 6-18 A sketch of a U-tube anti-rolling tank

An anti-rolling tank (ART) is basically a large flume or a U-tube tank fitted inside a ship containing fluid to produce a moment in antiphase (ideally) with the ship motion. As illustrated in **Figure 6-18** for a simple U-tube passive tank, the fluid inside the tank moves out of phase with the roll motion to produce a counter moment opposite to the roll excitation moment induced by the wave. However, it is not possible to achieve the desired goal at all frequencies of the roll motion, since the phase angle of the fluid movement is a function of the frequency. Therefore, it is customary to design a passive ART to produce a maximum stabilizing moment at the natural roll frequency of the ship, because the roll amplitude at resonance is usually the largest. Passive ARTs have some disadvantages. They occupy a considerable volume in a ship and reduce the metacentric height, consequently decreasing the stability of the ship. Furthermore, they tend to amplify the roll motions at low encounter frequencies and in the worst cases it may be necessary to block their functions completely. Active anti-rolling tanks have recently risen in popularity since the motions of the fluid can be controlled to allow roll stabilization in a comparatively wide range of motion frequencies. However, active ARTs are also heavy and take up much cargo space, and therefore they are used mostly in special purpose vessels, such as coastguard patrol ships.

5.3 Stabilizer Fins

As shown in **Figure 6-19**, active stabilizing fins are mounted on rotatable stocks at the turn of the bilge near the midship and moving in such a manner as to counteract rolling caused by waves. Stabilization is achieved through the moments created by lift forces of the fins which, as is widely known, vary with the angle of incidence and the forward speed of the fins (or the speed of water relative to the fins). Therefore, the angle of incidence of the fin is controlled by a device (e.g., gyroscope) sensitive to the roll motion of the ship to develop appropriate roll stabilizing moment. Let us consider, for example, a ship rolling to starboard (inclined to the right). The angle of incidence of the right fin must be modified so as to increase the lift and at the same time the left fin must be rotated to have the same incidence angle but in the

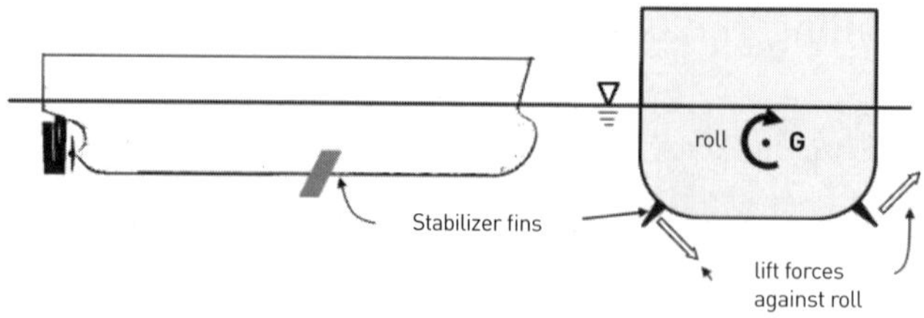

Figure 6-19 A sketch of stabilizer fins

opposite direction. This ensures the moment produced by the fins opposes the wave-excited roll moment. The vertical components of the lift will cancel, but the horizontal components add up to yield some sway force and yaw moment.

The performance of active stabilizer fins is dependent on the lifts produced by the fins which are proportional to the square of incident flow velocity. Therefore, active fins are effective for high speed ships. They can also be applied to control hull attitude (trim) and pitching of the ship simultaneously with rolling, if two pairs of fins are installed near the bow and stern.

6. Other Major Issues Associated with Ship Motions

Understanding the various phenomena arising from motions is important for safe and proper operation of the ship. This section discusses the major phenomena and problems related to ship motions.

6.1 Wave Loads

A ship moving in waves are subjected to wave loads. As shown in **Figure 6-20**, the pressure force acting on local area elements (segments) of the hull surface are called 'local loads' and the resultant force over the whole wetted surface of the ship is called 'global wave loads.' The pressure acting on the wetted surface of a ship moving forward in waves constantly changes with the relative motion between the ship and waves. The local loads are the results of impulsive or changing pressures acting on the wetted surface, such as due to slamming.

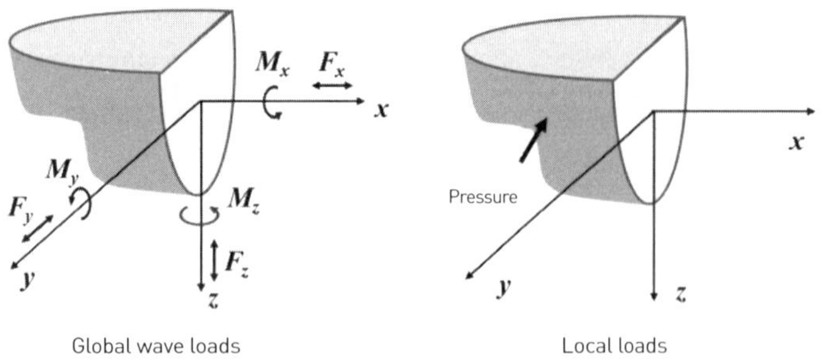

Figure 6-20 Wave loads acting on a ship

The global loads can be computed by a strip method, for example, in which the hull is sliced into several two-dimensional sections perpendicular to the x-axis. The force and moment acting on each section is calculated by integrating the local loads over the surface of the section. By integrating the sectional forces and moments over the entire length, the longitudinal distribution of the shear stress, bending moment and torque acting on the hull are found. These are known as the primary hull loads. While the local loads cause local damage to the hull, the primary hull loads cause the hull to deform or even to break if excessive. That is why the primary hull loads are used as critical information in structural design or in predicting and mitigating vibration and noise of a ship.

In fact, local loads are not as important as the primary hull loads in structural design of a ship, except the impact loads due to slamming. In principle, when two ships with identical principal dimensions but different hull forms are moving in the same waves, the ship experiencing larger motion experiences less wave loads. In other words, reduced ship motions provide better passenger comfort in return for higher burdens on the ship structure. Therefore, an optimum hull form will be a compromise between motion characteristics and structural strength.

6.1.1 Wave Loads Acting on Monohulls

The primary loads acting on a certain section of a monohull are shown in **Figure 6-20**. In general, when a ship sails in head seas (waves travelling in the opposite direction to the ship), both the vertical shear stress F_z and the vertical bending moment M_y reach their respective maximum when $\lambda/L \approx 1.0$, i.e. when the wavelength is almost the same as the ship's length. That is why the loads for $\lambda/L = 1.0$ are considered the most important parameter in the design of hull structures. Meanwhile, if a ship encounters quartering waves (moving in a direction approximately 45° or 135° to the ship's heading) the frequency of which is nearly the same as the natural roll frequency of the ship, then the horizontal shear stress F_y, the horizontal bending moment M_z and the torque M_x become more important in the range of λ/L value between about 0.3 to 0.4. Accidental ruptures of the hull by such loads have been occasionally reported on a ship whose deck is wide open, such as timber carriers.

6.1.2 Wave Loads Acting on Twin Hulls

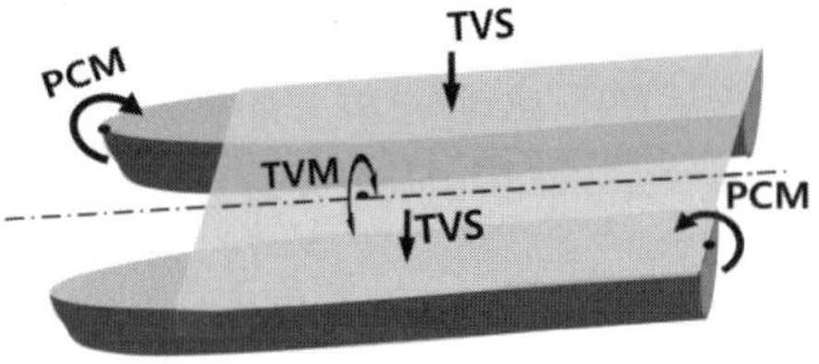

Figure 6-21 Additional wave loads acting on a catamaran

As shown in **Figure 6-21**, wave loads acting on catamarans consisting of two parallel hulls of equal size or small waterplane area twin hull (SWATH) vessels supported by two submerged bodies have three additional components to those acting on monohulls, namely, transverse vertical shear force (TVS), transverse vertical bending moment (TVB) and pitch coupling moment (PCM).

6.2 Added Resistance

In general, the resistance of a ship refers to the resistance at the design speed in calm water as a rule. However, motions add to the resistance and therefore the actual resistance of the ship will change over time according to the wave characteristics as illustrated in **Figure 6-22**. The difference between the mean resistance and the resistance in calm water is called 'added resistance' and amounts to 20 to 30% of the total resistance of a ship moving in high waves. Unlike wave-making resistance, which results from the work done by a ship in wave-generation, the added resistance occurs due to the motion of the hull relative to the incident waves. In other words, resistance of a ship increases in waves because of the diffraction and radiation of waves caused by the hull movements. Therefore, the added resistance in waves is proportional to the square of the relative motion amplitude and hence to the square of the wave amplitude. It is plain that the peak values of the added resistance will occur when the relative motions are at the maximum. The speed of a ship often fails to reach the design speed in waves for a given engine power because of the added resistance and reduced propeller efficiency due to waves, and the resistance of wind accompanying the waves. This phenomenon is known as the involuntary speed loss in waves and differs from the speed loss by resistance increase due to current or marine fouling.

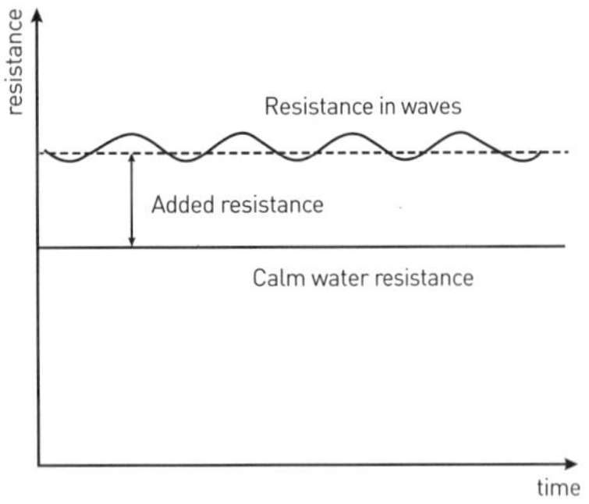

Figure6-22 Added resistance in waves

6.3 Slamming and Deck Wetness

Excessive relative motions between a ship and the water surface can lead to slamming or deck wetness as shown in **Figure 9-3**. Slamming is an undesirable impact mostly occurring when the bow of the ship is exposed in the air and collides with the water surface as it re-enters the water. The local impact loads on the hull surface colliding with the water may cause elasto-plastic deformation of the primary members of the ship structure and can eventually lead to severe structural damages. Even impacts which in themselves do not pose a serious risk can lead to fatigue failures of the structure, if the hull is pounded repeatedly for a considerable period of time. Deck wetness may occur anywhere on the deck but severe wetness usually occurs at the bow, from slamming, since some of the water thrown up by the slamming and scooped up by the submerged bow will be shipped onto the deck due to the forward speed of the ship.

The shipped water poses threats to the deck fittings, cargo and crew, and can even cause the ship to capsize due to the weight of the water on deck in extreme cases. Therefore, ship operators must reduce the speed or adjust the direction of travel in high seas to avoid slamming or deck wetness. Fundamentally, however, a hull structure must be designed to have sufficient strength to resist such impact loads.

6.4 Broaching-to

When a ship is poised on the steep forefront of a high wave in following or quartering sea conditions, the ship can be accelerated to ride on the wave. This phenomenon most frequently occurs when the ship's heading is within $\pm 15°$ and the wave celerity is almost the same or a little bit higher than the ship speed. This is known as surf-riding. In this situation the so-called broaching-to phenomenon may occur, which can lead to the ship capsizing as a result of a sudden change of the ship's heading and unexpected large heeling.

According to the International Maritime Organization's (IMO's) Maritime Safety Committee, surf-riding and broaching-to may occur when the heading θ is within $\pm 45°$ and the ship speed (in knots) is higher than $1.8 \cos \theta$ times the square root of the ship's length (in meters). Therefore, small boats are more vulnerable to surf-riding and possible broaching-to. To avoid this problem the ship speed, the course or both should be taken out of the dangerous region.

6.5 Hydroelastic Behavior of Ships

As previously stated, a ship moving in still water or in waves is regarded as a rigid body and elastic vibrations of the ship are ignored when dealing with the main ship motions. However, if the ship experiences impulsive loads through slamming, the elastic behavior of the ship must be taken into account along with the rigid-body motions. This is called the hydroelastic behavior of a ship, typically represented by the springing and whipping phenomena, and is an area where further developments are expected in the field of ship motion.

Recent advancements in computational fluid and structural dynamics have drawn academic attention to such fluid-structure interaction (FSI) problems.

6.6 Nonlinear Motions and Three-dimensional Methods

The aforementioned methods for analysis of ship motion are based on the assumption that a ship is subjected to incident waves of small amplitudes and, accordingly, the induced ship motions are also small. More specifically, linear theories assume that the wetted surface (submerged surface) of the moving ship does not change in size or shape, if both amplitudes of the ship motion and the wave are small.

However, as the amplitudes of the wave and the motion increase, changes in the wetted surface can no longer be ignored. Linear theories cannot deal with such cases, and in order to carry out more accurate analyses of large ship motions it is necessary to introduce non-linear analysis methods.

Furthermore, the two-dimensional strip methods cannot be applied to general bodies which are not slender like ships; a good example is offshore structures. Therefore, to deal with the motions of such bodies, three-dimensional methods are required to calculate hydrodynamic forces acting on the body as a whole instead of two-dimensional cross-sections. It can be said, in conclusion, that ultimately theoretical hydrodynamics in ship motions may be required to perform three-dimensional analyses on nonlinear ship motions by taking the hydroelasticity into account.

7. Ship Maneuverability

7.1 Importance of Ship Maneuverability

The recent rapid increases in marine traffic and ship speed have led to a high frequency of accidents of collisions and groundings. These accidents have become major causes of marine pollution due to oil spills as well as of injuries and property losses. Water and air pollution on land also worsens the sea environment and the self-purification action of the sea is very important for cleaning and global recirculation of the air and water. A clean sea is, therefore, mankind's last environmental recourse and an important resource that we must hand down to the next generation.

Realizing this, IMO has tightened various regulations concerning conservation of marine environment. One of the actions that it has taken is to make it mandatory to design and build ships of good maneuverability. As a result, although a ship may be delivered even though it fails to achieve a contract speed as long as appropriate penalties are paid, the ship can never be delivered if its maneuverability does not meet the IMO standards. The standards are divided largely into dynamic stability, and course-keeping, initial turning/course-changing, yaw checking, turning and stopping abilities as shown in **Table 6-2** [ITTC, 2002].

Table 6-2 Overview of IMO standards for ship maneuverability

Measure of Maneuverability	Criteria and Standard	Maneuver	IMO Standard
Turning Ability	Tactical Diameter: TD	Turning Circle	$TD < 5L$
	Advance: Ad		$Ad < 4.5L$
Course Changing and Yaw Checking Ability	First Overshoot Angle: $\alpha 10_1$	10°/10° Zigzag test	$\alpha_{10,1} \leq \begin{cases} 10.0° & \text{if } L/V < 10s \\ 5.0° + 0.5°\,L/V & \text{if } 10s < L/V < 30s \\ 20.0° & \text{if } L/V > 30s \end{cases}$
	Second Overshoot Angle: $\alpha 10_2$		$\alpha_{10,2} \leq \begin{cases} 25.0° & \text{if } L/V < 10s \\ 17.5° + 0.75°\,L/V & \text{if } 10s < L/V < 30s \\ 20.0° & \text{if } L/V > 30s \end{cases}$
	First Overshoot Angle: $\alpha 20_1$	20°/20° Zigzag test	$\alpha_{20,1} \leq 25°$
Initial Turning Ability	Distance traveled before 10° course change: l_{10}	10°/10° Zigzag test	$l_{10} \leq 2.5L$
Stopping Ability	Track Reach	Crash stop	$TR < 15L^{(1)}$

L: ship length (m), V: test speed (m/s), $^{(1)}$ for large, low powered vessels, $TR < 20L$.

The course-keeping ability is an ability of a ship to keep a predetermined course without excessive rudder changes. An acceptable course-keeping ability can therefore be achieved even for a ship which is marginally unstable. The initial turning ability describes the change of heading per distance traveled when a turn is initiated, and the yaw checking ability is a measure of the ship's response to counter-rudder, i.e. rudder deflected to the opposite direction, when already in a turn. The turning ability is the ability to turn the ship at a maximum design rudder angle and is measured by the advance at 90° change of heading and the tactical diameter at 180° change of heading. The stopping ability measures how long it takes for the ship to stop with the engine at full astern.

7.2 Dynamic Stability of a Ship

A ship is designed to run a straight course when its rudder angle is brought to the neutral position, which may not be equal to zero degree for a single screw ship since its propeller rotates in one direction. The stability of a ship is defined as an ability to return to a state of equilibrium when momentarily disturbed by an external force or moment without any corrective rudder. The control-fixed motion stabilities of a ship are usually classified into three types as shown in **Figure 6-23**. Straight-line stability of a ship implies that the final path of the ship, after a disturbance, is straight. The direction of the path, however, generally differs from the old one. A ship is said to be directionally stable if the new path is parallel to the initial path after a disturbance. If the ship returns to its original path, the ship is said to have positional motion stability.

A ship that is directionally stable or has positional motion stability tends to return to an equilibrium position so that it experiences difficulty in making a maneuver. By contrast, an unstable ship is often easy to maneuver but requires more energy spent to keep the predetermined course. Therefore, a compromise between stability and maneuverability must be made. Straight-line stability is desirable for maneuvering of a ship, especially for manual steering, but it must be borne in minds that a degree of automatic control enables proper maneuvering of even an unstable ship.

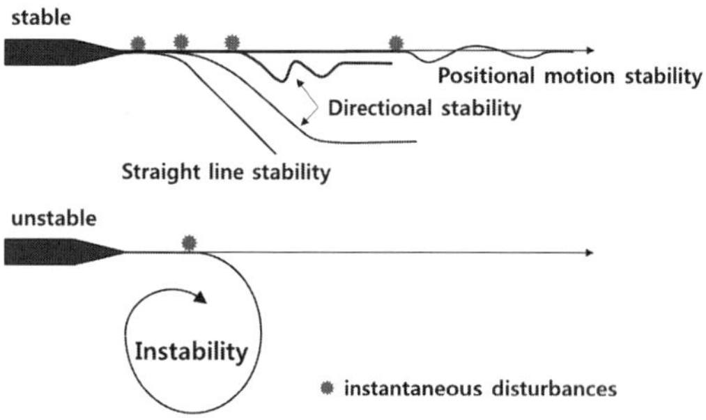

Figure 6-23 Dynamic stability

7.3 Definitive Maneuvers

Maneuvering performance of a ship is judged by using a number of maneuvering criteria, as shown in **Table 6-2**, which have been devised to establish the basic stability and control characteristics of a ship.

7.3.1 Dynamic Stability: Direct and Reverse Spiral Tests

Direct spiral tests are performed to identify the directional stability characteristics of a ship. In the tests, rudder angles are changed by small increments and steady yaw rates are measured as a function of the rudder angle.

For a directionally stable ship, only one yaw rate can result for any given rudder angle, whereas for an unstable ship there are regions where more than one yaw rate for a given rudder angle exist as shown in **Figure 6-24**. The ship, unstable in straight line motion tends to continue turning against its rudder in the region bounded by the dotted lines in **Figure 6-24**(b) where the rudder angle is small. A further increase in rudder angle beyond the point results in the yaw rate of the ship abruptly changing its sign and the ship suddenly starts to turn in the opposite direction.

The reverse spiral test is an alternative to the direct spiral test in which the ship is steered at a constant yaw rate and the mean rudder angle required to produce the rate is measured. This is difficult to achieve manually and consequently an autopilot is usually used. Results for a directionally stable ship are similar to those from direct spiral tests. For an unstable ship, the shape of the hysteresis loop is still identifiable with the dashed curve in **Figure 6-24**(b).

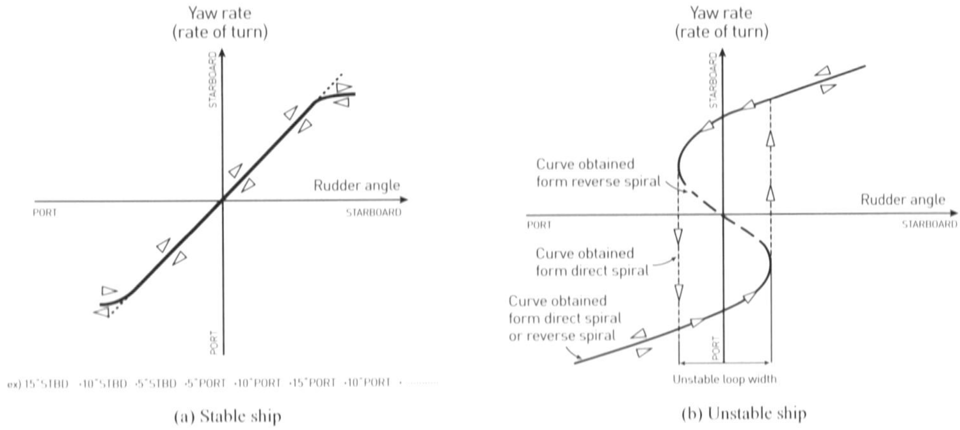

(a) Stable ship (b) Unstable ship

Figure 6-24 Relationship between yaw rate and rudder angle for stable and unstable ships

7.3.2 Initial Turning/course Changing and Yaw Checking Ability: Zigzag Tests

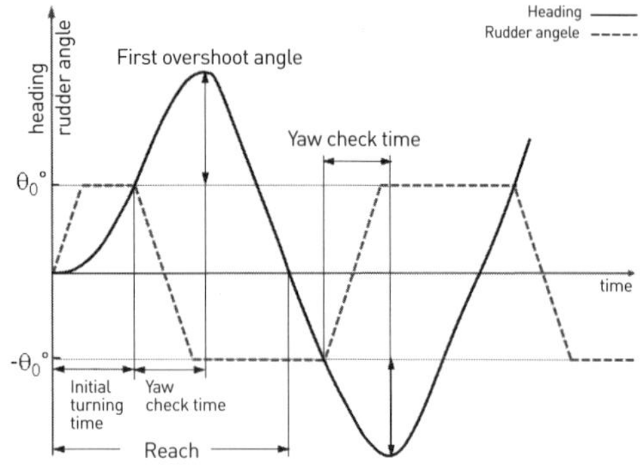

Figure 6-25 A result of zigzag tests (Z maneuver)

Next to the spiral tests in importance are the zigzag tests also known as the Z maneuver. The results of the tests indicate the capability of the rudder to control the ship. Overshoot angles shown in **Figure 6-25** are indicative of course changing and yaw checking abilities of the ship's rudder. The zigzag tests along with speed trials are one of the most important tests to confirm performance of a newly launched ship and must be conducted in a sea area before delivery. Therefore, IMO has established guidelines on the limits of overshoot angles of a ship and recommends verifying them through 10°/10° and 20°/20° zigzag maneuvers. For instance, during a 10°/10° zigzag maneuver test, the rudder is set at 10° starboard while the ship turns starboard. If the ship's heading (yaw angle) reaches 10° starboard, the rudder is turned to 10° port. The ship will fail to respond immediately and will continue to turn starboard despite the sudden change in the rudder angle. After a while the ship begins to turn to port and the ship's heading at that point minus the heading at the point when the rudder was put over to port is defined as the overshoot angle. There are standards on the acceptable first and second overshoot angles for the 10°/10° zigzag test, whereas only the first overshoot angles are specified for 20°/20° zigzag tests. If an overshoot angle exceeds the limit, the ship is judged to have a poor course-changing ability.

7.3.3 Turning Ability: Turning Circle Tests

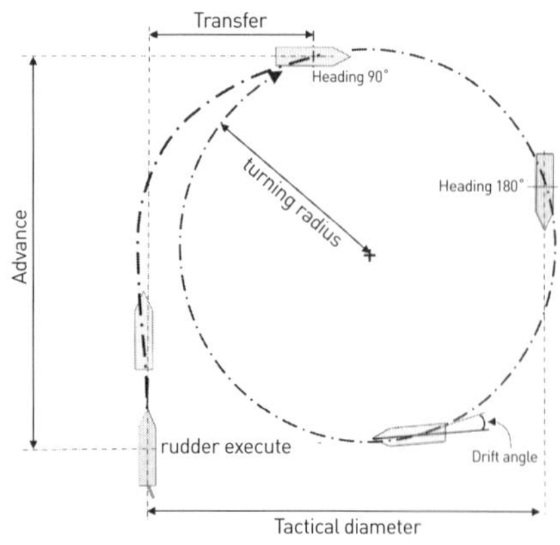

Figure 6-26 Turning circle maneuver

The turning ability of a ship is the measure of its ability to turn at the maximum rudder design angle (not exceeding 35°) at the test speed or other means of directional control. As shown in **Fig 6-26**, the key numerical information obtained in a turning circle maneuver includes advance, a distance travelled by the ship in the direction of the original approach while the ship goes through 90° change of heading from the time of the rudder execute, transfer which is the distance travelled by the ship in the transverse direction of the original approach while the ship goes through 90°

heading, and tactical diameter which is similar to transfer but at 180° change of heading. In addition, the speed lost in a turn and maximum roll angle, as well as the peak and final yaw rates, should be recorded. The steady turning ability of a ship can be simply understood as the yaw rate at a given rudder angle. Suppose that a ship A turns with the yaw rate, or yaw angular speed, of 3°/sec at a rudder angle of 5° while another ship B yaws at the rate of 4°/sec at the same rudder angle. Then the tactical diameter of B will be shorter than that of A and B is considered to have better turning ability than A.

7.3.4 Stopping Ability: Crash Stop Tests

A ship is often forced to come to a stop when the risk of imminent collision or grounding is foreseen during a voyage. Stopping ability of a ship is measured by the track reach, the distance the ship travels along its track from the moment 'full astern' order is given until the ship is stopped dead in the water, and the head reach, the distance along the original approach direction of the ship during the same duration as shown in **Figure 6-27**. The IMO criteria in stopping ability are given in **Table 6-2**.

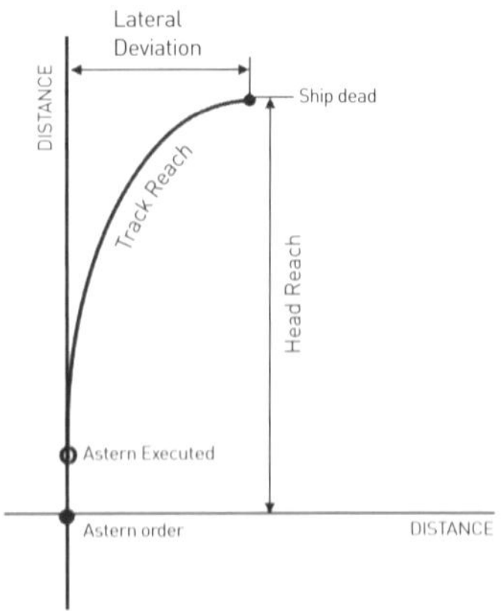

Figure 6-27 Crash stop

8. Evaluation of Ship Maneuverability

8.1 General Consideration

As discussed earlier, the motions directly related to ship maneuverability such as surge, sway and yaw occur on a horizontal plane, and therefore the volume and shape of the submerged part of the ship, and consequently the magnitude and distribution of the buoyancy, does not change during these motions. In the absence of the restoring mechanism for these motions, motion control is obtained through the actions of rudders in most cases.

There is a time delay between a rudder deflection and the ship's response. This makes a ship behave differently depending on the turning rate of its rudder: if a rudder oscillates at a low rate, the ship may respond well to the rudder changes but may not respond at all if the rate increases above a certain limit. In other words, a ship's response depends on the turning rate of its rudder and such a characteristic may bring on the risks that make the ship's voyage dangerous.

Stopping of a ship takes considerable time, since a ship, unlike automobiles, does not rely on ground friction or brakes, but use fluid resistance acting between the ship and water to reduce speed. Therefore, even with the propeller full astern, stopping distance could be too long for a ship in emergency to avoid collision or grounding.

The forces acting on a ship while maneuvering can be divided largely into those acting on the hull below the water surface and those acting above.

1) Forces acting on the hull below the water surface
   - Forces induced by ship motion
   - Propeller thrust
   - Rudder force
   - Current force, mooring force, etc.

2) Forces acting on the hull above the water
   - Wind force, etc.

Clearly the most important among those are the forces induced by the ship motion, the thrust produced by the rotation of the propeller, and the rudder force. Once these forces are known, it is possible to estimate the maneuverability of the ship. The thrust of the propeller and rudder force in uniform flow can be predicted with high accuracy these days. On the other hand, the thrust of a propeller working in open water is different from the force generated by the propeller placed close to the stern. Therefore, further studies are required to incorporate the effects of wake and hull-propeller-rudder interaction. As for the rudder, considerations must also include the effect of propeller slipstream. Some of the existing prediction formulae incorporate such effects, and they are considered reliable in a practical sense. The forces acting on the hull during the ship's maneuvering motion, however, are difficult to predict accurately, because they are functions of too many variables such as the principal dimensions, shape and speed of the ship and appendages.

8.2 Forces Arising from Ship Motions

This section discusses the methods for finding forces exerted on a ship during maneuvering. This is the most important process in the study of ship maneuverability. As pointed out a number of times already, ship motions relevant to maneuvering lack inherent restoring mechanisms and are mostly governed by hydrodynamic forces. However, as water is a viscous fluid, it is extremely difficult to calculate theoretically the hydrodynamic forces acting on the hull. There are, nonetheless, some empirical and analytical methods frequently used to approximate the force as discussed below.

8.2.1 Model Tests

The hydrodynamic forces and their derivatives acting on a scaled model of a ship are measure for various drift angles, rates of turn, and towing speeds. Planar motion mechanism (PMM) tests are conducted by towing a model ship with its heading either fixed at a given angle (drift angle) to the direction of the travel or oscillating in a certain pattern. The oscillation of the drift angle is controlled so as to emulate pure sway motion (by transversely reciprocating the bow and stern concurrently) or pure yaw (by transversely reciprocating the bow and stern with a certain phase angle). It is shown in the first diagram of **Figure 6-28**.

Rotating arm tests (CMT) are performed with a model mounted at a given radial location of an arm rotating at a constant rate (the second diagram of **Figure 6-28**). As shown in **Figure 6-29**, the test results can then curve-fitted using the least square method, for example, and be expressed as empirical formulae. The use of hydrodynamic forces and coefficients obtained through these tests, however, are limited to the tested ship and its geometrically similar hull forms. New tests must be performed to find the forces and hydrodynamic derivatives for any other ships.

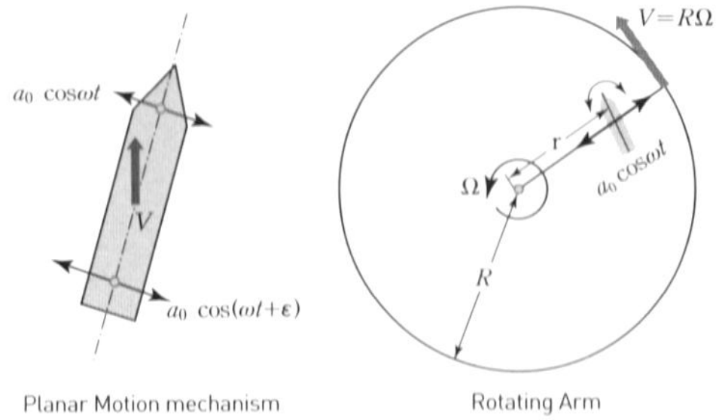

Figure 6-28 Planar motion and rotating arm test mechanisms

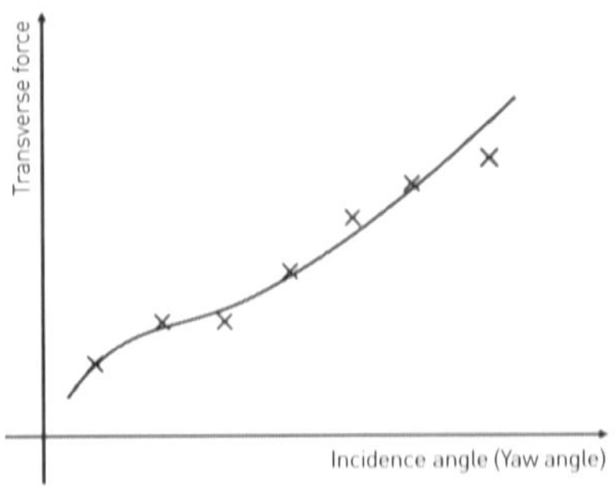

Figure 6-29 Least square curve fitting for construction of an empirical formula

8.2.2 MMG Models

In 1972, the mathematical modeling group (MMG), consisting of Japanese maneuverability experts, suggested a mathematical model to estimate hydrodynamic forces and their derivatives. The MMG models have been made public and over the ensuing decades the MMG models have been widely used to obtain the hydrodynamic forces and their derivatives of many ship types and models as functions of major parameters. Such parameters include the ship's yaw rate, speed, length, beam, draft and block coefficient (C_B) among others. The models are simple and easy to use, but they do have limitations including the failure to consider the variations in local shape geometry (e.g., installation of appendages or minor changes in bow shapes).

8.2.3 CFD-based Calculations

Since ships maneuver very slowly, viscous forces prevail over inertial forces, and analytic methods find it difficult to deal with these forces in a closed form. However, recent development in computer technology has led to some successful efforts to find numerical methods to calculate the viscous flow field around a ship while maneuvering. However, the calculations require considerable resources and the solutions thus obtained have yet to be validated. Nevertheless, the CFD methods are expected to become of practical use in the foreseeable future.

8.3 Maneuvering Simulation and Sea Trials

As stated already, IMO has become stricter regarding the maneuverability standards ever since marine accidents have been recognized as a primary cause of marine pollution as well as injuries and damages. In addition, ship-owners are increasingly interested in predicting the maneuverability of their ships in advance at an early stage of design. It is a common practice for the owners to require the shipyards to ensure that the maneuverability of their ships meet the IMO standards from the very beginning of the design process. To comply with the requirements, the shipyard normally examines the maneuverability of the ship being designed at an early design stage through computer simulations. The simulations should therefore yield estimations of all the characteristic performance parameters required by the IMO standards, such as turning, course changing and stopping abilities, based only on principal dimensions and the operating conditions. Ship designers then must adjust the ship's principal dimensions, rudder area and rudder shape, among others, to comply with the IMO maneuverability standards.

Upon the launch of a newly built ship, the shipyard should conduct a sea trial to verify its performance. The sea trial has two primary goals: one is to determine whether the ship achieves the design speed and the other is to determine whether the ship meets the IMO maneuverability standards, although the checklist includes noise and vibration levels, motion characteristics in waves and so on. As pointed out earlier, maneuverability of a ship is predicted for calm waters but it will surely be affected by current, wind, and waves in real sea conditions. Therefore, sea trials must be conducted in sea conditions as calm as possible. The effects of wave, current, and wind on ship maneuverability remain to be investigated more thoroughly in the future.

9. Control of Ship Maneuvering

Ship control collectively refers to the methods used to steer a ship on a desired course and heading. In order to achieve this it is essential that the current position of the ship is determined. By comparing the current position to a position at a short time past, the sailors are able to determine if the ship is on a desired course. Of course, any deviation to the course can be corrected by applying rudder forces.

In the past ships sailing in an open ocean had to rely on celestial navigation which uses the apparent positions of celestial bodies in the sky. For these purpose vast nautical almanacs, detailing their positions at each hour of the year, had to be published at regular intervals. Those sailing in sight of land are able to fix their position by measuring relative angles to key landmarks appearing on a map. This is known as geonavigation.

More recently, the radio direction finders (RDF), which use directions to the known radio beacons, became a useful navigation tool for ships and aircraft. However, modern navigation usually relies on global positioning systems (GPS) which uses a series of satellites observable from any place on the earth. The introduction of these electronic methods has allowed the use of autopilots which adjust ship's heading to maintain its predetermined course. Reliable and effective autopilots have also freed the helmsmen from the relatively mundane task of having to watch the ship's course constantly and making adjustments to the rudder continually while the ship is on a straight line course. There are, however, a few important points to consider when designing an autopilot. If a ship's heading is automatically corrected as soon as the ship is seen to be off course, the ship may navigate along the shortest path, but its fuel efficiency can suffer, because changing the heading and adjusting rudder deflection consumes energy and increases ship resistance. Therefore, the design of an autopilot must be optimized for the following three factors:

- the increase in travel distance due to being off-course;
- the ship motion arising from steering and the subsequent increase in resistance; and
- the increase in resistance caused by the rudder deflection.

As previously stated, a ship responds very slowly to steering and its maneuverability depends on the turning rate of its rudder, even when steered to the same angle. For example, if a rudder is turned 10° starboard and then 10° to port in one second, the ship may not respond; if the same action is taken over five minutes, the ship may respond nicely to the rudder deflection. The response to steering varies from ship to ship and, therefore, the characteristics must be incorporated when designing a ship control system. Details of control theories and evaluation of control performance are beyond the scope of this book, but interested readers are referred to the many publications on this important subject.

References

1. ITTC, 2002. Recommended Procedures and Guidelines: Maneuvering Committee. In: *23rd International Towing Tank Conference*. Venice, Italy 8~14 September, 2002. ITTC: Venice.
2. Lloyd, A.R.J.M., 1989. *Seakeeping*, Chichester: Ellis Hormwood Ltd. (distributed by Halasted Press; a division of John Wiley & Sons).
3. Newman, J.N., 1977. *Marine Hydrodynamics*. Cambridge: MIT Press.
4. Song, M.J. et al., 2011. Numerical analysis and validation of weakly nonlinear ship motions and structural loads on a modern container ship, *Ocean Engineering*. 38, pp.77-87.
5. The Society of Naval Architects and Marine Engineers (SNAME), 1971. *Principles of Naval Architecture*. Translated to Korean by Yim, S.J., 1991. Seoul: Daehan Textbooks Co., Ltd.

CHAPTER 7

Propulsion Plant

1. Development of Marine Propulsion Systems
2. Types of Propulsion Machinery
3. Shafting System for Ship Propulsion
4. Fuels
5. Future Technologies for Power and Propulsion Plant

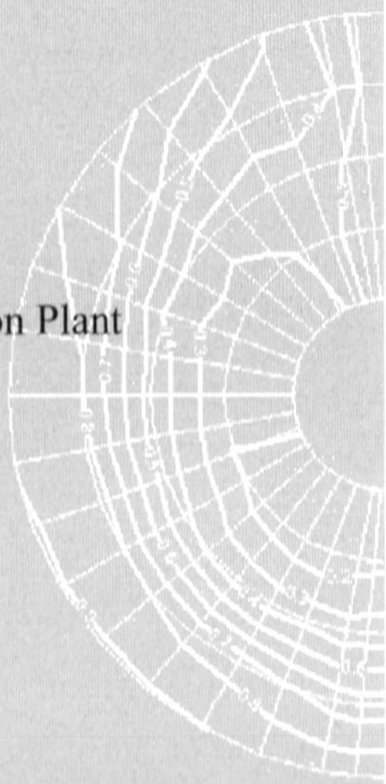

1. Development of Marine Propulsion Systems

Marine propulsion systems are the mechanisms or systems used to generate thrust to move a ship on or in water and can be divided into human, wind, and mechanical powered systems. The history of marine propulsion spans from the age of oars and sails to the modern mechanical engines. Oars were used to assist propulsion from primitive times until the 10th century and sails were used until the introduction and wide acceptance of mechanical propulsion systems in the 19th century which continues to evolve to this day. Most modern ships are now propelled by mechanical systems consisting of a motor or engine turning a propeller or jet drives, while paddles and sails are still used on some special ships and smaller boats.

1.1 Beginning

The first contraption ever used in the history of mankind for travelling on the water must have been propelled by hand strokes. It is interesting to conjecture how the early man came to fashion an oar-like device. What is certain is that oars and steering devices were soon developed and were used universally. For example, the prehistoric petroglyphs at Bangudae, located upstream of the Taehwa-gang River, Ulsan, Korea, illustrates a whaler with 18 people on board, implying that the people may have used oars back then. Later, the invention of a kind of rowlock, or thole pins, attaching the oar to the hull, might have considerably improved propulsive efficiency of the oars. This allowed many people to join forces to propel a ship and enabled travel across seas. Western oars are impulse propulsion systems, which generate propulsive force by the reaction to water being pushed backward. Unlike western oars, traditional Korean oars exploit lifting forces generated on them while being moved left and right in water at the right angles to the ship's direction of travel, and can be considered to be a form of lift propulsion systems which include modern screw propellers.

1.2 Wind-powered Sailing Ships

It is not exactly clear when sails were first used, but it can be presumed that people started using sails from prehistoric times together with oars. Ships painted in ancient Egyptian murals or Viking ships used in Northern Europe are the most prominent examples. In Korea, a thousand-year-old ship, excavated in Wando Island, and figures of ships engraved on a copper mirror of the Goryeo dynasty are evidences of early use of sails. Even at an earlier time, the Korean Admiral Jang Bogo was recorded to have pioneered a shipping route extending from China to Japan through Cheonghaejin, a naval base near Wando Island, and Korean sailing ships were active on this route.

Early sailing ships were equipped with a single mast. However, in the 15th and 16th centuries, during the rise of colonial explorations, the need for larger and sturdier ocean-going ships led to the development of three-mast ships to harness wind power more efficiently. Later the increased cargo transport requirement to and from colonies led to the advent of large warships with guns mounted and large sailing ships servicing fixed routes. In the Far East, meanwhile, two-mast naval ships of the Joseon dynasty fought against the Japanese invasion of Korea in 1592 and the Korean delegate dispatched to Japan after the war used a two-mast ship.

1.3 Advent of the Steamship

The year 1786 marked the beginning of steamships when John Fitch built a 14-meter-long ship and successfully conducted a sea trial. Nevertheless, it is still a matter of debate as to which was the first successful steamer. William Symington's ship, the *Charlotte Dundas*, built in 1801 and propelled by a paddle wheel, is regarded as the world's first practical steam boat. John Stevens built a screw-driven steamer in 1802 and then produced a new steamer equipped with an improved steam engine and two screw propellers which successfully crossed the Hudson River in 1804. The steamship *Clermont*, built by Robert Fulton in 1807, successfully carried passengers from New York City to Albany, covering a 240-kilometer distance in 32 hours and it is credited as the first successful commercial steamer by many. The age of commercial steamships had begun.

Early steamships used steam reciprocating engines driving paddle wheels. In 1811, the *New Orleans*, a side-wheeler built by Fulton and Livingston and the first steamboat on the 'western waters' of the United States, took to the waters of the Mississippi River. In 1814, there were only 20 steamers in New Orleans, but barely twenty years later, the number rose to about 1,200, playing a critical role in the development of the central region of the United States.

1.4 Steam Turbine

The origin of steam turbine dates back to the 1st century, when the Greek mathematician Hero (or Heron) of Alexandria published a well-recognized description of a steam-powered device called an aeolipile (also called Heron's engine). In 1480, Leonardo Da Vinci also proposed the steam turbine principle. In 1882, Swedish engineer Karl Gustaf Patrik de Laval built an impulse steam turbine. Further advances followed, and in 1890 a nozzle was used to increase the speed of the steam entering the turbine blades. This is now known as a de Laval nozzle and was of great importance in rocket design. A Laval's turbine (patented in 1883) attained a speed of 42,000 revolutions per minute and such high rotational speeds demanded that he also design new approaches to reduction gearing, which are still in use today. In 1893, he built and operated a reversible turbine for marine use.

In 1884, Charles A. Parsons, a British engineer, invented a multistage steam turbine to regulate the rotational speed and succeeded in linking a power generator through a reduction gear. In 1894, Parsons launched *Turbinia*, the first turbine-powered steamship, as shown in **Figure 7-1**. *Turbinia* later achieved a maximum speed of 34.5 knots with improvement in its propulsion system to overcome propeller cavitation. With the advent of diesel engines, however, competition for domination in the marine engine market began to hot up.

Figure 7-1 The first steam turbine-powered steamship *Turbinia* (Source: http://en.wikipedia.org/wiki/Turbinia)

1.5 Diesel Engine: the 'Best' Marine Propulsion System

The diesel engine was invented by Dr. Rudolf Diesel in 1897 and it developed rapidly since then. In 1911, a medium-speed, four-stroke engine made by B&W Diesel was successfully installed on the *Selandia*, the first ocean-going diesel powered-ship; and the success led more ship-owners to build ships powered by diesel engines. In 1914, less than 300 ships with a total tonnage of 235,000 tons were powered by diesel engines. Ten years later, the number grew to about 2,000 vessels with 2 million tons, and in 1940 the number rose to approximately 8,000 vessels totalling 18 million tons.

After World War I, diesel engine-powered ships occupied about 25% of the world tonnage and increased to 60% by 1939. The marine diesel engine has made remarkable progress over the past century with the development of a fuel injection system, the application of a supercharger, the use of low quality fuels, the development of very large engines and the advent of an electronically controlled engine, among others.

Recently, LNG carriers have begun to use the dual fuel (gas-oil) engine instead of the steam turbine, but the diesel engine remains the main power source for propulsion of almost all commercial ships. The output of diesel engines has also reached 100,000 horsepower.

2. Types of Propulsion Machinery

2.1 Diesel Engines

A diesel engine is a type of reciprocating internal combustion engine that uses the heat of compression to ignite diesel or fuel, and hence is known as a compression-ignition engine. There are three speed groups in diesel engines; high-, medium- and low-speed engines classified by the rotational speed (RPM). An average high-speed engine has a rotational speed ranging from 1,000 to 2,400 rpm and produces up to 3,500 horsepower. A medium-speed engine has a speed from 400 to 1,000 rpm and produces up to 10,000 horsepower. Lastly, a low-speed engine is designed to have the same RPM as the optimum RPM of a ship's propeller and often has a thermal efficiency exceeding 50%.

All heavy-duty diesel engines are low-speed engines and have a rotational speed below 200 rpm. The heavy-duty engine produces between 5,000 and 45,000 horsepower on average, but the latest engine is capable of producing up to 100,000 horsepower and much more economical than any other propulsion systems. Low-speed engines are normally used for the propulsion of large container ships or tankers, whereas medium- and high-speed engines are used for the propulsion of car ferries or fishing vessels and are also widely used as auxiliary power units.

2.1.1 High-speed Engines

An average high-speed engine operates at 1,000 rpm or higher and is widely used as main propulsion machinery for small ships, and auxiliary engines for main or emergency generators of large ships. **Figure 7-2** shows a typical high-speed engine.

Figure 7-2 A typical high-speed engine
(Courtesy of MTU, source: https://www.google.co.kr/)

2.1.2 Medium-speed Engines

An average medium-speed engine produces 10,000 horsepower or less at 400 to 1,000 rpm. They are commonly used for the propulsion of small ships and special ships such as cruise ships, passenger and car ferries, and Ro-Ro (roll-on roll-off) ships. Medium-speed engines dominate generator markets for seagoing ships, and only those ships that require less output use high-speed, four-stroke engines. **Figure 7-3** is an example of a medium-speed engine and **Figure 7-4** represents output of the various types of HiMSEN Engines, a range of medium-speed engines developed by the Hyundai Heavy Industries of Korea.

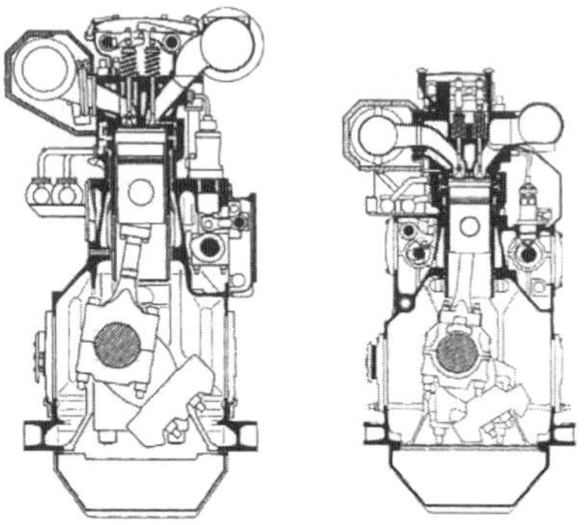

Figure 7-3 A typical medium-speed engine
(Courtesy of MTU)

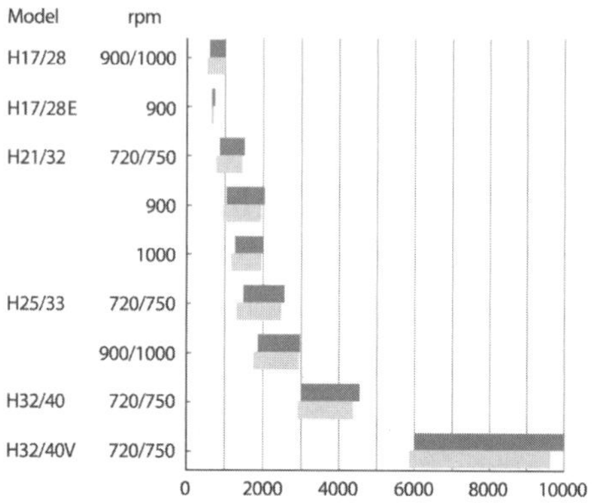

Figure 7-4 Characteristics of medium-speed engines
(Courtesy of HHI)

2.1.3 Low-speed Engines

An average low-speed engine operates at about 200 rpm to keep a piston speed below 4 m/s. The leading low-speed engine manufacturers include MAN Diesel & Turbo of Germany,

Figure 7-5 A typical low-speed engine
(Courtesy of Wärtsilä-Hyundai, source: http://hugin.info/131481/R/)

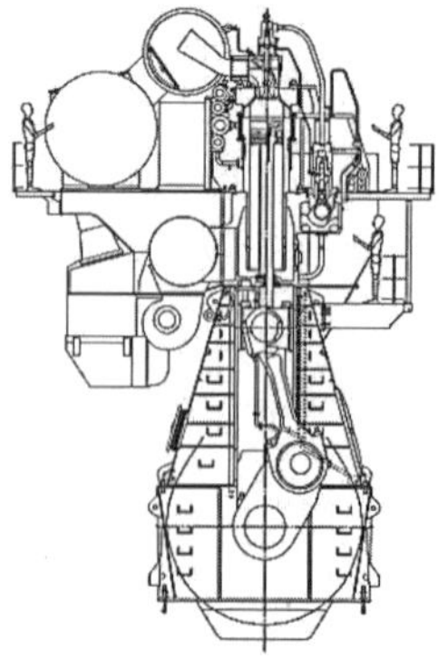

Figure 7-6 A typical low-speed engine
(Courtesy of Man B &W, source: http://jeb.biologist.org/content)

manufacturer of the MAN B&W MC/ME engines, Wärtsilä of Finland manufacturing the Wärtsilä RT engines, and Mitsubishi of Japan producing the UE engines. The MAN B&W engines produce between 1,100 kW and 97,300 kW with a series of four to 14 cylinders. **Figures 7-5** and **7-6** are examples of low-speed engines.

2.2 Electronically Controlled Diesel Engines

2.2.1 Introduction

There have been two great milestones since the introduction of the diesel engine: the development of a supercharger and an electronically controlled system. While the supercharger made huge improvements in engine performance, the electronically controlled system provided a turning point in ensuring flexible engine controls.

The camshaft has been a useful piece of hardware in controlling fuel, exhaust and air-start valves, and in driving fuel injection and exhaust valves, and air-start distributors. However, when controlling a combustion process in a variety of ways for different purposes and loading conditions, the camshaft has been a huge obstacle. This is because changes cannot be made to the shape or position of a camshaft when it is operating. **Figure 7-7** is an example of a fuel injection system driven by a conventional camshaft.

In the electronically controlled engine, the mechanical control mechanisms of the camshaft are replaced by computers and software; and the drive function with a hydraulic pump, an accumulator (e.g., common rail) and a control valve. These changes may vary depending on the manufacturer and type of the engines. For instance, the computer is used to control the fuel, exhaust and air-start valves, or a cylinder greasing system of a two-cycle engine; for four-cycle engines, the control is limited to fuel valves. The electronic control of the combustion process provides a number of advantages, such as reducing exhaust gas emissions and operating costs, and improving operational performance at slow speeds, but also poses new problems, such as verifying the reliability of modified or added hardware and software.

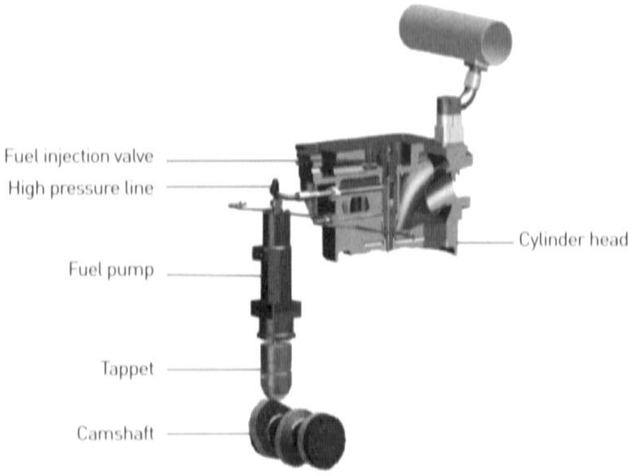

Figure 7-7 A typical fuel injection system driven by a conventional camshaft

2.2.2 Conventional Diesel Engines vs Electronically Controlled Engines

Figure 7-8 compares the outline drawings of a conventional diesel engine and an electronically controlled engine, both made by Sulzer (currently Wärtsilä). The electronically controlled engine does not have the exhaust valve-driving system, a fuel pump, a camshaft, a reversing servomotor, a fuel link system, an actuator, a VIT system, a starting air distributor and a camshaft driving system, which are featured in the conventional engine. Instead, it has a common rail platform, a control system, a supply unit and a related piping system. The electronically controlled engine was developed to allow fine-tuning to maintain an optimum state of combustion and is an advanced system designed to cope with environmental issues facing the 21st century. Furthermore, more efforts are needed to utilize efficiently various electrical or electronic systems and computer software and enhance problem-solving abilities.

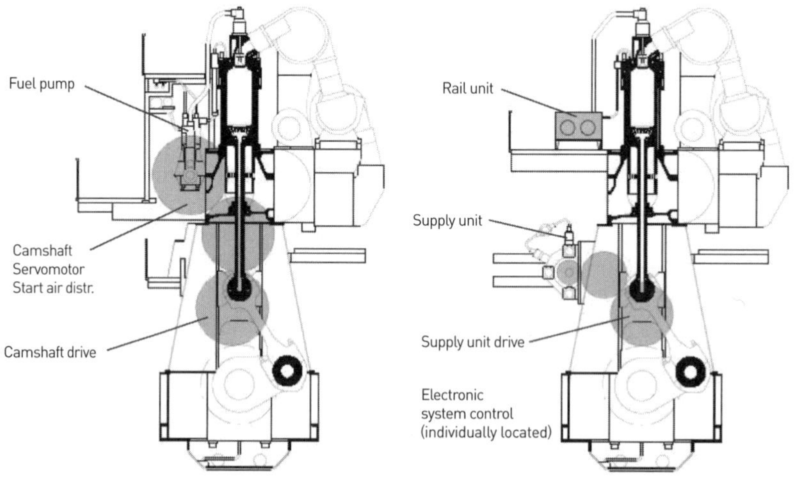

(a) Conventional engine (b) Electronically controlled engine

Figure 7-8 Difference between a conventional diesel engine and an electronically controlled engine
(Courtesy of Wärtsilä-Hyundai)

2.3 Gas Turbines

A gas turbine refers to a rotary engine which extracts energy from the flow of combustion gas. A gas turbine consists of a compressor, a turbine and a combustion chamber and is driven by the force of expanding high-temperature, high-pressure gas flow, generated when the fuel mixed with compressed air is burnt in the combustion chamber. The purpose of the gas turbine determines the design so that the most desirable energy form, of shaft torque, thrust and compressed air is maximized.

Gas turbines have been mostly used on naval ships, but they also have potential for merchant ships. Gas turbine manufacturers are trying to break into the shipping industry, as evidenced by the turbines installed on container ships, small gas carriers and the Finnjet – a ferry which operated in the Baltic Sea in the 1970s. However, rising fuel prices and improvement of diesel engines, such as increased power-ratio and improved ability to combust low-quality heavy fuel, are among problems that remain to be resolved. **Figure 7-9** shows a marine gas turbine.

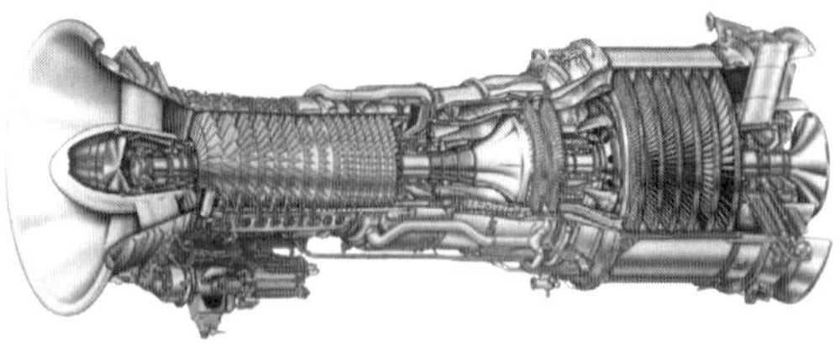

Figure 7-9 A large gas turbine for high-speed ferries, GE LM 2500 (Source: http://www.google.co.kr/)

2.4 Steam Turbines

2.4.1 Outline of Steam Turbines

A steam turbine is a device that extracts thermal energy from pressurized steam and uses it to do mechanical work on a rotating output shaft. An impulse turbine is driven by momentum change, occurring when high-temperature, high-pressure steam obtained from a boiler or a nuclear reactor passes through nozzles turning the steam flow into high-speed jets which are directed to collide with the turbine blades and change their course. A reaction turbine is driven primarily by a reactive force produced when this steam accelerates and changes its course through the blades and the exhaust.

2.4.2 Classification of Steam Turbines

Steam turbines can be classified by the type of steam flow, steam direction, operating conditions, number of stages and turbine axis.

1) Impulse turbines

The impulse turbine is driven by a force produced from the momentum of high-speed steam jet expanding at the nozzle and changing its course after impacting with the turbine blades.

2) Reaction turbines

The reaction turbine is turned by reactive forces, as the name implies, rather than by a direct push or impulse. It has no nozzles and, instead, the convergent spaces between the moving blades mounted on the shaft function as nozzles. Fixed or stationary blades of the same shape as the moving blades fastened to the casing guide the steam into the moving blade system to produce the reactive force on the moving blades. The force is produced when the steam changes direction and increases in velocity as it expands through the moving blades. A reaction turbine is rotated primarily by reactive force but also to some extent by direct impulse.

3) Impulse-reaction turbines

Figure 7-10 An impulse-reaction turbine mounted on a ship
(Courtesy of Wärtsilä-Hyundai)

The impulse-reaction turbine takes advantage of both types by combining impulse and reaction stages. This turbine derives its rotation from the impulse of the steam striking the turbine blades at the first stage by recovering energy through temperature and pressure drop. Most of the energy recovered from the impulse stage and the remaining reactive force is recovered by changing the steam flow direction. An installation of impulse-reaction turbine is shown in **Figure 7-10**.

4) Axial turbines

In an axial turbine the steam flows parallel to the shaft of the turbine to impact rotor blades.

5) Radial turbines

In a radial turbine the steam flows outward from the shaft to the casing and is usually a reaction turbine having both fixed (stator) and moving blades. For a radial turbine, the flow is accelerated and smoothly orientated at 90°, by the stator, towards the moving blades on which its direction changes and speed is reduced to drive the rotor.

6) Single-casing turbines

A single-casing turbine has one casing and is appropriate for small and medium-capacity turbines.

7) Multi-casing turbines

A multi-casing turbine consists of two or more high-, medium- and low-pressure casings. Steam works by expanding in each of the casings.

2.4.3 Key Components of a Steam Turbine

In general, turbines are composed of a few stages arranged on the same driving shaft. The assembly varies, depending on the turbine type, but turbines share the following key components: foundation, casings, rotors and blades, nozzles, bearings, shaft and shaft glands (**Figure 7-11**).

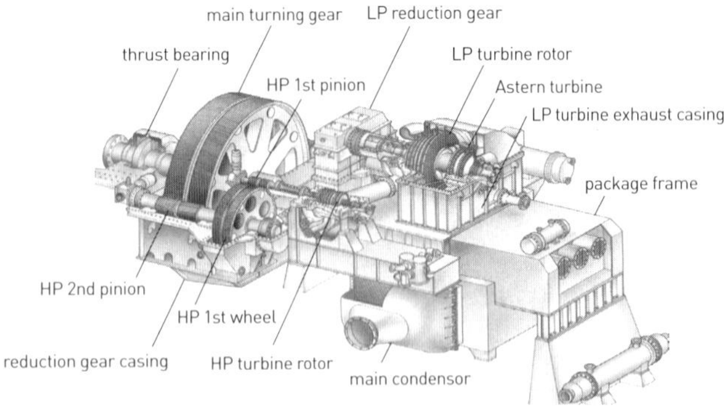

Figure 7-11 Key components of a steam turbine

The high-pressure steam, produced in a boiler, goes through steam and regulator valves, which are opened or closed by a speed governor, expands in a steam chest, and reaches an exhaust air chamber after each stage. Since the volume of steam markedly increases as the steam is expanded, there must be enough space as a passageway. The size of nozzles and moving blades becomes greater towards the low-pressure stage. **Figure 7-12** is the schematic diagram of a turbine control system.

1) Rotor

A rotor consists of a turbine shaft, a wheel and blades, and the shaft is supported by bearings installed outside the casing. The blades are fixed to the rotor wheel alternately with nozzles; creating a ring-shaped steam passage.

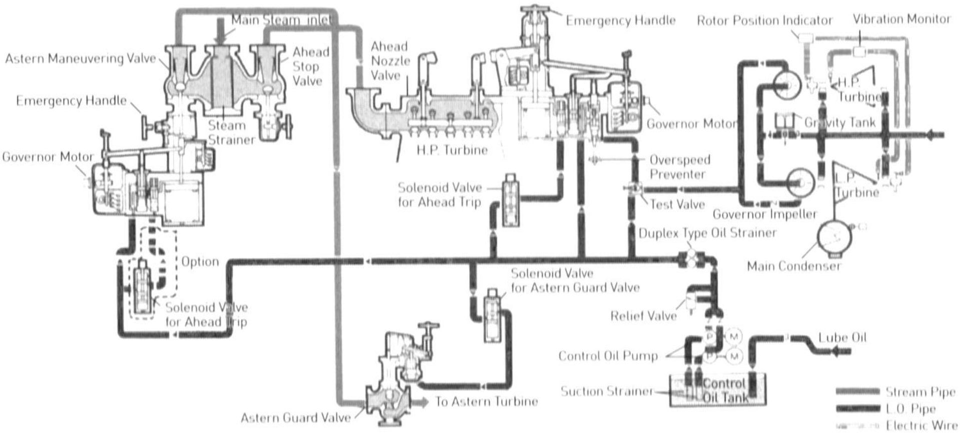

Figure 7-12 A schematic diagram of the turbine control system

2) Moving blades

Since the blades are fixed to the wheel or rotor and rotate fast while converting the kinetic energy from steam into torque, they must endure high-temperature, high-pressure steam. The blades are, therefore, made primarily of chrome (Cr) or chrome-molybdenum (Cr-Mo) alloy steel. When using low-pressure steam, blades made of manganese-copper (Mn-Cu) alloy or copper-nickel (Cu-Ni) alloy can be used when the temperature of the steam is lower than 315℃; copper-zinc alloy may be used if the temperature is below 220℃.

3) Nozzles

Nozzles are designed mainly to convert heat energy of the steam into kinetic energy by accelerating the steam and additionally to guide the steam to the blades. Nozzles are integrated with a diaphragm, which is split into two parts and supported by the casing. Depending on the cross-section, nozzles are divided into convergent, divergent, and parallel nozzles. A nozzle consists of an inlet, a throat and an outlet, and comes in a variety of shapes, but usually convergent or convergent-divergent nozzles are used.

4) Shaft gland

A shaft pierces the casing of a high-pressure turbine and makes a gap between the shaft and the casing through which high pressurized steam leaks, and shaft glands are used to minimize this gap. Shaft glands come in various forms, but are divided into labyrinth and carbon packing seal. Labyrinth packing seal creates a very small opening between a shaft and packing strips machined and installed on a casing around the shaft. The packing exploits the pressure drop occurring when steam is leaked through the small openings, and, therefore, the steam pressure becomes closer to the atmospheric pressure as the steam passes through several packing strips successively. This type of packing is used for high-pressure sealing, instead of carbon packing that is prone to damage under high temperature and high pressure.

Carbon packing is based on the same principle as labyrinth packing. Carbon packing blocks are used for low temperature and low pressure sealing by pressing down on a shaft using springs; they are prevented from spinning by a key. The use of carbon packing, however, does not guarantee a complete block of steam leakage from a turbine. When the pressure between the shaft and casing drops below the atmospheric pressure in a low-pressure turbine, air may get into the condenser through the turbine and it must be blocked off using sealing steam.

5) Bearings

Two types of bearings are used for steam turbines: main and thrust bearings. The main bearing supports the weight of the rotor and maintains the correct radial clearance between the rotor and casing.

Bearings are usually divided into sleeve, ball and roller bearings. Among those, sleeve bearings are normally used to prevent an excessive load per unit area, ensure a smooth flow of lubricant for cooling and lubrication, and facilitate the installation and repair of the bearings.

Thrust bearings receive force acting in the axial direction of the turbine shaft and hold the shaft firmly to maintain the designated position. Thrust bearings are installed at the front and back of collars fixed to the shaft with Babbitt metal-coated shoes and there is only a small clearance between the collar and the shoe to maintain the correct shaft position. Bearings designed to transmit high thrust have a number of collars fixed to the shaft and an appropriate number of shoes installed.

2.5 Dual Fuel Engines

A dual fuel (DF) engine is an engine that can run on either diesel fuel or liquid natural gas. **Figure 7-13** is an example of dual fuel engine. When operated in LNG mode, the engine emits negligible amount of noxious gas because it burns clean fuel. Since the primary component is methane, the fuel used by the engine is the most effective hydrocarbon fuel in terms of energy density per carbon atom, and has the largest amount of hydrogen per unit energy among all fossil fuels. Compared with heavy fuel oils (HFO) or diesel fuel, LNG can reduce carbon dioxide emissions by about 20% if operated in the same environment. It also reduces noxious emissions by 72%, benzene by up to 97%, and almost completely eliminates sulfuric oxide and soot (PM: particulate matter).

Furthermore, there is no visible smoke and no sludge or lead emissions. Natural gas is a very safe fuel because it combusts well in engines; it is lighter than air and has a high ignition

Figure 7-13 A dual fuel engine, Wärtsilä 6LDF
(Source: http://hugin.info/131481/R/)

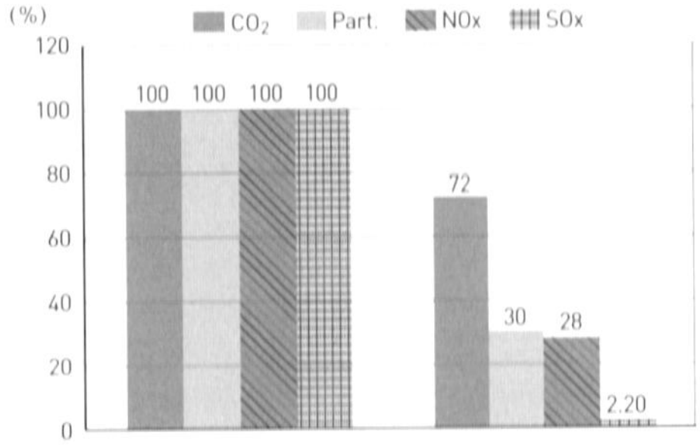

Figure 7-14 Reduction in noxious gas emission
(Courtesy of Wärtsilä Hyundai)

temperature. **Figure 7-14** compares the results of noxious gas emissions in diesel and LNG modes.

The dual fuel four-stroke engines made by Wärtsilä can be operated in either gas or diesel mode. In gas mode, the engine is operated in a lean-burn Otto cycle by combusting a mixture of air and rarified gas in the combustion chamber. Lean-burn refers to a combustion process that provides the cylinder with more air than what is needed to completely burn the fuel and helps to improve fuel efficiency.

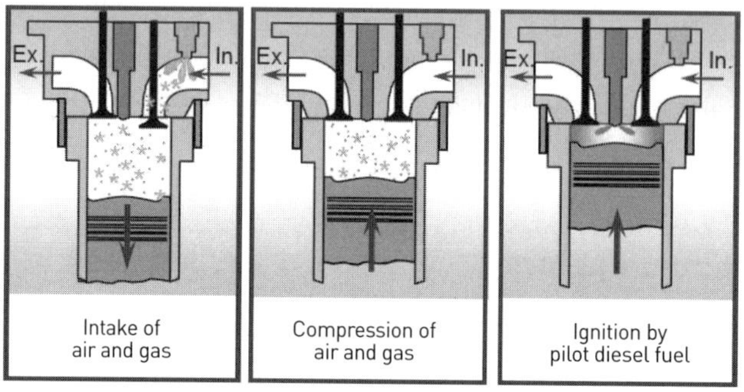

Figure 7-15 Ignition of fuel in DF gas mode (Courtesy of Wärtsilä-Hyundai)

Gas is fed into the cylinder through the suction air vent in the suction stroke. **Figure 7-15** shows the combustion process in DF gas mode. The gas mixture is ignited using a small amount of diesel fuel which is sprayed into the combustion chamber. To reduce nitrogen oxide emissions, only a very small amount of pilot diesel fuel should be used. Therefore, the DF engine made by the 'micro-pilot injection system' sprays less than 1% of the diesel fuel consumed in an ordinary diesel engine under normal load conditions. **Figure 7-16** is

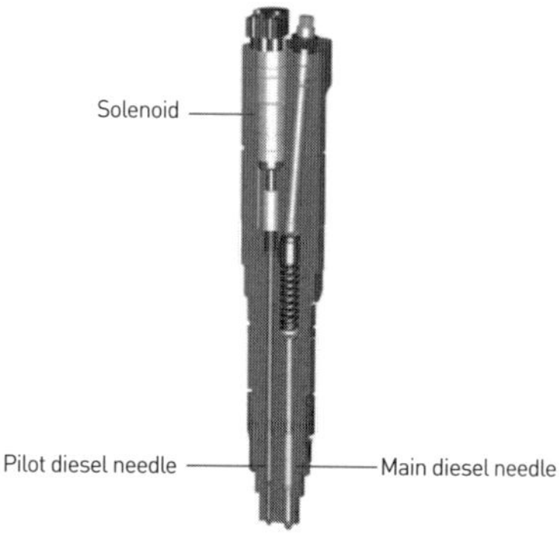

Figure 7-16 A dual fuel injection valve
(Courtesy of Wärtsilä-Hyundai)

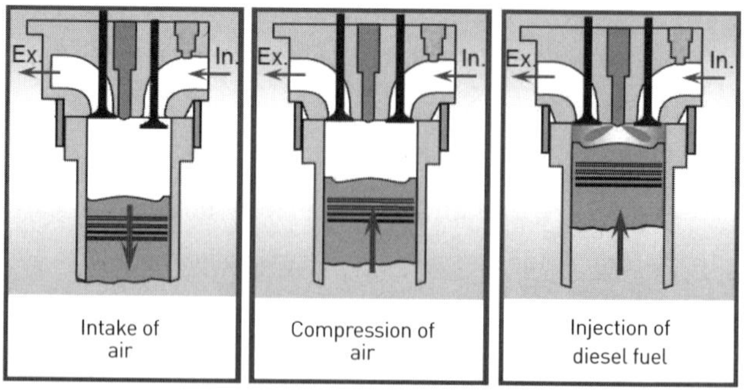

Figure 7-17 Dual fuel Diesel mode
(Courtesy of Wärtsilä-Hyundai)

an example of the micro-pilot nozzle and **Figure 7-17** shows the combustion process in DF diesel mode.

When a DF engine operates in gas mode, combustion must be precisely controlled to prevent knocking or ignition failure, and Wärtsilä proposed a complete electronic control system of pilot fuel spraying and gas injection into each cylinder head as the only way to achieve this goal.

Figure 7-18 is an example of an electronically controlled fuel injection system used on a cylinder head. The overall air-gas ratio is regulated using an exhaust valve, and the exhaust gas is partially made to bypass the turbine's supercharger. This system ensures that the air-gas ratio remains unchanged, despite the variation in surrounding conditions, such as temperature. **Figure 7-19** represents an air-fuel ratio of a typical DF turbine.

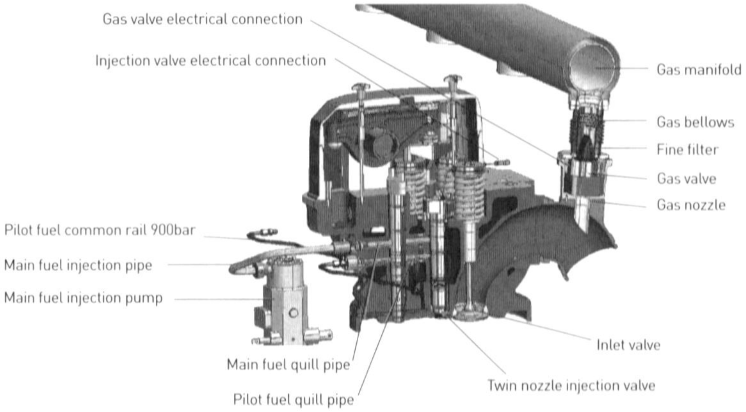

Figure 7-18 Fuel injection system in a cylinder head (Courtesy of Wärtsilä-Hyundai)

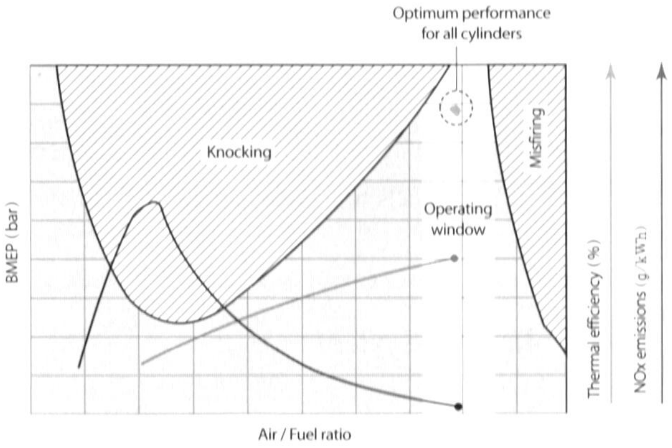

Figure 7-19 Air-fuel ratios of a lean-burn engine (Courtesy of Wärtsilä-Hyundai)

3. Shafting System for Ship Propulsion

If a marine propeller shaft is simply aligned in a straight line, loads on the bearings supporting the shaft cannot be evenly distributed, causing such problems as abnormal wear in shaft bearings, no load or overheating in intermediate bearings, and wear or damage in a reduction gear. Under the shaft alignment theory, the distribution of loads in straight-line alignment and the reaction influence number, or load variation when each bearing is changed by unit displacement, are calculated to adjust the alignment of bearings both vertically and fore-and-aft so that the shaft alignment appears close to a natural curve in order to ensure that the shaft is stable and insensitive to any disturbance.

The importance of studies on propeller shaft alignment was first recognized in the U.S. navy ships in the late 1950s. Since then theories began to develop on how to determine the ideal position of each bearing. The findings of these studies were extended to merchant ships, and more research was conducted from the late 1960s to the early 1970s, resulting in a number of theories related to various aspects.

The recent development of very large ships has called for a huge increase in power and this has led to an increase in damage to the aft bearing of the main engine and aft stern-tube bearings caused by the misalignment of shafts, especially when the ships are equipped with large, low-speed two-stroke engines. Such a tendency can also result in the deformation of double-bottom beneath the engine room or the main engine bed, due to draft change or thermal expansion. While the increased power requires a shaft with a higher stiffness, the use of high-tensile steel in hull construction has made structural members thinner, leading to easier deformation of stern. Therefore, very large ships require more precise shaft alignment than the existing smaller ships.

To optimize shaft alignment during the design stage, an analysis should be made of the bearing reaction forces under given loading conditions, bending moments caused by the propeller thrust or deformation of the main engine, and elastic deformation of bearings acting perpendicular to the loads. Academia, shipyards, and classification societies have conducted numerous research projects to reduce damage caused by shaft misalignment.

3.1 Structure of Shafting System

Figure 7-20 outlines the structure of a shafting system. The propeller shaft ① is supported by bearings in two locations ⑭ in the stern tube ②. To prevent seawater from flooding the ship (aft end), and to prevent oil in the tube from leaking into the hull (forward end), shaft glands (③,④) are installed. The propeller shaft and the intermediate shaft ⑤ are connected by the shaft coupling ⑥. For convenience in manufacturing and to make installation and assembly easier, the intermediate shaft is made into several pieces of appropriate lengths and is connected with couplings. The intermediate shaft is supported and driven by sliding bearings ⑫.

Since the engine ⑨ is installed near the stern of the ship on most ships, the propeller shaft becomes relatively short and usually only a single piece of intermediate shaft is used. The thrust produced by the propeller is transmitted to the hull through the thrust shaft ⑦ and the thrust collar ⑧.

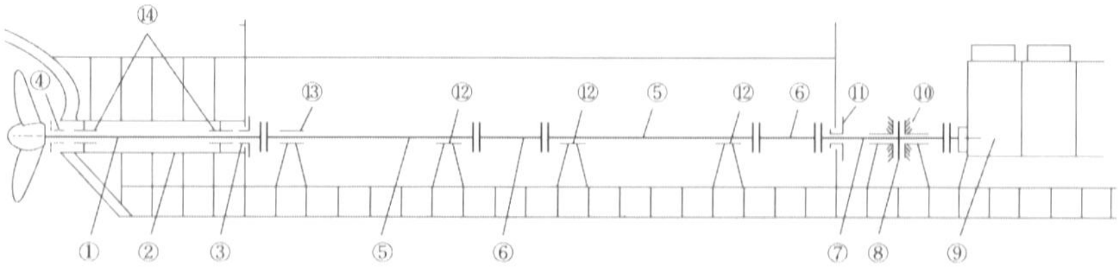

Figure 7-20 Outline structure of ship shafting system

3.2 Considerations for Propeller Shaft Alignment

1) Expected significant events in shaft alignment during operation

- Deformation of the double bottom caused by draft or route changes;
- Variation in bearing reaction forces due to thermal expansion in the diesel engine, turbine or reduction gear, arising from the temperature difference between stationary and operating modes (**Figure 7-21** shows an example);

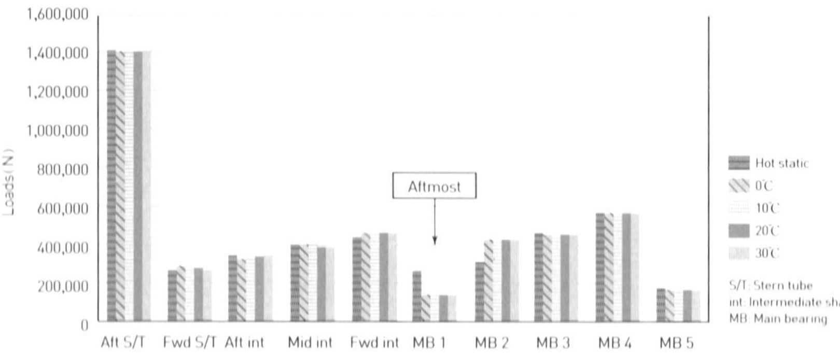

Figure 7-21 A typical change in bearing reaction forces with temperature variationt

- Shifts in the effective supports of stern tube bearings due to the impact of a bending moment caused by the thrust produced when the shaft rotates;
- A bending moment on the misplaced thrust pad due to misalignment of the center of propeller thrust and the shaft;
- An additional force produced by electromagnetic resistance of the shaft generator, which is using a power take-off (PTO) system on the propeller shaft;
- Wear in bearings of a seawater lubrication system using lignum vitae;
- An unbalanced inertial force caused by the rotation of a shaft or propeller.

2) Additional requirements for shaft design

Despite the foregoing problems, the shafting system has to meet the following requirements:

- Bearings at both ends of a reduction gear should have the same reaction force.
- All shaft bearings should have the same surface pressure.
- The direction and magnitude of each bearing load should be invariable and uniform under any operating conditions.
- The distribution of loads on the stern tube bearings should be as uniform as possible and no local concentration of loads at the edge of bearings.

For turbines or diesel engines featuring reduction gears, a reduction gear system is relatively short to allow the lowering of the shaft bearings or reduction system so that the average slope of the shaft centerline becomes parallel to the bearing centerline. Since low-speed diesel engines are too long to be aligned to have corresponding slope, they are bored to have the same slope as that of the stern tube bearings, or the same slope as that of the outside diameter of bush bearings. The latter method is recommended, as it allows precision machining.

In addition, to prevent local load concentration on the edges of the stern tube bearings made of white metal, the rear of the bearing is partially bored to have a single or dual slope over about 1/3 of its entire length.

4. Fuels

Fuels for internal combustion engines can be divided into liquid and gas fuels.

4.1 Liquid Fuels

4.1.1 Types

Liquid fuels for internal combustion engines can be subdivided as follows:

- Petroleum-based fuels: gasoline, kerosene, diesel, heavy oils, etc.
- Coal-based fuels: benzol, gasoline converted from coal liquefaction, gasoline, kerosene, diesel, heavy oils, etc.
- Vegetable-based fuels: ethyl alcohol and methyl alcohol.

4.1.2 The Process of Producing Fuels for Internal Combustion Engines

Most fuels used in internal combustion engines are petroleum-based and they are all hydrocarbons. In general, hydrocarbons types are aliphatic, aromatic, and naphthenic. Of them, aliphatic hydrocarbons are highly combustible, aromatic hydrocarbons are slightly combustible and naphthenic hydrocarbons are somewhere in between.

Petroleum, or crude oil, is a complicated mixture of many kinds of hydrocarbons. However, the component ratio of a crude oil is indicated by its origin. Crude oil also contains small amounts of sulfur, water and ash. When sulfur is burned, it becomes sulfur dioxide or sulfur trioxide and the latter is combined with water to produce sulfuric acid (H_2SO_4), corroding the engine. The fuel with a high content of sulfur, therefore, is inappropriate to use and requires special treatment.

Various grades of fuels are produced by refining crude oil in a distillation system. When the crude oil is heated, light oils with a lower boiling point evaporate first and are cooled to liquefy. The crude oil is separated successively into a variety of oil products according to their evaporation temperatures. This is called fractional distillation and the products of its process are shown in **Table 7-1** together with average properties of each product.

Table 7-1 Products of crude oil and their characteristics

Product Name	Evaporation Temperature		Density	low calorific power	theoretical amount of air	
	Paraffin oils	Naphthenic oils		[kcal/kg], [kJ/kg]	Nm^3/kg	Kg/kg
Gasoline	150°C or less	125°C or less	0.60~0.74	[10500], [44.00]	11.44	14.79
Kerosene	150°C ~300°C	125°C ~275°C	0.74~0.82	[10300], [43.16]	11.37	14.70
Diesel oil	300°C ~350°C	275°C ~325°C	0.82~0.88	[10170], [42.61]	11.00	14.22
Heavy oil	350°C or higher	325°C or higher	0.89 or more	[9900], [41.48]	10.72	13.86
Lubricant			-	-	-	-
Residues	Pitch	Asphalt	-	-	-	-

• Nm^3: normal cubic meter

1) Gasoline

Gasoline is a colorless, transparent fuel that is highly volatile and flammable and used as a fuel for gasoline engines. Gasoline is produced in the following ways:

① Straight-run gasoline

Straight-run gasoline is derived from direct distillation of crude oil.

② Natural gas gasoline

The gasoline is obtained through the compression and liquefaction of casing-head natural gas, rich in hydrocarbon vapors, that is taken without processing from an oil well. Furthermore, absorption gasoline is obtained by using an oil to absorb the natural or refinery gas containing the gasoline and then distilling it from the oil.

③ Cracked gasoline

Cracked gasoline is a gasoline manufactured by heating crude petroleum distillation fractions or residues under pressure, or by heating with or without pressure in the presence of a catalyst, so that heavier hydrocarbons are broken into others, some of which distill in the gasoline range.

④ Reformed gasoline

Reformed gasoline is a lower-thermal-value fuel gas made by a catalytic or thermal reforming process of high-thermal-value natural and refinery gases.

2) Kerosene

Kerosene was originally a trademark of a lamp company in the US and the fuel kerosene is usually called paraffin in the UK. Kerosene, a thin, clear liquid formed from hydrocarbons, is obtained from the fractional distillation of petroleum. Both volatile and flammable oils are inferior to gasoline and used to fuel kerosene engines. Otherwise, it can be used to fuel oil stoves. It comes in various colors from colorless to yellowish. Compared to gasoline, kerosene undergoes slower combustion thereby easily depositing soot inside the cylinder, and produces a foul odor. Many kerosene engines use gasoline to start up and use kerosene after the engine is warmed up.

3) Diesel oil

Diesel is also known as light oil, neutral oil or gas oil. Diesel-powered engines generally have a better fuel economy than equivalent gasoline engines and produce less greenhouse gas emissions. In addition, they are also used as cleaning or cutting oils for machinery. Unless an appropriate pre-heater is in place to use light fuel oil for an electrical ignition engine, it results in poor combustion performance.

4) Heavy oil

Heavy oils, also known as bunker or fuel oil, are thick and dark brown and usually refer to petroleum products which are obtained by mixing the residue of the atmospheric distillation of crude oil and diesel oils. If the dominant component is diesel oils, it is called Bunker 'A' fuel oil; 'C' oil, if the residue is the dominant component; and 'B' oil, if the two exist in equal ratios. Viscosity increases from Bunker A to B to C oils and so does the sulfur content. Heavy oils are used as fuel by ships, industry, and large-scale heating and power-production installations.

Bunker A fuel costs the most and is usually 1.7 to 2.0 times as expensive as the C fuel, despite price fluctuations. Bunker A generates heat of 10,200kcal per kg and C generates about 9,700kcal per kg. Though the amount of Bunker C needed is about 5% more than that of Bunker A, Bunker C is more commonly used because it costs less. Therefore, Bunker C is used for large diesel engines with high fuel consumption, while Bunker B is used to start a ship or operate it at low loads.

Since heavy oils are high in sulfur, the sulfur deposited on the inner surface of a cylinder after combustion causes corrosion and wears in the cylinder liner and ring. This problem is more commonly encountered with Bunker C than the other fuel oils.

4.2 Fuel Gas

The typical gases used for internal combustion engines are petroleum-based fuels and can be largely divided into liquefied natural gas (LNG) and liquefied petroleum gas (LPG).

1) Liquefied natural gas (LNG)

Liquefied natural gas is liquefied dry gas for easier transportation and storage and usually called natural gas. Dry gas refers to the component which does not turn into a liquid at normal temperatures among gases obtained from underground or deep-water gas fields (oil deposits). Its primary component is methane (CH_4).

2) Liquefied petroleum gas (LPG)

Liquefied petroleum gas is a natural gas that comes from oil fields or a by-product of oil refineries. Its primary components are propane, propylene, butane and butylenes.

4.3 Bio-fuels

Bio-fuels refer to fuels that are derived from biomass which includes all living organisms and their metabolic waste by-products. Unlike fossil fuels, bio-fuels are a type of renewable energy sources. These bio-fuels can be divided into biodiesel, which is a vegetable oil-based fuel, and bio-alcohol, which is produced through the fermentation of sugarcane or corn.

The advantages of bio-fuels are: firstly, unlike fossil fuels, plant-based bio-fuels have a carbon cycle and do not increase the level of carbon dioxide in the air; secondly, biodiesel, for instance, generates far smaller amounts of hydrocarbons that cause smog than ordinary diesel. Bio-alcohol also generates less exhaust gas. The drawbacks include the high risk of corrosion in engines due to high pH levels and the need for larger fuel tanks due to lower energy content per unit volume.

1) Biodiesel

When plant-based oil is mixed and reacts with methanol, it becomes an ester and generates biodiesel and glycerol. Glycerol can be separated from biodiesel using the difference in specific gravity and used for other purposes, and biodiesel can be used as automotive or marine diesel fuel.

2) Bio-alcohol

Bio-alcohol is obtained through the fermentation of carbohydrates (starch or sugar content) in sugarcane or corn. Bio-alcohol is ethanol with the same elements as ordinary alcohol, but different types of alcohol such as butanol or propanol can also be produced. In Brazil, a country with high sugarcane production, bio-alcohol is actively used as a substitute for gasoline to fuel motor vehicles.

5. Future Technologies for Power and Propulsion Plant

This section discusses equipment that is expected to be used as marine propulsion systems in the future. **Figure 7-22** shows prospects for the development of zero-emission ship propulsion systems as envisaged by NYK, a Japanese shipping company.

① Present (as of 2010)

As shown in **Figure 7-22**, most commonly used is the conventional system where the main diesel engine is directly coupled to the propeller shaft. A relatively new electric system combines a diesel generator and an electric motor.

② Near future (projected for 2020)

The 2020s will see a decline in the conventional system with the main engine coupled to the propeller, which will be replaced by an electric system combining biodiesel generators and electric motors, or an electric system combining fuel cells that use methanol or LNG and electric motors. Additionally, an electric system that combines solar batteries and electric motors will come into use.

③ Age of zero emissions (~2050)

By the 2050s, the conventional system with the diesel engine coupled to the propeller will completely disappear. In its place, an electric system combining fuel cells which use H_2 from a variety of sources and electric motors, and an electric system combining solar batteries and electric motors will dominate as marine propulsion system.

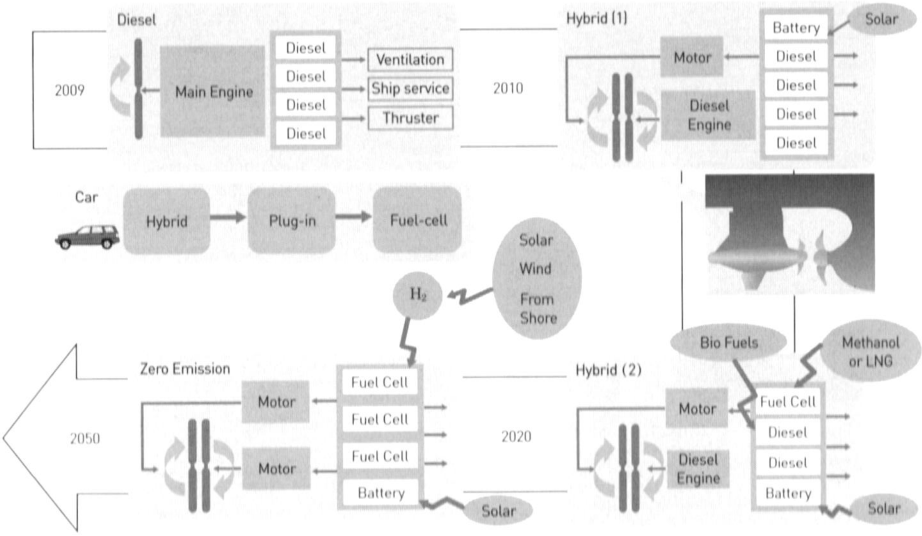

Figure 7-22 A concept of futuristic eco-friendly ship propulsion system
(Source: http://www.nyk.com)

5.1 Future Technologies for Power Units

5.1.1 Waste Heat Recovery System

1) General waste heat recovery systems

A waste heat recovery system (WHRS) refers to an energy recovery heat exchanger that recovers heat from hot exhaust of a ship's main engine. The recovered heat is used to preheat the incoming gas or to drive a power or steam turbine generator. The WHRS gained attention along with the development of fuel efficient ships during the first oil crisis in 1973.

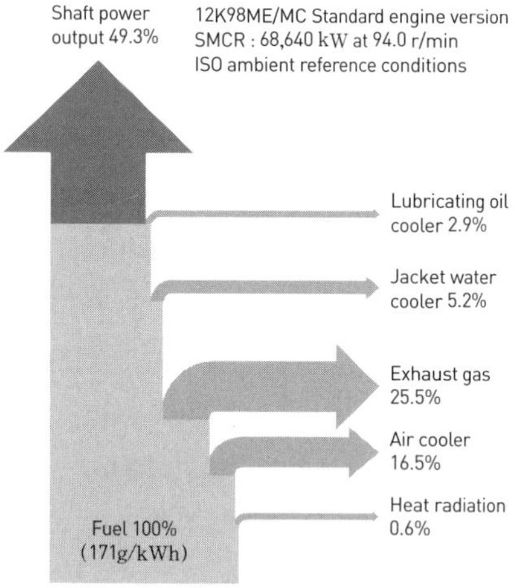

Figure 7-23 Heat balance diagram of a diesel engine

Continuous technological innovations in the past have allowed modern main engines for ships to achieve a thermal efficiency of up to 50%; 25% of the energy content is wasted as heat discharged in exhaust gases. **Figure 7-23** represents the heat balance diagram of the MAN Diesel Engine (12K98ME/MC). The figure shows that, of the 100% fuel consumption, the energy transmitted to the shaft is 49.3% of the fuel energy content, while 25.5% is discharged in exhaust and 16.5% is consumed by an air cooler. There is, therefore, an urgent need to develop a technology to exploit waste heat discharged in exhaust gases.

The waste heat recovery systems for ships that have been developed so far drive power turbines using

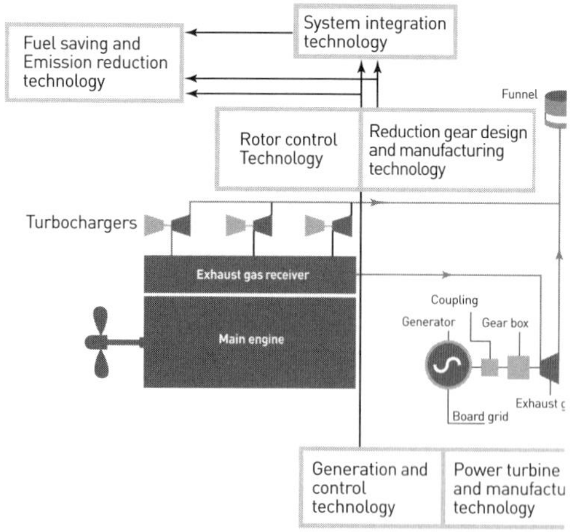

Figure 7-24 Schematic diagram of the waste heat recovery system

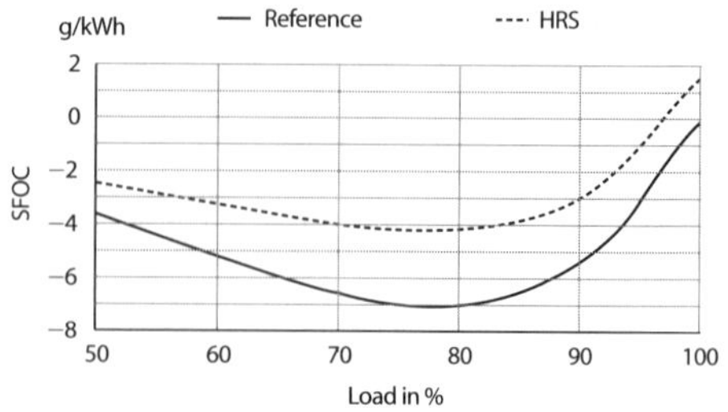

Figure 7-25 Comparison of fuel consumption rate between engines with and without waste heat recovery system

latent waste heat in exhaust gases, and then the turbine drives a generator. Through this means higher fuel efficiency and lower fuel consumption rate is achieved. **Figure 7-24** is a schematic diagram of the waste heat recovery system where a power turbine is driven. Its primary components are the main engine, a super charger (turbocharger), a power turbine, a reducer and a generator.

Figure 7-25 compares the fuel consumption of existing main engines for ships after using a waste heat recovery system. The waste heat recovery system not only saves fuel, but also reduces exhaust gas emissions.

2) Types of waste heat recovery systems for ships

Waste heat recovery systems for ships can be classified into power turbine (PT), steam turbine (ST), and combined power-and-steam turbine (combined PT and ST).

(1) Power turbine

Figure 7-26 shows a system using a power turbine (PT). The power turbine uses the turning power produced from part of the exhaust gases supplied from the manifold to the supercharger. The power turbine system can use up to 14% of the heat contained in exhaust gases for power generation and produce about 5% of the shaft power.

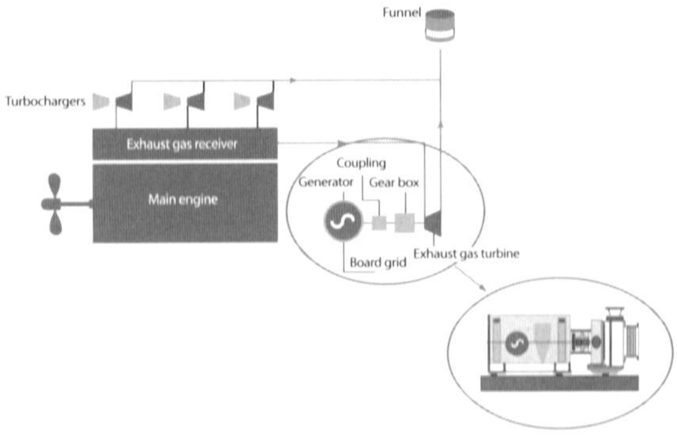

Figure 7-26 Schematic diagram of a power turbine generator

A waste heat recovery system using a power turbine consists of an exhaust gas turbine, a gearbox (reduction gear), a generator and a coupling that connects the generator and gearbox.

(2) Steam turbine

As shown in **Figure 7-27**, the steam turbine (ST) is driven for power generation by the steam produced in the economizer that transfers heat from the exhaust gas. Depending on the type of steam produced by the economizer and on how the steam is used, steam

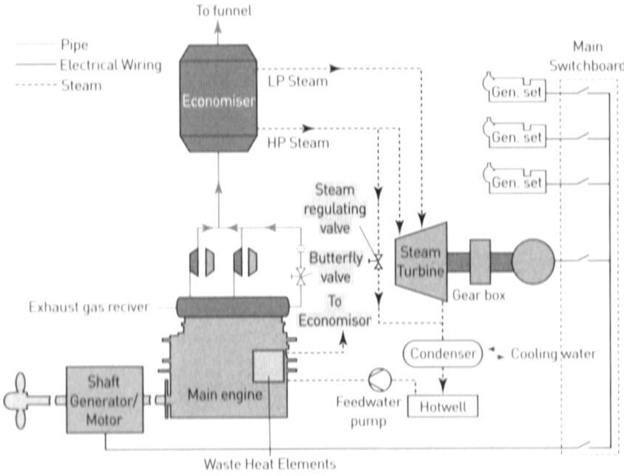

Figure 7-27 Schematic diagram of a steam turbine generator

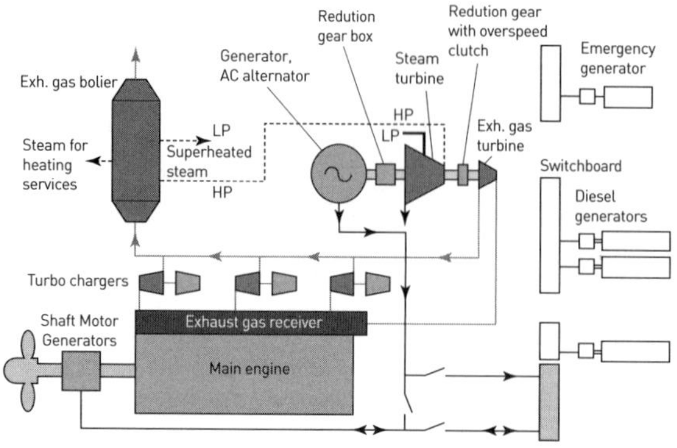

Figure 7-28 Schematic diagram of a power-steam turbine generator

turbines can be classified into single- and dual-steam pressure systems.

Figure 7-27 is an example of the single steam pressure turbine system. It supplies feed water from the hot well to scavenge air coolers of main engine with a feed water pump and to the exhaust gas economizer, and then generates steam in the evaporator. The steam then passes through the super heater to produce high-temperature, high-pressure steam to run the steam turbine.

(3) Power-steam turbine

Figure 7-28 illustrates a waste heat recovery system with a combination of power and steam turbines. Exhaust gases discharged from the main engine pass through the supercharger and then the economizer to produce low- and high-pressure steam. The exhaust gases bypassing the supercharger are supplied to the exhaust gas turbine and generate power using the steam and power turbines.

5.1.2. Fuel Cells

Fuel cells are next-generation clean generators that produce energy of high efficiency (at least 40%) and high power density by directly converting chemical energy to electric energy. This conversion takes place through the oxidization or reduction process. Cells use hydrogen as fuel and oxygen as an oxidizer. Since the only by-product is water, fuel cells produce hardly any pollutants and are highly efficient and eco-friendly, as they provide higher generation efficiency than existing technologies.

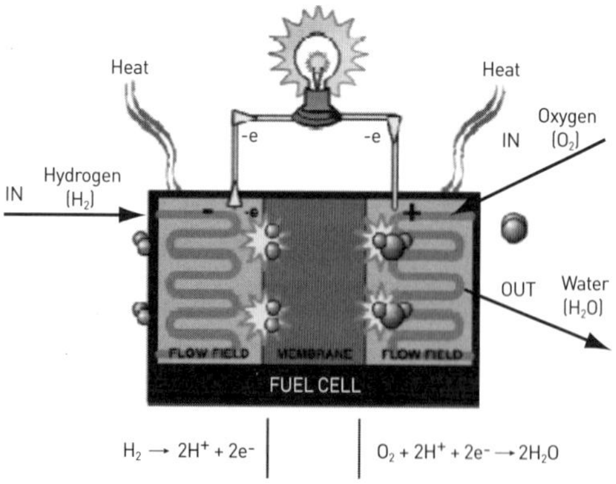

Figure 7-29 Mechanism of electricity generation using a solid polymer fuel cell

1) The mechanism of power generation using fuel cells

A fuel cell generates electricity when, at the anode, hydrogen is oxidized by a catalyst and is split into hydrogen ions (H^+) and electrons (e^-). The hydrogen ions, then, move through the electrolyte membrane, and the electrons move to the cathode through an external circuit and generate water, heat, and electricity when reacting with oxygen molecules at the cathode. **Figure 7-29** illustrates the mechanism of electricity generation using a solid polymer fuel cell.

2) Classification of fuel cells

Fuel cells are classified into five types by the kind of electrolyte, through which ions move between fuel and cathode, or they are divided into low- and high-temperature types by the operating temperature.

Fuel cells can produce electricity of from a few to some tens of megawatts, depending on the operating temperature and capacity. They thus have a wide variety of applications almost anywhere conventional energy source, such as petroleum, gas and electricity, are currently used. The capacity of fuel cells can be wide-ranging, from micro power for mobile devices to computers, marine buoys, home appliances, automobiles, aircraft, submarines, ships and large commercial power generation systems. **Figure 7-30** shows the scope of application of fuel cells.

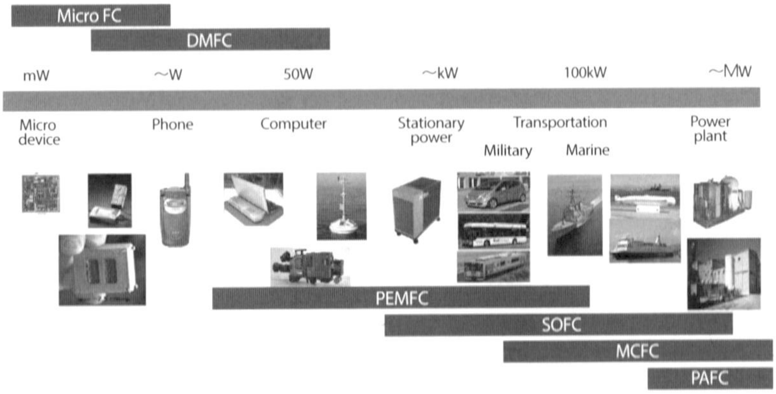

Figure 7-30 Application fields of various types of fuel cells

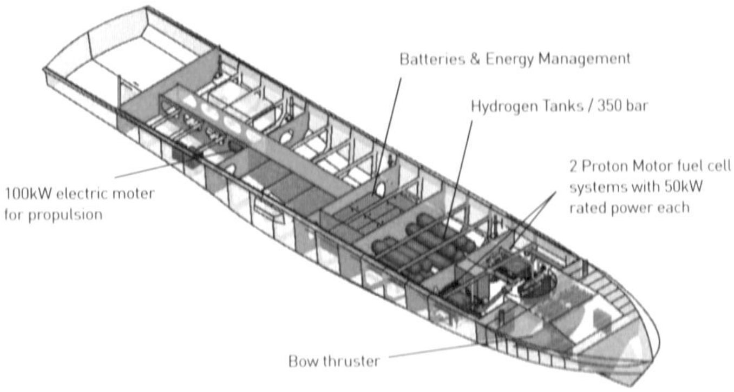

Figure 7-31 Marine fuel cell system for small ships (Source: Korean Register of Shipping)

3) Marine fuel cell systems

Marine fuel cell systems can substitute generators and engines of ships. **Figure 7-31** represents the structure of a fuel cell system for small ships generating about 100kW. The system consists of a high-pressure hydrogen tank, a battery, fuel cells, and an electric motor.

5.1.3 Wind Energy

Since ancient times wind has been a major source of power for propelling ships, but the development of large vessels equipped with propellers has made fossil fuels the chief source of power for propulsion. However, the sharp rise in oil prices and global warming has made it mandatory to reduce marine carbon dioxide emissions. Therefore, efforts are being made to use wind power to reduce fuel costs and carbon dioxide emissions in shipping.

1) Overview of wind energy

A ship uses wind power in roughly two ways: it uses wind power as a direct or indirect source of propulsion. The former includes conventional sails, airfoils, and kites, as well as a Flettner rotor using the Magnus effect, an aerodynamic phenomenon that produces lift perpendicular to the direction of the wind blowing over a rotating cylinder. When a ship travels at 15 knots across the North Atlantic Ocean under ideal weather conditions, it can save fuel by up to 15% at the speed of 15 knots, and even up to 44% at 10 knots, using this direct technology.

A ship can indirectly use wind power by installing a wind turbine on it to generate power during operation or while moored. Using an auxiliary power unit can reduce fuel costs. Wind turbines are divided into horizontal-axis wind turbines (HAWT) and vertical-axis wind turbines (VAWT). Commercial horizontal-axis turbines usually have a capacity of 5MW and larger turbines are more efficient. On the other hand, vertical-axis turbines are only suitable for small capacity generators, since high capacity generators are too large.

2) Examples of using wind energy

Figure 7-32 A ship with Flettner rotors
(Source: //www.google.co.kr/)

(1) Flettner rotor

It was in 1924 that the first attempt was made to mount a rotor on a ship using the Magnus effect. However, the attempt went almost unnoticed due to low oil prices at the time. With the recent increase in oil prices, however, shipping industry began to consider applying this technology to ship propulsion, and ENERCON, a German turbine manufacturer, built a ship featuring four 25-meter-tall Flettner rotors in 2009. **Figure 7-32** is an example of a ship featuring such Flettner rotors.

(2) Kites

Another way to propel a ship by wind is to use kites, and a German company SkySails is the leader in this technology. The company can make kites with an area of up to 640 m^2 and is planning to develop a large kite of up to 5,000 m^2. Kites can be deployed and recovered from a control panel installed on the ship's bridge. A report by the company claims that its kite can save engine power by 15% to 20% at 15 knots. **Figure 7-33** shows an example of a ship using a kite.

Figure 7-33 A kite helping propel a ship
(Source: //www.google.co.kr/)

(3) Wind profiles or wing sails

Wind profiles (or wing sails) are made of solid material instead of fabric, but they can produce thrust just like traditional fabric sails. Since wing sails are mounted on the deck of a ship and computer-controlled, they can generate more thrust than conventional sails. **Figure 7-34** is an example of a ship with airfoils installed on the deck like conventional sails.

Figure 7-34 Wing sails
(Source: //www.google.co.kr/)

It has been reported that, when wing sails are mounted on a RO-RO ship, they could produce up to 25% of the maximum engine power at the speed of 15 knots.

5.2 Future Technologies for Propulsion Systems

5.2.1 Diesel-electric Propulsion

A diesel-electric propulsion system is designed so that the electric power supplied to a motor driving a propeller shaft can be adjusted in response to the speed requirement and operating conditions. In general, four to five diesel generators with varying output are installed to optimize the output. Using the diesel-electric propulsion system saves electricity by 10% or more, allows flexible arrangement of cargo holds, and ensures flexible, effective operation and excellent load substitutability.

Once the system is adopted, it is known to enhance a ship's energy efficiency by 5 to 20%. As one of the most common ways of saving energy, the system has the following characteristics:

1) Overview of diesel-electric propulsion system

An electric propulsion system consists of diesel engines, generators and thrusters. Usually multiple medium-speed gas and dual fuel diesel engines are used to drive generators. These generators are connected to a high-voltage transmission system on the main switchboard and all loads, including the propulsion system, thruster, auxiliary machinery and ship systems, are powered through this power transmission system. The total load is distributed uniformly by the operating generators, and the ship's generation system is designed, taking into consideration total power requirement, operating mode, system flexibility, and emergency reserve power and so on as well as equipment costs.

Figures 7-35 and **7-36** are examples of two common electric propulsion systems. **Figure 7-35** shows two medium-speed motors connected to a single reduction gear to propel a ship with a single propeller, while **Figure 7-36** represents a twin-propeller system with each propeller directly connected to a low-speed motor.

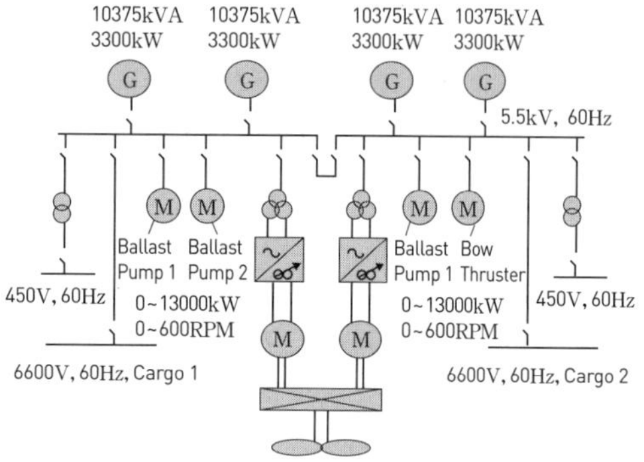

Figure 7-35 General wiring diagrams for a single propeller

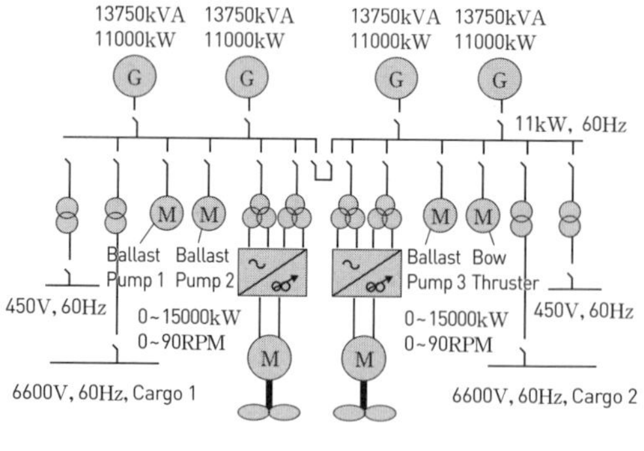

Figure 7-36 General wiring diagrams for twin propellers

2) Propulsive efficiency

An average dual fuel engine at a rated load has an efficiency of about 47%. Considering the fuel consumption and propeller shaft output, as well as 8 to 10% of transmission loss, the final efficiency is nearer 43%. The efficiency of a steam turbine propulsion system, which consists of a boiler, a steam turbine and a gear, does not exceed 30%. Once a small load is introduced while the system is working in manual mode or with speed limits, the efficiency gap between the steam turbine and electric propulsion becomes greater. **Figure 7-37** is a chart comparing the propulsion efficiency of electric and steam turbine propulsion systems.

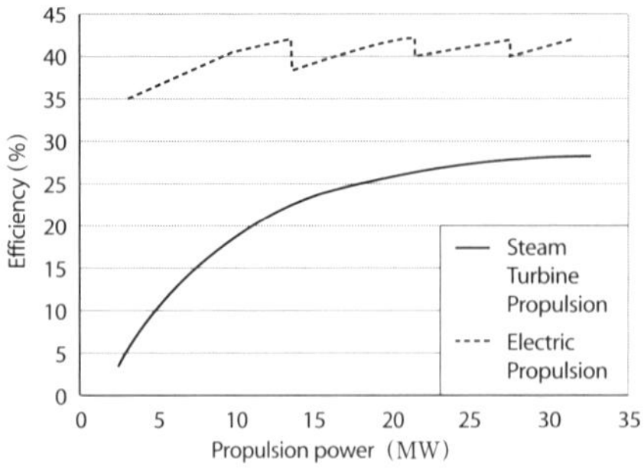

Figure 7-37 Comparison of the propulsion efficiency of electric and steam turbine propulsion systems
(Source: https://www.google.co.kr/)

5.2.2 Pod Propulsion Systems

Recently, the pod propulsion system using an electric motor has become the most popular marine propulsion system. This system features an AC motor housed in a 'pod' which is mounted on the outside of a ship's hull at a fixed position but allowed to rotate, or azimuth, freely. Because of this azimuthing ability the ship does not require rudders. **Figure 7-38** is an example of the pod propulsion system.

Figure 7-38 A pod propeller
(Source: https://www.google.co.kr/)

The original pod propulsion system was introduced by a shipyard and a manufacturer of marine electric machinery in Finland. A propulsor was installed protruding from the hull surface, and this could rotate to produce thrust in any direction, even rearward. It was sold with the name 'Z-peller.' This propulsion system was designed to drive a propeller using a gear coupled onto the motor in the ship. However, the modern pod propulsion system uses an integrated motor and propeller placed outside the hull.

Using this system, a motor does not have to be installed inside a ship and a long propeller shaft is no longer needed. It also eliminates the engine room, housing one or more large engines, and allows multiple generators to be scattered in different places to send the produced power to the pods, dramatically increasing flexibility in the general arrangement of the ship. Furthermore, it also permits the development of new hull forms, because the ship is freed from the constraints imposed by the conventional propulsion systems. The pod propulsion system, in particular, can be used to maneuver a ship in a congested or limited area like a port where lateral movements of the ship are necessary, since it produces thrust in any desired direction.

5.2.3 Magneto-hydrodynamic (MHD) Propulsion

Magneto-hydrodynamic propulsion is a system that does not require propellers. The system was developed based on Fleming's left-hand rule that an electrical current flowing perpendicular to a strong magnetic field causes the wire to experience a force perpendicular to the magnetic field and to the direction of the current. The reaction of a fluid accelerated by the aforementioned force propels a ship. The system is also known as a water-jet propulsion system using a magneto-hydrodynamic force instead of pumps. Since it does not have any moving parts, it does not experience vibration or cavitation. **Figures 7-39** and **7-40** show the "Yamato 1," a Japanese model ship using MHD propulsion and its underlying principle, respectively.

Since its first introduction in 1960, the U.S. has continued to develop the MHD propulsion system. In 1966, for the first time in the world, an MHD propulsion model test was performed. At the time, however, due to difficulty in producing strong enough magnetic fields, the test merely demonstrated that electromagnetic propulsion was theoretically possible. Compared with propeller propulsion, MHD propulsion was far less efficient and, as a consequence, it drew little or no attention among shipbuilding engineers.

Figure 7-39 Yamato 1, an MHD propulsion ship
(Source: https://www.google.co.kr/)

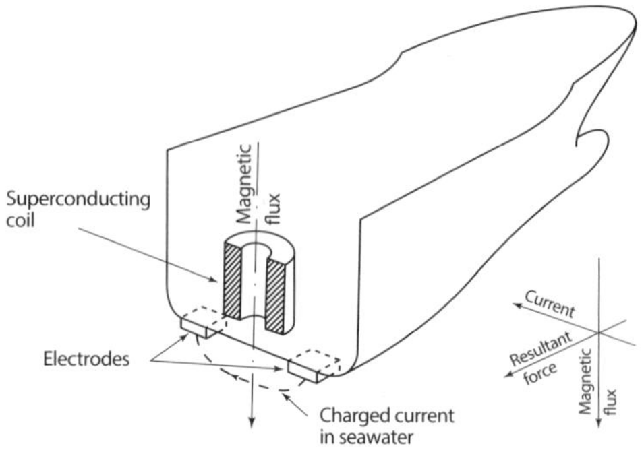

Figure 7-40 Principle of magneto-hydrodynamic propulsion

In the late 1970s, however, MHD propulsion had attracted renewed attention when it was suggested that superconductors can produce magnetic fields several times stronger than those of conventional magnets. In 1979 Japan developed the first MHD model ship in the world using a superconducting magnet. In 1985 The Ship and Ocean Foundation of Japan began a research and development program of superconducting electromagnetic propulsion ships and, in 1992, built a model ship called the Yamato 1 and ran a variety of tests on it at the Port of Kobe. It could, however, only achieve a propulsive efficiency far lower than those of conventional propellers, since the supplied electric power was mostly consumed in heating seawater passing through the system.

Commercialization of the system depends on the development of normal-temperature superconductors. Even if the development succeeds, there will still be many challenges, including the effect of high magnetic fields on the marine environment and improvement required in its propulsive efficiency. It is still a propulsion system that one can only dream of.

References

1. Cha, S.B., 2011. *Fuel Saving & CO_2 Reduction for MANB & Two-Stroke Engines* [Technical note]. Busan: MAN Diesel & Turbo Korea Ltd.

2. Energy & Environment Business Center, Korean Register of Shipping, 2010, *Special research on the development of energy-saving marine and marine carbon dioxide capture technologies* (in Korean). Seoul: the Ministry of Oceans and Fisheries.

3. Ikeda, R., 2007. *Ship Equipment* (in Japanese), Tokyo: Natsumesha Co. Ltd.

4. Hupli, J., 2008. *Technical review and development 50DF*, [Technical note]. Samho-eup, Jeollanam-do: Wartsila-Hyundai Engine Co., Ltd.

5. Jang, S.Y. et al., 2000. A review of electric propulsion system. In: Korean Institute of Power Electronics (KIPE), *Power Electronics Conference*. Gurye-gun, Jeollanam-do 19~22 July, 2000. KIPE: Seoul.

6. Jeon, H.J., 1986. *Marine power transmission systems* (in Korean), Seoul: Taehwa Publishing Company.

7. Jeon, H.J. & Choi, J.S., 1999. *Lectures on internal combustion Engines* (in Korean). Seoul: Hyosung Publishing Company.

8. Kim, J.S. et al., 2008. A study on the performance analysis for power converters of electric propulsionship. *Korean Marine Engineering Journal*, 32(8), pp.1278-1284.

9. Korea Maritime History Institute, *the era of steamship* (in Korean) [online]
Available at: http://www.seahistory.or.kr/Menu03/Menu03Sub02_5_1.htm
[Accessed 18 June 2015]

10. Korean Register of Shipping, 2004. *Electronically controlled diesel engines*. Seoul: Korea Register of Shipping.

11. Lee, D.C., 2011. *Marine diesel engines* (in Korean) [class handout], Marine Engine Lectures. Mokpo: Green Marine Equipment RIS Center, Mokpo National Maritime University.

12. MAN Diesels: Thermo efficiency system (TES) for reduction of fuel consumption and CO_2 emissions. 2005. Available at: http://www.mandieselturbo.de//files/news/filesof5055/P3339161.pdf
[Accessed 18 June 2015]

13. Pounder, 2004. *Marine diesel engines and gas turbines*. Translated from English by Jeon, H.J. & Lee, D.C., 2011, Seoul: Dongmyungsa.

14. Takisawa, S., 1995. *Frontier Technology in Evolution of Ship* (in Japanese), Tokyo: Seisando-shoten Publishing Co.

CHAPTER 8

General Arrangement

1. Introduction to Compartments and General Arrangement
2. Arrangement of Bulkheads
3. Arrangement of Machinery Spaces
4. Arrangement of Fuel Oil Tanks
5. Arrangement of Cargo Spaces by Ship Type
6. Arrangement of Deck Spaces
7. Layout of Accommodation Quarters
8. Layout of Navigation Bridge
9. General Arrangement of Passenger Ships

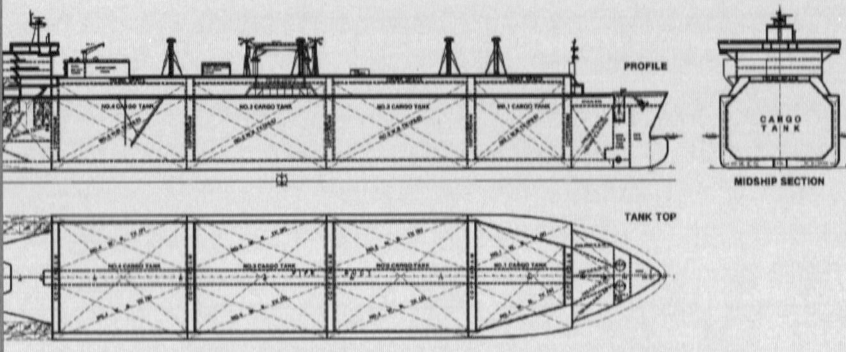

1. Introduction to Compartments and General Arrangement

1.1 General Description of Compartments

Major compartments in a ship include cargo holds, fore peak tank, aft peak tank, ballast tanks, fuel oil tanks, engine room, steering gear room, accommodation quarters (deck house), navigation bridge(wheel-house), stores, passageways for crew, piping and wiring ducts, cofferdams and void spaces, pump rooms, manifold spaces, and towing and mooring equipment spaces.

A cargo hold should be designed suitable for the cargo to be carried and be furnished with the equipment required by the relevant standards and regulations.

As shown in **Figures 8-1** and **8-2**, the cargo holds in crude oil carriers, bulk carriers and gas carriers are generally arranged between the collision bulkhead and the forward bulkhead of engine room. Furthermore, the cargo holds have the double bottom and double hull structure surrounded by ballast tanks or void spaces.

The engine room, accommodation quarters and the bow area in crude oil carriers and liquefied gas carriers are segregated from the cargo tanks by installing cofferdams fore and aft of the cargo holds as shown in **Figure 8-2**. Additionally, ballast tanks, fuel oil tanks and pump rooms when installed instead of cofferdams can be accepted as segregation spaces. Main characteristics of cargo holds and general arrangement for various ship types will be discussed in Section 5. of this chapter.

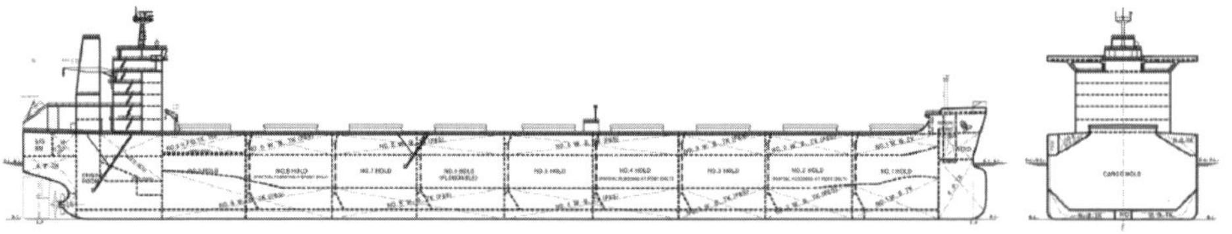

Figure 8-1 Arrangement of cargo holds in a bulk carrier

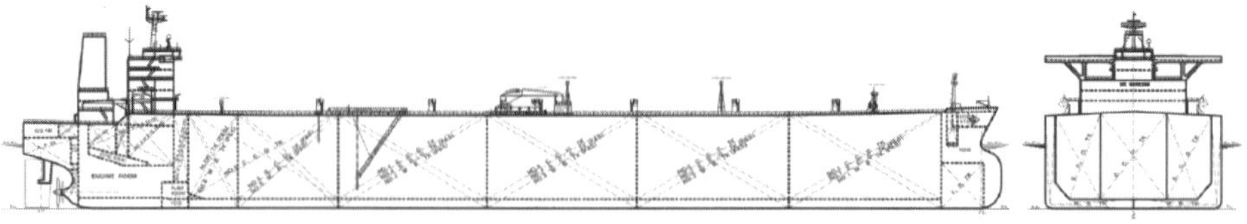

Figure 8-2 Arrangement of cargo holds in a crude oil carrier

1.2 Factors to Consider in General Arrangement Design

The functionality of the compartments is largely dependent on their arrangement. Therefore, an optimum arrangement of compartments in a ship is crucial in ensuring safety, high loading capacity and operational convenience of the ship as well as minimizing the construction cost, fuel consumption and other operational expenses. The general arrangement must be optimized during the preliminary design stage, or even earlier if possible. Relevant classification rules and other standards applied to ships must be taken into consideration, along with other factors discussed below.

1.2.1 Trim and Longitudinal Strength

During the structural design the trim and maximum bending moment are calculated for all loading conditions. Since the arrangement of compartments has a direct bearing on the ship's trim, it will inevitably influence the maximum longitudinal strength required. A ship in full load condition with evenly distributed load in all cargo holds must not have trim by the bow. It also must be designed not to have excessive trim in ballast condition; it should have adequate draft at the bow, but, more crucially, sufficient draft at the stern to ensure full propeller submergence.

Furthermore, during the preliminary design stage, the ship's maximum bending moment and maximum shear force should be predicted and reflected in the structural design. The relevant classification society approves the ship to have the maximum longitudinal strength as specified. Once the designed ship is built, the ship operator must always pay a close attention to the loading condition so that the maximum stress allowed is not exceeded.

1.2.2 Segregated Ballast Tanks

For crude oil carriers over 20,000 DWT and product oil carriers over 30,000 DWT, segregated ballast tanks must be installed as required by MARPOL 73/78 (International Convention for the Prevention of Pollution from Ships 1973, and 1978) Regulation 18.

The volume of the segregated ballast tanks must be decided so that, when they are filled with ballast water in light ship condition, the design ballast draft should be obtained. The ballast draft must be so that the midship draft (d_m) is greater than 2.0m + 0.02L, where L is the freeboard length as defined by the International Convention on Load Lines (ICLL), and the stern trim should not exceed 0.015L. Sufficient stern draft should be maintained at all times so that the propeller is completely submerged. The aforementioned rules for crude oil and product oil carriers are intended to prevent marine pollution caused by ballast water being loaded into and discharged from cargo tanks.

For bulk carriers, there is a separate regulation requiring enough capacity of ballast tanks to ensure safe operation rather than pollution prevention. For container ships or vehicle carriers there is no exclusive regulation related to ballast tanks, because those ships rarely operate without any cargo.

1.2.3 Damage Stability

Appropriate subdivision and arrangement of compartments is also imperative in meeting the damage stability requirements. The rules pertaining to damage stability include the International Convention on Load Lines (ICLL 1966), the Convention on the Prevention of Marine Pollution, the International Code for the Construction and Equipment of Ships Carrying Liquefied Gases in Bulk (IGC Code) and the Safety of Life at Sea (SOLAS 2009), and they are applied according to the ship type and its purpose.

To comply with the damage stability requirements, the ship's compartments may have to be divided into smaller spaces and their arrangement altered.

1.2.4 Safety and Dangerous Spaces

Dangerous spaces are tanks, holds and storage spaces designated for carrying dangerous cargo and fuel oil according to the provisions in the International Maritime Solid Bulk Cargoes (IMSBC) Code, International Bulk Chemical (IBC) Code, IGC Code, International Electro-technical Commission (IEC) 60092, International Association of Classification Societies (IACS) SC 79, International Maritime Dangerous Goods (IMDG) Code, and other relevant classification rules. Once designated as such, they must meet the requirements of provisions of ventilation, explosion-proofing and safety equipment. They must also be safely segregated from the adjacent safe spaces according to the methods set forth by the aforementioned rules.

For liquefied gas carriers and gas fuel-propelled ships, the designation of gas-danger zones and the protective arrangement of gas-safe zones must be carried out in accordance with the IGC Code and the standard of IEC 60092. Furthermore, the relevant provisions in the International Gas Fuelled Ship (IGF) Code, which is currently under development, should also be incorporated.

1.2.5 Ensuring Visibility from Navigation Bridge

SOLAS Regulation V/22 requires that adequate navigation bridge visibility should be provided from the steering stand regardless of draft, trim and loading conditions. The bridge visibility should be such that 'the sea surface forward of the bow to 10° on either side is not obscured by more than two ship lengths (2 × LOA), or 500m, whichever is less.' Therefore, the bridge height, trim, draft and cargo loading conditions must be taken into consideration. For container ships, in particular, maximum allowable container stack heights on deck which is in front of the bridge must be limited to ensure good visibility.

1.2.6 Air Draft

When there is a bridge or a similar structure with a limited clearance above water across any part of the routes that the ship is expected to pass, the air draft must be less than the clearance. In general, air draft is expressed as the distance from the waterline at the minimum draft or the ship's baseline to the uppermost point of the fixed structure of the ship. The number of accommodation floors, the height of the navigation bridge and the maximum height of the antenna mast should be taken into consideration in determining the air draft.

If absolutely necessary, the antenna mast can be made foldable to reduce the air draft temporarily. For bulk carriers the distance from the waterline to the upper part of hatch coaming is sometimes treated as air draft, and this is to accommodate the operation of cargo handling equipment in dedicated terminals. Usually the ship-owner decides what air draft is needed by judging from the expected shipping routes.

1.2.7 Forecastle Deck

The forecastle deck is used to provide enough space for towing and mooring equipment at the bow and to prevent the bow area from being flooded in rough seas. In addition, a forecastle deck house can be installed for freeboard assignment as specified by the ICLL or to secure some extra reserve buoyancy.

1.2.8 Arrangement of Accommodation Quarters and Funnel

Conventional ships with the engine room near the stern usually have accommodation quarters and funnels above the engine room and fuel tanks. Some of the very large container ships have the accommodation quarters well forward of the funnel and engine room spaces due partly to the protection requirement of the fuel oil tanks. Locating the accommodation block forward also improves the bridge visibility and helps to increase the above-deck loading capacity of containers. The space below such accommodation block can be utilized to locate the fuel oil tanks and they will have much smaller influence on trim, even with the inevitable variation in the amount of fuel carried in them during operation.

1.3 Rules Relevant to General Arrangement

Since many ships are required to operate on international shipping routes, they must meet all the conventions and codes set forth by the IMO. In addition, when using certain routes, such as canals, ships have to comply with the rules of each canal authority and the classification rules. They are required to meet not only the international standards, but also the national rules of the flag state and visiting countries. More details about this point can be found in Section 3. Design Constraints of Chapter 3.

2. Arrangement of Bulkheads

Bulkheads are partitions dividing the internal space of a ship according to the purposes of utilization. However, they are dominant structural strength members too and should be designed so that they can make maximum contributions to the structural strength of the ship while providing the optimum allocation of spaces for a given hull form. In particular, engine room bulkheads, bow and stern collision bulkheads, and some horizontal bulkheads must be installed according to the requirements of international standards.

2.1 Arrangement of Peak and Machinery Space Bulkheads

2.1.1 Engine Room Bulkheads

SOLAS requires that watertight bulkheads be installed to separate the engine room from the cargo holds and accommodation quarters, because it contains a variety of critical equipment. Piping arrangement to/from the cargo spaces is also limited in ships carrying flammable cargo, such as oil tankers.

2.1.2 Collision Bulkhead

SOLAS also requires that a watertight collision bulkhead be installed extending from the floor to the bulkhead deck. It should be situated aft of the fore perpendicular (FP) at a distance not less than $0.05L$ or 10m, whichever is smaller, and not more than $0.08L$ or $0.05L + 3$m, whichever is greater, unless permitted by the relevant authorities.

If any part of the ship below the load line protrudes forward from the FP, such as a bulbous bow, the distances described above are measured from either a point at a mid-length of such protrusion, at a distance 0.015L forward of the FP, or at a distance 3m forward of the FP, whichever is less.

2.1.3 After Peak Bulkhead

The after peak bulkhead is installed to prevent the engine room at the stern from being flooded due to any damage to the propeller shaft. Rules for the after peak bulkhead are not as strict as those for the bow collision bulkhead. While the collision bulkhead is required to extend to the bulkhead deck, the after peak bulkhead is allowed to be lower than the bulkhead deck, as long as it extends to a height above the summer load water line.

2.2 Arrangement of Horizontal Watertight Bulkheads and a Double Hull Structure

2.2.1 Double-bottom Bulkhead

For oil tankers, a double-bottom bulkhead in the cargo space of oil tankers is required to prevent marine pollution caused by oil spills. For other ship types, it is used to reduce the risk of sinking due to grounding.

Many ship types which are required to comply with the damage stability regulations specified by any conventions other than MARPOL are primarily required to survive collision accident. However, it was judged that there were insufficient provisions for grounding casualties, and consequently a mandatory requirement was introduced in SOLAS 2-1 Regulation 9 which requires a double-bottom bulkhead to be fitted at a certain height from the after peak bulkhead to the collision bulkhead. The 'certain height' refers to the height of the double bottom which must be at least 5% of the beam of the ship, but not less than 0.76m, although it does not need to exceed 2.0m.

2.2.2 Other Horizontal Watertight Bulkheads

For a dry cargo ship with Type B freeboard which has to satisfy the probabilistic damage stability requirements, an additional horizontal watertight bulkhead other than the freeboard deck may be installed in order to increase the survival rate in the event of the watertight compartment being flooded. This is more or less limited to car carriers with very high molded depth (D). For passenger ships, there are additional complex and strict rules for horizontal watertight bulkheads to achieve enhanced safety.

2.2.3 Double Hull Structure

The MARPOL 73/78, Regulation 19 requires that for all oil tankers, the clearance between the bulkheads forming cargo tanks and the hull plates should be at least w from the side and h from the bottom, according to the following formulae:

w(lateral clearance) = 0.5m + (DWT) / 20,000 or 2.0m, whichever is less but not less than 1.0m
h(bottom clearance) = (B) / 15 or 2.0m, whichever is less but not less than 1.0m

3. Arrangement of Machinery Spaces

3.1 Components of an Engine Room

An engine room contains power systems for the propulsion of the ship as illustrated in the sectional drawing in **Figure 8-3**. It accommodates one or more main engines, generator engines, auxiliary boilers, shafts, engine room cranes, pumps, air compressors, compressed air tanks, fuel oil and lubrication oil tanks, engine room fans, engine control room, workshops, engine room parts lockers, purifier room, passageways and other access facilities.

3.2 Criteria for the Arrangement of the Machinery Space

Since the size of the engine room is depended on ship type, hull form, main and auxiliary engines, and shafting system, it is one of the first factors that must be determined during the concept design stage. The length of the engine room on a typical stern-engine ship accounts for 11 to 13% of the entire ship length, and the length has a great influence on hull strength, cargo hold volume, trim and ship price. It is important to minimize the size (especially the length) of the engine room. When developing a new hull form, it should be possible to extend the limits of the engine room size to meet the owner's wishes.

The main factors that need to be considered when determining the location of the main engine include the capacity of the boiler and the generator engines and the location of the shaft generator, if used. However, the main engine should be located as far aft as possible to minimize the length of the propeller shaft. Efforts should be made in order to optimize the arrangement of the zones within the engine room as well. The main considerations in this task include the location of the main engine and the length and height of the engine room.

3.2.1 Location of Main Engine

In the engine room arrangement, the location of the main engine is determined by considering various elements as shown in **Figure 8-3**.

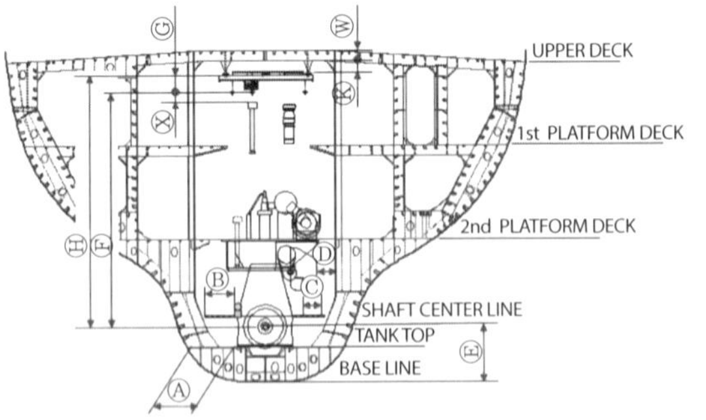

Ⓐ Space for plumbing and installation of side stopper
Ⓑ Passage beside turning gear
Ⓒ Passage under air cooler
Ⓓ Passage around main engine
Ⓔ Height from base line to shaft center line
Ⓕ Height from shaft center line to crane hook (It is determined according to the size of engine)
Ⓖ Required height for crane and installation of I-beam for crane operation
Ⓗ Height from shaft center line to underside of upper deck longitudinal in the engine room compartment
Ⓚ Space for installation of drain pipe from accommodation (250 mm)
Ⓦ Size of web beam for upper deck in the engine room compartment
Ⓧ Clearance space (150~200 mm)

Figure 8-3 Factors which determine the location of main engine

3.2.2 Length of Engine Room

In a conventional stern-engine ship, the length of engine room is determined by considering the elements shown in **Figure 8-4.**

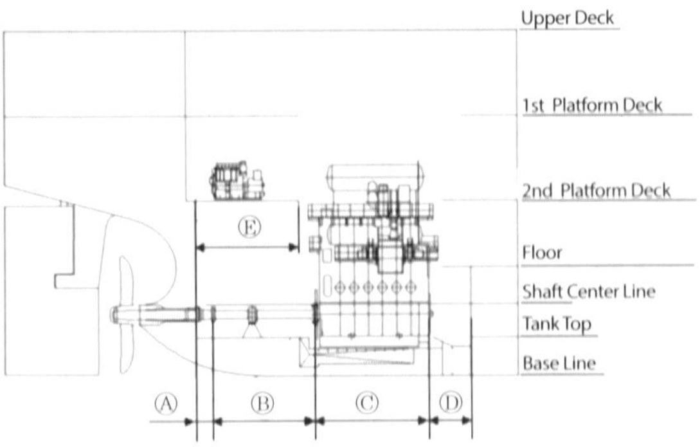

Ⓐ length from engine room after peak bulkhead to the end of propeller shaft
Ⓑ length of intermediate shaft – clearance for propeller shaft replacement (A+B)
Ⓒ length of main engine
Ⓓ length between main engine to engine room forward bulkhead
Ⓔ length for the installation of generator engine

Figure 8-4 Factors determining the length of engine room

3.2.3 Criteria for Determining the Deck Height of Engine Room

The number and height of decks arranged in engine room are determined by taking the following into consideration:

- The number of decks must be kept to a minimum, as they contribute to the ship's weight.
- The platform deck floor should be kept flat.
- The deck floor should be supported vertically with hull structural member in as consistent a manner as possible.
- The height of tank top and floor are determined by the shaft centerline and the type of the main engine.
- VLCCs and other ordinary commercial ships usually have two platform decks.
- Large container ships and LNG carriers often have three platform decks.
- Enough room should be provided for replacing the pistons of main engine.
- Enough room should be provided for replacing the pistons of generator engines.

4. Arrangement of Fuel Oil Tanks

Pollution caused by cargo oil is already regulated under MARPOL, but there are no international regulations for possible pollution caused by fuel oil. However, the increase in ship size and speed will lead to a dramatic increase in the amount of fuel oil carried by ships. The growing awareness of the fact that a collision or grounding can result in serious marine pollution led to the 2005 adoption of related provisions by the IMO, and to the mandatory protection of fuel oil tanks on every ship delivered or to be delivered since August 2010.

According to the rules, a ship should be fitted with a complete double hull structure throughout, or it needs to prove that the ship meets the allowable limits of oil spill in the event of a damage occurring.

4.1 Design Criteria for Fuel Oil Tanks

The double hull requirement of a fuel oil tank is almost the same for the cargo tank arrangement of oil tankers, but it requires relatively smaller clearances. For oil tankers larger than Panamax class, both the side and bottom clearances of the cargo tanks must be at least 2.0 meters, but they are required to be only 1.0m and 2.0m, respectively, for the fuel oil tanks. The amount of oil spill is calculated by adding the oil spill due to the lateral damage caused by collision and the bottom damage caused by grounding. The requirement is deemed to be satisfied if the total spill is less than the permissible spill amount for a given ship size. The calculations are made as follows:

4.1.1 Lateral (collision) Damage

① For each of the fuel oil tanks, the probability of damage in longitudinal, transverse or vertical directions are determined according to the location of the tanks;
② The tank capacity is multiplied by the damage probability;
③ Steps ① and ② are repeated for each fuel oil tank, and the results are summed.

4.1.2 Bottom Damage

① For each of the fuel tanks the probability of bottom damage is determined (bottom and lateral damages have different probability of occurrence even for the same location);
② The capacity of each tank and the probability are multiplied, and the results of all tanks are summed;
③ When calculating bottom damage, water pressure effect can be considered. For lateral damage, the amount of the spill is assumed to be the entire volume of the tank. However, for bottom damage, it can be assumed that a portion of the fuel oil remains in the tanks due to the external water pressure which results in a smaller amount of spill than in the lateral damage case. The maximum capacity of a fuel oil tank is limited to 2,500m^3 in all cases, whether it meets the double hull or maximum spill requirements.

5. Arrangement of Cargo Spaces by Ship Type

5.1 Bulk Carriers

A bulk carrier has five to nine cargo holds, depending on its size, as shown in **Figure 8-5**. The transverse bulkheads between cargo holds are usually constructed of corrugated plates, but sometimes they are constructed of flat plates suitably stiffened. Pipe ducts and a stool are installed at the bottom. Each hold has a hopper structure at each of the four corners, as illustrated in the cross section drawing in **Figure 8-5**. Both single and double hull structures are allowed. While the single hull structure provides the maximum cargo hold capacity, the double hull structure gives a smooth internal surface without any protrusions. Thus, it is good for handling cargo.

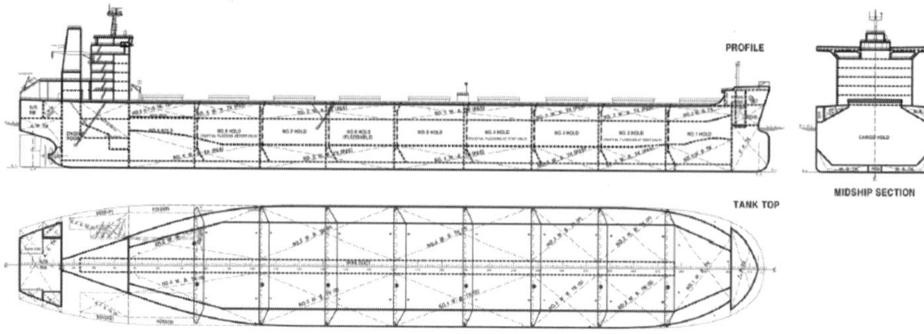

Figure 8-5 General arrangement of a typical bulk carrier

5.2 Container Ships

The meteoric rise of container ships since 1960s, especially since 1970s, is due largely to the fact that their turn-around time at ports can be minimized, because containerized cargo can be handled much more efficiently than break bulk cargos. Containers are carried in holds and on deck, and some ships have an open hold structure without any hatch cover, even though most modern container carriers do have hatch covers which are made strong enough to withstand the weight of the on-deck containers. Just like other ship types, economy of scale applies to container shipping as well. This has led to many Panamax class container ships being built which are the largest ships that can undertake the passage through the Panama Canal. Nevertheless, the advantages to be gained from larger ships are so great that many post-Panamax class ships appeared, an example of which is illustrated in **Figure 8-6**. The prospect of the new larger Panama Canal being built has also led to an increase in ship's breadth, length and container capacity. It has contributed to the growing number of new container ships with a carrying capacity of more than 11,000 TEU, although with diminished speed. Large ships have better fuel efficiency using low-speed engines, especially with a waste heat recovery system (WHRS) which reduces fuel consumption. Moreover, the application of 2-tier lashing bridges allows more tiers of containers to be carried on deck.

By placing fuel oil tanks in the midship cargo hold area or bulkhead spaces between cargo hold, the double skin protection requirements can be met. On the other hand, by adopting the twin-island deckhouse structure, in which the accommodation quarters are located at midship above the fuel oil tanks, consequently visibility from the bridge and cargo capacity, can be improved.

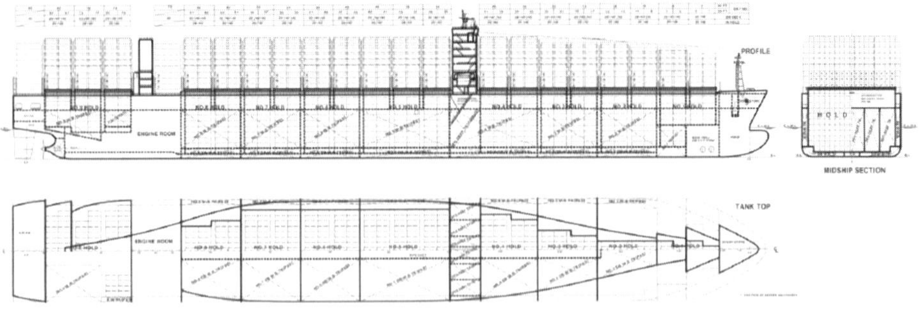

Figure 8-6 General arrangement of a large container ship

5.3 Crude Oil Carriers

A VLCC has two longitudinal bulkheads for each cargo hold as shown in **Figure 8-7**, but smaller Suezmax (160,000 DWT) and Aframax (118,000 DWT) ships can have a single longitudinal bulkhead to separate each cargo hold into port and starboard tanks. The cargo holds are segregated from safety-critical spaces like the engine room by a void space between them and the adjacent cargo holds. A crane for handling cargo manifolds and pipes is installed in the midship. Since this type of ship has a narrow and tall superstructure, thorough and careful reinforcement is required to ensure sufficient structural strength of the bridge support and to minimize vibration.

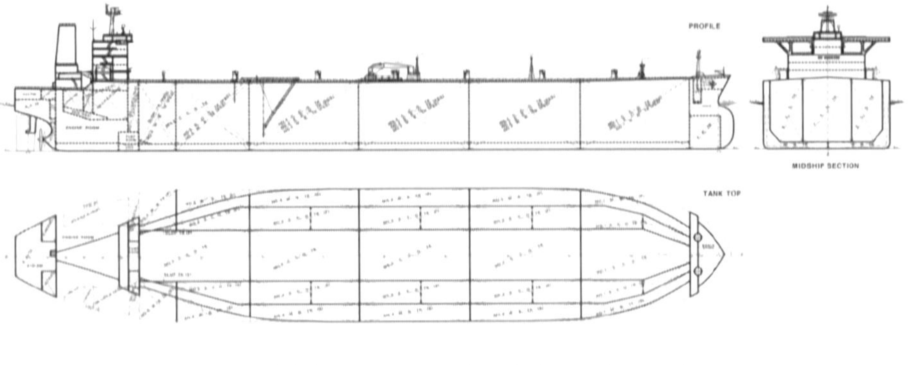

Figure 8-7 General arrangement of a typical crude oil carrier

5.4 Liquefied Natural Gas Carriers

Figure 8-8 presents the general arrangement of a liquefied natural gas (LNG) carrier. LNG carriers are classified into Moss, membrane, and self-supporting prismatic B (SPB) types, depending on the cargo containment system (CCS). A cargo hold should be designed by taking into consideration on the unique properties of LNG, the temperature of which reaches −162°C, and it must be fitted with insulating materials to minimize gas boil-off. Cofferdams are installed at both ends of a cargo hold, and a trunk deck is placed on top of the membrane tank to provide a double skin structure around the entire structure.

The cargo compressor room, electric motor room, vent mast and manifolds are arranged on the trunk deck. The passageways and wiring duct are connected to the forecastle of the ship along both port and starboard sides. It also features bow fuel oil tanks and bow thruster rooms and a pipe duct in the double bottom. Since the design draft is shallow, it allows a sunken poop deck structure.

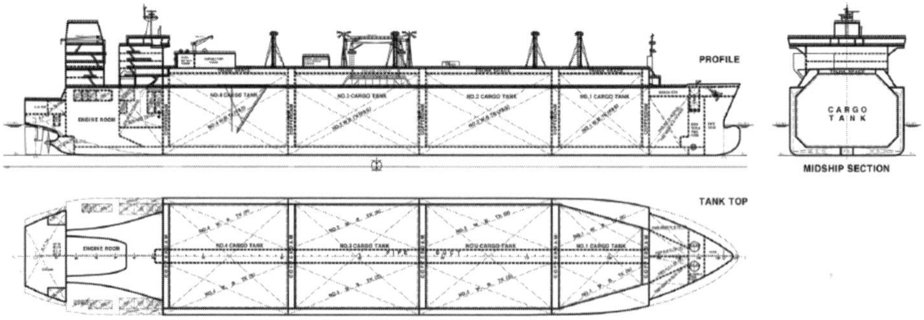

Figure 8-8 General arrangement of a typical membrane type LNG carrier

5.5 Car Carriers

Figure 8-9 shows a typical car carrier which is a type of Ro/Ro (roll-on/roll-off) ship from the way the cargo (vehicles) are loaded and unloaded. The car carrier is designed to transport trucks, buses, bulldozers and trailers as well as automobiles.

Unlike other merchant ships, car carriers have many decks, typically nine to thirteen. To allow Ro/Ro of cargo, external ramps are installed at stern, side and/or bow, and there are fixed or retractable internal ramps to allow movement between the car decks. In order to ensure efficient loading and unloading operations the ramps and structural pillars must be arranged taking into account the turning radii of vehicles and the on-board traffic flow. The advances in shipbuilding technology have led to the owners preferring one-pillar systems to two-pillar systems.

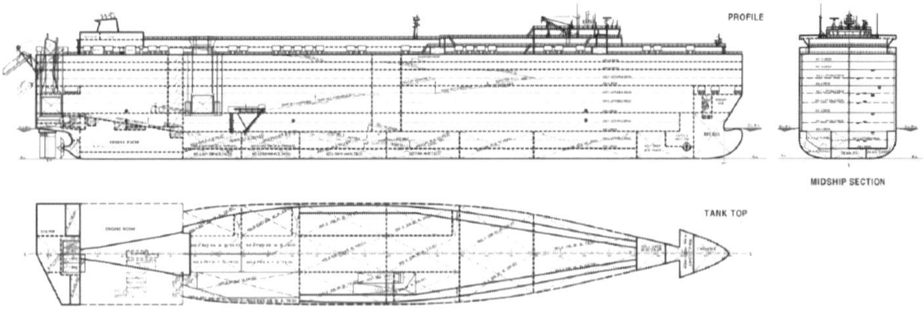

Figure 8-9 General arrangement of RO/RO carrier

6. Arrangement of Deck Spaces

6.1 Anchoring and Mooring System

6.1.1 Equipment number

A ship is subjected to external forces due to wind, waves and current while anchored or moored. It is also occasionally necessary to utilize the anchor for emergency turning and rapid deceleration. It is important, therefore, that the anchoring and mooring systems are made strong enough to withstand possible forces applied to them in all foreseen circumstances. The most immediate concern is to prevent any of the component parts failing, particularly anchors dragging or anchor chains/mooring ropes parting. Furthermore, sometimes ships may have to be towed, when disabled due to mechanical failures, accidents or any other causes, and the failure of any part of the towing system is highly dangerous, possibly resulting in serious damages to the ship being towed and the towing ship.

The size and number of anchors, anchor chains, mooring ropes, towing ropes and the deck equipment should be carefully determined. The equipment number is often used for this task and it is calculated empirically as shown below. The lift and braking capacity of the windlass can then be determined by taking the minimum breaking load and diameter of the anchor chain into consideration

$$\text{EQ. NO} = (\Delta)^{2/3} + 2BH + A/10$$

where EQ. NO: Equipment number

B: Beam

Δ: Summer displacement (m^3)

A: Lateral projected area above load water line (m^2)

H: Height from the load water line to the highest point of the deck house or cabin broader than $B/4$ and located on the highest through deck.

6.1.2 Anchoring System

An anchoring system is provided primarily for keeping a ship at a fixed location at sea but it is also utilized in emergency for rapidly reducing the ship's speed, for example, to avoid collision in congested waters. Anchors are also used for restraining a stranded ship to prevent further damage. The equipment and outfitting included in the anchoring system are: anchors, anchor chains, anchor windlasses, chain glands, chain pipes, hawser pipes and bell mouth. For more details on the arrangement of anchor system components, see Section 4. Deck Outfiting in Chapter 10.

6.1.3 Mooring System

A ship is a floating structure and therefore it must have sufficient equipment for safe mooring which can withstand the environmental loads. Mooring lines are used to restrain the ship against drifting away from the quayside or movement in the longitudinal direction along the quay. Breast lines prevent the ship's movements in transverse directions and spring lines inhibit the ship's fore-and-aft movement. Mooring equipment consists of mooring winches and capstans; and mooring outfitting includes side fair-leaders, chocks, bollards, bitts, deck stand rollers, emergency towing components and single-point mooring components. See Section 4. Deck Outfiting in Chapter 10 for further details of mooring systems and equipment.

6.2 Steering Gear Room

A ship must have a steering gear to control the rudder which is needed to keep the ship on its course or maneuver it. The steering gear needs to be installed in an easily accessible space as far removed from the engine room as possible. For ordinary commercial ships, the steering gear room is a separate space at the rear of the engine room, and it contains auxiliary machinery including steering gear, hydraulic systems for deck machinery, and other equipment. Steering equipment placed in the steering gear room includes an automatic steering system, hydraulic oil storage tanks and lubrication oil distribution pumps. In some cases an emergency fire pump room may be put in this space as well.

6.3 Boatswain Store

A 'boatswain store' or 'bosun store' is located below the forecastle deck and is used to store deck supplies, spare parts and equipment needed for repair and maintenance by the boatswain and his deck crew. Hydraulic motors for driving deck machinery and electric switchboards can also be installed here. Other equipment and facilities that can be found in a boatswain store include steel shelves, wooden grating for rope storage, Suez searchlights, ropes and spare accessories. **Figure 8-10** is a 3D drawing of the arrangement of equipments in the boatswain store which is useful to assess the store layout and identify any possible interference among the equipments.

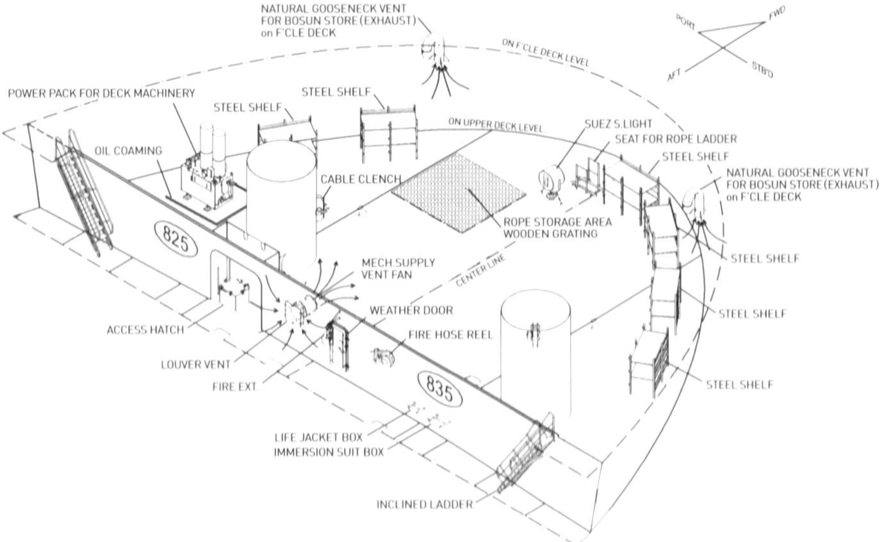

Figure 8-10 A 3D drawing of the layout of equipment and facilities in the boatswain store

6.4 Manifolds

Manifolds are pipe systems connecting the tanks on board to the storage tanks ashore, and are used for loading and discharging liquid cargo. It is usually located in the middle area of the ship. Manifolds on an oil tanker consist of cargo oil lines for loading/discharging oil, fuel oil lines, diesel lines, and a vapor line for venting the evaporated gas during loading and unloading.

The number and location of manifolds, clearance from the sheer strake, height from the deck, and spacing among manifolds depend on the ship size and cargo type, and are determined in accordance with the recommendations of the Oil Companies International Marine Forum (OCIMF). The OCIMF also provides detailed rules for the work platform and spill tanks underneath the manifolds. Since the other design details should also follow the OCIMF rules, the designer must carefully check the 'Rules for Manifolds and Accessories for Oil Tankers' when designing a manifold. If the cargo is liquefied gas, the manifolds must be designed in accordance with the 'Rules for Refrigeration Manifolds for Liquefied Gas Carriers.'

6.5 Lighting Equipment

The lighting system design determines the most suitable lighting source, fixture type, size and location, in order to achieve the optimum working conditions in a given space appropriate for the purpose of the space in terms of quality, intensity and direction of lighting. The first requirement for good illumination is that all persons within the room should be able to discern objects and other people clearly so that they can move and work without being hampered. The second requirement is to provide a comfortable working environment.

The design considerations of lighting equipment include brightness, distribution of brightness, glare, flicker, shadows, spectrum and mood; and the design parameters are location, equipment, cost and ease of repairs of lighting fixtures. Lighting fixtures that are commonly used are fluorescent, flood and incandescent lights.

6.6 Masts

Masts of a ship are tall posts where navigation and signal lights are installed. Masts normally found on a ship include a bow mast mounted on the forecastle deck; a radar mast, mounted atop the accommodation quarters; and a stern mast, mounted on the poop deck. Navigation and signal lights must be installed on the masts according to the International Regulations for Preventing Collisions at Sea (COLREG, 1972).

6.7 Pump Room

A pump room contains cargo pumps for moving cargo oil to/from cargo tanks, ballast water pumps for taking in and discharging ballast water, and a ballast water treatment system for sterilizing microorganisms in the ballast water.

Since the pump room is located close to a cargo tank and carries dangerous oils, it is considered a hazardous space. Therefore, it is placed in front of, and physically separated from the safety-critical engine room. The cargo pumps are driven by a steam turbine to enhance safety, but the ballast pumps are driven by electric motors.

7. Layout of Accommodation Quarters

Since accommodation quarters are spaces for the crew to rest and work, they should be arranged to provide maximum safety and comfort.

7.1 Overview of an Accommodation Plan

The accommodation block also contains a navigation bridge, navigation and communication equipment, cargo control room, fire control room and air-conditioning control rooms. An accommodation plan is a drawing that illustrates the arrangement of the rooms and cabins on each deck.

The layout of accommodation quarters must follow the owner's preferences but it must comply with the provisions of SOLAS, the International Labor Organization Conventions, ICLL, IMO Resolution A467 and other relevant classification rules. Moreover, the national regulations of the flag state must also be complied with.

7.2 Cabin Components

A cabin is bound by the hull, decorative and/or noise-isolating partitions and deck lining on the steel decks, and will have openings such as at least one entrance door and windows for natural lighting. Each cabin will be furnished with at least a bed, chairs, a desk, drawer units, a wardrobe or a locker and lighting, similar to a hotel room. Cabins for senior personnel are usually suites with a number of rooms.

7.3 Areas within Cabins

7.3.1 Accommodation Quarters

Cabins are allocated depending on the ranks of the crew on board which are summarised in **Table 8-1**.

Table 8-1 Rank of crew on board

Class	Grade	Person		
		Deck Department	Engine Department	Other Departments
Officer Class	Captain Class	Captain	Chief Engineer	
	Senior Class	Chief Officer	2nd Engineer	
	Junior Class	2nd Officer	3rd, 4th Engineer	Radio Operator
	Petty Class	Boatswain	No.1 Oiler	Chief Cook, Chief Steward
Crew	Crew Class	Able Seaman	Oiler	Cook
Other				Suez Crew, Repair Crew

7.3.2 Public Spaces

The 'public spaces' is a collective term for all common spaces provided in order to ensure crew's comfort. They vary considerably from ship to ship, but usually include dining rooms, lounges, smoking rooms, gymnasiums, swimming pools and offices.

7.3.3 Control Spaces

The control spaces are the places where control of navigation, fire and cargo handling takes place. A cargo control room, ballast control room, bridge and chart room, communications room and fire control room belong to this category.

7.3.4 Other Spaces

- Catering spaces: galleys and pantries for cooking food, and storing supplies, kitchen equipment and utensils.
- Sanitary spaces: include bathrooms, laundry rooms, dryer rooms and saunas.
- Passageway spaces: include stairs, corridors and stairways to the engine room and emergency exits from the engine room.
- Provision spaces: dry and refrigerated food stores.
- Machinery spaces: spaces for a variety of equipment, such as air-conditioner rooms, emergency generator rooms, and CO_2 rooms.

8. Layout of Navigation Bridge

8.1 Criteria for Equipment Layout in the Navigation Bridge

8.1.1 Overview

The wheel house and chart room are the hub of ship control. In control of any complex organization such as a ship, it is crucial for the controller to be able to gather information, appraise the situation, decide on appropriate responses and deliver orders in the most accurate, expeditious and logical way possible in any operating conditions. Therefore, the equipment should be laid out by considering functional and operational factors as well as aesthetics.

8.1.2 General Criteria for Arrangement

1) Monitoring

The basis of ship control in the wheel house is based on the information gathered through monitoring. While monitoring is important in every direction, it is especially important in the area extending from dead ahead to 90° starboard, because, according to the collision prevention rules, if a ship that detects another ship on its starboard side, and there is a risk of collision if both ships maintain the current courses, it is the 'give-way' ship. A give-way ship must take an early and decisive action to keep out of the way of the other ship. The monitoring station is often located slightly to the starboard side in the front center area of the wheel house and it should be designed to provide as wide a view of the surroundings as possible.

2) Traffic flow of personnel

There is a close spatial relationship between the locations of monitoring workstation, chart table, radar, steering wheel and steering stand. When the ship leaves and arrives at a port, the communication equipment, such as engine room telegraph, telephone, and order-delivering devices, is often used. Therefore, their functions and the relationship between them should be taken into consideration when arranging the equipment, so that the personnel movement can be made as direct and short as possible.

3) Harmonious room arrangement and aesthetics

The equipment installed indoors should be built-in and housed in a console or group panels to enhance the appearance.

8.1.3 Criteria for Equipment Layout

Once an overall structural drawing of the navigation bridge and a list of equipments is prepared, the equipments are arranged according to the criteria discussed below. **Figure 8-11** shows the actual layout of equipment in a typical wheel house.

Figure 8-11 Actual arrangement of equipment in a wheel house

1) Radar indicator

Since a radar indicator is closely related to the chart table, they are installed close together. When more than two indicators are installed, the owner may wish to have one of them to be installed near the monitoring workstation.

2) Steering stand

Since the steering stand is the reference location for the quartermaster, it should be installed on the centerline of the wheel house, and it is absolutely imperative to secure good visibility from this position.

3) Magnetic compass projection tube and voice tubes

The magnetic compass projection tube and voice tubes should be arranged in such a manner to allow the quartermaster at the steering stand a clear view of the gauges located at the front of the wheel house through the lower part of the projection tube. The location and height of the equipment should be determined so that the needles of the magnetic compass are clearly visible.

4) Chart table

The chart table should be wide enough to have two charts on it. Also, there should be enough room between the chart table and the back wall of the chart room for opening the chart drawers.

5) Bridge console

The bridge console of a ship can be installed either hard against the front wall of the wheel house or with a clearance from the front wall. Considering the position of the monitoring equipment, it is usually placed on the starboard side of the wheel house.

6) Bridge group panel

The centralized panel of bridge equipment is installed either stand-alone to the port side of the centerline or built into the partition between the wheel house and the chart room.

7) Gauge board

A gauge board consists of indicators for speed, depth, steering angle and main shaft rotation, a wind vane and an anemometer, and a chronometer, which is installed at a head-up position at the front of the wheel house where the quartermaster can clearly see the gauges. The board is usually installed on the centerline or slightly to the starboard side, depending on the ship-owner's preference.

8) Other equipment

The arrangement of equipment other than those described above is determined according to the purpose of each device.

8.1.4 Miscellaneous Considerations

① In determining the layout of any equipment, it is important to consider needs for repair and maintenance. In particular, manufacturer's instructions should be carefully read and followed, for example, the requirement to provide sufficient space for dispersion of heat from the equipment.

② The bottom of equipment installed on the ceiling must be sufficiently high so as not to cause any obstruction to the movement of personnel.

③ Alarms, buzzers and bells must be arranged within the audible distance of the crew on watch. If any signals with the same sound can cause confusion, they must be toned differently.

8.1.5 Applicable Rules and Regulations

The rules and regulations related to the arrangement of the navigation bridge include national regulations of the flag state in addition to the following:

- DNV GL Nautical Safety Rules: Applied when DNV GL additional class notations NAUT- are to be adopted. Deals with bridge design, bridge equipment, and displays and their layout.
- Panama Canal Rules (International Maritime Pilots Association, Aug. 1976): Deals with the requirements in navigation bridge designs, including visibility, windows, access and design of bridge wings.

9. General Arrangement of Passenger Ships

The general arrangement of passenger ships depends on the ship type, but it has the following common characteristics:

- Distinctive appearance and interior designs are important.
- It is recommended to segregate passenger and public spaces from crew spaces.
- Safety of passengers and crew is paramount.

All passenger ship-owners want to have their own unique features in the ships that they own or operate, and they demand an exterior appearance that is as distinctive and pleasing as the ship's interior design, since the competitiveness in this market is sought through differentiation in attractiveness. Much effort is made in order to satisfy the continuously changing passenger tastes while maintaining its distinctiveness, which makes the general arrangement design of passenger ships a very difficult task. The designer, therefore, plays an important role in examining, identifying and resolving any conflicting demands on general arrangement from various considerations. Furthermore, changes are frequently made to the ship design during the design and construction process.

To ensure the comfort and convenience of passengers and crew, the accommodation quarters and public spaces are afforded much more space than those required for the operation of the ship, such as engine and auxiliary machinery spaces. This is one of the reasons for the extravagant number of decks on passenger vessels. The ships will have swimming pools, shopping arcades, lounges, play rooms for children, casinos, theaters, large restaurants and nightclubs in addition to the cabins distributed to various decks. The ratio of passenger spaces to crew spaces serves as a measure of the ship's comfort: a higher ratio indicates more luxurious ship. The passenger and crew spaces are clearly separated, and they also use separate corridors, stairs and elevators. Ensuring smooth flow of passengers, cargo and services by providing appropriate access routes is another priority area to be considered in general arrangement design.

Since a passenger ship can carry a large number of people, it must satisfy strict safety requirements. It has main vertical zones (MVZ) of limited length designated to prevent the spread of fire. Furthermore, depending on the characteristics of each space, there are strict requirements for fire insulation for each deck or bulkhead. The ship has spaces separated by watertight bulkheads to provide protection against flooding, and it is subject to more strict stability standards than other types of commercial ships. Some regions of the world impose additional stability requirements (e.g., the Stockholm Agreement in northern Europe) in addition to the SOLAS requirements. Moreover, there are requirements for the provision of lifesaving appliances, other equipment and systems, and procedures for the evacuation of passengers.

More recently the safety regulations have become more stringent through the introduction of Safe Return to Port (SRtP) requirements. In this concept the ship itself is considered a lifeboat, and the regulation requires the ship to be able to return to the nearest harbour or a safe haven in the event of fire or flooding occurring. This means that passenger ships must have sufficient degree of redundancy and physical dispersion of propulsion, control and some service systems throughout the ship. This may impose many additional constraints in general arrangement. To put many concepts into practice within the given constraints it is important to have an effective coordination among system components. Therefore, each system has to be designed after identifying its influence on other components and in close consultation with other designers.

Based on these three characteristics discussed above, various trends have emerged in the general arrangement design of passenger ships. Aside from the unique design, the considerations of passenger comfort and convenience, and strict safety requirements, there are matters that cannot be effectively regulated by the existing rules and regulations. As a result, the concept of 'alternative design' is being adopted. The idea is that a design should be approved, if it can be proved that an equivalent or better level of safety can be achieved even though some of the rules and regulations cannot be met. The alternative design concept has actually been applied to some recently built passenger ships. With the continuous application of this practice in the future, the general arrangement of passenger ships is expected to evolve into a variety of styles.

References

1. CAD & Graphic, 2005. *PLM guide book*, Seoul: BB Media.

2. Erleben, K. et al., 2005. *Physics-based animation*. Rockland, MA: Charles River Media, Inc.

3. Kwon, Y.J., 2006. *The principles of ship design* (in Korean). Seoul: Dongmyungsa.

4. Lee, C.E., 2014. *Ship Design* (in Korean). Seoul: Cheongmungak [revised edition of original book published by Daehan Textbooks Co. Ltd.]

5. Park, M.G. & Kwon, Y.J., 1995. *The principles of ship design* (in Korean). Seoul: Hangukyigonghaksa.

6. Shin, J.G., 2006. *CAD, digital manufacturing and PLM* (in Korean). Seoul: Sigmapress.

7. The Society of Naval Architects and Marine Engineers (SNAME), 1971. *Principles of Naval Architecture*. Translated to Korean by Yim, S.J., 1991. Seoul: Daehan Textbooks Co., Ltd.

8. The Society of Naval Architects of Korea, 1993. *Introduction to Naval Architecture and Ocean Engineering* (in Korean). Seoul: Dongmyungsa.

9. The Society of Naval Architects of Korea, 1996. *Introduction to Ocean Engineering* (in Korean). Seoul: Dongmyungsa.

10. The Society of Naval Architects of Korea, 2011. *Introduction to Naval Architecture and Ocean Engineering* (in Korean). Seoul: GS Intervision.

11. Yoo, H.S. et al., 1997. *The element of Ocean Engineering* (in Korean). Seoul: Dongmyungsa.

CHAPTER 9

Ship Structures

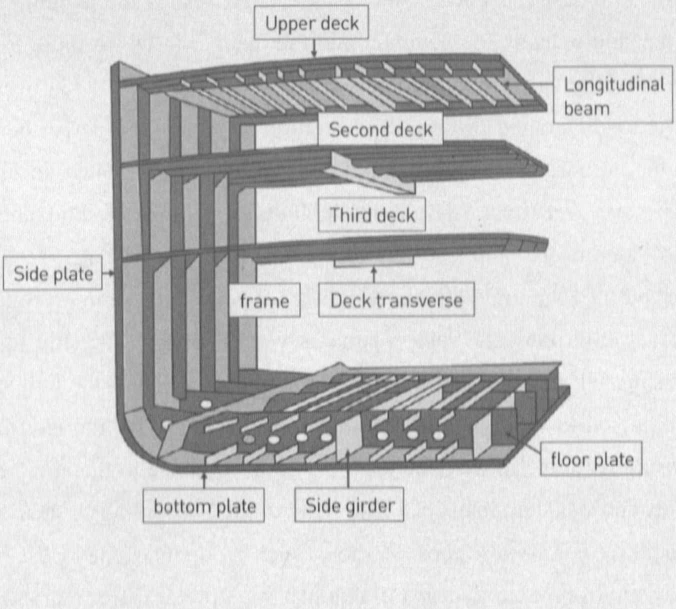

Ships are diverse not only in their external dimensions but also in internal structures. A ship's hull is usually built of plates and stiffeners and should be constructed strong enough to withstand the rigors of operating in heavy seas throughout its life.

Rolled steel, high tensile steel, aluminum alloy, timber, fiber reinforced plastic (FRP) and various other materials are used in constructing a ship's hull. Outer shells must be watertight, and the hull form is designed to provide the required displacement while presenting minimum resistance against forward motion and acceptable seakeeping characteristics. The space within the hull is divided into appropriate compartments to provide spaces for storage of the cargo, machinery and equipment necessary for the operation of the ship, and suitable accommodation space for the passengers and crew.

All the structural members, including the outer shell, deck and the internal subdivisions contribute to the structural strength of the ship.

1. External Forces Acting on the Hull

1.1 General

The first step in structural design is understanding the characteristics and magnitude of the external forces that the ship is likely to encounter and the manner in which these forces are applied to the structure of the hull. Here, the loads are examined in the context of practical structural design. The loads applied to the hull structure are many and it will be helpful to classify them for the purpose of analysis: static loads which do not change in time; and dynamic loads which vary over time. Static loads include the hydrostatic buoyancy force and gravitational force due to the ship's structural elements and cargo that it carries. The total buoyancy force and the total gravitational weight must be in balance in an equilibrium state. When an external environmental factor, such as waves, acts on the ship to disrupt this equilibrium, the ship will move in order to restore the balance. Thus, the hull structure is simultaneously subjected to the dynamic external forces induced by the environment, and inertia forces arising from the movements of the ship in response to the environmental excitation. Static loads and dynamic loads can also be classified according to their sources, such as hydrodynamic forces or mass. **Figure 9-1** shows such grouping of the loads.

The most important environmental factor influencing the structural loads is the waves.

Wind and current have significant effects on stability or mooring of the ship, but their effects on the overall hull structural loads are relatively mild. Therefore, it is usual to consider just the incident wave forces and the consequent ship motions in the structural design.

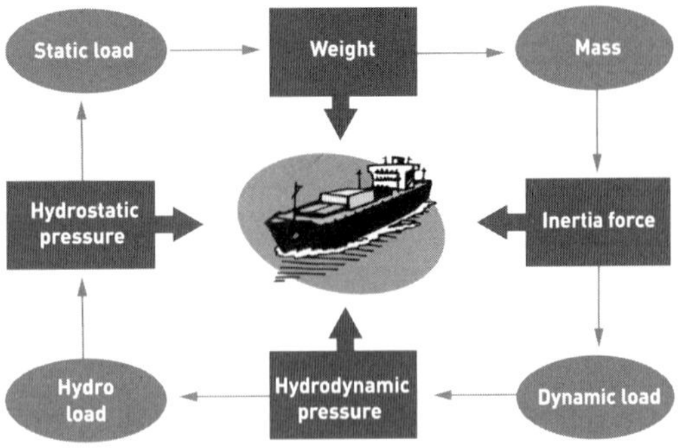

Figure 9-1 Loads exerted on hull

1.2 Static Load

The equilibrium state of the ship without waves could be expressed as

$$F_S = (-Mg) \tag{9-1}$$

Here, M refers to the total mass of the ship including the mass of the ship itself and all loaded cargo. The mass of the ship multiplied by the gravitational acceleration g is the ship's weight, and the weight of water displaced by the hull is equal to the buoyancy force F_S. When the two are equal in magnitude and are applied in line but in opposite directions, the ship reaches a static equilibrium state, since the sum of the forces will be zero and there is no moment.

The distribution of the net vertical forces in longitudinal direction can be found by subtracting the longitudinal weight distribution from the buoyancy force distribution. Integrating the net vertical forces in lengthwise distribution results in the shear force distribution. Integrating shear force distribution in turn produces the bending moment distribution. It is sufficient to consider the weight and buoyancy forces as simplified external forces, when an overall ship structural analysis is to be carried out directly. However, when a ship is idealized into a simple beam, or when local structural analysis is to be carried out, such as on midship section, shear force and bending moment distribution data is useful. As shown above, static shear force and bending moment used in such situations can easily be calculated by using simple programs if the form of the ship and weight distribution is known.

1.3 Dynamic Force

When there are waves, the dynamic component of the forces can be expressed as

$$F_D = (-Ma) \tag{9-2}$$

where F_D is the hydrodynamic pressure force acting on the hull surface through the interaction of the waves and the ship, and Ma on the right side of the equation is the inertia force. Hydrodynamic pressure force on the left side of the equation includes restoring force, added inertia force and influence of the friction caused by the movement of the ship. Thus, the six simultaneous equations expressing the ship's 6 degrees of freedom motions must be established in terms of acceleration, velocity and displacement, including the cross-coupling terms where appropriate, as has already been discussed in Chapter 6.

1.4 Design Loads

According to the unified regulations of IACS (International Association of Classification Societies), ship design loads are estimated by assuming 25 years as the design life. This is in contrast to the determination of safety factors of offshore structures based on its survivability of 100 years of lifespan. The latter is because these immobile structures, unlike ships, cannot actively evade heavy weather.

A design life of 25 years means that the hull structure must be designed to withstand the highest waves the ship is likely to encounter in 25 years. Therefore, probabilistic and statistical methods must be used to determine the design loads. In order to do so, appropriate assumptions on the probability distribution of the waves and response of the ship must be made.

There are two methods of calculating the maximum stress likely to be experienced by the hull: the equivalent design wave method and the spectral stress method. The equivalent design wave method calculates the maximum wave load statistically and applies that stress on the structural model. The underlying assumption is that the maximum stress applied to the hull can be calculated by applying maximum wave loads on various wave conditions.

On the other hand, the spectral stress method conducts structural analysis in waves of unit height. This is to get the transfer function of the stress for statistical processing with the ocean wave spectrum and the long term observational data to determine the maximum stress. It is mainly used in the fatigue life evaluation.

1.5 Impulse Loads

Generally, a hull structure experiences the highest wave load when the wave length is similar to the length of the ship. The period of the wave is comparatively long in such cases making the load vary slowly, and consequently, structural analysis can be conducted quasi-statically. Generally, the load applied by the waves is the most fundamental in the overall structural design, since it not only affects the ship globally, but also is the most important factor in hull deformation.

However, slamming on the bow or stern of a ship, and the accompanying green water, plus the sloshing caused by the oscillation of fluid in internal tanks, also causes intensive impact loads in rough sea conditions. Such impulse loads have localized effects on the hull structure, and are reviewed for local reinforcement after the overall design of a ship structure under the design wave condition is completed.

1.5.1 Slamming

Slamming is the phenomena of hydrodynamic impact on the hull surface generated, for example, when the bow of the ship reenters the wave surface after being lifted out of water due to excessive ship motion in heavy weather. There is also flare impact caused when the bow with a large flare angle slams down on the water surface. The first example is known as bottom slamming and the latter is called flare slamming.

Bottom slamming often occurs in ships with flat bottom such as oil tankers and bulk carriers, while flare slamming occurs more often in ships with large flares such as container ships. Bottom slamming at stern frequently occurs in ships with a very long flat transom stern as in container ships. Impact pressures like these may cause structural damage, such as deformations of the outside planking or buckling of the stiffeners. Impact pressures, such as due to slamming, may be evaluated directly by conducting dynamic structural analysis or static analysis based on equivalent static head.

Figure 9-2 shows the distribution of bottom slamming pressure used by the IACS unified regulations. The impact force caused by slamming is not only the main cause for the damages occurring to the outside plating of the bottom and flare, but also is a threat to the overall

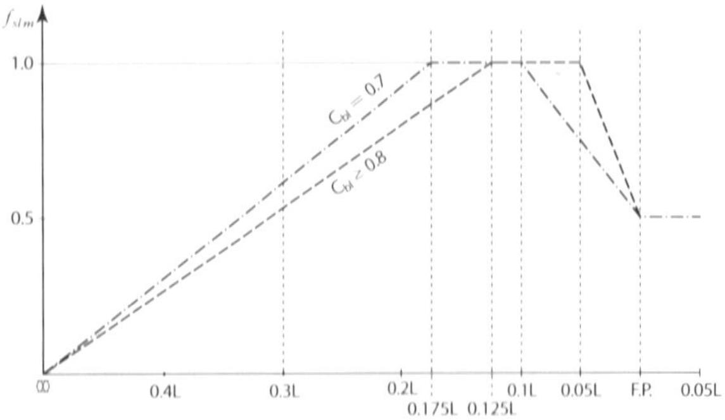

Figure 9-2 Distribution of bottom slamming pressure

structural safety of the hull. Large amplitude vibration in the vertical direction of the hull is called whipping and it can be caused by slamming. The effects of whipping caused by shock waves or pulsating gas bubbles in the event of nearby underwater explosions must be considered in the structural designs of naval vessels.

1.5.2 Sloshing

For a ship carrying liquid cargo in wide tanks, such as oil tankers or LNG carriers, the liquid cargo sloshes about in the tanks due to ship motion. Depending on the amount of liquid volume in the tank, this may produce a very large impact pressure inside the tank, particularly when the internal fluid motion and the hull motion resonate at certain frequencies. Sloshing is manifest in a variety of ways depending on the motion of the ship and the level of the liquid cargo, and therefore the tank structure must be designed to withstand such pressures by using model experiment or numerical analysis in diverse situations.

1.5.3 Springing

Springing is the phenomenon of constant vibrations caused by the resonance of high-frequency components of the ocean waves and the hull girder's lower mode vibration, regardless of the main engine rpm and state of the sea. Optimization of the hull structure and use of high tensile steel in recent years have reduced the rigidity of the hull girders and, as a consequence, their natural frequency, thereby increasing the possibility of springing. Structural reinforcements against whipping or springing are necessary, but precise evaluation on the hydroelastic behavior is a prerequisite for such a course of action.

1.5.4 Green water

Green water is a phenomenon in which a dark, unbroken wall of seawater is visible,

(a) A bulk carrier

(b) A naval vessel

Figure 9-3 Green water

unlike the white spray of water, in front of the bow when the bow enters the water at a high speed during a pitching motion, as illustrated in **Figure 9-3**(b). As the descending speed of bow is decreased, the wall of seawater collapses, causing the water to fall on the deck with large downward impact force. As the seawater continues to flow rapidly aft on the deck, it causes further impact on the deck structures or equipment, as illustrated in **Figure 9-3**(a). Breakwater is installed on the bow area to mitigate such effects.

1.6 Other Loads

There are many local loads other than the aforementioned examples. These include axial force and hydrodynamic pressure force caused by the propeller, vibration of the engines, mooring forces, and loads caused during loading and unloading of cargo. Also, there are loads caused by explosions, impact or flooding.

2. Hull Damage

Ships are used to create added value through transporting passenger or cargo to desired locations across rivers or oceans. Compared to that of ground transportation, operational environment for ships is usually much harsher. Therefore, careful consideration must be given to the structural strength of ships at the design stage to withstand such harsh environment. Marine accidents can cause not only damage to the ship hull but also result in serious consequences such as loss of human lives and pollution of the marine environment. Causes of the accidents are very hard to identify precisely due to the diverse and complicated mechanisms associated with each accident. Nevertheless knowledge regarding the types of damage is important in mitigating the consequences of accidents.

2.1 Midship Fracture

Damages at the midship region of the hull resulting in the ship breaking in half during a voyage are not common, but are one of the most catastrophic accidents when they do occur.

(a) The *Prestige* (b) The *Selendang Ayu*

Figure 9-4 Examples of ships broken into two

The primary causes of such accidents include grounding, impact of high waves, collision, and explosions of torpedoes or underwater mines.

Figure 9-4(a) shows the sinking of the oil tanker *Prestige*, which happened due to impact of high waves during a storm in 2002 off the coast of Spain. **Figure 9-4**(b) is the midship fracture accident of the bulk carrier *Selendang Ayu* which happened in Alaska in 2004. The *Prestige* broke up after the initial cracks caused in rough seas propagated gradually. *Selendang Ayu*'s hull could not withstand the pressure induced by the sea current after grounding and ended up in breaking due to an accumulation of fatigue damage.

When ships, especially oil tankers or bulk carriers, experience forces right angles to the longitudinal axis caused by the waves, the midship region receives a huge stress by the consequent bending moment and shear forces, resulting in cracks that can develop into fractures.

2.2 Fatigue Fractures

Fatigue fractures are observed in bridges, airplanes, machine parts or any structures that are repeatedly exposed to dynamic stresses. They also occur when the stress is lower than the yield strength or tensile strength of the material at a state of static load. Such fracture phenomenon is called 'fatigue fracture,' because it occurs after repetitive occurrences of stress and deformations.

Actually, most of the fractures that happen in machine parts are caused by fatigue. Many accidents in cars, airplanes, pumps and turbines are often attributable to fatigue fractures. Fatigue is an important cause of metal fracturing as evidenced by the fact that approximately 90% of all metal fractures are caused by fatigue. They are particularly dangerous type of fracture phenomenon, because they occur without any visible warning signs prior to the fracture happening, making it difficult to contain them from progressing to large-scale disasters. Therefore, consideration of fatigue strength is of the utmost importance at the design stage.

Fatigue cracks typically initiate at the locations where stress is concentrated, such as where there is an abrupt change in structural dimensions, defects of weld zones, and geometric discontinuity such as grooves or holes. These discontinuities are critical locations which can act as the nucleus of the fatigue fracture. Reduction or elimination of such structural discontinuity can reduce the likelihood of fatigue fractures.

Welding introduces some factors which can make the welded structures susceptible to fatigue fractures. However, fatigue life may be improved by grinding the weld toes to reduce the discontinuity of geometric forms, post weld heat treatment (PWHT) to release the welding residual stress, or surface treatment to generate compressive residual stress.

Figure 9-5 shows the connecting region of a side longitudinal frame and a transverse stiffener of a container ship. As can be seen, this is constructed using fillet welding which is most often used in shipbuilding. **Figure 9-5**(a) is an example of fracture developing after repeated loads on the weld toe region, causing a concentration of stress and the resultant cracks which then propagated. Fatigue fractures caused by stress concentration at the weld of insufficient penetration depth are shown in **Figure 9-5**(b).

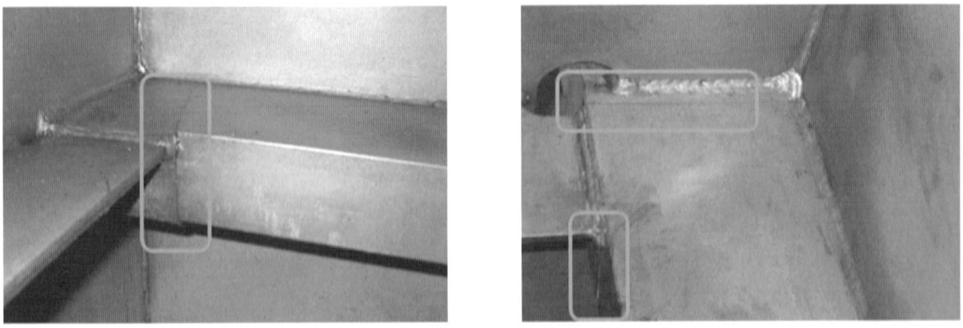

(a) Weld toe

(b) Shallow weld penetration

Figure 9-5 Fatigue fractures

2.3 Brittle Fractures

Brittle fracture is a phenomenon in which small defects in a material propagate rapidly to a total failure caused by a load on the structure. It often occurs in locations where stresses are concentrated, such as welding defects or structural discontinuities. However, brittle fracture may be caused even without noticeable initial defects. Owing to recent development in material technology, many kinds of ductile steel that can resist brittle fractures have been developed. Thus, the possibility of brittle fractures can be controlled to very low levels, if appropriate materials are selected for the operational temperature range.

Before examining brittle fracture, ductile fracture is discussed as a way of contrast.

2.3.1 Ductile Fractures

Ductile fractures are a physical phenomenon that is in contrast to brittle fractures. **Figure 9-6** shows a tensile test specimen of highly ductile aluminum.

A significant reduction of the cross sectional area and elongation of the specimen is clearly visible. This is a typical feature of ductile fracture in which the material absorbs much energy during large deformation before resulting in fracture. The cup-like shape of the one surface and conic shape of the opposite surface is characteristics of the fracture surface.

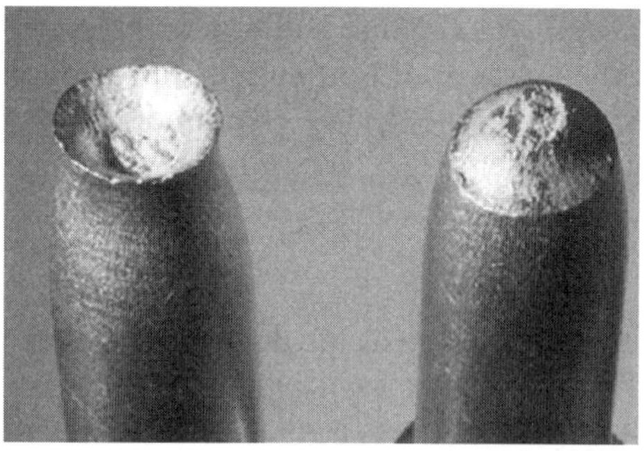
Figure 9-6 Ductile fracture (aluminum)

For this reason, ductile fractures are also called cup-and-cone fractures. The center of the internal surface of the specimen in **Figure 9-6** shows irregularly elongated fiber-like surfaces, which tells us that considerable plastic deformations have taken place before the fracture.

2.3.2 Brittle Fracture

Figure 9-7 shows a specimen of mild steel material after a tensile test that has undergone brittle fracture. This shows that there is almost no changes on the cross sectional area or the length prior to the fracture. Almost no elongation is observed in the specimen, and therefore it may be concluded that not much energy has been absorbed by the material through deformation up to the fracture taking place. A clean cross section and a metallic crystal surface are characteristics of this fracture, which often occurs at very low temperatures. Typical materials susceptible to brittle fractures include glass, concrete, rocks or iron with a high carbon content.

Figure 9-7 Brittle fracture (mild steel)

2.3.3 Examples of Brittle Fractures

During the Second World War, more than 500 T-2 Tankers of 16,000 DWT were mass produced using simplified continuous manufacturing processes which were hastily developed for wartime production based on welded construction methods. Such a T-2 Tanker was anchored at Portland Harbor in Oregon on January 16 in 1943. Suddenly the ship split into two at the midship and shot up from the water. **Figure 9-8** shows photographs of the severed midship part, clearly showing how the fracture occurred. Since then, this accident has been studied intensely as a typical example of welding defects and brittle fractures.

Figure 9-8 T-2 Accident of the Tanker SS *Schnectady* broken into two

Brittle fractures are caused by propagating cracks without plastic deformation. Such cracks are called unstable cracks, and they rapidly propagate once initiated even without increase in stress.

Figure 9-9 is an example of a brittle hull fracture as a result of rapidly propagated cracks developed at the midship of an oil tanker. Some cracks were formed at a structural discontinuity due to welding defects and repeated application of high level stresses caused by rough weather conditions, and then certain factors caused the cracks to propagate rapidly to split the oil tanker completely into two.

Figure 9-9 Brittle fracture in an oil tanker

Brittle fractures are more dangerous than ductile fractures because of their two main characteristics defined earlier. Firstly, ductile fractures occur after considerable plastic deformations, which give personnel time to take preventive measures. However, brittle fractures occur abruptly without any warning, rendering any preventive measures impossible. Secondly, ductile materials generally have high capacity to absorb much strain energy during the progress of ductile fractures, while brittle material absorbs little strain energy due to the lack of plastic deformation.

2.4 Localized Damages

2.4.1 Localized Damages Due to Collisions

In most of the collision accidents, the bow of the assailant ship collides into the side of the damaged ship. **Figure 9-10** shows the damaged parts of the Egyptian carrier vessel *Wadi Halfa* after its collision with a British cargo ship *Scot Isles*. As the picture shows, the damage outline of the damaged ship matches the shape of the bulbous bow of the assailant ship. Moreover, the bulbous bow of the assailant ship is relatively rigid, resulting in minor damages, while the damaged ship which received the impact experienced greater damages.

Figure 9-11 shows the result of a collision accident between a British tanker *Amenity*

(a) *Scot Isles* (damaged ship) (b) *Wadi Halfa* (assailant ship)

Figure 9-10 Damaged regions after collision 1

(a) *Tor Dania* (damaged ship) (b) *Amenity* (assailant ship)

Figure 9-11 Damaged regions after collision 2

and the Norwegian roll-on/roll-off vessel[1] *Tor Dania*. This illustrates a different form of damage in that the assailant ship suffered greater damages. This is largely due to the fact that *Amenity* did not have a bulbous bow and, consequently, its bow area was relatively weak. Nevertheless, it is clearly seen that the bow form has left an imprint on the damaged region of the other ship when collision occurred.

2.4.2 Localized Damage Due to Grounding

When a ship is driven onto sand bars or ashore for whatever reason, the sand erosion caused by tidal currents and tidal difference can often give rise to a heeling moment and cause the ship to list and this can eventually lead to capsizing. During this process the rudder or propeller protruding from the protective hull may also be damaged.

When a ship runs aground on a reef, the structural damage is more direct. If the difference between the depth of the reef and the draft of the ship is small, large lengthwise damages can be anticipated despite rather shallow penetration depth on the ship's bottom. In **Figure 9-12**(a) severe damages that the *Sea Empress* suffered in a grounding accident is shown, including bottom plates torn off and inner structural members severely mangled almost beyond recognition. These were caused by repeated forward and reverse maneuvers of the ship in an effort to escape from the reef. It is easy to see that, when a ship runs aground at a high speed in waters much shallower than the ship's draft, the bow structure can be severely damaged often accompanied by upward bending as can be seen in **Figure 9-12**(b).

(a) *Sea Empress*

(b) *Maersk Kendal*

Figure 9-12 Damaged regions after grounding

1 Ships designed to carry wheeled cargo, such as automobiles, trucks, trailers and railroad cars that are driven on and off the ship on their own wheels.

3. Characteristics of Ship Structure

The bow and stern of a ship's hull usually form smooth streamline shapes in order to minimize water resistance. In terms of strength, however, the hull is often treated as a long prismatic box-shaped girder. The outer hull is a completely watertight structure, composed of shell members, such as bottom plates, side shells, deck plates and so on. Subdivisions of the internal space are made in longitudinal and/or transverse directions, according to space and safety requirements.

3.1 Hull Structural Members

Ships navigate through rough seas carrying cargo and various machineries on board. Therefore, ships should be able to withstand their loads, while resisting against the water pressure applied from the outside of the hull. In order to do so, various structural members are required. Much of a conventional ship's shell is flat and therefore is built up by welding flat plates together. In contrast, however, the bow and the stern are usually highly curved and it is necessary to form these shell plates by using various curved surface production processes.

Plates that make up the shell have relatively weak stiffness against normally applied loads. So, the plates must be reinforced by frames and beams that run in the longitudinal and/or transverse directions such as shown in **Figures 9-13** to **9-15**. Additionally, pillars are installed between the decks to support the loads of the cargo. To cope with large bending loads applied to a hull, large beams or girders are arranged in the longitudinal or transverse direction. On the other hand, ships often have areas where high stresses occur locally, especially where members are connected at right angles or at structural discontinuities. Such areas run a high risk of failure due to stress concentration, and therefore local reinforcements or stiffeners like brackets must be installed.

3.2 Characteristics of Longitudinal Direction

Most ships are long and slender, and therefore the most obvious structural strength requirements are against longitudinal bending moments and shear forces on the cross sections. These loads acting on the ship's hull vary longitudinally. Consequently, the structural characteristics of the hull also vary in the longitudinal direction. Typical locations and structural characteristics on general merchant ships are as follows.

3.2.1 Midship Cargo Holds

The middle part of the hull is where cargo holds are situated. Most ships have a double bottom structure in this region and the inner bottom plate constitutes the bottom of the cargo holds. The shape and the structure of the cargo holds vary depending on the characteristics of the cargo.

3.2.2 Bow

The bow structure must be able to withstand water impact pressure, since it comes into contact with the waves directly. It also has a higher probability of direct impact from collision with other ships, floating materials and shore structures, and therefore the bow structure must be designed solid enough to minimize the consequences from such collisions. To protect the contents of the foremost cargo hold from such accidents collision bulkheads (forepeak bulkheads) must be installed at a certain distance away from the fore end of the ship so that the cargo hold is isolated to a degree. In the space between stem and collision bulkhead, anchor chain lockers and ballast tanks are placed. In the forecastle, windlasses and various other machineries required for the operation of the ship are arranged and sometimes bow thrusters are placed under the waterline.

3.2.3 Stern

The stern region is where the equipment related to the propulsion of the ship, such as the propeller and propeller shaft, is situated. Equipment related to the ship's control and operation, such as rudders, steering and mooring gear, is also usually placed here, and an after peak bulkhead generally divides the engine room from the stern. The propulsion system and main engines are the major excitation sources of the shipboard vibration and noise. Structural design for minimizing vibration and noise level is mandatory, since the deckhouse is often situated here containing the wheelhouse, work space for the deck officers and crew accommodation.

3.2.4 Engine Room

The main engine is usually located in the stern area to minimize the space occupied by the machines and the length of the propulsion shaft which connects the engine to the propeller, and this practice tends to simplify the related structures. The bottom structure of the engine room requires additional reinforcements to support the heavy main engine and other machinery located there.

3.2.5 Structure Above Upper Deck

Structures above the upper deck that are formed from the extensions of the side shells of the ship are called the superstructure. When the width of superstructure is smaller than the width of the ship obtained by continuous upward extension of the side shells, it is called a trunk or a deck house. A superstructure, trunk or deckhouse is installed only at necessary locations rather than throughout the whole length of the ship. Therefore, they do not contribute to the longitudinal strength of the ship structure. On the contrary, they do necessitate additional reinforcements due to the possibility of stress concentration as a result of structural discontinuity. A deck house takes the form of conventional buildings, and is used as an accommodation space for the crew members or a wheelhouse. Since it does not contribute to the overall strength of the ship, light materials can be used to build it.

3.3 Framing system of the Hull Structure

The outer shell of the hull is made of flat panels and they require stiffening using frames. There are three types of framing system according to the arrangements of the stiffeners: longitudinal framing system, transverse framing system and combined framing system.

3.3.1 Transverse Framing System

In this system the major structural parts, such as the deck, sides and bottom plates, are reinforced with stiffeners arranged transversely as shown in **Figure 9-13**. The resulting structure is simple and construction is relatively easy. Since the longitudinal stiffeners do not penetrate the transverse web frames, there is no need to cut out slot holes in them. There are fewer structural members protruding into the cargo hold, which makes it suitable for small to medium sized ships, such as small general cargo ships, fishing vessels and so on. As the ship becomes larger and longer, however, the longitudinal strength of the hull girder becomes more critical. In order to compensate for this weakness, the shell plating has to become thicker, making this system of framing unsuitable for larger cargo ships.

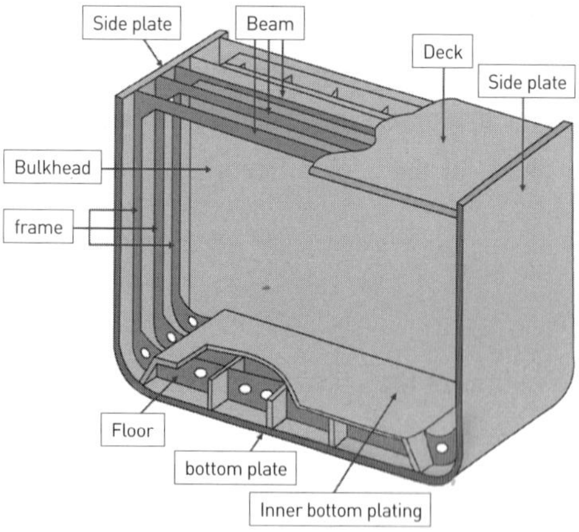

Figure 9-13 Transverse framing system

3.3.2 Longitudinal Framing System

In this approach the major structural parts are reinforced by longitudinal stiffeners, as illustrated in **Figure 9-14**. The transverse strength of the hull is reinforced by large transverse web frames. It has two key advantages over the transverse system: superior longitudinal strength and a lighter hull structure. However, hold spaces tend to be cluttered by many intruded structural elements resulting in a complex hold space, and many slot holes are needed in the transverse web frames to allow the longitudinal stiffeners to go through them. Therefore, this framing system is not appropriate for general cargo ships. However, most ships that transport liquid cargo, such as oil tankers, naval vessels and large merchant ships that have relatively large length-to-beam ratio adopt this framing system.

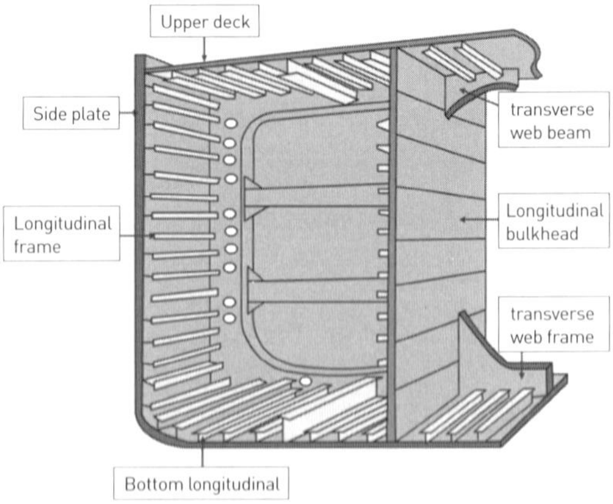

Figure 9-14 Longitudinal framing system

3.3.3 Combined Framing System

This is a hybrid of the above two framing systems. The deck and bottom structures adopt a longitudinal framing system, while a transverse framing system is used for the side structures. **Figure 9-15** shows an example of such a system. Cargo holds are relatively less cluttered, while the longitudinal strength of the ship is also secured. Cargo holds of a bulk carrier are a typical example of the combined framing system. Even in ships of longitudinal framing system, local transverse framing is often used at bow and stern regions that do not experience large bending moments and bottom of the engine rooms in which large concentrated loads are applied.

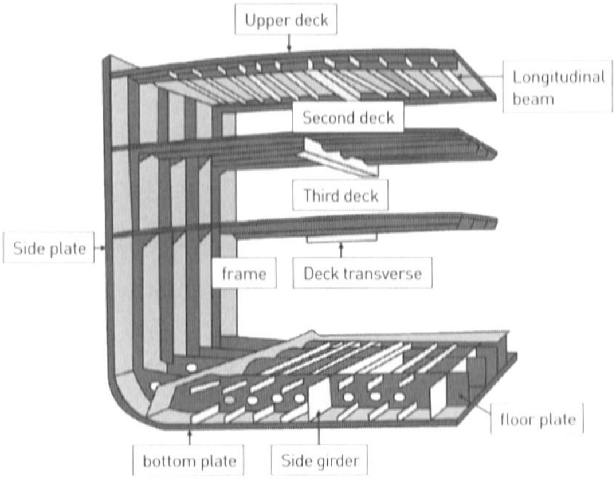

Figure 9-15 Combined framing system

4. Structural Strength of the Hull

The total weight of the hull including the cargo and the total buoyancy of the ship is in equilibrium when the ship is afloat. However, the longitudinal distribution of the load and the buoyancy is not balanced in most locations. This imbalance between the load and the buoyancy is the cause of shear force and bending moment on the hull structure. The effects of the imbalance become more significant when the ship is fully loaded with cargo and/or passengers and encounters rough seas.

Therefore, it is very important to evaluate the safety of the hull structure for the given sea route. In order to do so, the forces that act on the hull in a given loading condition must be examined. There are forces that try to bend the hull in the longitudinal direction, forces that try to deform the cross sections and forces that try to twist and warp the hull even in still water. Thus, reviews on these forces are in order.

4.1 Longitudinal Strength

When a ship is floating in still water in equilibrium encounters a train of waves, the distribution of the buoyancy along the hull changes in proportional to the height of the wave surface from the still water surface. Therefore, the difference between the buoyancy and gravitational force distribution along the longitudinal direction of the hull varies in time. Provided the still water bending moment is not greater than these changes, tensile stress and compressive stress alternatively appear on the deck and bottom, due to the periodic changes of bending moment as shown in **Figure 9-16**.

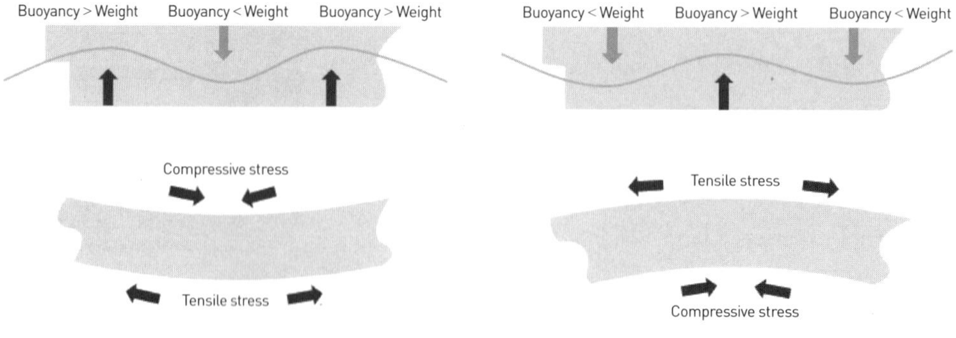

Figure 9-16 Longitudinal effects of waves on the hull structure

If the tensile or compressive stress that acts on any structural members exceeds the allowable stress of the material, that member will experience plastic deformation or shear. Thus, the hull structure must be arranged to resist the applied loads without experiencing such damage. Such design exercise is known as the longitudinal strength design.

In the longitudinal framing system, it is primarily the longitudinal frames that contribute to the longitudinal strength, while the transverse bulkheads and large web frames cover the transverse strength required. The transverse girders of the ship's bottom that connects to the web frames on the sides and the deck girders form large circumferentially closed ring type structures, and they reinforce the ship's transverse strength.

The weight/buoyancy distribution and waves are the dominating factors in determining the load the hull receives. As stated briefly above, still water bending moment arises from the uneven longitudinal distribution of the difference between weight and buoyancy forces, which can be easily estimated. However, wave bending moment is more complex and contains some uncertainties due to the stochastic nature of waves.

4.1.1 Still Water Bending Moment

If an object of a uniform cross section and density is afloat in still water, the difference of buoyancy force and the weight will be zero throughout its length, causing no bending moment. However, if the equilibrium between the buoyancy and the weight distribution is broken partially by adding some mass in one part of the object in such a way as to cause no trim, the weight distribution will be larger than the buoyancy distribution at the location with the added weight, while the buoyancy is larger than the weight in other areas. As a result, a bending moment appears.

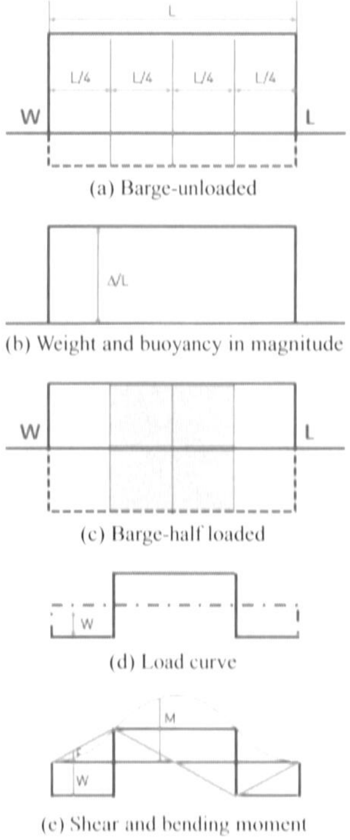

Figure 9-17 Bending moment on a barge in still water

When a rectangular barge of uniform density in longitudinal direction is afloat in still water as shown in **Figure 9-17**(a), it is in static equilibrium so that the weight and the buoyancy per unit length in any locations throughout its length are equal to each other in magnitude but acting in opposite direction as illustrated in **Figure 9-17**(b).

However, when cargo is loaded on the central two shaded segments as shown in **Figure 9-17**(c), the barge will experience a the parallel sinkage until the resulting increase of buoyancy is identical with the weight of the loaded cargo, and then its static equilibrium is restored. The area under the weight distribution curve and the buoyancy distribution curve must be equal, but since the buoyancy is applied uniformly throughout the length of the barge while the weight is greater in the midship, The load curve from the difference between weight and buoyancy will be distributed as shown in **Figure 9-17**(d).

Figure 9-17(e) shows a shear force diagram, which results from integrating the load distribution on the barge. Starting from 0 at the aft end of the barge, it linearly increases until the sign of the load curve changes to negative when it starts to decrease linearly, and then switches back to linear increase when the sign of the load curve changes back to positive, resulting in 0 at the forward end. The diagram also shows the bending moment curve, which can be obtained from integrating the shear force diagram. The relationships between the various curves, which appear in the process of acquiring bending moment curve, are summarized as follows.

- The area under the buoyancy curve is equal to the area under the weight curve.
- The centroid of area under the buoyancy curve and the centroid of the area under the weight curve are vertically in line.
- The net area under the load curve, which is the difference between weight curve and buoyancy curve, is equal to 0.
- The maximum and minimum values of the shear force occur at all intersecting points of the load curve and the base line.
- The maximum bending moment occurs when the shear force is equal to 0.
- The values of shear force curve and bending moment curve at each of the end points are 0.

4.1.2 Wave Bending Moment

Once the hull form is determined, the distribution of the buoyancy acting on the hull in the waves can be estimated based on the longitudinal wave profiles. However, weight distribution is dependent on the longitudinal arrangement of the machines and distribution of cargo, and therefore the weight and center locations of each component must be calculated. If the waves are assumed to be at a standstill, still water bending moment calculations proposed earlier may be used to calculate the wave bending moment.

Actual sea state consists of propagating waves of various wavelengths in diverse directions, and sometimes extremely large waves overlap spontaneously. Moreover, the relative position of the waves and the ship is constantly changing, rendering accurate calculation of the bending moment of the ship impossible. Consequently, it has been the usual practice to determine the strength requirements of a ship hull by assuming that the ship is a girder floating at a static equilibrium on hypothetical stationary standard waves.

However, this method has the disadvantage of missing the influences of inertial forces arising from the motion of the hull.

Fortunately, however, a simple equation involving major dimensions of the ship, such as length, width, depth and displacement, has been recommended by IACS for estimating the bending moment and shear force. This simplified relation has been derived from direct analysis of many ships and statistical analysis of the operational records of these ships at sea.

When more precise analysis is required, the direct analysis explained in 5.2.3 may be applied. Let us assume that the actual waves of the ocean could be expressed as superposition of various component waves. If the effects of all the component waves can be calculated, the total hull response to the spectral sea can be estimated by superposition. The effects of component waves can be calculated by numerical methods, such as strip method or more advanced computational methods (computational fluid dynamics), or obtained from model experiments. In this way all dynamic effects of inertia in response to buoyant forces and weight are automatically included in the calculations.

In addition, substantial amount of real-time measurement data of the stresses experienced by the hull girder in actual operations has been accumulated. Statistical information from these data allows the bending moment to be predicted with enough precision and confidence to be used in design.

4.1.3 Section Modulus of Hull Girders and Stresses

Having calculated the bending moment, stress on the upper deck or the bottom plate can be calculated by applying the simple beam theory, assuming the hull structure as a single girder. In other words, the stress on the upper deck or bottom plate is expressed in **Equation 9-3** using bending moment M_B and section modulus SM of the hull cross section.

$$\sigma = \frac{M_B}{SM} \tag{9-3}$$

Theoretically, the lengthwise variation of the section modulus must be expressed as a curve for exact analysis of the rigidity of the ship. According to load line regulations, all continuous longitudinal members under the deck must be included in the calculation of the section modulus, excluding decks or girders installed for the purpose of support. Generally only those members effective against tension and compression are included in the section modulus calculation.

A curve of longitudinal stress distribution can be produced using the bending moment diagram and the section modulus curve. The bending stress is highest near the midship, and, therefore, normally only the midship region of length 0.4L for general cargo ships and 0.5L for oil tankers needs to be investigated. Since the bending moment decreases at a faster rate than section modulus outside this region, stress generally decreases towards bow and stern.

4.2 Transverse Strength

Consider a hull structure composed of the bottom, two sides, a 'tween deck and the upper deck. On the upper deck, loads from the cargo are evenly distributed, and water pressure is applied from the outside on the sides and the bottom. Inside the ship, the weight of the structure and the cargo stored inside are distributed as shown in **Figure 9-18**(a).

The bottom and the decks are supported by the sides, and the sides are supported by the decks and the bottom. Thus, it is reasonably obvious that the bottom, decks, and both sides will deform bending inward. Therefore, the cross sectional structure of the hull must have sufficient transverse strength to withstand such loads.

If the port and starboard level of water surface is not even as in **Figure 9-18**(b) due to incident waves or asymmetrical distribution of the cargo or machinery, the deformation may also occur asymmetrically. The phenomenon that causes asymmetrical deformation on a cross section is called raking. Of course, the hull must withstand raking loads as well.

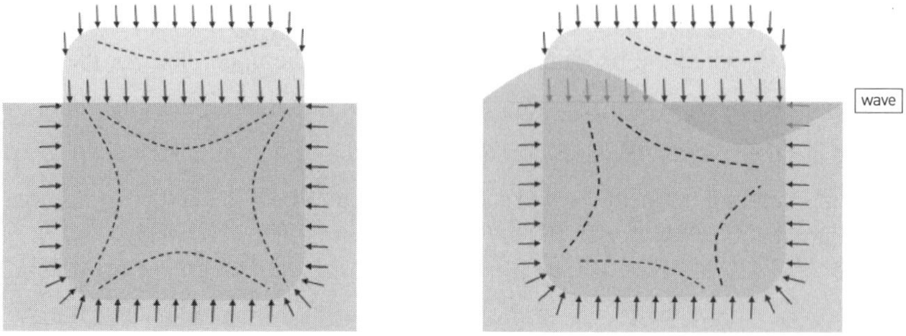

(a) Deck load and bottom and side loads

(b) Raking

Figure 9-18 Transverse effects of waves on a cross section of a ship

4.3 Torsional Strength

When a ship is inclined in the transverse direction due to constant strong wind from the side as shown in **Figure 9-19**(a), the force of the wind to incline the ship and the distribution of the hydrostatic pressure to restore the ship is at equilibrium. However, the longitudinal distribution of the wind pressure on the hull is not equal to the longitudinal distribution of the restoring force of the ship, causing torsional moment on the hull.

Another example is shown in **Figure 9-19**(b), where the ship is advancing at an oblique angle to the direction of the waves. In the fore body of the ship, a moment is generated by a wave crest on the starboard side to restore the hull in a counter-clockwise direction, while the moment in the aft body could be to incline the ship further (clockwise) due to a wave crest on the port side.

In this condition the hull experiences a torsional moment. Such torsional moments cause shear deformation on some structures such as transverse bulkheads. Generally, torsional moments are at their maximum when the wave crests or troughs are amidships in quartering waves, the wavelength of which, measured in the longitudinal direction of the hull, is similar to the length of the hull.

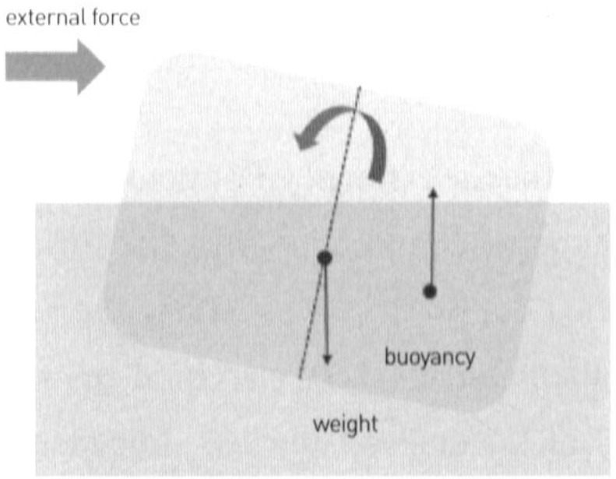

(a) Twisting force caused by inclination

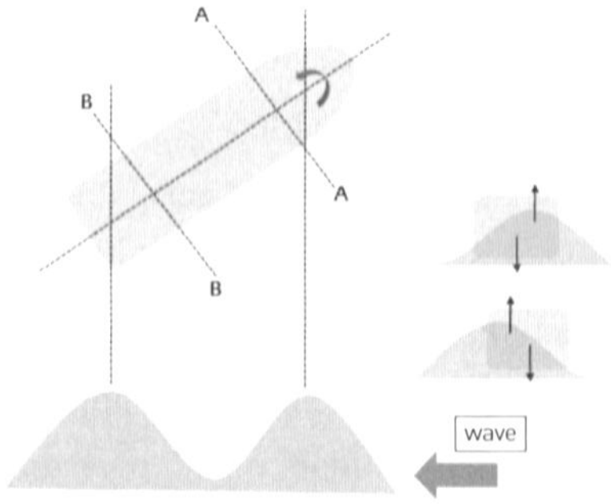

(b) Twisting force caused by waves

Figure 9-19 Torsional strength

4.4 Local Strength

In addition to the longitudinal, transverse and torsional strength, a ship's structure must have sufficient local strength to withstand the following loads:

- slamming, whipping, springing due to ship motion;
- green water due to ship motion;
- wind pressure in stormy weather;
- intrinsic residual stresses of welded structure;
- water pressure due to flooding;
- dynamic load due periodic movement of the liquid and bulk cargo;
- impulsive force due to embarkation contact, grounding or collision; and
- forces due to the hull's vibration, etc.

When any of the above phenomena occurs, local damage may result on the decks or panels being subjected to the loads directly and pillars or other internal support members to which the loads are transmitted. Also, heavy equipment and excessive or incorrectly stowed cargo can cause damage to some structural members.

5. Structural Analysis of the Hull

Various forms of deformations occur on the ship's hull due to many types of external forces acting on the hull explained earlier. Corresponding to such deformations, ship structures must be designed to secure sufficient strength without structural damage on the ship throughout its life.

The structural design of a ship is based on the assessment of structural safety against various loads that the ship will commonly experience throughout its life: in other words, structural analysis. This process must be carried out by strictly satisfying the regulations of the recognized authority with applications of the latest advanced structural analysis methods, and the shipbuilders must prove the validity of the analysis process if needed. Generally, design methods based on the comparative strength concept, which allows for equal or more strength through comparison with existing ships which have proved their structural safety via accumulated voyage data, are most often used.

In designing new types of ships that have never been built, rational structural analysis methods, based on the absolute strength design concept, could be used. This design process contains the statistical analysis of the wave loads and finite element structural analysis. Recently, applications of such direct structural analysis methods have become more widely used in evaluations of structural safety of general merchant ships.

5.1 Strength Assessment of Structures

In addition to the load data, information on the mechanical characteristics of structural materials is essential for assessing the strength structural strength. Of course, the design data such as structural configurations and scantlings are also indispensable. After the basic information has been prepared, actual analysis is conducted to simulate structural behavior through the mathematical and numerical modeling process. If reasonable analysis methods have not yet been developed, or additional data is needed, scale model experiments or onboard tests may be conducted. When the analysis results are obtained, they are compared with the safety assessment criteria.

5.1.1 Characteristics of the Materials

Various material properties, representing strength and stiffness, are necessary for structural analysis. Mechanical characteristics of materials are always determined through experiments, and material manufacturers, like steel mills, generally provide material test results for mechanical properties. Data for conventional materials which are mass-produced could easily be obtained from various handbooks that have been published already. When special materials are applied in specific environments, material tests may be required, if the characteristics of the materials are not easily obtained.

5.1.2 Evaluation of the Loads

It is natural that the larger the external forces applied on a structure, the bigger and stronger the structural members must be designed. The types, characteristics and assessment procedures of loads acting on the ship's hull are as illustrated in **Figure 9-1**.

5.1.3 Structural Analysis Method

Once the appropriate structural analysis method for the target problem has been determined, the structure has to be modeled according to the characteristics of the analysis method. At this stage, various technological judgments and assumptions should be made to simplify the problem. Predicted loads are applied to the structural analysis model prepared through such processes to calculate the structural behavior. The obtained results of the analysis are considered and assessed carefully and rationally. Therefore, the possibilities of damage evolution on the structural members are estimated for all the load cases, and reasonable reinforcement plans are proposed as needed for re-design to secure the safety of the structure.

5.1.4 Finite Element Method

According to the characteristics of loading and purpose of analysis, various methods can be applied to predict the structural responses accurately, such as deformation or stress. Actual hull structures are so complex that there is no means of perfectly simulating the hull's structural behavior as yet. In the past, theoretical analyses have been used on simple structures. However, a powerful and efficient numerical analysis method called finite

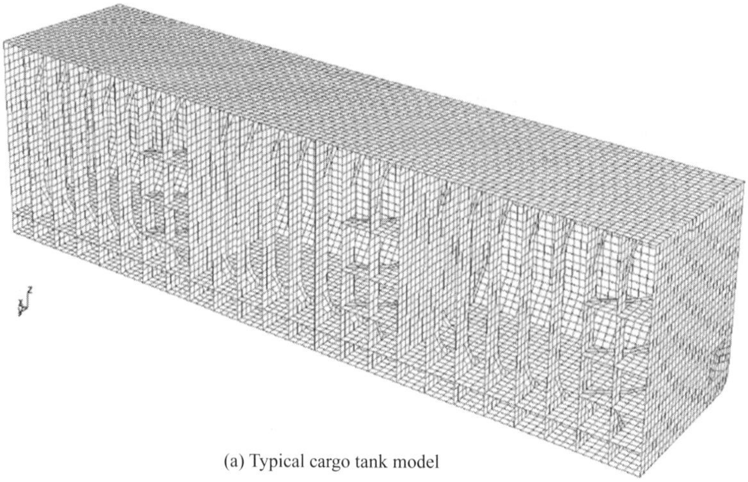

(a) Typical cargo tank model

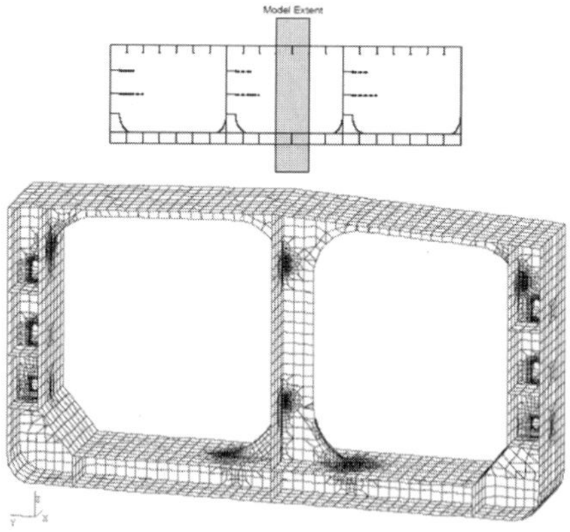

(b) Detailed calculation model of transverse web frames for one spacing

Figure 9-20 Finite element modeling of an oil tanker for structural analysis

element method (FEM) was introduced in the early 1960s. In this method the target structure is divided into numerous discrete elements so as to model it very closely to the actual structure. Since then, FEM has been developed as a strong structural analysis tool, thanks to the meteoric development of computing capabilities and the memory capacity of computers, which enable global analysis of hull structures of interest as illustrated in **Figure 9-20**(a), in addition to the detailed structural analysis on stress concentration regions as in **Figure 9-20**(b) with detailed mesh divisions when and where needed.

In FEM the geometry of the target structure is modeled using finite numbers of nodal points and elements. The governing equation, expressed as a high-order differential equation, is converted into a set of linear simultaneous equations for easier solution. Solutions converge as the number of elements increase, but, of course, time required for computation, memory capacity and man-hours required for modeling, and so on may make using a large number of elements impractical.

In order to acquire guaranteed objective reliability of the structural analysis, shipbuilders, classification societies, research institutions, and universities are using commercial codes that have been validated worldwide, such as MSC/NASTRAN, ANSYS, ABAQUS, MARC, etc.

With the advent of FEM, precise analysis of most of static and dynamic structural behavior has become possible, such as natural and forced vibration analysis and impact analysis, in addition to general linear elastic analysis of hull structure. This change has brought huge advancements in the design and safety assessment of hull structures. Various effective non-linear structural analysis methods, such as large deformation elasto-plastic analysis, ultimate strength analysis of hull girders, and thermal stress analysis which has experienced many difficulties until recently, have all been proposed and are accepted gradually, thanks to the relentless improvements of computing performance.

The hydro-elasticity problems have remained the most challenging topic. Recently, however, various attempts have been made in the fluid-structure interaction analysis. Notable examples are ALE[2]-based hydro-codes[3] applied to the fluid impact response problems, such as sloshing and slamming phenomena. To analyze the fluid-structure interaction at a reasonable level in complex, 3-dimensional flow, however, dramatic improvements in the analysis algorithm and computing power are still needed, because the analysis of fluid domain takes an overwhelming amount of computation time.

5.2 Structural Design and Safety Assessment of the Hull

Structures of the ship's hull are so complex that the design based on the rational load estimation and accurate structural analysis is practically impossible. Although much information about actual ocean waves has been gathered and accumulated, much uncertainty still remains and the only rational way of estimating the wave load currently available is

2 Arbitrary Lagrangian and Eulerian Method: ALE solves the fluid equations on an arbitrarily moving grid and shows efficient capability in fluid-structure interaction problems, where the fluid mesh near the structure undergoes large deformations and unacceptable distortion.

3 It is loosely defined as a finite element analysis code for solving large deformation, finite strain transient problems that occur on a short time scale. It can deal with highly nonlinear structural problems such as impact, crash, rupture and fluid-structure interactions. Generally explicit solvers are adopted for reducing the huge computation time. The representative commercial hydro-codes are MSC.Dytran, LS-DYNA and so on.

based on statistical treatment. In the process of the structural idealization and modeling of a target, it is highly probable that the subjective view point of the analyzer intervenes and the material properties used are also rarely as uniform as theoretically assumed. Also, in the shipbuilding process, all members cannot be produced precisely as specified by the construction drawings. Therefore, the structural strength also carries a certain amount of uncertainty.

For such reasons, statistical characteristics cannot be ignored in the estimation of the hull structural responses. In the past, it was customary to deal with the statistical characteristics using concepts such as allowable stresses and safety factors based on experiences of similar vessels. However, more efforts are being made now to apply statistical theories and reliability analysis methods for more rational structural design.

5.2.1 Concepts of Hull Structural Design

Structural design of a hull is a process of deriving an optimum compromise between two conflicting requirements.

Firstly, the ship must be designed to minimize the building and operating costs by minimizing the weight of the ship, while satisfying the functional requirements such as deadweight, speed and so on. Secondly, the ship structure must be designed to guarantee the safety of the passengers and crew members in addition to the prevention of ocean pollution. To these end national governments, international organizations and classification societies are striving to develop and manage structural design standards.

In other words, hull structural design is a process in which the arrangements and scantlings of structural members are determined to withstand loads predictable in all possible situations, while pursuing the minimum costs for construction and maintenance.

5.2.2 Structural Analysis and Safety Assessment

As discussed earlier, the initial structural design process of the hull focusses on scantlings of local members and longitudinal strength of the hull girder. Once these initial decisions are made, various structural analysis methods, mentioned earlier, can be used to conduct a more detailed safety assessment. The results of the structural analysis are then compared with the safety assessment criteria of a classification society. In order to minimize the possibility of eventual structural failures, detailed considerations of various failure mechanisms are necessary.

5.2.3 Direct Analysis of the Hull Structure

The most important external force that should be considered in ship structural design is the wave load. Wave load is an essential factor that causes the longitudinal structural responses directly relevant to the safety of the hull girder, but the accurate evaluation of it is not easy, due to many uncertainties. Therefore, each classification society has proposed its own simplified empirical formulae for the estimation of the wave load. In designing a new vessel that has never been built before, however, such empirical formulae cannot guarantee adequate structural strength.

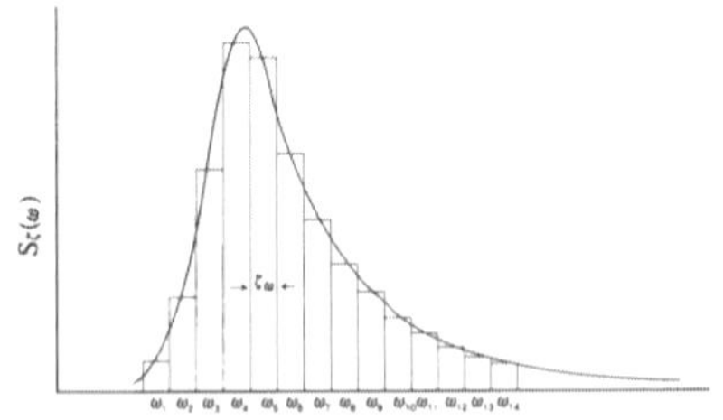

(a) Wave spectrum

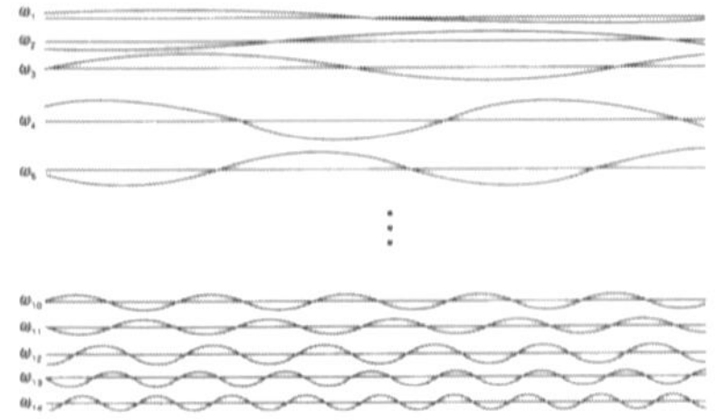

(b) Regular wave component

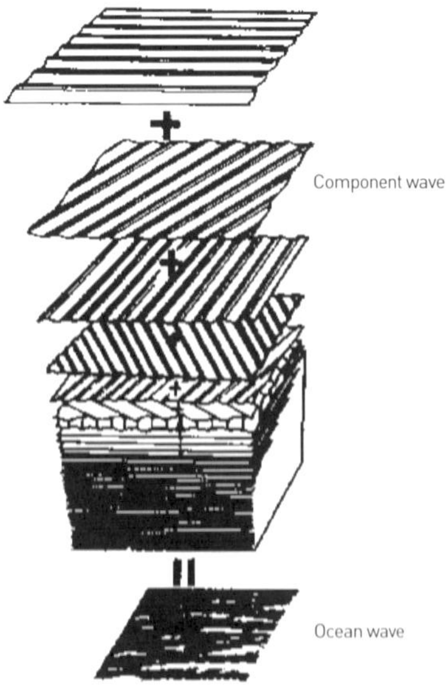

(d) Component waves forming Ocean waves

(c) Irregular ocean wave

Figure 9-21 Concepts of component waves forming ocean waves and spectrum

Thus, hull response to the waves, such as wave bending moment, shear force, acceleration and water pressure, is calculated first from the results of seakeeping analysis[4] based on hydrodynamic theories. Statistical characteristics of ocean waves obtained from the wave

4 It is an accurate hydrodynamic analysis for calculating the behavior of ship motion in waves including velocity/acceleration, pressure acting on hull surfaces, hull girder bending responses and so on. In the past, 2-D strip theory was used widely. Recently, however, the 3-D panel method, which can model the hull surface realistically, is generalized.

spectrum[5] and ocean wave statistics[6] observed in the sea route are considered. Wave spectrum expressed in **Figure 9-21**(a) assumes that the ocean waves consist of regular wave components and shows energy contributions of component waves in diagram. If the regular wave components shown in **Figure 9-21**(b) are superposed linearly, an instance of irregular ocean waves possessing the given statistical characteristics can be obtained as in **Figure 9-21**(c). The heading directions of the component waves, which are diverse in the actual ocean, as in **Figure 9-21**(d), shall be considered stochastically. Therefore, structural response of a ship's hull needs to be assessed statistically from the short-term and long-term perspectives. Such detailed processes of ship structural analysis are called the direct analysis method.

5.2.4 Structural Safety Assessment Criteria

Safe hull structures allow the ship to carry out its intended mission without suffering structural damages under even the worst loads predicted. Thus, sufficient understanding of the damages of materials is necessary to assess structural safety rationally. In general, the failure modes of materials are classified into the following four types:

1) Yield

The results of material tests show that the materials manifest elastic behavior in a relatively small range of applied stress. However, when the stress level exceeds the yield stress to cause plastic deformation, not only does it result in permanent residual deformation but also reaches the tensile strength to increase the possibility of sudden unstable fracture. Thus, excluding some special cases, most of the structures are designed so that the stress level of the members is controlled under the yield stress of material, eliminating the possibility of plastic deformation.

In the case of large structures, such as ships, there is almost no possibility of total failure even when small scale yielding occurs locally on a limited area, because the stress is redistributed to neighboring members. Nevertheless, to maintain the members free from yielding is a very significant criterion in the structural design standards.

5 Refer to the explanation on wave spectra in section 1.2.2 of Ch.6 and Pierson-Moskowitz spectrum shown in Figure 9-21.

6 Wave occurrence probability data for sea state i.e. wave height, period, wind velocity, direction, etc. They are collected and rearranged statistically at worldwide sea routes based on the observations by officers of merchant ships, weather ships or automated wave measurement systems.

2) Buckling collapse

When the level of the axial compressive stress on long and slender members, such as pillars, or in-plane compressive stress on thin plate/shells reaches a certain threshold, known as critical stress, sudden lateral deformation occurs followed by immediate unstable collapse. Such a phenomenon is called buckling, and is a very important failure mode to be considered in structural design.

Usually in slender or thin structural members, the level of critical stress is lower than yield stress. Such cases are called elastic buckling, because buckling occurs suddenly in the elastic state of members. Safety against elastic buckling stresses for simple members can be easily predicted using various theoretical formulae of critical stress.

For shorter columns, however, the critical stress increases and becomes higher than the yield point, possibly leading to plastic buckling. Computations for such cases are highly complex and are not used except in special cases. The level of critical stress is not related to the strength of the materials, such as yield stress or tensile strength, but proportional to the stiffness of the structure, such as modulus of elasticity of the material and bending rigidity of cross sections of columns or thickness of members.

3) Fatigue fracture

As explained in section 2.2, when fluctuating cyclic loads are repetitively applied to the structure, damage may occur even when the stress level is lower than the yield stress or tensile strength. Every time a cyclic load is applied, irreversible micro-damages occur, which accumulate in the material ultimately to cause cracks long enough to be visible. Once the visible crack is created, the stress is concentrated on the tip of the crack and propagates finally to result in the fracture of the member. This failure phenomenon is called a fatigue fracture. It is often observed in high speed rotary machines, structures with severe vibrations and transportation vehicles such as airplanes, ships and cars. Fatigue fracture is divided into two types: low cycle fatigue and high cycle fatigue, according to the level of stress applied to the members and the number of loading cycles until failure occurs.

Low cycle fatigue fractures occur due to plastic deformations caused by repetitive application of stress higher than the yield stress level. Cutting nails or wires by repetitively folding them sideways is taking advantage of this failure mode. The number of applications to fracture is relatively small (less than 103~104 times) and hard to analyze theoretically yet, but it is not considered to be an important failure mode because it is observed rarely in ship structures.

Compared to low cycle fatigue, high cycle fatigue fractures occur due to low level of cyclic loads more frequently. In the early stages, microscopic cracks are created and grow to lead to a sudden, unstable fracture. Before the fracture the deformation due to cracks is not easily observable, and so it is hard to make preventive repairs. Therefore, there is often a great number of casualties and property damage caused by high cycle fractures, as in the collapse of Seong-su Bridge over the River Han in 1994.

The fatigue life of structures is thought to be the sum of crack initiation life (leading up to the first crack) and crack propagation life (the time it takes for the crack to propagate until the final break through). The propagation phenomenon of cracks is not clearly investigated yet, and consequently complex theoretical and experimental approaches are needed to predict crack propagation life. In general, crack initiation life takes up most of the fatigue life and much experimental data regarding crack initiation life (S-N curve)[7] have been published. Therefore, simple fatigue life estimation methods, based on S-N data, are conventionally used. **Figure 9-22** shows a typical example of an S-N curve which has a distinctive shape for each material, but the characteristics of the curves are basically the same.

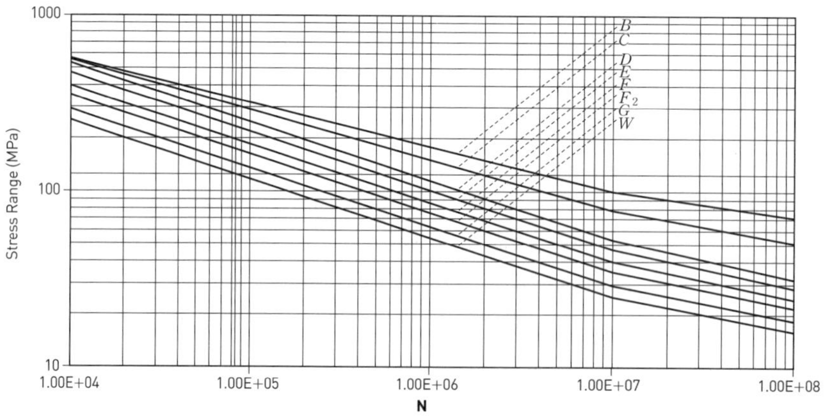

Figure 9-22 S-N curves used in the design of a double-skin oil tanker

To prevent fatigue fractures, detailed fatigue strength assessment must be performed during structural design. Quality control in the shipbuilding process, including welding, is very important, and ships in service also must be inspected regularly for any sign of fatigue cracks.

It has been observed recently that, fatigue damages in ship structures are on the increase. This is a wake-up call for more careful assessment of stress concentration regions during design. Classification societies are also preparing technical manuals for fatigue fractures.

7 S-N curve is a graph of stress amplitude versus the number of cycles to failure determined by fatigue tests for material specimen under cyclic harmonic loadings.

4) Brittle fractures

Even materials with high ductility at room temperatures show brittle characteristics at low temperatures. When a structure experiences an impact in a low temperature environment, small defects in the material can rapidly propagate to result in ruptures. This phenomenon is called a brittle fracture. Brittle fractures are initiated not only from the points of stress concentration due to notches or welding defects, but also from the fatigue cracks – even when there are no initial defects.

Recent development of material technology has produced structural materials with high resistance to fractures. As a result, the probability of occurrence of brittle fractures can be controlled to a very low level by selecting appropriate materials for the operating temperature.

6. Vibration and Noise

Ship vibration and noise induced by onboard machinery (especially the main engine and propeller), fluid movement in the cargo holds and pipes, and ocean waves, all reduce habitability and comfort of passengers and crew leading to a reduction in their work efficiency. In addition, excessive vibration can cause malfunction or damage of onboard equipment and machinery and fatigue damage of the hull structure in severe cases.

For the design and construction of high quality ships with low vibration and noise, one can consult the *Control of Ship Vibration and Noise* published by the Korean Register of Shipping. The vibration and noise of ships is normally assessed according to the contract specifications between the owner and the shipbuilder, and governed by the relevant IMO codes, classification rules and standards of the International Organization for Standardization (ISO). Efforts have to be made to reduce vibration and noise during the design stage through analytical and empirical methods, but compliance to the specifications, codes and rules are tested through measurements in the sea trials after construction.

If excessive vibration and noise is found during the sea trials, it must be rectified using some corrective countermeasures. However, such work is expensive and time-consuming, and in the worst cases, can result in the delay of ship delivery. Thus, prevention of vibration and noise during design is more effective and important than remedial measures.

6.1 Ship Vibration

All objects, including ships, have unique natural frequencies determined by their inertia and rigidity. Moreover, as the frequency of excitation force approaches the natural frequency of the object, resonance can occur in which the magnitude of the vibration becomes much larger than can be normally expected from the magnitude of the excitation force. The preventive design of ship vibration is normally done by investigating the possibility of the ship structures and onboard equipment resonating with major excitation forces induced by the propulsion system and auxiliary machinery. The vibratory responses of ship structures to the known excitation forces are examined through forced vibration analyses.

6.1.1 Excitation Sources of Ship Vibration

Ship vibration is induced by various sources including propeller, main and auxiliary machinery and waves as illustrated in **Figure 9-23**. The magnitude of ship vibration is proportional to the magnitude of excitation forces, and therefore it is important to design or select a propeller and the machinery to minimize the excitation forces. Furthermore, resonance of the ship structures with the major excitation sources must be avoided as much as possible.

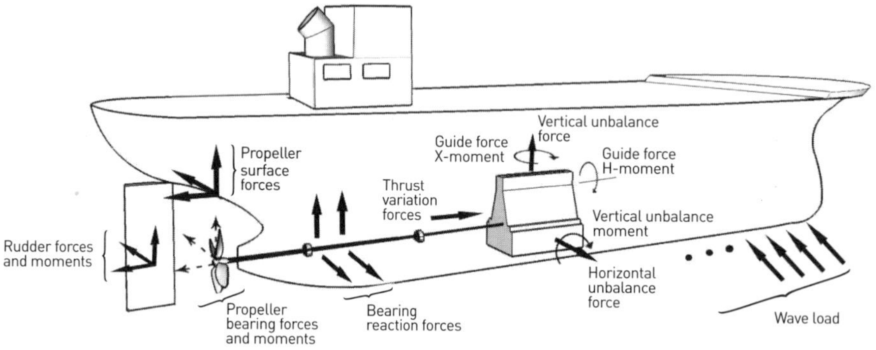

Figure 9-23 Major vibration excitation forces in a ship

1) Propeller-induced vibration

Propeller excitation force is a kind of hydrodynamic forces caused by the rotation of the propeller blades. It can be divided into propeller surface force and propeller bearing force. The propeller surface force manifests itself as pressure fluctuation acting on the hull surface above the propeller. The vortex and cavitation induced during the rotation of propeller blades generate this force. The surface force may cause vibrations in the hull girders, superstructures and local structures in the aft body.

Propeller bearing force varies periodically. Moments are caused from fluctuating pressure distribution on the propeller blades rotating in a non-uniform wake field. This force and moment are then transferred to the hull structures through the propeller shaft and bearings. Propeller bearing force may cause vibrations in the propeller shaft, hull girder, superstructures and local structures in the aft body.

2) Main engine-induced vibration

Almost all ships use turbines or diesel engines as their main engine. In the case of turbines, vibration excitation force is negligible. Diesel engines, on the other hand, have rotating and reciprocating mass of the pistons and periodic changes of combustion gas pressure in the cylinder, leading to unbalanced forces and moments which cause vibration of the hull girder structure, double bottom and main engine body. The components of the excitation force, having the same frequency as the rotating speed of the main engine, can be easily compensated for by installing counter weights on the crankshaft.

However, higher-order components of excitation forces cannot be easily dealt with. Low-speed diesel engines with 6 or fewer cylinders, in particular, usually generate a large, second order vertical unbalanced moment, and can cause severe vibrations in the hull girder if resonant.

The excitation forces and moments are produced by periodic changes of combustion gas pressure in the cylinders and consequently their frequencies are higher compared to those of unbalanced excitation forces. They excite torsional and longitudinal vibration of propeller shaf directly, and cause vibration in the main engine block, superstructure, engine room structure and so forth.

3) Wave-induced vibration

Hydrodynamic impact, due to bow slamming or green water on deck while operating in rough seas, can cause hull girder vibrations called whipping. Low-draft ships and high-speed ships are more susceptible to this. Whipping can induce damage to local structures or excessive stress in hull girder structures.

Furthermore, ship structures can experience continuous vertical hull girder vibrations, unrelated to main engine speed, even under favorable sea conditions. This is known as springing, and is often observed in large vessels such as ultra large crude oil carriers and container carriers.

4) Other excitation sources

Excitation forces induced by machinery, such as diesel generators, pumps and compressors, may cause vibrations in the machinery itself or its support structures. Moreover, water hammering phenomenon experienced in the rapid on-off operation of duct valves and surging phenomenon of pumps experienced in the non-uniform irregular in-flow of fluids may cause vibrations in ducts and pipes. Sloshing, which is the periodic motion of the free surface of fluid in a partially-filled tank, may cause vibration in the tank structure.

6.1.2 Types of Ship Vibration

Ship vibration is often categorized into hull girder vibration, superstructure vibration, various local structure vibration, propeller shaft vibration and vibration of the main engine block.

1) Hull girder vibration

Hull girder vibration refers to a beam-like vibration of the hull structure as shown in **Figure 9-24**. The modes of hull girder vibrations are classified into vertical vibration, horizontal vibration, torsional vibration and longitudinal vibration, according to the direction of the vibratory displacement.

Because the depth-to-breadth ratio of conventional ships is less than 1, the vertical rigidity of usual ship structures is less than the horizontal one. Furthermore, vertical excitation forces

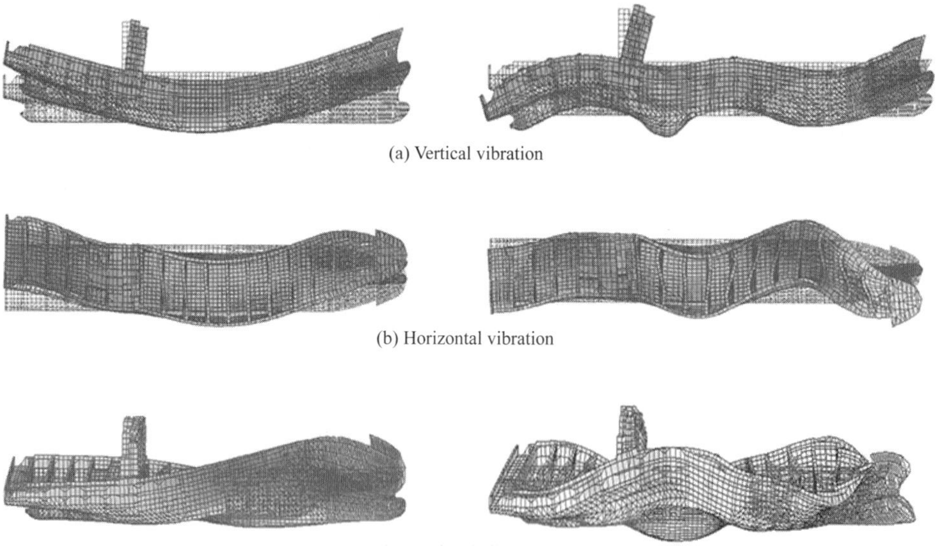

(a) Vertical vibration

(b) Horizontal vibration

(c) Torsional vibration

Figure 9-24 Natural modes of vibration in hull girder[8]

8 Mode shapes shown in Figures 9-24, 9-25, 9-26 and 9-28 are exaggerated representations of real vibration shapes.

are normally larger compared to the horizontal excitation forces. Therefore, vertical vibration is the most important mode to prevent in the hull girder design. Horizontal and torsional vibrations of a hull girder normally appear in a coupled form, because the vertical position of mass center, i.e. center of horizontal behavior, is different from that of shear center, i.e. center of torsional behavior. Torsional vibration is important in ships with open sections such as container carriers which have relatively weak torsional rigidity. Longitudinal vibration generally is not considered in vibration preventive design, because conventional ships have enough rigidity and the longitudinal excitation forces are usually small.

Excessive hull girder vibration can impair comfort and habitability for the crew and passengers, and can damage the hull girder structure in severe cases.

2) Superstructure vibration

The superstructure, which houses accommodation and work area for the crew and funnel system for the main engine, can experience vibrations due to bending and shear deformation of the superstructure itself, the rotational motion caused by elastic behavior of its supporting structures and hull girder vibration. Typical representative vibration mode shapes of superstructure are shown in **Figure 9-25**.

Excessive superstructure vibration, normally induced by the resonance with the main engine and propeller excitation forces, may reduce crew comfort and cause damage to onboard equipment and structures. As the allowable vibration level of the superstructure is normally included in the contract specification, it is very important to control and reduce the superstructure vibration.

3) Vibration of local structures

Local vibration in decks, engine room, double bottom structures, masts, funnels and

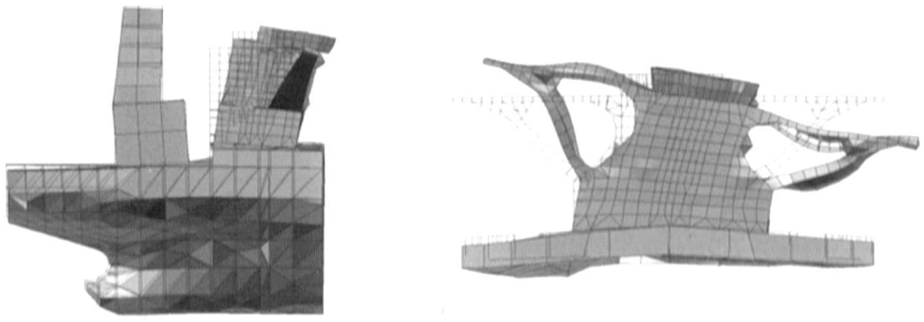

(a) Longitudinal vibration in superstructure

(b) Lateral vibration in superstructure

Figure 9-25 Natural modes of vibration in superstructure[8] propeller blades can be much more severe than hull girder vibration. It can occur by direct excitation from adjacent vibration sources or vibratory energy transmitted through hull girder and adjacent structures. Large vibration of a local structure can cause not only vibratory fatigue damage of the structure itself but also malfunction of equipment installed on the structure. In particular, control of local vibration and rationalization of mast structures supporting precise electronic equipment and navigation bridge wings accessed by watchkeeping officers and pilots is very important. **Figure 9-26** shows mode shapes obtained by free vibration analysis of the radar mast and a deck.

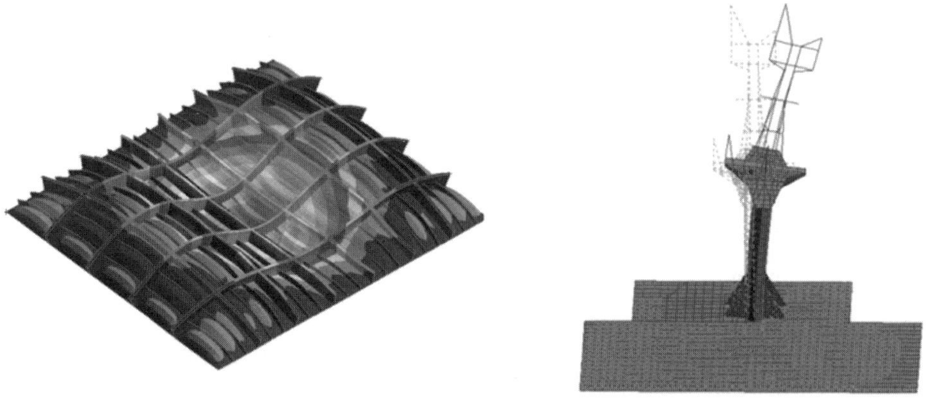

(a) Deck structure (b) Radar mast

Figure 9-26 Examples of natural vibration mode in local structures[8]

4) Propulsion shaft vibration

Vibration of the propulsion shaft connecting the main engine and propeller are classified into torsional vibration, longitudinal vibration and lateral vibration as shown in **Figure 9-27**. The longitudinal and torsional vibration of the shaft can be coupled.

The control and reduction of propulsion shaft vibration is very important for safe operation of ships, because the vibration can damage the shaft or its bearings. Thus, classification societies specify allowable dynamic stress level of shaft induced by its torsional vibration.

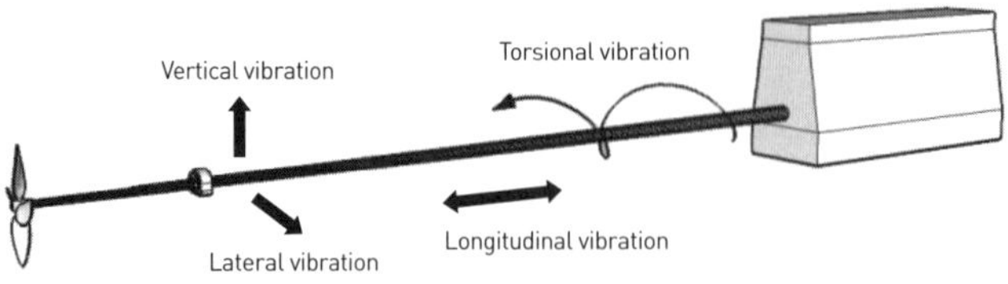

Figure 9-27 Types of vibration in propulsion shaft

5) Diesel engine vibration

Diesel engines are used widely as the main engines of ships. These not only generate excitation forces, but vibrate themselves and can cause damage to their own structure or parts such as equipment and pipes connected to the engines. The vibration of diesel engines could be classified into horizontal H-type vibration in a transverse direction, X-type vibration of torsional mode on horizontal plane, and L-type vibration in the longitudinal direction, as shown in **Figure 9-28**.

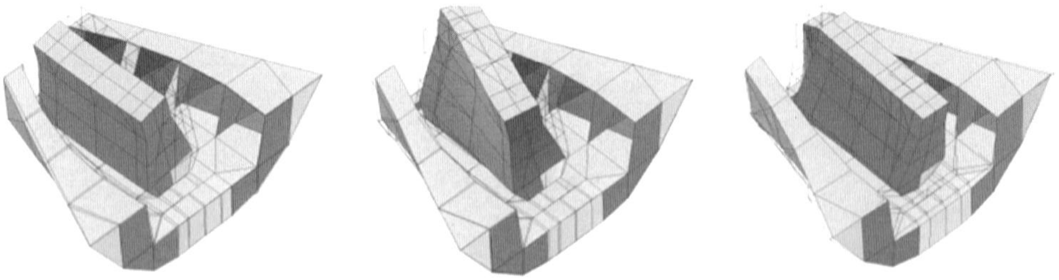

(a) H-type vibration (b) X-type vibration (C) L-type vibration

Figure 9-28 Natural vibration modes of diesel engines[8]

6) Other vibrations

Ships can experience other kinds of vibration, such as vibration induced by auxiliary machinery, onboard equipment vibration, fluid-induced vibration in the pipes and cargo holds, wave-induced vibration and so forth. Such vibrations in severe cases may cause damage to the hull structure, onboard equipment, pipes and cargo holds.

6.1.3 Assessment and Control of Ship Vibrations

To control the magnitude of ship vibration to less than the tolerable limits, preventive methods in the design process and countermeasures based on measurements during or after construction are used. An example of such a vibration control procedure is shown in **Figure 9-29**.

To assess the ship vibration responses objectively, the ISO and classification societies provide standards and rules to evaluate habitability of the crew and passengers, vibration fatigue of the hull structure and the safe operation and performance of shipboard machinery and equipment against environmental vibration. These standards are revised according to the needs in line with technological development and social/environmental changes. Representative standards and rules for ship vibration assessment are presented in **Table 9-1**.

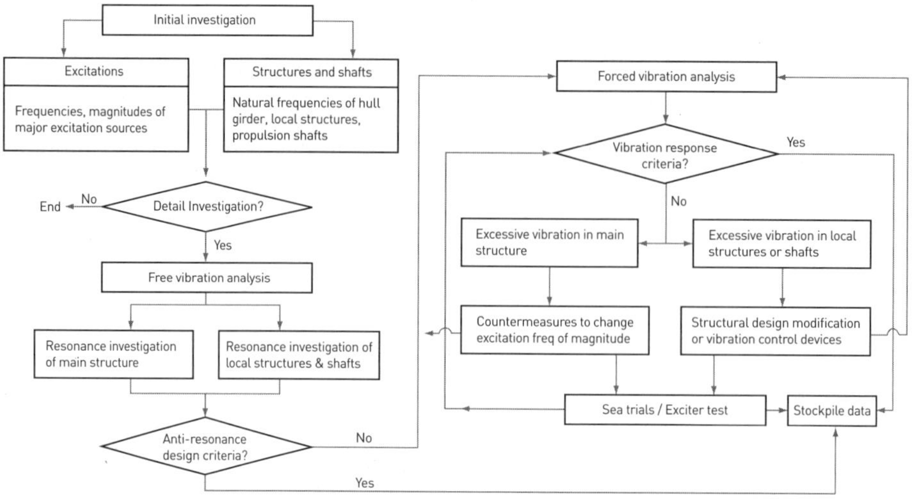

Figure 9-29 Procedures for ship vibration control

Table 9-1 Examples of standards, rules and guidelines for assessing ship vibration

Evaluation items		Relevant standards, rules and guidelines*
Habitability		ISO 6954 (12 and 16), DNV GL rule: Part 5 Ch. 12 (3)
Ship structures		ISO 20283-2 (17), Rules and guidelines of classification societies
Reciprocal engines and propulsion system		Diesel engines: Maker's and MESJ guidelines Propulsion shaft: Rules of classification societies
Onboard equipment	Vibration	ISO 10816 (13), ISO 7919 (14), DNV GL rule: Part 6 Ch.15 (4)
	Vibration resistance	ISO 10055 (15), DNV GL standards for certification (2)

• Numbers in parentheses are reference numbers.

6.2 Ship Noise

Excessive noise is unpleasant and in extreme cases can harm the hearing of passengers and crew after prolonged exposure. Noise control is one of the key technologies in design and construction of ships, because noise level onboard ships should be below the tolerable limit as specified by the contract or as regulated by the IMO.

6.2.1 Noise Sources and Propagation Paths

There are various sources of noise on board a ship, including main and auxiliary machines, propellers, heating, ventilation and air conditioning (HVAC) systems, fluid flow inside pipes and wave impact. The major sources are main engines, generators and propellers, their noise can permeate the whole ship, while auxiliary machines such as air compressors and pumps are local noise sources.

Some noises are propagated through the air (airborne noises) and some are transmitted in the form of structural vibration (structure-borne noises). In general, airborne noise affects the source space and its adjacent spaces sharing walls or decks, but has little influence on the spaces separated from the source by two or more decks or walls, because airborne noise is greatly attenuated during transmission through walls or decks.

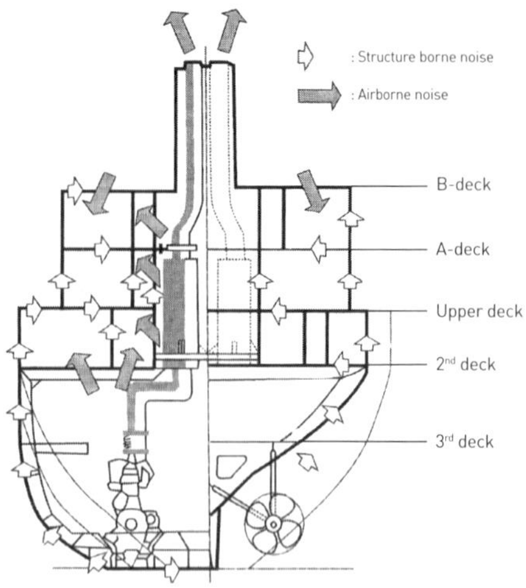

Figure 9-30 Transmission paths of airborne and structure-borne noise

On the other hand, structure-borne noise propagated in the form of structural vibration is easily transmitted far away from the sources as illustrated in **Figure 9-30**, because ship structure is light but has continuity. The noise induced by HVAC systems is both airborne and structure-borne, and are propagated through ducts to rooms regardless of the layout. Thus, the noise levels of rooms far away from noise sources are mainly determined by the structure-borne and HVAC system noise components.

6.2.2 Control and Assessment of Ship Noise

Shipboard noise is controlled by applying preventive measures during the design process and remedial countermeasures when found excessive after construction. Noise reduction techniques include using sound absorbing and insulating materials, floating floors to reduce the propagation of structure-borne noise, resilient mount to reduce the transmission of equipment noise, and installation of silencers. However these measures increase the cost of construction and weight of the ship, as well as taking up valuable ship space.

Therefore, it is often more satisfactory to control the noise level through effective design of general arrangement with minimal application of sound reduction treatments. For instance, residential areas and other spaces which need to satisfy high level of noise criteria are placed as far away from the noise sources as possible. In addition, it is necessary to specify an allowable sound power level of equipment. Also, shipbuilders and ship-owners can demand related documents and certificates in their procurement process, thus encouraging the manufacturers to produce low-noise equipment. This will ultimately improve the ship's quality by minimizing the noise from sources. **Figure 9-31** shows an example of a procedure to control ship noise.

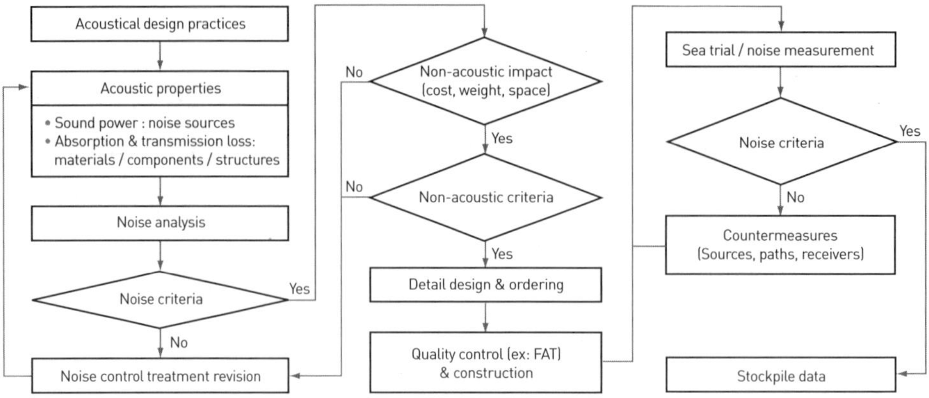

Figure 9-31 Procedures for ship noise control

IMO and classification societies provide codes and rules to ensure habitability for crew and passengers, to enhance work efficiency and to protect hearing of persons exposed to the noise over prolonged periods. IMO provisions most widely applied to general merchant ships are summarized in **Table 9-2**. On the other hand, stricter noise criteria are applied to some ships and offshore structures, including passenger ships, drillships and FPSOs (floating, production, storage and offloading), requiring higher level of protection against noise. DNV GL rules, UK HSE regulations and Australian AS 2254 standards are examples of regulations requiring more stringent level of noise control.

Table 9-2 IMO noise level limits

Designation of rooms	Noise level limits, dB(A)		
	Resolution A.468(XII) since 1981 (18)	Resolution MSC.337(91) from July, 2014 (19)	
	≥1,600 GT	1,600 up to 10,000 GT	≥10,000 GT
Machinery spaces, not continuously manned	110	110	110
Workshops	85	85	85
Machinery control rooms, galleys	75	75	75
Navigation bridge, chartrooms	65	65	65
Offices, mess rooms	65	65	60
Cabins, hospitals, radio rooms	60	60	55

References

1. Department of Energy, 1990. *Offshore installations: guidance on design, construction and certification-Annex 52*, 4th Edition. London: HMSO.

2. DnV GL, 1999. *Rules for classification of SHIPS/HSLC/MOU-Part 4 Chapter 5 Instrumentation and automation*, [online]
Available at: https://exchange.dnv.com/publishing/rulesmou/1999-01/MOU-Part%204%20Chapter%205.pdf
[Accessed 19 June 2015]

3. DnV GL, 2003. *Rules for ships-Part 5 Chapter 12 Comfort class*, [online]
Available at: https://exchange.dnv.com/publishing/rulesship/2009-07/ts512.pdf
[Accessed 19 June 2015]

4. DnV GL, 2011. *Rules for classification of ships-Part 6 Chapter 15 Vibration class*, [online]
Available at: https://exchange.dnv.com/servicedocuments/dnvts615_2014-01%20(2).pdf
[Accessed 19 June 2015]

5. DnV GL, 2014. *Classification notes no.30.7, Fatigue assessment of ship structures*. (supersedes Classification notes no.30.7, June 2010), [online]
Available at: https://exchange.dnv.com/publishing/cn/2014-04/CN30-7.pdf
[Accessed 19 June 2015]

6. Hertzberg, R.W., 1989. *Deformation and fracture mechanics of engineering materials*. 3rd ed. NewYork: John Wiley & Sons.

7. Hugh, O.F., 1983. *Ship structural design: arationally-based, computer-aided, optimization approach*. New York: Wiley-Interscience. (5th edition published by SNAME in 2005).

8. Hwang, J.H. & Yim, S.J., 1989. *Merchant Ship Designs* (in Korean). Seoul: Munundang.

9. Hydro Lance corporation, *Superfast, smooth, half the fuel, greater safety*, [online]
Available at:http://www.hydrolance.net/RO-RO-container-FastShips.htm
[Accessed 19 June 2015]

10. IMO, 1981. Resolution A. 468(XII) *Code on Noise Levels on Board Ships*. London: IMO Publishing.

11. IMO, 2012. Resolution A. MSC. 337(91) *Code on Noise Levels on Board Ships*. London: IMO Publishing.

12. ISO, 1984. ISO 6954:1984 *Mechanical vibration and shock-Guidelines for the overall evaluation of vibration in merchant ships*. Geneva: ISO. (This standard has been revised by ISO 6954:2000).

13. ISO, 1995. ISO 10816-1: 1995 *Mechanical vibration-Evaluation of machine vibration by measurements on non-rotating parts-Part 1: General guidelines*. Geneva: ISO.

14. ISO, 1996. ISO 7919-1:1996 *Mechanical vibration of non-reciprocating machines-Measurements on rotating shafts and evaluation criteria-Part 1: General guidelines*. Geneva: ISO.

15. ISO, 1996. ISO 10055:1996 *Mechanical vibration-Vibration testing equipments for shipboard equipment and machinery components*. Geneva: ISO.

16. ISO, 2000. ISO 6954: 2000 *Mechanical vibration and shock-Guidelines for the measurement, reporting and evaluation of vibration with regard to habitability on passenger and merchant ships*. Geneva: ISO.

17. ISO, 2008. ISO 20283-2: 2008 *Mechanical vibration-Measurement of vibration on ships-Part 2: Structural vibration*. Geneva: ISO

18. Korea Maritime Institute, 2003. *Trends of strengthening vessel safety following Prestige incident and its ripple effects* (in Korean). Current issues in maritime and fishery. Seoul: KMI.
Available at: http://www.kmi.re.kr under (research data/discontinued/Analysis of current issues in maritime

and fishery/2003-01).
[Accessed 19 June 2015]

19. Korea Register of shipping, 1997. *Control manual of shipping vibration and noise* (in Korean). Seoul: Korea Register of shipping.

20. Korea Register of shipping, 2010. *Classification society and steel ship regulations* (in Korean). Seoul: Korea Register of shipping.

21. Korea Register of shipping, 2014. *Control manual of shipping Vibration and noise* (in Korean), 3rd edition. Seoul: Korea Register of shipping.

22. Lee, J.S., 1981. *Ship structural mechanics* (in Korean). Seoul: Korea Maritime Research Institute.

23. Lee, S.M. et al., 2010. Current status and perspective of ship vibration technology (in Korean). In: Korean Society for Noise and Vibration Engineering. *Spring meeting*. Jeju, 12~14 May 2010. Seoul: Korean Society for Noise and Vibration Engineering.

24. Marine Accident Investigation Branch (MAIB), 2009. *Accident investigation report 10/2009 Collision between general cargo vessel Scot Isles and bulk carrier WadiHalfa*. Southampton: MAIB.
Available at: https://assets.digital.cabinet-office.gov.uk/media/547c6fe7e5274a429000005b ScotIslesWadi HalfaReport.pdf
[Accessed 19 June 2015]

25. Park, S.K., 1998. *Introduction to knowledge of Naval Architecture* (in Korean). Seoul: Dongmyungsa.

26. Standards Australia, 1988, AS 2254-1988 *Acoustics-Recommended noise levels for various areas of occupancy in vessels and offshore mobile platforms*. Sydney: Australia.

27. Terasawa, K. ed., 1971. *Strength of ship hull structure* (in Japanese). Tokyo: Kaibundo Publishing Co.

28. The Society of Naval Architects and Marine Engineers (SNAME), 1971. *Principles of Naval Architecture.* Translated to Korean by Yim S.J., 1991. Seoul: Daehan Textbooks Co., Ltd.

29. The Society of Naval Architects and Marine Engineers (SNAME), 1983. *Design guide for shipboard airborne noise control, Panel M-27 Machinery noise.* T&R Bulletins and Reports, T&R 3-37. Jersey City: SNAME.

30. The Society of Naval Architects of Korea, 1996. *Shipbuilding Engineering* (in Korean). Seoul: Dongmyungsa.

31. The Society of Naval Architects of Korea ed., 2011. *Introduction to Naval Architecture and Ocean Engineering* (in Korean). Seoul: GS Intervision.

32. Wikimedia, 2004. *JUNEAU, Alaska (Dec. 19, 2004)-An over-flight photo taken Dec. 19 show the bow and stern sections of the 378-foot freighter SelendangAyu near Skan Bay. Unified Command photo.* {{PD-USCG}}
Available at https://commons.wikimedia.org/wiki/File:Selendang_Ayu.jpg
[Accessed 19 June 2015]

CHAPTER 10

Outfitting

CHAPTER 9
CHAPTER 10
CHAPTER 11
CHAPTER 12
CHAPTER 13
CHAPTER 14
CHAPTER 15
CHAPTER 16

1. Steering Gear and Rudders

1.1 Steering Gear

Steering gear is equipment installed in the after body of a ship to operate its rudder(s), and is controlled by signals from the steering wheel or the autopilot in the wheel house.

1.1.1 Rotary Vane Type Steering Gear

This type of steering gear has a circular stationary housing with stator vanes fixed radially inward toward rotor and a rotor with outward radial vanes attached to it. The rotor is located in the middle of the stationary housing, and thus the space between the stator and the rotor is divided by the vanes. The rudder is controlled by the rotor which in turn is controlled by high pressure hydraulic oil in the spaces between the vanes. A typical rotary vane type gear is shown in **Figure 10-1**.

(a) Rotary vane type steering gear

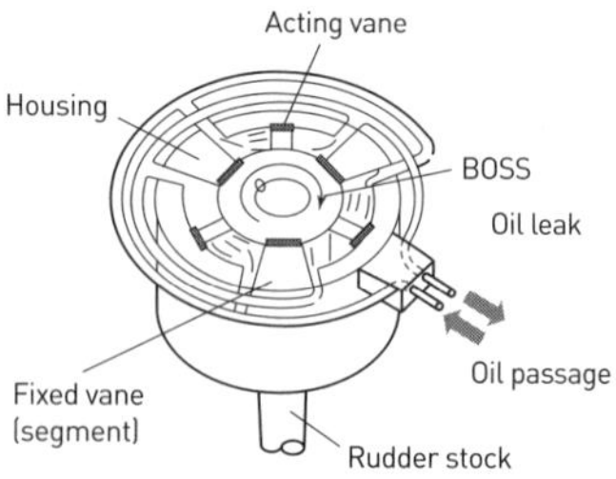

(b) Composition of actuator fot rotary vane type steering gear

Figure 10-1 Rotary vane type steering gear

1.1.2 Ram-cylinder Type Steering Gear

This type is also referred to as a Rapson slide type. Hydraulic cylinders are arranged fore and aft of the rudder stock. The hydraulic rams operated by hydraulic pressure push/pull the tiller which then rotates the rudder stock. A typical ram-cylinder type is shown in **Figure 10-2**.

(a) Ram-Cylinder type steering gear

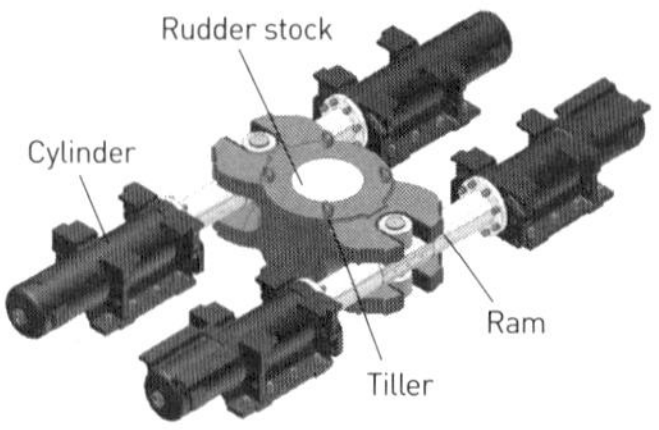

(b) Ram-Cylinder type

Figure 10-2 Ram-cylinder type steering gear

1.2 Rudder System

The rudder is installed at the stern behind the propeller. When water flows past the rudder, the pressure difference between the face and the back surface of the rudder blade generates a lift force. This lift force is used to steer the ship. A typical horn-type rudder and its installation arrangement are shown in **Figure 10-3**.

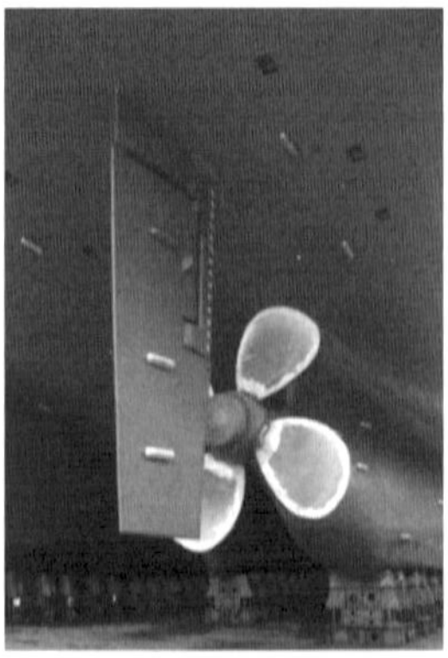

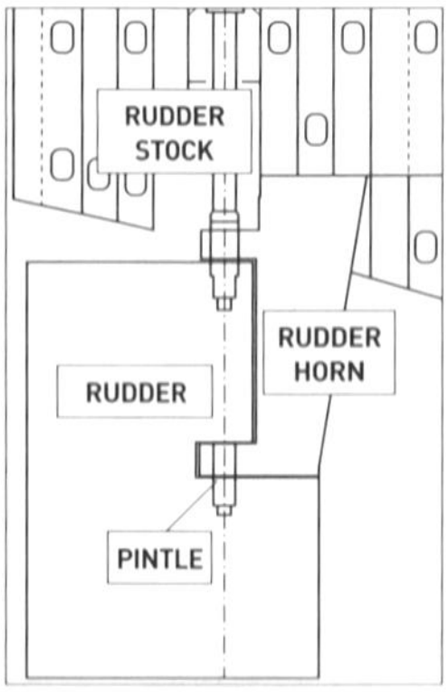

Figure 10-3 An example of semi-balanced rudder in position

1) Rudder types by the installation features

(1) Rudder with heel

This type of rudder sits on the heel gudgeon as well as the upper pintle. The heel is formed by extending a stern bottom frame as shown in **Figure 10-4**(a).

(2) Semi-balanced rudder/semi-spade rudder

The semi-balanced or semi-spade rudder shown in **Figure 10-4**(b) is connected to the rudder horn; the pintle shaft and rudder stock act as a hinge to rotate the rudder, as shown in **Figure 10-3**. The rudder below the pintle is balanced to some degree so that smaller forces are required to turn the rudder.

(3) Balanced rudder/full-spade rudder

The rudder is hung from the hull by the stock which is near the center of effort of the rudder blade to minimize the rotational torque.

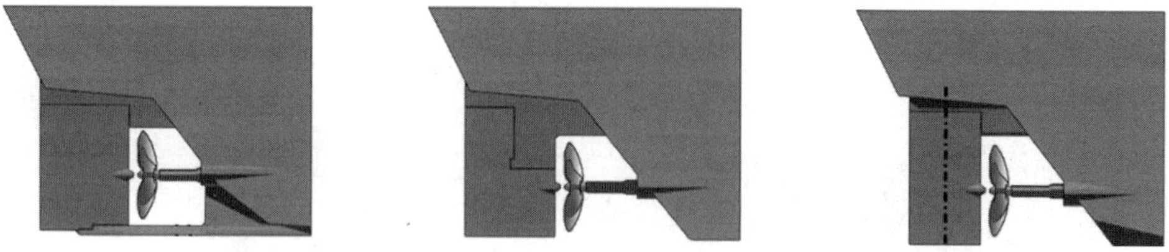

(a) Rudder with heel (b) Semi-spade rudder (c) Spade rudder

Figure 10-4 Rudder types by the installation features

2) Rudder types by the specific section shape

(1) Rudder with twisted leading edge

Rudders with twisted leading edge have asymmetric sections as shown in **Figure 10-5**. The leading edge of the section is shifted slightly to one side in order to straighten the rotational flow component in the propeller wake so that the rudder can absorb some of its kinetic energy. Thus, the leading edges of the rudder in the upper-half and lower-half are made slightly twisted in opposite directions when observed from an upstream angle. The blade twist is effective not only in absorbing rotational energy, but also in gaining thrust force components from the rudder btlade.

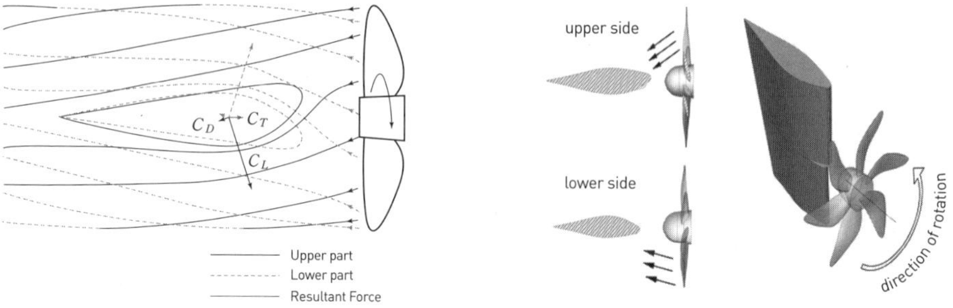

Figure 10-5 Rudder with twisted leading edge

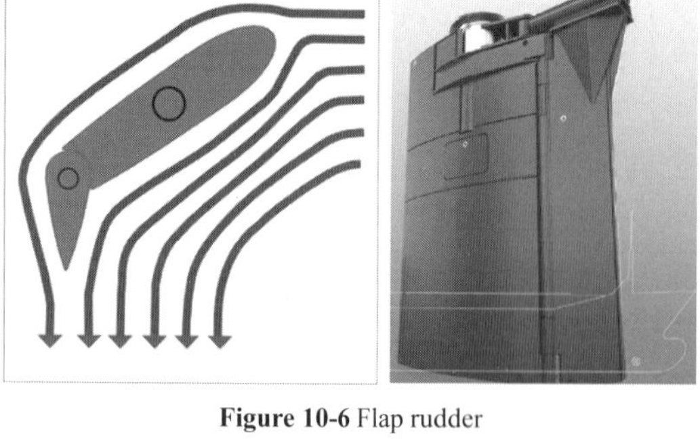

Figure 10-6 Flap rudder

(2) Flap rudder

A flap rudder consists of a main blade and an additive blade, so-called a flap, as shown in **Figure 10-6**. When the rudder angle is altered, the angular change of the flap is amplified by a mechanical link, generating a bigger lift. This type of rudder improves the controllability and turning force, compared to conventional rudders, especially at low speed.

(3) Schilling rudder

The section of the schilling rudder spreads out like a fish tail at the trailing edge as shown in **Figure 10-7**. This section shape was developed in 1970s to improve the controllability of slow barge-shaped ships operating in inland waterways. The speed loss experienced when applied to ocean-going vessels used to be a factor against its wide application, but this problem has since been ironed out and it can now be used on all vessels, regardless of their size, from tug boats to VLCCs. This type of rudder improves the course keeping stability, reduces steering gear vibration at large rudder angles, and improves the controllability in going astern.

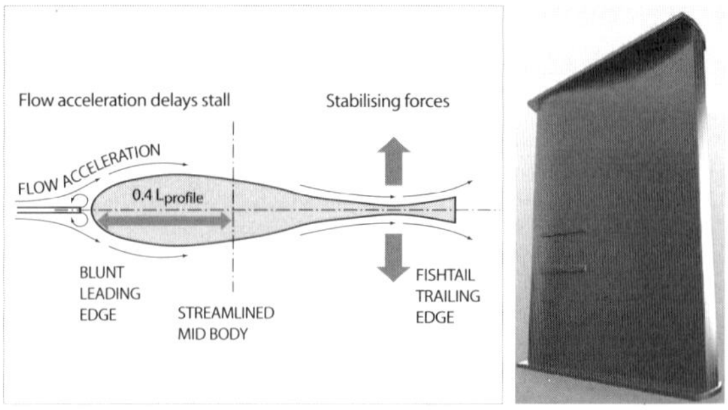

Figure 10-7 Shilling rudder

1.3 Autopilot

Autopilot is an important method of steering a ship and its schematic is shown in **Figure 10-8**. Autopilot may be based on adaptive control or PID control.

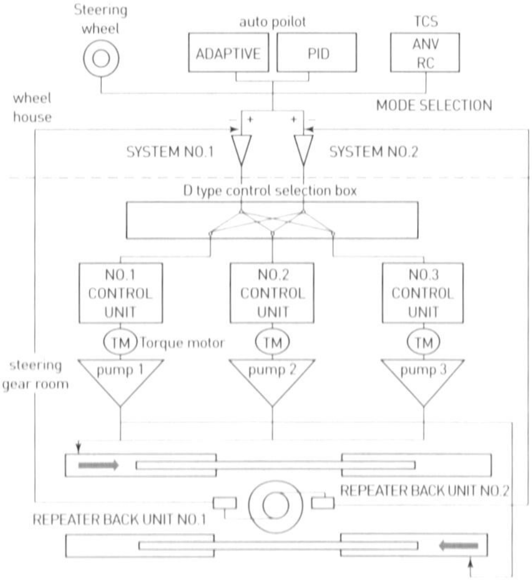

Figure 10-8 Auto pilot control flow chart

1) Adaptive control

Dynamic response characteristics of the ship in typical operating conditions, such as loading condition, draft, speed, wind direction wind speed are modeled with MRACS (model reference adaptive control system). Based on this model, the rudder angles required to maintain an optimum navigational route are calculated and commands are issued. Weather data is processed with a Kalman filter to avoid excessive rudder movements due to the bow yawing in bad weather. In this way it can reduce the dependence on skillful quartermasters and prevent increase in fuel consumption due to excessive and frequent rudder movements often made by inexperienced quartermasters.

2) PID (proportional integral differential) control

Required ship's heading is given to the controller as the target. The controller then receives the signal of the actual ship's heading from a gyrocompass for example. It analyzes this feedback signal and, taking into account various other factors, such as ship's maneuvering dynamics and weather condition, computes the rudder angle required. The main heading change command is obtained by summing a quantity proportional to the error in heading angle, a quantity proportional to the rate of change of the error and another term proportional to the integral of the error (hence the name PID controller). The differential term is proportional to the yaw rate and is included to reduce excessive rudder movements by anticipating the heading angle. The integral term removes long-term heading drift due to a bias in the system.

2. Navigation and Radio System

This class of equipment consists of the navigation system, radio system and intercommunication system. The radio system and navigation system are regulated by Chapters four and five of SOLAS and the essential equipment required by SOLAS should comply with IMO standards.

2.1 Navigation System

1) Radar (radio detection and ranging)

Radar detects objects around the ship by transmitting radio waves and receiving their reflections. It is essential for navigation especially at night and in poor visibility. Radar consists of 'X-band' and 'S-band' in principle and it is installed in the wheel house.

ARPA (automatic radar plotting aid) detects, displays, memorizes and tracks objects around the ship. Using these functions, an ARPA determines if any object ahead, for example a reef or another ship, presents a hazard of collision. If it decides that there is a risk of collision, it sounds an alarm to alert the navigator. ARPA is used by connecting to the existing radar, either S-band or X-band.

2) ECDIS (electronic chart display and information system)

ECDIS displays digital hydrographic chart on a monitor and is also capable of tracking navigation routes.

3) AIS (automatic identification system)

Originally developed for airplanes, AIS was introduced to ships following the Braer accident in Britain in 1993. It is used for collision prevention and VTS control, and transmits and receives the ship details, ship type, position, heading, ship speed and other information relevant to navigational safety.

4) DGPS (differential global positioning system)

Ships at sea can measure their position and speed precisely with DGPS by using 24 satellites orbiting the earth twice a day at an altitude of 20,200 kilometers, of which 21 are actually used at any one time while 3 are kept in reserve. GPS can be relied upon to provide precise location information at any time in all conditions anywhere on earth.

5) Gyro compass

A gyro compass measures the true heading of the ship, based on the 'true north' indicated by the axis of a gyroscope spinning at a high speed.

6) Magnetic compass

A magnetic compass measures the ship's heading relative to the datum of terrestrial magnetic north. Just like the meridians linking the earth's two poles, the magnetic meridians link geomagnetic north and south poles. The magnetic poles are not coincident with the geographical poles, and this causes the two meridians to be at an angle to each other, and this angle is referred to as a deviation. Furthermore, there is another deviation due to the error of the compass itself which varies depending on the heading and the ship. This error is the difference between the earth's magnetic north and the compass north and it is known as magnetic deviation. The information needed to correct the magnetic deviation is obtained during the sea trials.

7) Rudder angle indicating system

The rudder angle is electrically displayed on the indicators installed in the wheel house and other necessary places such as steering gear room.

8) Echo sounder system

An echo sounder sends out acoustic signals and measures the time taken for their reflections to come back. This is then used to determine the water depth under the ship.

9) Speed log system

The speed log system measures ship speed and there are a few types depending on the method of speed measurement: Doppler speed log, acoustic speed log and electromagnetic speed log.

10) Master clock system

A chronometer and a crystal clock are used as the master clock system. The ship's chronometer utilizes the regular electrical signals obtained from electronically excited vibration of quartz. An acceptable error range of a crystal clock for navigation is usually less than ± 0.2 seconds per day within a temperature range of $-10°C$ to $+50°C$. Unlike chronometers, crystal clocks are not affected by temperature and humidity, because they are electrically activated, and they are also convenient as they do not need to be wound. Furthermore, a crystal clock can be connected to other slave clocks throughout the ship.

11) Wind measurement

Wind speed is measured with an anemometer and its direction with anemoscope. An anemometer can be of either windmill type or cup type.

12) VDR (voyage data recorder)

A VDR is equivalent to an airplane black box and it records, maintains and manages various data in real time during a voyage. Voyage data, engine conditions, navigational information, weather information, etc. are converted to digital forms and then transmitted to the main memory for storage. The stored data can be accessed at any time and output in any form when required. The mandatory installation of VDR has been in discussion since the capsize of Estonia, a Ro-Ro passenger ferry, in the Baltic Sea in 1994 with a loss of 852 lives. VDR records are very useful in investigation of marine accidents so that similar accidents can be avoided in the future.

13) Whistle system

A whistle system carried on board includes all acoustic signal devices which can sound short or long blasts, referred to as whistles in the regulations, using steam, compressed air or electricity. In order to diversify the tone of acoustic signals, ranges of signal frequencies are assigned to various ship sizes as shown in **Table 10-1**.

Table 10-1 Acoustic signals

Length of ship	Frequency of acoustic signal	Description of the tone
Over 200 m	70~ 200 Hz	Heavy and deep sound
Over 75m ~ under 200m	130~350 Hz	Loud and bright sound
Under 75m	250~700 Hz	High and urgent sound

14) Weather facsimile

A weather facsimile receives radio wave signals of weather charts, newspapers, fishing information and so on from wireless stations around the world and reproduces them in their original form. This device uses frequencies from 3 to 24 MHz.

2.2 Composition of Communication Equipment

1) Radio telephony

The essential communication equipment on board a ship includes MF/HF radio telephones for communication with adjacent ships or shore stations within 100 nautical miles and VHF radio telephones the operational distance of which is from 20 to 30 nautical miles. The usual communication channels include coastal stations, satellites and adjacent ships as illustrated in **Figure 10-9**.

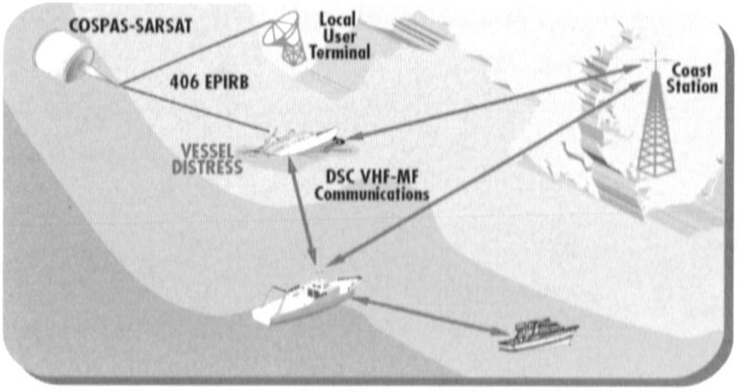

Figure 10-9 Channels of communication

Equipment required by GMDSS (Global Maritime Distress and Safety System) includes portable VHF/FM two-way radio telephones with more than two frequencies including 156.8MHz FM for communication between ships in distress, rescue ships and rescue airplanes.

UHF radio communication systems operating in frequencies between 300 and 3000 MHz are also in use. These radio systems are portable devices for communicating within a ship, between adjacent ships or between ship and shore, and are used in loading/unloading, rescue operations and other on-board operations.

2) NAVTEX (navigational telex) receiver

The NAVTEX receiver uses the international common frequency of 518 kHz. It automatically receives MSI (marine safety information) through NAVTEX. The information contains weather and any warnings in effect, and is supplied by coastal countries for the benefit of vessels operating nearby. It is possible to select the option of receiving only the necessary information. A NAVTEX coastal country gathers navigational warnings, weather warnings and information on search and rescue, and broadcasts the information either on its own or jointly with other countries to the relevant sea areas. Maritime safety information for ships sailing in international waters or areas where NAVTEX is not available can be accessed via narrow band direct printing (NBDP) through its eight designated short wave frequencies or the high-performance enhanced group calling (EGC) system of INMARSAT.

3) EPIRB (emergency position indication radio beacon)

EPIRB is required by GMDSS. This system transmits distress signals of 406 MHz through six satellites including COSPAS satellites at an altitude of 1000 kilometers and SARSAT satellites at 850 kilometers. It can be used throughout the world, as it uses polar orbit satellites. However, it can take up to about 70 minutes from transmission from a beacon to reception on land, because the satellite orbit is low at1000 kilometers.

In the event of distress at sea, on land or in air, this system is manually or automatically activated to transmit radio signals. When COSPAS-SARSAT satellites detect the signals, the location of the distress can be determined within an error range of one to two miles. The information is then passed on to a nearby rescue coordination center (RCC) for rapid search and rescue. If a ship sinks by four meters from the datum or heels to more than 45 degrees, EPIRB should be automatically activated, and it should also be possible to be activated manually.

4) Lifeboat search and rescue location devices

Search and rescue location devices are required by GMDSS and they are carried on the lifeboats in which people abandon ship. In the event of difficulties at sea, these devices show the location of the lifeboats on the 9 GHz (X-band) radar display or AIS of other ships passing nearby, and is consequently known also as search and rescue radar transponder (SART).

5) SSAS (ship security alert system)

SSAS was developed to transmit security alerts and status reports to shore stations in the event of pirate or terrorist attack and other dangerous situations.

6) Satellite communication system (INMARSAT-C)

INMARSAT-C is essential equipment on board ships. It uses small and light terminals to provide accumulated data services. Voice service is not provided.

7) Satellite communication system FLEET-F

FLEET-F was developed to combine all existing communication services except INMARSAT-C. This maritime satellite communication system uses a global beam in support of distress/safety communications, and there are FLEET F77, F55 and F33 services. FLEET F77 has provided commercial services since April 2002, but it will stop services in 2015 and FB500 will replace it. Telephony, facsimiles and data (up to a maximum of 64 kbps) services are provided and MISDN (Mobile Integrated Service Data Network) or MPDS (Mobile Packet Data Service) can be chosen.

8) Satellite communication system Fleet Broadband 500 (FB)

Fleet Broadband 500 was originally developed as the ISDN basis of the INMARSAT services. However, as the improved frequency band has become the broadband basis, Fleet Broadband marine service has been newly developed. FB uses three INMARSAT-4 (I-4) satellites to provide communication services to ships in the Pacific, Indian and Atlantic Oceans.

2.3 Intercommunication System

Equipment in this category includes on-board command system which uses microphones and speakers, magnetic telephones which can call or communicate without a separate power supply, automatic exchange telephones, and CAS (communal aerial system) for radio and television.

CCTV systems are used to provide specific visual information to designated users. Their usage includes observation and alarm monitoring of auto-pilot system, and various equipment including main and auxiliary engines. CCTV systems are also used as security equipment and to observe key machinery and loading/unloading equipment.

3. Loading / Unloading System

3.1 Equipment of Cargo Handling Systems

Onboard cargo handling equipment is used to load or unload cargo and consists of cranes, hoists, winches, masts, posts, stays and derrick booms. The performance of the cargo handling system determines how fast a ship can load and unload its cargo in ports (port turnaround time). Loading/unloading is carried out while the ship is anchored or moored at a quay or buoy. Sometimes, a barge is tethered alongside and loading/unloading is carried out using the on-board handling systems.

The typical cargo handling systems are derricks, deck cranes and gantry cranes, but conveyor systems are sometimes used for ships carrying bulk cargos such as ore and coal.

1) Derricks

As shown in **Figure 10-10**, two derrick booms are installed on two derrick posts and loading/unloading is carried out by controlling the cargo ropes with winches. There are two ways of operating the derricks. In the first the booms are fixed and vertical and sideways movement of the cargo is controlled by the two cargo lines with the winches. The second method uses the winches only for lifting and lowering the cargo, while the sideways movement is achieved by rotating the derrick booms around the posts. Derricks are good for handling general or bale cargo, but poor for arranging cargo in cargo holds.

2) Deck cranes

There are four types of deck cranes: cranes fixed on the ship's center line, side-to-side mobile cranes, port or starboard fixed cranes, and fore and aft mobile cranes. These have better cargo handling capability than derricks. They handle all cargos within its turning

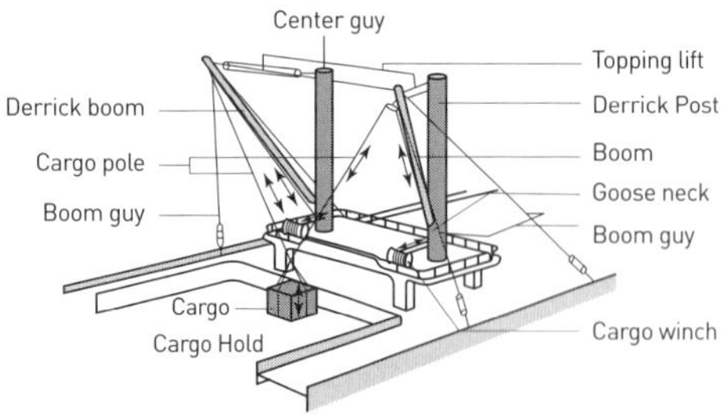

(a) Derrick boom fixed type

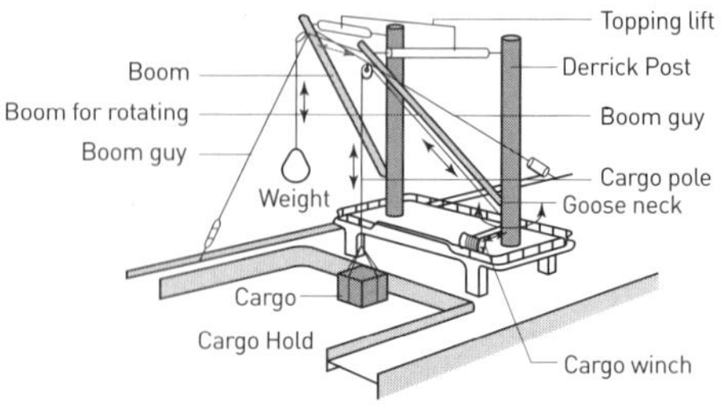

(b) Derrick boom rotating type

Figure 10-10 Cargo handling system (derricks)

radius as long as they are within the maximum safe working weight. Deck cranes are easy to get ready for loading/unloading, secure for sailing, and adjust for cargo weight.

3) Gantry cranes

Gantry cranes move along rails on deck. They are the best cranes to handle large bundle cargos such as bundled timber and newsprint papers. If all superstructures are placed towards the stern during general arrangement, the gantry crane can be made to move throughout the cargo area without hindrance and therefore high cargo handling efficiency can be achieved.

3.2 Embarkation Ramps

1) Gangway ladders, accommodation ladders

Gangway ladders are for boarding and disembarking in ports. They are installed at convenient locations near accommodation blocks least affected by waves, making sure that they do not interfere with the operations of cargo handling systems and lifesaving equipment. They are usually installed on both sides of the ship, but oil tankers may have it only on one side.

The standard slope of a gangway ladder is 55 degrees to the water surface in the ballast condition. The whole length of gangway ladders is generally more than six meters, and they could be made foldable, if the difference between light load draft and full load draft is more than 2.5 meters. The width of gangway ladders should be at least 600 mm between the handrails and the effective width of the treads should be not less than 500 mm. The strength of the ladder should be such that it can withstand the load of a person of 75 kilograms weight on every other tread with a safety factor of 5 to 8. It must also be able to withstand the collision impact of an embarkation boat.

A wharf ladder is shown in **Figure 10-11**. Details concerning types, materials, size and fabrication methods of these ladders used as passageways between ships and shore are specified in KS V 2622 and KS V 2623. Light ladders are preferable as they are easier to put away at sea and deploy when required.

Figure 10-11 Gangway ladders

Figure 10-12 An example of a gangway ladder used as pilot ladder

2) Pilot ladders

Rope ladders are often used for pilot ladders and they are also used for access route between the ship and small boats such as rescue craft. Details of types, materials, size and fabrication methods for pilot ladders are specified in KS V 2625. The International Maritime Pilot's Association (IMPA) prohibits pilot ladders over 9 m in length and recommends the use of gangway ladders instead as illustrated in **Figure 10-12**.

4. Deck Outfitting

4.1 Mooring Arrangement

A ship is restrained to one location against the drift forces due to wind, waves and tide by deploying anchors or mooring to a buoy or quay. The anchoring system consists of anchor chains, windlasses, chain compressors, chain clenches and anchors as shown in **Figure 10-13**.

1) Chains

Chains are made of θ-shaped steel elements linked together and their unit length is 27.5 m. The stud in the middle of each link provides transverse support and it enhances the strength of the chain and prevents the chain being twisted. Large ships have anchor chains made of 10 to 15 lengths at both port and starboard, but smaller ships may use wire or fiber ropes instead of chains.

2) Mooring winches

Mooring winches are equipment used for pulling up chains and anchors, and there are hydraulic, electric and pneumatic types. They have to deal with heavy loads and the pulling speed should be at least 9 m/min. Mechanical band brakes are often used for their braking system.

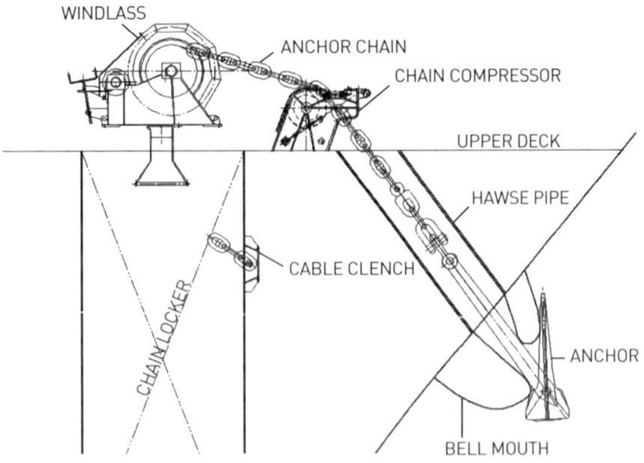

Figure 10-13 Mooring system

3) Chain compressors

When a ship is anchored, the chain compressors prevent direct delivery of the load on the chains to mooring winches, and while the ship is on the way they fix chains by pressing down on them to prevent anchors from falling. They are installed between mooring winches and hawse pipes.

4) Cable clenches

Cable clenches tie the end of chains to the ship, and are located in chain lockers in merchant ships. They are designed to detach the chains easily by simply removing a fixing pin in emergency.

5) Anchors

All ships carry anchors as essential equipment and they act as fixed points to tie the ship to at sea. The main purpose of anchors is to restrain the ship at a fixed point, but they can also be used for an aid in emergency stopping or turning, as they provide high resistance when partially lowered in the water.

When a ship is stranded, anchors can be deployed at suitable places using external help to improve stability. They can also be used to pull a ship off reefs using mooring winches. In order to be effective in their primary function anchors should be made heavy enough and designed to dig into the seabed easily. Various anchors have been developed over the years, but they can be roughly divided into stock anchors and stockless anchors. Stockless anchors often used in general merchant ships can be dropped in any attitude. The weight of the chain and the anchor itself pushes the fluke down, and therefore, as soon as the chain pulls on this type of anchor, the fluke digs into the ground as shown in the illustrations on the left of **Figure 10-14**. The harder it is pulled, the more it digs in, increasing the anchoring force.

When the angle of the line at anchor is greater than, say about 12°, as when the anchor is raised, the fluke comes out of the seabed easily with very little force as shown by the illustrations on the right of **Figure 10-14**.

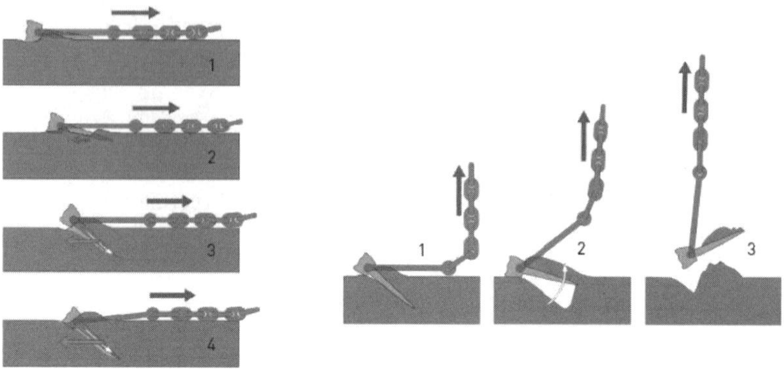

Figure 10-14 The action of anchor

4.2 Mooring Equipment

During loading/unloading operation of cargo in ports, the ship must be secured against the forces due to wind, waves and tide using sufficient and safe mooring arrangement. Considering the severity of damages that can occur due to unexpected drift, the mooring arrangement is very important, but surprisingly simple methods are used for mooring.

Mooring a ship means tying it to buoys or dolphins of quays with mooring ropes. Similar equipment is used in single point mooring (SPM), arrangement in a dense mooring area and mooring for bunkering at sea.

When moored, a ship is restrained from moving away from the quay or fore-and-aft movement. If the mooring fittings of the ship and shore are ideally arranged, two mooring cables are sufficient to restrain the ship in the transverse direction and two more are enough to restrain the fore and aft movement. However, the arrangement of fittings is not perfect in most cases, and so additional cables are necessary at bow and stern.

To moor a ship safely the following five rules apply:

① Arrange transverse cables at right angles to the longitudinal direction of the ship to prevent it from moving away from the quay.

② Arrange the lines restraining the fore-and-aft movement to be as parallel to the ship's side as possible.

③ Reduce the angle between mooring lines and the quay parallel as far as possible.

④ Arrange mooring lines symmetrical fore and aft as far as is possible.

⑤ Use ropes of the same material and size for all mooring lines.

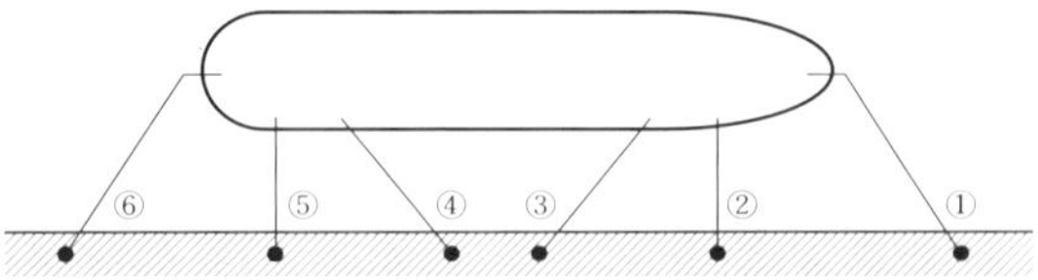

Figure10-15 The arrangement of mooring lines

4.3 Mooring Fittings

The number and size of mooring lines must be in accordance with the provisions of IACS rules, especially those given in Chapter 10 of Common Structural Rules for Bulk Carriers. Open or closed chocks are used to change the direction of mooring lines coming into the ship from outside and they also protect the ropes. Ships using the Panama Canal must be equipped with Panama chocks approved by the Canal Authority as well. One-post bollards and bitts with one or two posts must be provided on deck to tie mooring lines. Roller fairleads, deck stand rollers, universal fairleads, and horizontal rollers and guides are used to control and change the direction of mooring lines, while protecting the ship and the lines.

Figure 10-16 SPM of oil tanker

4.4 Single Point Mooring and Emergency Towing

1) Single point mooring

The arrangement shown in **Figure 10-16** is the mooring method required by the Oil Companies International Marine Forum (OCIMF), and is often used by oil tankers. In this method the ship does not come alongside a quay, but moor to a buoy with loading/unloading facilities. Tankers must have chain stoppers and fairleads in the bow and must be equipped with winches or capstans of over 15 tons load capacity which can handle 150 m length of 80 mm rope.

2) Emergency towing

IMO issued a requirement for all oil tankers with deadweight of more than 20,000 tons capacity built on or after 1 January 1996 to strengthen the bow and stern for mooring. It requires these structures to withstand the load of at least 1000 kN for tankers of smaller than 50,000 DWT, and at least 2000 kN for tankers larger than 50,000 DWT.

5. Lifesaving System

5.1 Marine Lifesaving Appliances

Lifesaving appliances consist of equipment to help people evacuate from ships in difficulty and equipment used to rescue people from the sea. Lifeboats and life rafts are survival craft for helping evacuation, and rescue boats are typical appliances for rescuing people from the sea. Life buoys and life jackets are personal lifesaving appliances and they must be equipped with various signaling and communication equipment in accordance with the provisions given in SOLAS and LSA (Lifesaving Appliances) CODE.

5.2 Survival Craft

1) Life boats

Lifeboats are used for evacuating a ship in emergency and they have self-propelling ability. They must be designed and manufactured to have sufficient stability and freeboard and appropriately compartmented to operate in rough seas when fully loaded with the allowable number of people and requisite equipment. It must be possible to launch the lifeboats safely at up to 10 degrees trim and 20 degrees heel of the stricken ship. A life boat should have sufficient structural strength, and, even if it suffers penetration damage below the waterline, they must not sink and have sufficient residual static stability.

A lifeboat should have about 40 kinds of equipment and provisions on board in readiness for emergencies, including emergency rations, drinking water, compasses, fishing gear, survival guides, distress signals, etc. necessary for survival for a certain period of time. The types of lifeboats include partially enclosed, fully enclosed, self-contained air support and fire proof. They may also be categorized as free-fall type and conventional type based on launching methods as shown in **Figure 10-17**.

(a) Partially enclosed type (b) Fully enclosed type (c) Free fall type (d) Conventional type

Figure 10-17 Types of life boats

2) Rescue boats

Rescue boats must be highly mobile and maneuverable to rescue people in distress at sea and an example is shown in **Figure 10-18**. A rescue boat should be able to accommodate at least six people, including five sitting and one lying down. Its speed should be not less than six knots and it should be able to maintain this speed for more than four hours when fully loaded with people and equipment. Furthermore, a rescue boat should be able to control a group of life rafts and be able to tow the largest life raft on board the ship when fully loaded with people and equipment at a speed of more than two knots.

A rescue boat must be equipped with unsinkable oars, floatable water bailers, effective compasses with a light-emitting object or proper illumination, sufficient length of mooring ropes, water-tight Morse code signaling lamps, first aid equipment, radar reflectors, warming equipment, and portable fire extinguishers.

Figure 10-18 A rescue boat

3) Life rafts

Life rafts are also used for emergency evacuation from a ship. They are used if it is not possible to use a lifeboat or a lifeboat is too far away to reach in time. There are two ways of launching: lowering with davits after boarding; and boarding after deploying it on the sea. All life rafts should be made to be capable of staying afloat for at least thirty

Figure 10-19 A life raft

days in all sea conditions. Just like lifeboats a life raft should be provided with emergency rations, drinking water, compasses, fishing gear, survival guides, distress signals and so on necessary for survival for a certain period of time. **Figure 10-19** shows a typical life raft with mandatory equipment and provisions displayed in front of it.

4) Embarkation rope ladders

Rope ladders are used for evacuation if the ship is trimmed excessively or heeled to more than 20°. These ladders should be installed at life boat launching stations or nearby positions. However, the competent authority may permit such ladders to be replaced by approved devices which can afford access to the survival craft when waterborne, provided that there is one or more embarkation ladders on each side of the ship.

5. 3 Personal Lifesaving Appliances

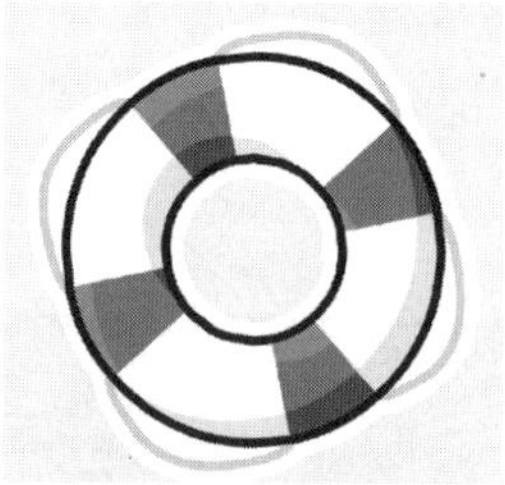

Figure 10-20 A lifebuoy

1) Lifebuoys

A lifebuoy is made of cork, kapok, balsa or any other similar material formed into a doughnut shape. This is covered with cotton and four hand straps are tied around it as shown in **Figure 10-20**. Their size varies a little depending on the manufacturers, but there are two types based on their weight and purpose.

(1) General life buoys

This type of buoy is used where quick release is not required. Its mass must be not less than 2.5 kg.

(2) Quick release lifebuoys

These buoys have a quick release mechanism and are equipped with a self-activating smoke signals or self-igniting lights. The mass of these buoys must be not less than 4 kg and are usually installed on both sides of the wheelhouse. The number of lifebuoys required depends on the ship length as specified in SOLAS chapter III Reg.32. The buoy is equipped with self-igniting light with or without self-activating smoke signalers as shown in **Figure 10-21** and lifelines, and their number and arrangement is also specified in SOLAS.

Figure 10-21 A self-activating smoke signaller

2) Life jackets

A life jacket is made by covering the core kapok panels with cotton and made into a shape of vest as shown in **Figure 10-22**, and is also known as life vests. In an emergency, people must board a lifeboat or life raft wearing the life jacket, and wait for rescue. Even if they cannot board such craft, they can still stay afloat with their heads above water by wearing a life jacket.

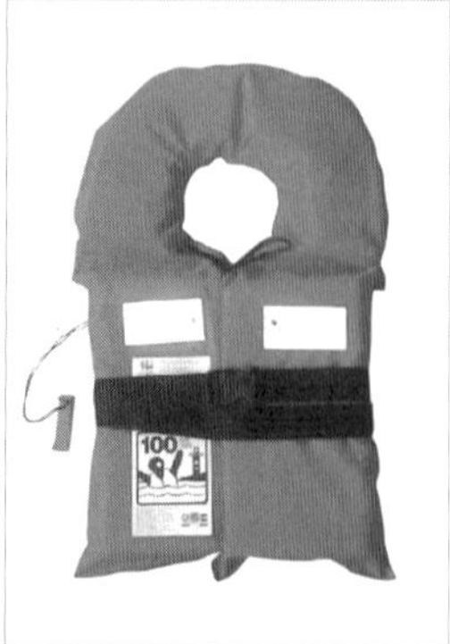

Figure 10-22 A life jackets

The number of life jackets must be at least the same as the number of people on board including the ship-owner and pilots, and the life jackets must be stored in each cabin. In addition, extra life jackets must be provided for the crew on duty in the bridge, bow and engine room. The locations for these extra life jackets can differ slightly depending on the relevant competent authority or classification society.

Life jackets should be made wearable by adults who weigh up to 140 kg and have a chest size of more than 1,750 mm, and life jackets for children and babies must be specially made and supplied. Life jackets should have a torch, a whistle and hand loops as specified in the LSA Code.

3) Immersion suits

Immersion suits provide protection from the cold sea water and are worn by rescue personnel. Ships operating exclusively in warm climates are exempt from the requirement to have them on board.

4) Lifeline throwing appliances

Ships of more than 500 GT operating on international routes are required by law to have line throwing appliances onboard. They are used to throw lifelines to ships or people in distress. These appliances are operated by a rocket or bullet pulling a thin and long rope known as a shot line. Four throwing appliances and four lifelines each more than 230 m long are formed into a unit and stored in the wheelhouse.

5.4 Distress Signal Equipment

This equipment is used to signal the location of distress, and the quantity and type should be in accordance with the provisions of SOLAS or national regulations. They are stored in the wheelhouse. The dimensions, quantity and method of use of the signal equipment are specified in the LSA code. There is a variety of equipment, including rocket parachute signals, rocket signals, manual flame signals and self-activating smoke signals.

6. Piping Arrangements

6.1 Ballast System

1) Definition and purpose

Ballast is heavy material loaded in addition to any cargo in order to secure necessary stability and proper trim necessary for safe and efficient sailing. Normally seawater is used for ballast, and the necessary tanks to carry it are provided at the bow, stern and both sides of the ship. The operational condition with no cargo but minimum amount of ballast is known as ballast condition.

2) Ballast piping arrangement

It is necessary to have an entirely separate piping and pumping system to deal with the liquid ballast. The ballast system consists of sea chests, strainers, pumps, transfer pipes, valves and ballast tanks.

(1) Double main line type

Two main pipes are arranged into each ballast tank and manifold service pipes are arranged in service tanks. Tanks are arranged on both sides, and two main lines are duplicated for redundancy.

(2) Manifold type

Ballast tanks are independently controlled in this type. It is used when the pipe size is small and the number of tanks is small.

(3) Ring main type

When the two main pipes of the double main line type are connected to each other, a ring main is formed. It can be changed to the double main line type by closing the isolating valve at the joint. This type is often used in bulk carriers with a duct keel.

(4) Duct keel type

This type uses the duct keel of a bulk carrier as a piping duct. By using the isolating valves it can operate as any of the double main line type, the manifold type or the ring main type.

3) Ballast pipe and piping components

Because seawater is used as ballast, ballast pipes and piping components are usually galvanized and the tanks are painted. Shipyards may deviate from the normal standards at the request from the owner to use GRE (glass reinforced epoxy) pipes which have good corrosion resistance and conductivity instead. Valves inside ballast tanks are remotely controlled from the ballast control room (BCR) or cargo control room (CCR).

6.2 Bilge System

1) Bilge

Bilge is sewage water from ships, including drain water from equipment and pipes, seawater on decks, and water mixed with oil.

2) Bilge piping

Bilge pumps are installed in engine rooms and bilge pipes are arranged in areas where bilge occurs. Engine room bilge mixed with various oils is stored and then processed or discharged on shore according to regulations. The bilge from water on deck is sometimes discharged through eductors.

(1) Cargo hold bilge

When the main bilge inlet pipes are placed in the keel duct, screw down non-return (SDNR) check valves must be installed in the branch inlet pipe from each cargo hold.

(2) Engine room bilge

The engine room bilge pumps are used for fire-fighting and other general purpose as well. Piping arrangement must be made so that oily bilge can be discharged ashore.

(3) Steering gear room bilge

The piping should be designed so that the steering gear room bilge, consisting of sea water that comes through the hatch and the oily bilge from various equipment in the room, can be discharged into the engine room bilge system by gravity.

(4) Boatswain's store and bow area bilge

The bilge in the boatswain's stores, bow spaces and chain lockers consists mostly of seawater ingress. Therefore the piping should be configured to discharge this bilge directly through eductors.

6.3 Liquid Cargo System

1) Liquid cargo

Some ships carry liquid cargos, such as chemical products, oil products, crude oil and liquefied gas, and they are handled through pipes. The basic components of such liquid cargo system include pumps, pipelines, valves and tanks. Depending on the amount and the physical properties of the cargos, the system is either placed close to each other or segregated. An inert gas system is used to prevent the vapor produced during loading and unloading operations from exploding. A tank cleaning system is used to remove traces of the prior cargo when a different type of cargo is to be loaded. Heating devices are sometimes used to reduce viscosity of some cargos for easy pumping operations.

2) Cargo handling piping

(1) Exclusive pumps

These pumps are installed in ships carrying a variety of types of cargo which need to be segregated, such as chemical carriers. Each cargo tank has its own pumps and piping system. These pumps should be arranged in accordance with segregation methods specified in various relevant regulations. **Figure 10-23** shows only one side of a normally port-starboard symmetrical piping system.

(2) Multi-purpose pumps

These pumps are used in ships carrying a single or similar type of cargos, such as crude carriers and product carriers. Pumps are arranged in a pump room and each is used commonly depending on the piping arrangement or for handling similar types of products. **Figure 10-24** shows an example piping system for three multi-purpose cargo pumps.

(3) Piping system for cargo handling and its components

Cargo loading/unloading pipes on chemical tankers is usually made of stainless steel. Especially for toxic or acid cargo, stainless steel is legally required. In this case welded connections are normally used with flange connections being allowed in some exceptional cases. General structural steel is used in product tankers or crude oil tankers, but the owner may opt for a different material in the case of product carriers.

When designing the cargo pipes, design pressures and materials required in regulations are decided on the basis of the suction head and the flow rate of the pumps. The material of gaskets used in the cargo piping is chosen from NBR, VITON, Teflon and so on to suit the characteristics of the cargo.

6.4 Valves

Valves connect pipes and are used to control flow rate and shut off the flow. Some are safety valves with an ability to control pressure. Typical forms and characteristics of valves used in the piping system of ships are as follows:

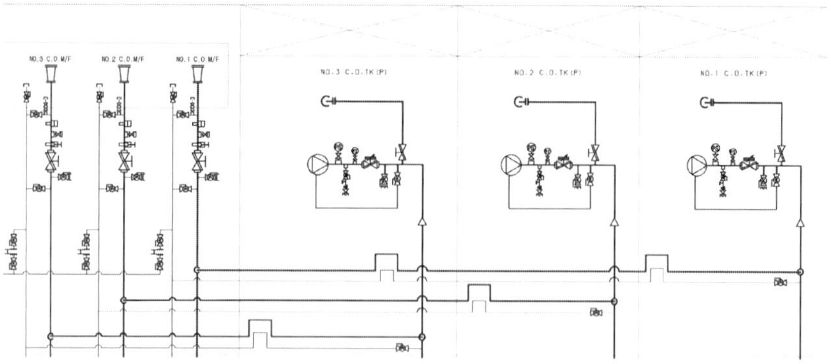

Figure 10-23 Exclusive pumping and piping system

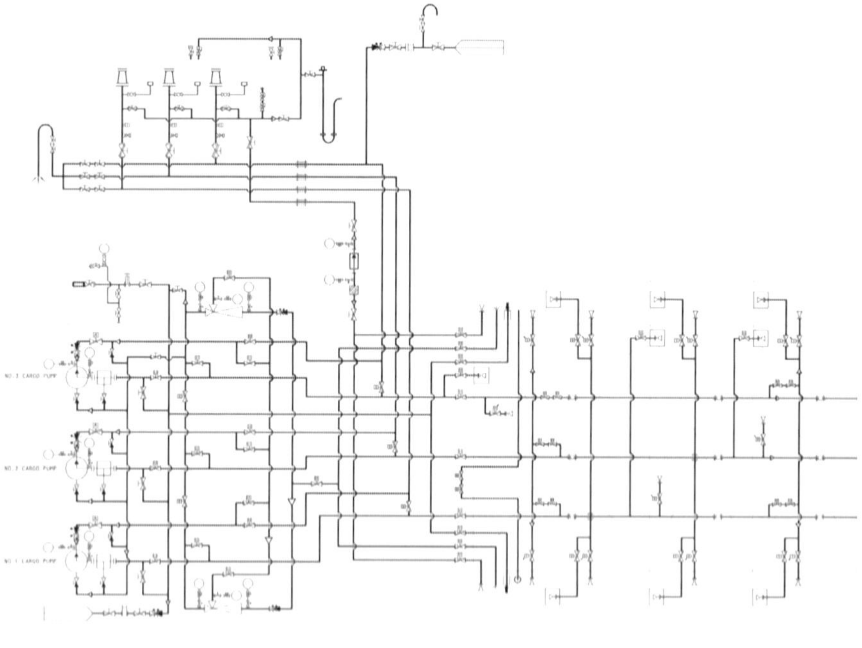

Figure 10-24 Multi-purpose cargo pumping and piping system

1) Globe valves

The fluid flows upward inside the valve and exits into the pipeline through a gap in the valve, forming an 'S' shape path. The gap can be controlled and this determines the flow rate. This type suffers from high pressure loss and turbulence, and so is unsuitable for low pressure flow. The inlet and outlet can be in the same direction or crossed at right angles. Some globe valves are used with a hose connected to the outlet pipes, and some are non-return valves.

2) Gate valves

These valves are opened or closed by lifting or lowering a gate out of or into the path of the fluid. They are unsuitable for regulating flow and usually used either fully closed or fully open.

Gate valves have the advantage of low frictional loss when fully open, but vibration can occur when partially open. They are used where foreign matters in the fluid are liable to get stuck in partially open valves.

3) Spherical valves

Sometimes called ball valves, their active part is a sphere with a hole in the middle. By turning the sphere, the hole can be aligned with the flow direction to open. Since they are simple and can be made small, they are often used in piping system for oil residue processing or tank cleaning.

4) Butterfly valves

The main closing mechanism is a flat disc which can be turned to open or close the flow. It can be used to regulate the flow. These valves are usually used in large pipes, but since these light and simple valves take up little space, they are often used in piping for cargo handling, ballast and cargo hold cleaning.

5) Small valves

Many small diameter valves are used for a variety of instrumentation, such as piping for gauges and sample-taking.

7. Fire Safety

7.1 Fire Safety Facilities

These are installed to ensure the safety of crew and passengers from fires which can occur at any time during sailing or cargo handling. SOLAS provides detailed fire regulations.

7.2 Terminologies

1) International fire test code

The regulations for fire testing are given in the International Code for the Application of Fire Test Procedures (FTP Code) as revised in 1996. SOLAS made it mandatory for ships built on or after 1st July1998 to use only products which pass the tests specified in this Code.

2) Standard fire test

Specimens of bulkheads, decks, ceiling linings, partitions, doors and so on must be tested in a test furnace by exposing them to test flames in accordance with the standard period and time-temperature curve as specified in the FTP Code. **Table10-2** summarizes the standards required and fire test conditions for various divisions and fire grades.

3) A-class divisions

The A-class divisions include bulkheads and decks satisfying the following conditions:

① Built of steel or equivalent materials suitably stiffened;

Table 10-2 Standard fire test

<table>
<tr><td rowspan="2">Standard fire test</td><td>Time(min)</td><td>0</td><td>5</td><td>10</td><td>15</td><td>30</td><td>60</td><td rowspan="2">Material</td></tr>
<tr><td>Temperature in test furnace(°C)</td><td>-</td><td>+556</td><td>+659</td><td>+718</td><td>+821</td><td>+925</td></tr>
<tr><td rowspan="3">A-class divisions (bulkheads and decks)</td><td>Fire Grade</td><td>A-0</td><td>-</td><td>-</td><td>A-15</td><td>A-30</td><td>A-60</td><td rowspan="3">Steel or other equivalent material.
Must be insulated to meet the fire grade.</td></tr>
<tr><td>Temperature of reverse side(°C)</td><td colspan="6">The average temperature should not exceed 140°C above the original temperature.
(The temperature should not exceed 180°C above the original temperature at any point.)</td></tr>
<tr><td>Fire integrity</td><td colspan="6">To be capable of preventing the passage of smoke and flame to the end of the one-hour standard fire test.</td></tr>
<tr><td rowspan="3">B-class divisions</td><td>Fire Grade</td><td>B-0</td><td>-</td><td>-</td><td>B-15</td><td>B-30</td><td>-</td><td rowspan="3">Approved non-combustible material</td></tr>
<tr><td>Temperature of reverse side(°C)</td><td colspan="6">The average temperature should not exceed 140°Cabove the original temperature. (The temperature should not exceed 225°C above the original temperature at any point.)</td></tr>
<tr><td>Fire integrity</td><td colspan="6">To be capable of preventing the passage of flame to the end of the 30 min test period.</td></tr>
<tr><td>C-class divisions</td><td>Fire integrity</td><td colspan="6">No particular requirement specified for heat flux and fire integrity.</td><td>Approved non-combustible material</td></tr>
</table>

② Built to be capable of preventing the passage of smoke and flame to the end of 60 minutes of a standard fire test; and

③ Insulated with approved non-combustible materials such that the average temperature of the unexposed side will not rise more than140°C above the original temperature, nor will the temperature, at any one point, including any joint, rise more than 180°C above the original temperature, within the time intervals specified for the relevant fire-grades.

4) B-class divisions

The B-class divisions include bulkheads, decks, ceiling materials and interior materials satisfying the following conditions:

① Constructed of approved non-combustible materials;

② Built to prevent the passage of flames during at least 30 minutes of the standard test; and

③ Insulated such that the average temperature of the unexposed side will not rise by more than 140°C, nor will the temperature at any one point, including any joint, rise by more than 225°C above the original temperature, after the time periods specified for the relevant fire grades.

5) C-class divisions

The C-class divisions are products constructed of approved non-combustible materials, but do not need to satisfy any requirements concerning heat flux and integrity.

6) Fire integrity of bulkheads and decks

Bulkheads and decks which form boundaries of various compartments should satisfy the fire integrity defined in **Tables 10-3**, **10-4**, **10-5** and **10-6**.

7) Compartments

- Control stations: spaces containing emergency power and ship's radio equipment, steering gear room, chart room, fire control stations, spaces containing central fire alarm equipment, control room for propulsion machinery, etc.
- Corridors: corridors and lobbies for passengers and crew.
- Accommodation spaces: public spaces, toilets, cabins, offices, hospitals, cinemas, recreation spaces, food stores, etc.
- Stairways: interior stairs, elevators, enclosed emergency escape trunks and escalators.
- Low risk service spaces: laundries, drying rooms, warehouse or storage area having less than 4m^2 of deck area not containing flammable liquid.

- Category A machinery spaces: places which contain any of the following machines and trunks to these places
  - internal combustion engines for activating main propulsion machinery;

Table 10-3 Fire integrity of bulkheads separating adjacent spaces (cargo ships excluding tankers)

SPACES		(1)	(2)	(3)	(4)	(5)	(6)	(7)	(8)	(9)	(10)	(11)
CONTROL STATIONS	(1)	A-0[e]	A-0	A-60	A-0	A-15	A-60	A-15	A-60	A-60	*	A-60
CORRIDORS	(2)		C	B-0	B-0 A-0[c]	B-0	A-60	A-0	A-0	A-0	*	A-30
ACCOMMODATION SPACES	(3)			C[a, b]	B-0 A-0[c]	B-0	A-60	A-0	A-0	A-0	*	A-30
STAIRWAYS	(4)				B-0 A-0[c]	B-0 A-0[c]	A-60	A-0	A-0	A-0	*	A-30
SERMCE SPACES (LOW RISK)	(5)					C	A-60	A-0	A-0	A-0	*	A-0
MACHINERY SPACES OF CATEGORY A	(6)						*	A-0	A-0[g]	A-60	*	A-60[f]
OTHER MACHINERY SPACES	(7)							A-0[d]	A-0	A-0	*	A-0
CARGO SPACES	(8)								*	A-0	*	A-0
SERMCE SPACES (HIGH RISK)	(9)									A-0[d]	*	A-30
OPEN DECKS	(10)										-	A-0
RO / RO CARGO SPACES	(11)											*[h]

Table 10-4 Fire integrity of bulkheads separating adjacent spaces (tankers)

SPACES		(1)	(2)	(3)	(4)	(5)	(6)	(7)	(8)	(9)	(10)
CONTROL STATIONS	(1)	A-0[c]	A-0	A-60	A-0	A-15	A-60	A-15	A-60	A-60	*
CORRIDORS	(2)		C	B-0	B-0 A-0[a]	B-0	A-60	A-0	A-60	A-0	*
ACCOMMODATION SPACES	(3)			C	B-0 A-0[a]	B-0	A-60	A-0	A-60	A-0	*
STAIRWAYS	(4)				B-0 A-0[a]	B-0 A-0[a]	A-60	A-0	A-60	A-0	*
SERMCE SPACES (LOW RISK)	(5)					C	A-60	A-0	A-60	A-0	*
MACHINERY SPACES OF CATEGORY A	(6)						*	A-0	A-0[d]	A-60	*
OTHER MACHINERY SPACES	(7)							A-0[b]	A-0	A-0	*
CARGO SPACES	(8)								*	A-60	*
SERMCE SPACES (HIGH RISK)	(9)									A-0[b]	*
OPEN DECKS	(10)										-

Table 10-5 Fire integrity of decks separating adjacent spaces (cargo ships excluding tankers)

SPACE		(1)	(2)	(3)	(4)	(5)	(6)	(7)	(8)	(9)	(10)	(11)
CONTROL STATIONS	(1)	A-0	A-0	A-0	A-0	A-0	A-60	A-0	A-0	A-0	*	A-60
CORRIDORS	(2)	A-0	*	*	A-0	*	A-60	A-0	A-0	A-0	*	A-30
ACCOMMODATION SPACES	(3)	A-60	A-0	*	A-0	*	A-60	A-0	A-0	A-0	*	A-30
STAIRWAYS	(4)	A-0	A-0	A-0	*	A-0	A-60	A-0	A-0	A-0	*	A-30
SERMCE SPACES (LOW RISK)	(5)	A-15	A-0	A-0	A-0	*	A-60	A-0	A-0	A-0	*	A-0
MACHINERY SPACES OF CATEGORY A	(6)	A-60	A-60	A-60	A-60	A-60	*	A-60[f]	A-30	A-60	*	A-60
OTHER MACHINERY SPACES	(7)	A-15	A-0	A-0	A-0	A-0	A-0	*	A-0	A-0	*	A-0
CARGO SPACES	(8)	A-60	A-0	A-0	A-0	A-0	A-0	A-0	*	A-0	*	A-0
SERMCE SPACES (HIGH RISK)	(9)	A-60	A-0	A-0	A-0	A-0	A-60	A-0	A-0	A-0[d]	*	A-30
OPEN DECKS	(10)	*	*	*	*	*	*	*	*	*	-	*
RO / RO CARGO SPACES	(11)	A-60	A-30	A-30	A-30	A-0	A-60	A-0	A-0	A-30		*[h]

Table 10-6 Fire integrity of decks separating adjacent spaces (tankers)

SPACE		(1)	(2)	(3)	(4)	(5)	(6)	(7)	(8)	(9)	(10)
CONTROL STATIONS	(1)	A-0	A-0	A-0	A-0	A-0	A-60	A-0	-	A-0	*
CORRIDORS	(2)	A-0	*	*	A-0	*	A-60	A-0	-	A-0	*
ACCOMMODATION SPACES	(3)	A-60	A-0	*	A-0	*	A-60	A-0	-	A-0	*
STAIRWAYS	(4)	A-0	A-0	A-0	*	A-0	A-60	A-0	-	A-0	*
SERMCE SPACES (LOW RISK)	(5)	A-15	A-0	A-0	A-0	*	A-60	A-0	-	A-0	*
MACHINERY SPACES OF CATEGORY A	(6)	A-60	A-60	A-60	A-60	A-60	*	A-60[a]	A-0	A-60	*
OTHER MACHINERY SPACES	(7)	A-15	A-0	A-0	A-0	A-0	A-0	*	A-0	A-0	*
CARGO PUMP ROOMS	(8)	-	-	-	-	-	A-0[d]	A-0	*	-	*
SERMCE SPACES (HIGH RISK)	(9)	A-60	A-0	A-0	A-0	A-0	A-60	A-0	-	A-0[b]	*
OPEN DECKS	(10)	*	*	*	*	*	*	*	*	*	-

- internal combustion engines the total output of which is 375 kW or more for purposes other than main propulsion; and
- oil-fired boilers and equipment handling or storing fuel oil.

- Other machinery spaces: spaces containing propulsion machinery other than those contained in the category A machinery spaces, boilers, fuel oil equipment, steam engines, internal combustion engines, generators, main electrical equipment, oil fuel filling stations, refrigeration machinery, anti-roll systems, ventilators, air conditioning machinery, other similar places and trunks to these places.
- Cargo spaces: spaces for cargo and trunks to them, and includes spaces containing cargo pumps on oil tankers.
- High risk service spaces: galleys, pantries, saunas, paint stores having a deck area of more than $4m^2$ for paint, stores for flammable liquids and workshops separated from machinery spaces.
- Open decks: open spaces on decks.
- RO/RO cargo spaces: spaces where roll-on/roll-off cargos are stored.

8) Methods of fire protection

Accommodation and service spaces are required to be protected by one of the following methods, but often I-C method is used. It is compulsory to use I-C in all oil tankers.

- I-C: All compartment bulkheads should be non-combustible B-class or C-class divisions.
- II-C: Wherever there is a risk of fire, automatic sprinklers equipped with fire detection and alarm devices must be installed, regardless of the form of the internal bulkheads.
- III-C: Wherever there is a risk of fire, fire detection and alarm devices must be installed, regardless of the form of the internal bulkheads. Equipment for detecting fire and warning should be installed in all places which have the possibility of fire outbreak, regardless of the class of inside compartment bulkheads. The deck area of the spaces neighboring accommodation areas or A-class or B-class divisions must not exceed $50\ m^2$ under any circumstances.
- Common rule: Fixed fire and smoke detectors and alarm devices should be installed in all corridors, stairways and escape routes in accommodation spaces.

9) Cofferdams

A cofferdam is a void space between two watertight bulkheads or decks. They are usually used to separate adjacent compartments.

10) Draught stop

Close-fitting draught stops are installed in the air spaces enclosed behind ceilings, paneling or linings in accommodation, work or control areas to prevent smoke from spreading. Draught stops are made of 1 mm thick steel sheets and should be spaced not more than 14m apart. In the vertical direction such enclosed air spaces, including those behind stairways, trunks and so on must be closed at each deck.

11) Prevention of heat transmission

Heat insulation treatment should be applied to bulkheads and decks crossing an insulated bulkhead or deck for a distance of at least 450 mm from the point of intersection including the thickness of the insulation material in the case of steel and aluminum structures. This is to prevent the heat of fire from being transmitted along walls or other structures in contact with thermal barriers.

7.3 Miscellaneous Points in the Fire Regulations and Codes

1) Pantries

Spaces containing cooking equipment, including coffee makers of more than 5 kW, toasters, dishwashers, microwave ovens, water boilers, electric hotplates with surface temperature of more than 150°C, and electric hotplates of more than 2 kW used for cooking for two or more people, are classed as pantries(SOLAS Ch. II-2, Reg. 3.45).

2) Inert gas generating room and incinerator room

Spaces containing inert gas generators and incinerators are included in Category A machinery spaces (SOLAS Ch. II-2, Reg.3.31).

3) Emergency fire-fighting pump room

When a room is within a Category A machinery space or adjacent to the space containing the main fire-fighting pumps, its boundaries must be insulated to the standard of fire grade A-60 as applied to control stations (SOLAS Ch. II-2, Reg. 10.2.2.3.2.1).

4) Spaces producing oil vapors

In spaces where oil can permeate the surface of thermal insulation must be covered with metal plates or glass fiber products to prevent oil vapors from permeating (SOLAS Ch. II-2, Reg. 4.4.3).

5) Space classification

If the common bulkhead between two adjacent spaces has a communication opening of more than 30% of the bulkhead surface area, the two spaces will be treated as one and its classification will be the one having the more stringent boundary requirements (SOLAS Ch. II-2, Reg. 9.2.2.4.2).

6) Electrical equipment rooms

Electrical equipment rooms containing automatic telephone exchange, spaces for air conditioning ducts, etc. should be included in category (7) Other Machinery Spaces (SOLAS Ch. II-2, Reg. 9.2.3.3.2).

7) Sauna rooms

Sauna rooms are Class A spaces, and it may include changing rooms, showers and toilets. The saunas shall be insulated to ‘A-60’ standard against other spaces except those inside of the perimeter and spaces of categories (5), (9) and (10). The distance from the hot surfaces to combustible materials must be at least 500 mm or the combustible materials must be protected. The sauna door shall open outwards by pushing, and electrically heated ovens shall be provided with a timer (SOLAS Ch. II-2, Reg. 9.2.2.3.4).

8) Means of escape

Details of means of escape for various ship types are given in SOLAS (Ch. II-2, Reg. 13), and a gist of some of them is as follows:

- The length of dead-end corridors should not be more than 7 m.
- Doors opening to the stairs or passageways used for escape should be at least 700 mm wide and must comply with the Fire Safety Systems Code.
- Category A machinery spaces must be provided with two escape routes.
- From machinery spaces other than those of Category A, two escape routes shall be provided. However, a single escape route will be accepted for spaces which are entered only occasionally and for spaces where the maximum travel distance to the door is not more than 5 m.

9) Protection of stairways

Protection of stairways is specified in SOLAS (Ch. II-2, Reg. 9.2.2.5).

- One end of a stairway connecting two decks must be protected by connecting to an enclosure formed by B-0 class divisions with a self-closing door.
- Both ends of an elevator connecting two decks must be contained within an enclosure

formed by A-0 class divisions with positive closing doors.

- Lifts used to carry meals and foods between decks are treated as elevators.

10) Bulkheads between cargo areas and Category A machinery spaces

Bulkheads between cargo spaces and Category A machinery spaces must be insulated to A-60 class standard, unless all dangerous goods are stowed at least 3 m horizontally away from such bulkheads (SOLAS Ch. II-2, Reg. 19.3.8).

11) Boundaries between accommodation spaces and cargo areas

Restrictions and heat insulation of bulkheads between accommodation spaces and cargo areas are specified in SOLAS (Ch. II-2, Reg. 4.5.2)

① Access doors to work spaces can be installed at a distance less than 5 m from the boundary bulkheads between accommodation areas and cargo areas. However, if an access door has to be installed at a distance of less than 3 m from such bulkheads, they have to be insulated to A-60 class standard.

② Windows and side scuttles on the sides of the superstructures and deckhouses within the limits specified above must be of the fixed (non-opening) type. Such windows and side scuttles, except wheelhouse windows, shall be constructed to A-60 class standard.

③ Fire insulations to A-60 class standard are required only up to the bridge deck.

12) Emergency generator rooms

Emergency generator rooms containing transitional source of emergency power and the emergency switchboard must not be contiguous to the boundaries of Category A machinery spaces but must be segregated from them by cofferdams (SOLAS Ch. II-2, Reg. 43.1.3).

8. Air Conditioning Facility

8.1 Air Conditioning Facility

Air conditioning maintains the temperature, humidity, cleanliness and flow of air in a given space in an optimal condition for its use. Therefore, the air conditioning differs by size and purpose of each space. Different types of spaces have different goal of air conditioning.

Examples are spaces used by relatively few people, such as houses, offices and cabins; public facilities used by many people, such as department stores and theaters; and hotels and hospitals. Nevertheless, the common objective of air conditioning of all these facilities is to provide comfortable environment for the users.

On the other hand, industrial air conditioning in factories, laboratories and similar facilities aims to provide stable production and research environment by keeping constant humidity, constant temperature and clean air, rather than pleasant environment for people.

8.2 Characteristics of Air

1) Type of air

Characteristics of air depend on the moisture content: dry air which does not contain any moisture at all; moist air, which contains some moisture; and saturated air, which contains a maximum amount of moisture that is possible for the temperature and pressure. As the temperature of air rises, saturation pressure also rises, and the air is able to contain more moisture. Conversely, if the temperature decreases, the saturation pressure decreases, and the maximum amount of moisture the air can contain also reduces.

2) Air conditioning load based on air characteristics

(1) Cooling load

In order to maintain the target temperature and humidity in a room in summer, the external heat must be blocked from permeating into the room, and at the same time some heat may have to be removed from the room. The heat to be removed from the room is called sensible heat.

When the air temperature drops, water condensation occurs and the load used in removing such condensation is called latent heat. The cooling load, therefore, consists of sensible heat and latent heat.

(2) Heating load

Similarly, in order to maintain the target temperature and humidity in a room in winter the heat and moisture dissipated into the external air must be replenished. The heating load therefore consists of sensible heat loss due to heat escape and latent heat loss due to moisture loss.

8.3 Air Conditioning Types

1) Single duct type

Since air has low specific heat, the duct has to have a large section area enough to handle a large volume of air. Low speed ducts have air flow speed of less than 15 with pressure of about 50 to 75 mm Aq, and, therefore, are used in places where there is no restriction in the cross section area of ducts, such as a structure with many compartments or single large volume spaces. High speed ducts, on the other hand, have air flow of more than 15 m/s with the pressure about 150 to 200 mm Aq. They are used when twice the air flow of low speed ducts is required. Although the cross section area of high speed ducts can be reduced, the cost of equipment is high due to higher power consumption, and some measures for reducing noise are needed.

2) Dual duct type

This type sends hot and cold air generated with cooling and heating equipment in a central machinery room via two separate ducts. The two streams of air are mixed to produce the required temperature before being supplied to the target rooms.

8.4 Selection of Equipment

To determine the cooling load and choose appropriate equipment, internal and external heat sources and their temperature need to be identified and heat conduction due to temperature gradient also has to be considered.

① **Table 10-7** on next page shows the number of minimum air changes per hour required for various types of spaces in general merchant ships.

② Condition for balance of air supply

The volume of air supplied mechanically and naturally has to be equal to the sum of the volume of air expelled, both mechanically and naturally exhausted, for each deck.

③ Calculation for choosing the equipment

In order to find the total load for satisfying the temperature conditions required within the ship, heat sources and loads of each compartment are calculated. The maximum volume of air required is taken to be either the volume needed for minimum air changes or circulation air volume, whichever is greater. The circulating air volume and the cooling capacity are then determined by considering the condition of air supply balance for each deck.

Table 10-7 The minimum number of air changes per hour for each type of space

Compartment	Air conditioning	Mechanical	
		Supply	Exhaust
Cabin	6		
Public spaces	8		
Office spaces	8		
W/House & Chart space	12(*)		
Hospital	8		
Navigation Locker	8(*)		
Private toilet			10
Engine changing room	10(*)		15
Ship's laundry room	15(*)		15
Deck changing room	10(*)		10
Galley	6(*)	15	30
Pantry	6(*)		15
Dry provision store	6(*)		
Fire station	5(*)		

8.5 Configuration of Equipment

1) Air conditioning equipment

The functions of main air conditioning sections are as follows.

- Air-mixing section: Fresh air from the outside and the air being re-circulated within accommodation spaces are mixed with manual or automatic dampers.
- Air-cleaning section: This section comes after the air-mixing section, and here fine dust in the air is filtered out. The filters must be easy to clean and replace.
- Air-heating section: Air heaters are made of copper tubes or aluminum fin coils and they heat surrounding air by circulating hot water, steam or oil through the pipes.
- Air-humidifying section: The humidity of air is maintained at an appropriate level by spraying steam or water.
- Air-cooling section: Air is cooled through heat exchange between refrigeration gases

which have been made to pass through expansion valves in copper pipes or aluminum fin coils and the air.

- Dehumidifying section: When air is cooled, water condensation develops. The moisture thus produced is removed here so as to prevent it from flowing into the air-supplying section.
- Air-supplying section: Air is supplied to the accommodation spaces from here. Double sided suction centrifugal fans are often used for this purpose.
- Air-discharging section: Air is expelled from this section. Silencers are often used to reduce the noise of air blowers.

2) Cooling system

Figure 10-25 shows a condensing unit of air conditioning system. When a refrigerant is vaporized in an evaporator, it absorbs heat from the surroundings and thereby lowering the temperature. The vaporized refrigerant is then compressed, thus raising its temperature. The high pressure refrigerant vapor is sent to a condenser where heat exchange takes place between the refrigerant and the surrounding air. The vapor is cooled sufficiently to return to its liquid state once again. It is then pumped to the evaporator, thus completing a cycle, which is repeated again and again. Elements of a cooling system are as follows:

Figure 10-25 Condensing units

- Compressor: Compressors of an open reciprocating type are used in normal merchant ships. The compressor draws the low-temperature, low-pressure refrigerant vapor from the evaporator, compresses it into high-temperature, high-pressure gas. The gas is then released into the discharge line. In this process the evaporator pressure is kept at a certain level.
- Condenser: The condenser extracts heat from the refrigerant to the outside air. The gas

is then cooled sufficiently to return to its liquid state. The liquid refrigerant then flows from the condenser into the liquid line

- Evaporator: The low temperature, low pressure supersaturated refrigerant gas is evaporated to become low temperature, low pressure gas by discharging it through an expansion valve, thus absorbing surrounding heat. If heat is absorbed excessively, the surface temperature of the evaporator will fall below freezing point and frost will accumulate on the surface of the cooler. Such frost must be removed, because the accumulated surface frost reduces the efficiency of the cooler and the volume of air supply.
- Receiver: The condensed liquid refrigerant is stored in the receiver temporarily before being sent to the expansion valve.
- Expansion valve: The high temperature, high pressure liquid refrigerant from the receiver is turned into low temperature, low pressure liquid as it passes through the expansion valve and expands. Thus the valve allows evaporation to take place more easily and also controls the flow rate of the refrigerant.
- Dryer: It removes moisture from air, since such moisture content in the cooling system can have adverse effects on equipment. Desiccants, which change color when water is absorbed, are used, and they must be changed regularly. It is a good practice to remove as much moisture as possible from the system with a vacuum pump before using the dryer first time in the trial.
- Liquid separator: The liquid separator is installed at the entrance of the compressor. It protects the compressor by collecting liquid refrigerant and preventing it from entering the compressor. Part of the refrigerant can reach the separator in a liquid state when the load is very low or there is a severe load fluctuation and the refrigerant is insufficiently evaporated.

8.6 Ventilation

Changing the air in a room with air from the outside by supplying or extracting air either naturally or through mechanical means using air flow, temperature and humidity is known as ventilation or air-circulation. The main purpose of ventilation in cargo and other spaces, excluding cabins and machinery spaces, is to supply fresh air, remove harmful gases, control temperature and dehumidify.

1) Ventilation type

(1) Natural ventilation

Natural ventilation relies on wind direction and speed, and air convection due to temperature difference between internal and external air. The natural ventilation requires no special equipment and no electric power. However, it is difficult to obtain enough pressure and to secure uniform and large volume of air flow. Furthermore, it is difficult to control the effects of wind direction, wind speed, waves and spray in some weather conditions.

(2) Forced ventilation

In this method air is supplied or discharged by mechanical means. Mechanical power can be used for both supplying and discharging air. Alternatively, mechanical power can be used only in either supplying or discharging air, allowing discharging or air-intake to take place naturally as a consequence.

(3) Dehumidifying ventilation

This ventilation is often required in ships which carry papers, forest products or cement. Fresh air is drawn in and dehumidified by heating with electric heaters or steam before sending it to cargo holds.

2) Ventilation plan

The usages of the target space and the reasons for ventilation have to be understood before the ventilation plan can be made.

① Oxygenating ventilation: This ventilation aims to supply oxygen to spaces that require it, such as engine rooms and diesel generator rooms.

② Dehumidifying ventilation: This ventilation aims to remove moisture content of the air in order to prevent cargo damage or protect crew health.

③ Temperature controlling ventilation: This ventilation supplies cooling air to thruster rooms, steering gear rooms and other spaces which contain various machines to prevent overheating. It also supplies cold or hot air to maintain comfortable working and living environment.

④ Gas removal ventilation: This ventilation aims to remove explosive or poisonous gases generated from dangerous materials in cargo pump rooms, paint storage, rooms in which carbon dioxide is generated and so on.

⑤ Deodorizing ventilation: This ventilation aims to remove unpleasant smells from galleys, restaurants, toilets, lockers and crew spaces.

3) Determination of ventilation volume

The ventilation volume should be determined by identifying the factors which necessitate the ventilation, taking into account usage and conditions of the space concerned. The volume required to satisfy the need for each factor is calculated and the highest value of them all is taken. However, the volume can usually be obtained by either considering the number of air-changes required or calculating the amount of heat generated and absorbed in the room which can lead to the volume of air required to maintain the temperature of the room at an appropriate level.

(1) Ventilation volume from number of air-changes

Since the minimum number of air-changes required is specified according to the usage of the space, the required ventilation volume per minute can be obtained by multiplying the volume of the space with the number of air-changes per minute.

(2) Ventilation volume from heat to be supplied

The temperature within an enclosed space is determined by the heat generated within it and heat transmitted through the boundary walls. The volume of air of certain temperature which needs to be supplied in order to maintain the air within such a space artificially at a required temperature can then be calculated.

(3) Type and characteristics of ventilating machines

An axial blower blows air using the lift and centrifugal forces generated by the rotation of the impeller blades, and the direction of air flow is the same as that of the axis of its impeller. If the rotation direction of an axial blower is reversed, air flow will reverse as well. Axial blowers include small household devices such as electric fans as well as large machines used in ventilation on board ships. A centrifugal blower, on the other hand, pushes out air at right angles to the axis by raising the air speed and pressure using the centrifugal force generated by rotating blades. Even when the direction of rotation is changed, the air flow does not change its direction.

(4) Type and character of ventilators

A number of different types of ventilators are used in ships as follows.

- Mushroom type: The height of this type is low and it can be opened/closed from inside the room. This type is installed where it receives little interference from waves.
- Gooseneck type: It is easy to keep out seawater and rainwater and is often used in narrow compartments, storage lockers and so on.
- Wall attachment type: This type is usually installed in navigation spaces.
- Cowl head type: This type is installed in engine rooms, air conditioner rooms and so on.

9. Pollution Prevention Facilities

9.1 Oil Leakage

1) Devices for monitoring and controlling oil discharge

MARPOL 73/78 Annex1, Reg. 15(a) makes it mandatory to install devices for monitoring and controlling oil discharge. These devices monitor the amount of oil contained in ballast water and tank washing water, and record the total amount of water discharged and the amount of discharge per nautical mile in liters.

It is also mandatory to install a recording device which can continuously record the oil concentration and discharge rate against time and date. If the instantaneous discharge rate of oil exceeds a permissible value, the discharge of water-oil mixture should be stopped immediately and completely.

If the monitoring/controlling device malfunctions, discharge should be stopped and such malfunction event must be recorded in the oil discharge register. An alternative manual device must also be installed which can be used in the event of the automatic equipment malfunctioning. There are class A and class B monitoring devices. The class A devices have a mechanical interlocking mechanism and a discharge control function, and are required for oil tankers of deadweight 4000 tons or more. Class B devices, on the other hand, are installed in oil tankers with GT more than 150 tons and deadweight up to 4000 tons.

2) Configuration of the monitoring device

① Oil ingredient densitometers to measure the oil content in discharge water in PPM unit.

② Flow meters to measure discharge rate at sea.

③ A speed log to indicate ship speed in knots.

④ A sampling device to transfer typical specimen of the discharge to the oil ingredient densitometer.

⑤ A control device to stop discharge to the outside of the ship.

⑥ A cut-out device to stop discharge to the outside of the ship in the event of the monitoring devices malfunctioning.

3) Configuration of the control device

① A processing device to convert the concentration of discharged oil, discharge flow rate and the ship speed to discharge volume per nautical mile and total discharge volume.

② A device to send a signal to the discharge control mechanism in class A devices.

③ A data recording device.

④ Display devices to show current operational status.

⑤ A manual auxiliary device to be used in the event of the main monitoring device malfunctioning.

⑥ A devices to send a cut-out signal to the control device in order to stop the discharge to the outside of the ship in the event of the class A device not functioning perfectly.

4) Sampling device

① Isolation valves must be installed near each sensor in a sampling device. However, if sensors are installed in cargo oil pipes, two isolating valves should be installed in series in the sampling pipe. One of these two isolating valves can be remotely controlled.

② Sampling sensors must be projected into the discharge pipe at least by a quarter of the diameter of the discharge pipe.

③ Samples which are returned to a slop tank must not be made to drop directly into it. In the case of tanks where an inert gas system is installed, a sealing device of appropriate height must be installed in the pipes connected to slop tanks.

5) Condition of discharge cut-off warning

① If the instantaneous discharge rate of oil ingredients is more than thirty liters per nautical mile.

② If the total discharge volume of oil ingredients reaches 1/30,000 of cargo loaded in the previous voyage.

③ If a malfunction occurs including power loss, loss of sampling function, serious malfunction of measuring or recording device, input signals to any sensor exceeding the effective capacity of the device and so on.

9.2 Ballast Water Treatment

From about 3 to 5 billion tons of ballast water is used every year worldwide and more than 3000 species of organisms move with the ballast water. Foreign species invade and viruses spread through this means, and IMO Regulation B-3 is concerned about controlling the movement of foreign species. The regulation for mandatory ballast treatment facilities has been phased in since 2009, and ships contravening this requirement are not allowed to enter ports.

1) Approval of ship ballast water treatment (BWT) plant

Depending on whether active substances are used or not, the BWT plant should be approved through either of the two channels below:

① Approval by government: Competent government authorities perform shore and on-board functional and environmental tests to ascertain that it conforms to the IMO D-2 discharge standard.

② Approval by IMO (only treatment plants which use active substances): Toxicity tests must be undertaken in addition to shore and on-board tests to obtain initial and final IMO approval of treatment plants which use active substances.

2) Type of ballast water treatment

Ballast water treatment can take place either in the ballast water intake/discharge pipes or in the ballast tanks. The treatment methods include mechanical and physical methods, such as filtration, cyclonic separation, ultraviolet radiation, heat and ultrasound, and chemical and biocide methods, such as insecticides, chlorination, metal ionization and ozonization.

Most treatment systems use two or three-stage processes where the first stage is usually separation through filtration followed disinfection/sterilization. At this stage some systems use a combination of two or more methods. **Table10-8** shows the merits and shortcomings of some popular treatment methods.

Table 10-8 Merits and shortcomings of sterlization methods

Methods		Merits	Shortcomings
Physical treatment	Filtration	- continuous treatment possible - simple devices	- difficult to remove very small particles - high basic treatment costs - maintenance of filtration devices required
	Centrifugation	- simple devices - no cleaning process	- low energy efficiency - may not kill all organisms
	Ultraviolet	- high efficiency - environment-friendly	- low penetrability and efficiency due to short wave length - high risk of corrosion due to ultraviolet devices being exposed to seawater
	Heat	- environment-friendly - potentially high price competitiveness	- difficult to use in large ships - efficient only at high seawater temperature
	Electric shock	-short treatment time	- high likelihood of leakage through body
	Ultrasonic wave	- a new technology with high potential	- still in the development stage.

Chemical treatment	Insecticide	- possible to treat harmful microorganisms	- not environment-friendly - storage facilities needed for chemicals
	Chlorine treatment	- high treatment efficiency	- treated ballast water may contain carcinogens
	Metal ion	- high efficiency	- some microorganisms tolerant to copper or silver can survive
	Ozone	- simple installation - good for treating microorganisms	- comparatively high installation costs

9.3 Tank Cleaning

When crude oil is transported by a ship, elements of wax, asphalt and sand contained in crude oil tend to settle at the bottom or other structures within the tanks. If such sediment builds up during a few voyages, it would be difficult to recover the oil from the bottom, and this inevitably leads to higher oil residues in the tanks. Furthermore, this will also result in reduced tank capacity and increased pollution risk.

For these reasons tanks are cleaned regularly to remove the sediment and to minimize oil residue. Cleaning is done either with water or crude oil.

1) Water cleaning

This conventional method uses seawater to remove deposits in cargo tanks. Seawater is heated to about 80°C, before being jetted at a high pressure (10 kg/cm^2) to remove the deposits. However, water used to clean tanks is contaminated with oil, so it cannot be discharged directly into the sea. Water is separated out from this mixture in a slop tank by allowing the oil to float on top before being discharged. This method is known as LOT (load-on-top) and had been the main method for minimizing oil pollution before crude oil cleaning method was introduced.

2) Crude oil cleaning

This method returns the cargo to the original state by dissolving sediments with crude oil itself during unloading. A portion of crude oil cargo is diverted to fixed cleaning pipelines and thence to the tank washing device for this purpose.

(1) Single stage cleaning method

This method is similar to water cleaning and begins when about one meter of oil remains in the tank. When the bottom of the tank is exposed or nearly exposed, upper, middle and lower parts of the tank are simultaneously cleaned. This method is often used in ships where non-programmable fixed type cleaners are installed.

(2) Two-stage cleaning method

It is the most popular method and the cleaning is carried out for upper and lower parts in two stages. It is possible to use programmable cleaners and requires higher manpower than the single stage cleaning.

(3) Multi-stage cleaning method

This method is used in ships which have programmable fixed cleaners. Cleaning is carried out in many stages, as the oil discharge progresses and oil level gradually goes down – usually in three stages. The tanks can be cleaned before the crude oil in the tank coagulates when dealing with heated crude oil. However, cleaning oil must be discharged before it coagulates, every time cleaning is halted.

- Upper part cleaning: When the oil level is lowered by several meters, the upper part is cleaned. The jet spray from the nozzles must be directed at main structures like horizontal girders.
- Middle part cleaning: When about two thirds of the loaded crude oil is discharged, the middle part is cleaned. Nozzles should be directed downwards at about 30 to 70 degrees toward main structures.
- Lower part cleaning: When one meter of oil is remaining, the lower part is cleaned. The trim should be kept to a maximum to assist the process of simultaneous washing and recovering oil residue. The number of washing devices required is determined according to the volume of oil to be collected. The number of cleanings is decided by many factors.

3) Shadow analysis table

The number and the locations of tank washing devices are decided by considering the internal structure of tanks as regulated by the IMO. The effectiveness of washing has to be graphically analyzed and the area that the jet from the nozzles cannot reach directly must be no more than 10% of horizontal surface and no more than 15% of vertical surface.

10. Accommodation Outfitting

10.1 Cabin Design

Cabins are the living spaces for passengers and crew who are confined inside a ship. Therefore, compared to shore accommodation facilities, there are more things to consider in designing cabins which may require more sophisticated technology and facilities. People cook, rest and enjoy recreation together as a small group, and so the accommodation spaces should be comfortable while providing privacy. The materials are selected in accordance with the owner's wishes, but fire protection and safety provisions must comply with the requirements of the relevant classification society and government departments.

The design includes accommodation arrangement, plan colors in the accommodation block, fire protection plan of all areas including engine room and cargo holds, steel outfitting design of accommodation, engine room and surrounding areas of cargo holds, cabin furniture arrangement, cabin facility arrangement, cargo hold insulation designs of LPG carriers and ethylene carriers, air condition designs for accommodation spaces, noise measuring and noise reduction designs of accommodation and engine room, vibration analysis and reduction of radar masts, and so on. Specifically, the superstructure area shown in **Figure 10-26** will be examined in more detail here.

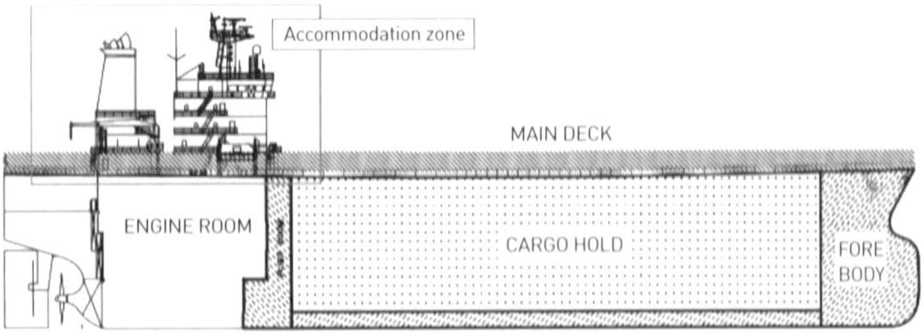

Figure 10-26 Accommodation design zone

10.2 ACA (Accommodation Compartment Arrangement)

Accommodation arrangement deals with the arrangement of cabins, wheel house, communication equipment rooms, cargo control stations, fire control stations and air conditioner rooms. ACA drawings show the plan of these spaces for each deck.

Figure 10-27 An example of private space

Figure 10-28 An example of public space

Figure 10-29 A typical control space

Figure 10-30 A galley

1) Subdivision of cabins

It is customary to plan private areas such as bedrooms and offices, depending on the rank: captains, senior officers, officers and ratings. **Figure 10-27** shows an example of private area of a senior officer and **Figure 10-28** shows an example of public area for the convenience and benefit of the whole crew. Dining rooms are provided separately for officers, crew and on-duty personnel. Lounges are usually provided separately for officers and ratings, but gymnasiums and swimming pools are often provided for joint use. Offices must be provided for deck department and engine department, but separate offices are also provided for the captain, the chief engineer and the purser.

2) Operation control room

An operation control room refers to a space dedicated to the tasks of general control of fire monitoring and fire-fighting, cargo handling and related equipment. An example of such a space is shown in **Figure 10-29**. In tankers cargo control rooms are usually arranged in the accommodation block, and in bulk carriers ballast control rooms are arranged in the accommodation block. Wheel house and chart rooms are required for navigational control of the ship and a communication room is needed for radio communications equipment. Furthermore, a space is required to control fire monitoring and fire-fighting equipment.

3) Galleys

Galleys have cooking utensils to be used for cooking food for passengers and crew, and an example is shown in **Figure 10-30**. Cooking facilities differ depending on nationality, but their arrangement is subject to rules and regulations. There are galleys used to cook food for crew and a small kitchen used for serving the food and cooking for on-duty personnel.

4) Sanitation facilities

These include private and public toilets, laundry rooms, drying rooms and sauna rooms.

5) Passageways

These include emergency escape routes from engine rooms to open decks, stairways and passages to move between decks and stairways connecting engine rooms and cabins.

6) Food stores

Meats, fish, vegetables, grains and beverages for use during the voyage are separately stored. This area is divided into cold stores and dry stores.

10.3 Main Cabin Outfitting

1) Doors

Doors connect the cabin to the accommodation block. Appropriate standard of divisions must be used in accordance with the fire regulations: class-A division capable of preventing spread of smoke and flame for 60 minutes in a standard fire test; class-B division capable of preventing spread of smoke and flame for 30 minutes in a standard fire test; and class-C division made of approved non-combustible materials.

Hollow watertight metal doors should be installed in places directly open to the outside to prevent seawater or rainwater ingress in any sea state. The wheel house must have sliding doors which can be opened and closed easily even in strong winds and the extent of opening should also be easily controllable.

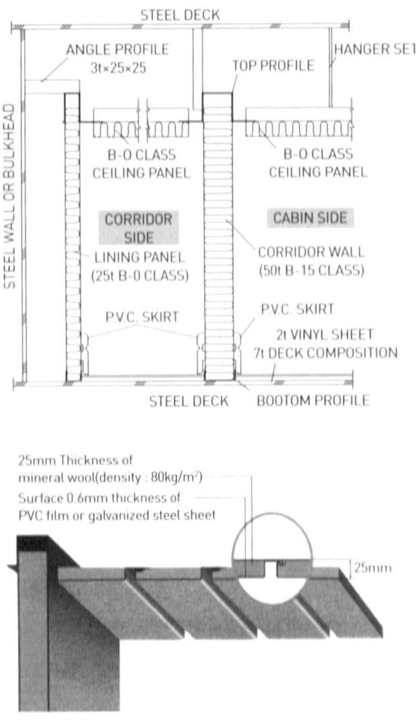

Figure 10-31 Wall & ceiling coverings

2) Interior finishing

Bedrooms, offices, common passageways and so on must be finished to a sufficiently high standard so that the crew can enjoy pleasant and safe living environment on board. Boundary walls and ceilings must be finished using fire-retardant covering materials approved by the competent authorities. **Figure10-31** shows an example of interior finish.

3) Deck covering

Decks are covered with synthetic rubber latex, vinyl tiles, carpets and noise proof insulation materials in order to minimize noise and vibration and provide insulation, good aesthetics and non-slip surface.

4) Lighting windows

Circular or rectangular windows, either fixed or openable, are installed for illumination and lighting. In the case of oil tankers, class-A fire windows should be installed on the front walls or ship sides within 3 m from front walls of cabin. When rectangular windows are installed, they are usually of 400 mm width and 600 mm height. However, if the owner requests alternative size of window, the thickness of the glass must be calculated according to the provisions in ISO 5779. Circular windows of 350 mm or 400 mm diameter are used and their thickness must be decided in accordance with ISO 5780.

Electrically heated windows are installed in the wheel house of ships sailing in cold regions. When water is frozen or snow or ice accumulates on these windows, ice can be melted by the electric heating lines to secure adequate visibility. Double glazed windows filled with argon gas are installed in engine control rooms and they also reduce noise from the engine rooms.

5) Steel outfitting

The steel outfitting attached to the outside of deckhouses includes steel radar masts, steel doors, inclined ladders, vertical ladders, hand rails, verandahs, etc. **Figure 10-32** shows a typical mast used in ships.

Figure 10-32 Radar mast system

6) Fire- and noise-proofing

In order to provide a pleasant living environment for the crew, protection of human life and property, noise-proofing and insulation, classes A, B and C divisions are applied in accordance with the fire regulations.

7) Furniture

Strong and good quality furniture suitable narrow spaces should be provided and appropriately arranged in the rooms.

8) Miscellaneous spaces

There should be a space for the crew's changing room and lockers. Other storage spaces include deck stores, paint and electrical equipment stores, cleaning equipment lockers, custom goods stores, laundry stores, etc.

11. Electricity and Lighting

11.1 Electric Power Sources

Ship's power sources can be diesel engines, turbine engines or the main engines driving generators, and they must be able to provide sufficient power for the operation of the ship. More than two generators must be installed so that necessary power can be supplied even in the event of one of them malfunctioning. Emergency generators are installed to supply electricity to essential equipment, such as steering gear, fire-fighting pumps, lighting, communications, navigation and radio equipment in an emergency of a shutdown of the main power supply due to an accident or malfunction of the main generators.

450 V 60 Hz three phase AC generators are normally used in general merchant ships, and the capacity of generators reaches thousands of kVA. In the case of ships like container ships and drillships, which have to contend with heavy load demand, 6.6 kV or 11 kV 60 Hz three phase AC generators are more commonly used.

440 V AC is used in motors for various auxiliary machines and loading/unloading equipment, and 220 V is used for lighting, communications, navigation and radio equipment. 110 V is also used from time to time, in which case transformers are used to drop the

voltage. The 24 V DC batteries with sufficient capacity are used as uninterrupted power source (UPS) to provide emergency power to communication, navigation, radio and warning equipment during a blackout.

11.2 Configuration of Power System

A symmetrical bus system is often used as the power supply method of choice to improve the sustainability of power supply to the main loads which has become critically important due to the rapid development of automation of ships. In this system the generator control console is placed at the center and the supply distribution switchboards are arranged on either side of it. The essential propulsion auxiliaries are divided into two groups, each feeding a separate distribution switchboard, so that the ship is perfectly capable of operating with just one switchboard operating.

A high-capacity isolating switch must be installed between the main switchboard and the emergency switchboard, so that the latter can be supplied from the main power source, whenever it is possible, as shown in **Figure 10-33**. When the power from the main source is cut, the switch is thrown open so that the power from the emergency power sources can be supplied instead. Components of the power system on board a ship and their functions are as follows.

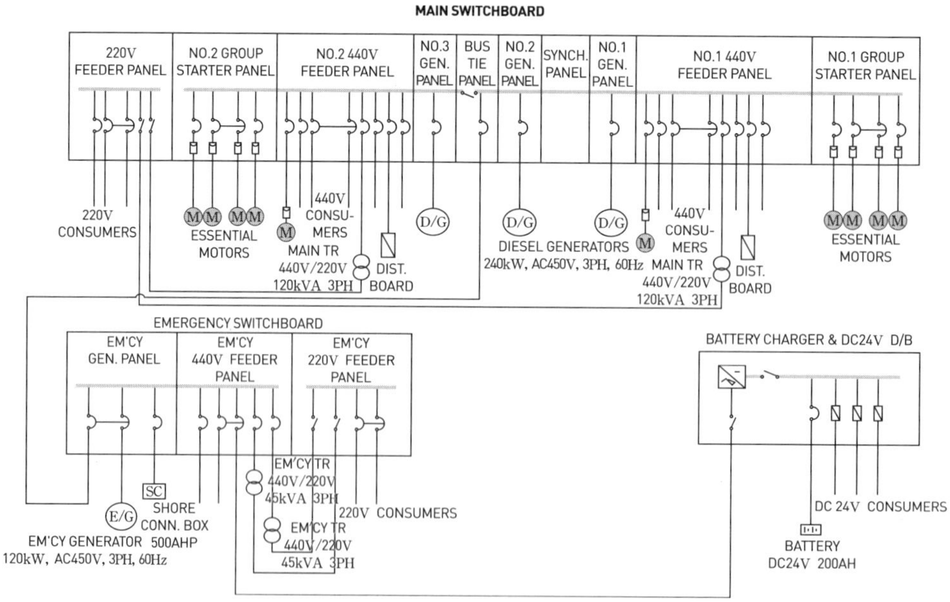

Figure 10-33 Power system single line diagram

① The main switchboard not only checks, controls and protects electricity produced by the main generators but also transforms, receives, and distributes it.

- Generator control consoles control the generator engines to produce and receive electricity and monitor the state of generators through various instruments.
- Synchronized operation control panels operate generators independently or in parallel and distribute the loads to each generator in response to varying power demand, and thus contribute to smooth supply of electricity in varying operational conditions.
- The normal operation of the main power line consoles is through two power lines connected to it. If one line malfunctions, a circuit breaker will be operated to supply power only to the intact power line. This isolates the malfunctioning line so that repair can be carried out.
- 440V AC power supply control consoles consist of a generator control console and a synchronized operation control panel. They control the use of electricity necessary for ship operation by supplying the power to functional start control consoles for various motors and distribution panels.
- 220V AC control consoles transform the 440V AC to 220V AC and then supply it to equipment which requires it.

② A main transformers changes 440V AC from the main power source to 220V AC and supplies it through switchboards to lighting, communication, controls and navigation systems.

③ Functional starting control consoles control start-up devices of pumps and fans, equipment essential for the propulsion system, cooling system and so on. These consoles, in connection with the main switchboard, are installed in the engine control room, and have the functions of starting and braking of pumps, overload protection, automatic standby and starting, automatic recovery from blackout and so on.

④ The emergency switchboard monitors and controls the power from the emergency generators, and transforms, receives and distributes the power, when the main power supply is disrupted due to accidents or malfunctions of the main generator.

- Emergency generator console controls the generator engines and receives the generated electricity. It also has various instruments with which the status of the generators is monitored continuously.
- A 440V AC emergency power supply control console provides the electricity received from emergency generator control consoles and supplies it to various electric appliances which require emergency 440V AC power.
- A 220V AC emergency power supply control console transforms the 440V AC power received from the emergency 440V console, and distributes the resulting 220V AC.

⑤ An emergency transformer can transform 440V AC emergency power to 220V and transmits it through an emergency switchboard to emergency lighting, communications, controls and navigation systems.

⑥ When the ship is moored to a shore facility for a long time, electricity from the shore can be received through shore power supply terminal boxes.

⑦ Functional starting control consoles are used to support the operation of various pumps and blowers, and they are often installed near the machines they serve.

⑧ The electricity from the main power source is distributed at the distribution switchboards according to power demand, using circuit breakers.

⑨ Battery charging and discharging equipment charges or discharges 24V DC storage batteries and supplies power to other equipment as required.

11.3 Lighting Equipment

Lighting design aims to determine light sources and their type, size and location in order to realize a light environment appropriate for the usage of the given space. Lighting in engine rooms, accommodation spaces, deck areas and other spaces is subject to rules and regulations. The places which require emergency lighting are specified by SOLAS, and they include all mustering stations, lifeboat stations, all passageways, stairs and entrances/exits, personnel elevators, propeller shaft and rudder equipment, all control stations, main and emergency switchboards, fire-fighting stations and lockers for firefighting equipment.

11.3.1 Navigation lights

There are many different kinds of lights and signal equipment used in ship navigation. **Figure 10-34** shows the arrangement of navigation lights deployed to be seen by the others in the area so as to prevent collision. Navigation lights are usually used at night but they can also be used during the daylight hours, if the visibility is limited due to rain, dense fog,

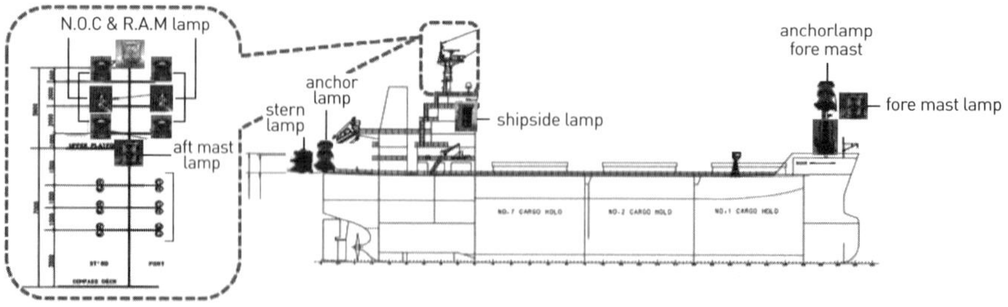

Figure 10-34 Arrangement of navigation lights

smoke or any other reason. The name and usage of navigation lights are as follows.

1) Foremast light and after mast lamp

These white lamps indicate the presence of a ship, and are visible from far away even when the side lamps are not visible. If the distance between two mast lamps is observed carefully for a time, a person on another ship can estimate its travel direction and heading change. For this reason, they are sometimes referred to as direction indicating lamps.

2) Side lamps

These lamps indicate the ship's heading, since a green lamp is installed on the starboard side while a red lamp on the port side. If the side lamps are carefully observed, another ship can estimate its heading and travel direction, thus providing basic information for determining the relative positions of the two ships. The lights are made visible only from a limited range of angles to prevent confusion arising and their backs are blocked off to prevent any light leaking towards the stern.

3) Anchor lamp

A ship at anchor does not have mobility, and therefore it is important for other ships to identify a ship at anchor. The anchoring lamp is a white light visible from 360°.

4) Stern lamp

It is a white light displayed at the stern for the benefit of another ship which may wish to overtake. This lamp is required to be located as near to the stern as possible so that it and the mast lamp indicate the size of the ship.

5) N.O.C. (Not Under Command) & R.A.M. (Restricted Ability to Maneuver) lights

These lights indicate the ship has limited ability to maneuver or is unable to maneuver, and are placed where they can be easily noticed by other ships so that they can take evasive actions. N.O.C. lights consist of two red lights placed at least 2m apart vertically in line. R.A.M. lights are made up of a red light – white light – red light placed at least 2m apart each vertically in line. All these lights must be placed where they can be easily seen from all directions. However, the restriction angle is allowed up to only 6°.

References

1. International Association of Classification Societies (IACS), 2012. *Common Structural Rules (CSR) for Bulk carriers*. London: International Association of Classification Societies Ltd.
   Available at: http://www.iacs.org.uk/publications/CommonRulesDoc.aspx?pageid=4& sectionid=2&linkid=1 [Accessed 20 June 2015]

2. International Maritime Organization (IMO), 2009. *International Convention for the Safety of Life at Sea* (SOLAS), Consolidated Edition 2009. London: IMO Publishing.

3. International Maritime Pilots Association (IMPA), 2008. *IMPA guidanceon the rigging of ladders for pilot transfer*. London: Maritime International Secretariat Services Ltd.

4. Park, S.K., 1998. *Introduction to Shipbuilding Technology* (in Korean). Seoul: Dongmyungsa.

5. The Korea Shipbuilders' Association (KOSHIPA), 1992. *Design guideline for accommodation outfitting* (in Korean). Seoul: KOSHIPA.

6. The Korea Shipbuilders' Association (KOSHIPA), 1995. *Design guideline for shippiping system* (in Korean). Seoul: KOSHIPA.

7. The Korea Shipbuilders' Association (KOSHIPA), 1996. *Design guideline for shippiping system* (in Korean). Seoul: KOSHIPA.

8. The Korea Shipbuilders' Association (KOSHIPA), 1995~1998. *Standard fittings* (in Korean). Seoul: KOSHIPA.

9. The Korea Shipbuilders' Association (KOSHIPA), 1998. *Design guideline for electrical outfitting* (in Korean). Seoul: KOSHIPA.

10. The Korea Shipbuilders' Association (KOSHIPA), 2000. *Design guideline for ship ventilation* (in Korean). Seoul: KOSHIPA.

11. The Society of Naval Architects of Korea, 2000. *Ship equipment* (in Korean). Seoul: Dongmyungsa.

CHAPTER 11

Ship Production Technology

1. Production Design

1.1 Introduction

The production design is a process of generating drawings to be used for ship construction, based on the structure and arrangement specified in the detail design and adapted to suit the capability and facilities of the shipyard for efficient building of the ship. The drawings include those for nesting, fabrication, details of component pieces and fitting-up, as outlined in **Figure 11-1**.

Ships are built according to construction drawings to suit the yard arrangement, equipment available, capability of the yard and building schedule. Each detail piece such as a plate or profile is fabricated by cutting and forming, and then joined with other pieces according to the construction drawings. For this reason, some shipyards consider production design as a process of ship production.

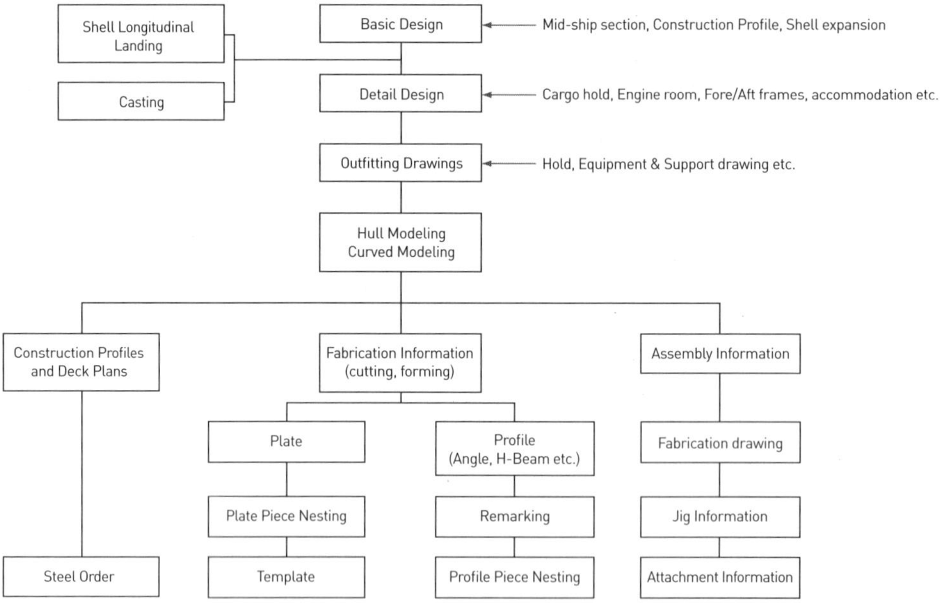

Figure 11-1 Production design process

1.2 Sequence of Production Design

In production design, the ship's structural members are grouped by its configuration, shape and working sequence, and then those of the same thickness are arranged as tightly as possible on a plate of the given thickness or a little thicker so that material scrap can be minimized. This process is called nesting. The fabrication drawings of all individual parts are then produced and the information is sent to the fabrication shop together with a list of necessary steel materials. It is absolutely critical that all parts are made and assembled exactly as designed. Therefore, the fabrication drawings must be clear, unambiguous and comprehensive to avoid misunderstanding or confusion by the workers.

1.2.1 Three-Dimensional Modeling of Hull Units

The two-dimensional detail drawings of the ship structure and arrangement are converted into three-dimensional models using CAD/CAM systems. **Figure 11-2** shows an example of such a model of an engine room block, and it is easy to see that visualizing the structure in this way will help the workers a great deal in understanding what needs to be done.

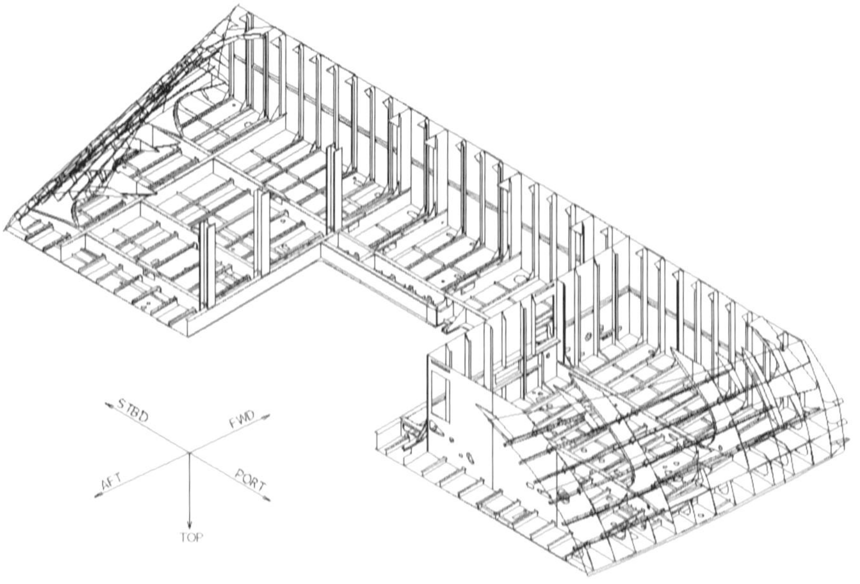

Figure 11-2 An example of three-dimensional CAD modeling (an engine room block)

1.2.2 Assembly Drawings

Assembly drawings are prepared for the assembly shops to construct the structures with the available facilities, and for this reason, it is important to plan the assembly work appropriate for the shop arrangement, characteristics of equipment, work schedule and working conditions of the yard. A typical assembly drawing contains information of the allowable tolerances for assembling and erection of blocks, welding methods, welding leg lengths, block assembly sequence, part numbers for the assembly and so on. Such information is displayed in accordance with the yard's standard practice, and an example is shown in **Figure 11-3**. Additional assembly information includes arrangement and height of pin jigs on working tables for fabricating curved blocks. The lengths and angles of attachment between shell plates and internal structural members are also provided.

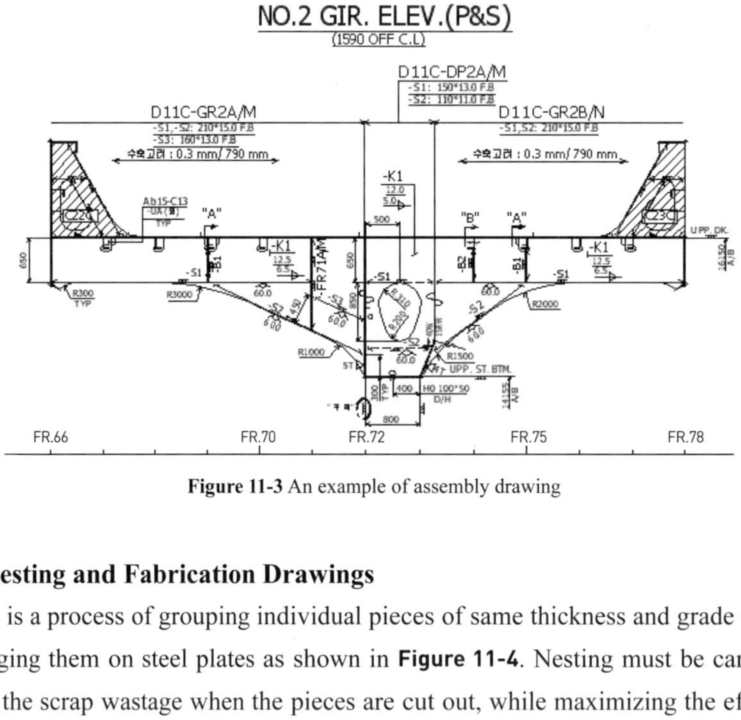

Figure 11-3 An example of assembly drawing

1.2.3 Nesting and Fabrication Drawings

Nesting is a process of grouping individual pieces of same thickness and grade of material and arranging them on steel plates as shown in **Figure 11-4**. Nesting must be carried out to minimize the scrap wastage when the pieces are cut out, while maximizing the efficiency of the cutting work. On occasion, therefore, it is allowed to cut some pieces from a plate thicker or having higher grades than designed, if it gives a better utilization of plates.

The next step is extracting numerical control (NC) cutting data from the nesting drawings and transferring them to the cutting machines in the fabrication shop. Of the pieces, those requiring curve forming are sent to curve forming presses or line heating facility. Drawings for cutting and forming in the production process are referred to as fabrication drawings.

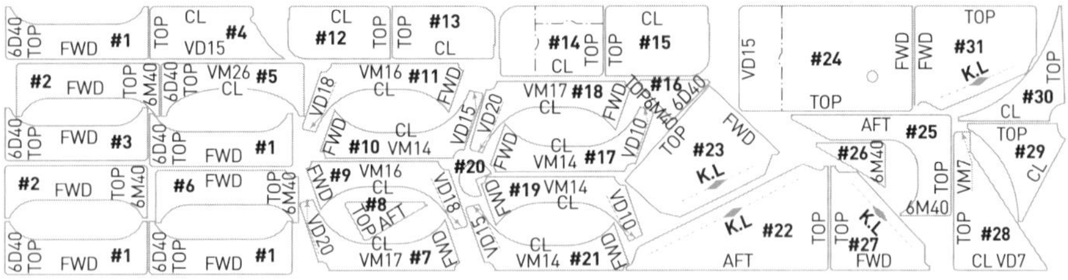

Figure 11-4 An example of nesting

1.2.4 Pre-Nesting, Procurement and Work Order

Steel for shipbuilding must be approved by the relevant classification society and is produced to order. The steel materials to be used for building a particular ship must be estimated well in advance of the scheduled cutting date to allow sufficient time for delivery and labeling of the procured steel.

The pre-nesting design is a process of determining the quantity, size and thickness of steel plates and profiles necessary for building a ship and its outcome directly affects the construction cost. Therefore, an effort must be made in the process to minimize the amount to be procured and fabrication work involved, and this of course implies minimizing the scraps. Once procured, a list of required steel material is produced based on the results of nesting. The work orders are then issued to the fabrication shop for cutting in accordance with this list.

1.3 Computerization of Production Design

The first CAD/CAM system introduced to shipbuilding by Korean shipyards was the VIKING system from Sweden for ship design and production in the 1970s and was used for hull construction only. Nowadays, most shipyards have established unified design systems of hull and outfitting design from the basic design to production design by applying 3D CAD systems such as TRIBON, AVEVA MARINE, GSCAD among others. The systems are connected to NC machines to automate the fabrication processes too.

Recent developments in CAD technology have enabled simulations based on three-dimensional modeling to be used for wide ranging applications, including checking interference between modules, studying vehicle movements in RO-RO ships, examining erection and turnover of blocks and planning production processes.

2. Shipbuilding Facilities

2.1 Shipyard Layout

2.1.1. Requirements for the Layout

Shipbuilding progresses from material procurement, fabrication, assembly and erection to launching and outfitting work in that order based on the designs. The process ends when the ship is delivered to the owner after having gone through the performance verification processes of various tests and sea trials.

The shop layout of a shipyard depends on a great variety of factors related to the production process and fabrication methods in a broad sense. In principle, however, a shipyard layout must be planned to:

① assure profitability in its operation;
② provide a safe and clean working environment;
③ guarantee product quality;
④ provide rational work process as a production system;
⑤ be able to be supervised, managed, repaired and maintained;
⑥ avoid pollution and environmental hazards; and
⑦ be ready for future expansion of the facilities.

2.1.2 Principles of Yard Layout

Shipbuilding is a typical production-to-order industry where a huge variety of parts are produced in small batches before being assembled. The parts may often require non-uniform work sequence, but the layout must be made to allow the individual parts to flow as smoothly as possible. The most efficient layout is based on the work flow group technology, and it is beneficial to adopt it even if partially. The next important points are to utilize the machines as efficiently as possible and to achieve a balance in the capacity of all the production processes involved.

Construction of a ship's hull calls for frequent transportation of heavy units and the layout must be carefully arranged to allow smooth workflow between consecutive processes with special emphasis on the reduction of travel distances and simplification of the work flow paths. For example, outfitting shops which used to be placed near the outfitting piers or docks are nowadays located near a hull fabrication shop, because pre-outfitting at the assembly stage has become the norm. Care must be taken to leave sufficient room for future expansions, and the scale of various shop buildings must be standardized wherever possible. The yard layout is not immutable but should be allowed to evolve continuously with the advance in fabrication methods. Therefore, ceaseless efforts must be made to assess and improve the layout by expecting the future needs and development, even if the present one appears to be working satisfactorily.

2.1.3 Examples of Yard Layout

There are three typical shipyard layout patterns, viz. I-type, L-type and U-type as shown in **Figure 11-5**. **Figure 11-6** shows an example of the most common U-type arrangement.

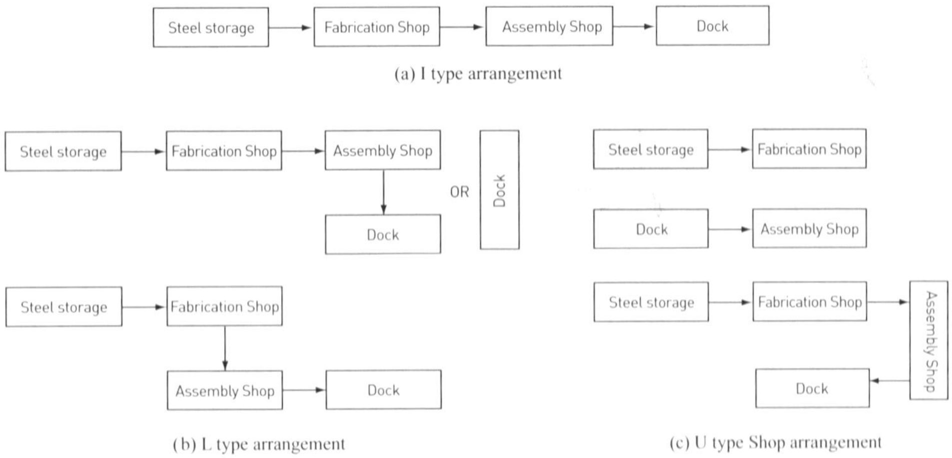

Figure 11-5 Typical patterns of yard arrangement

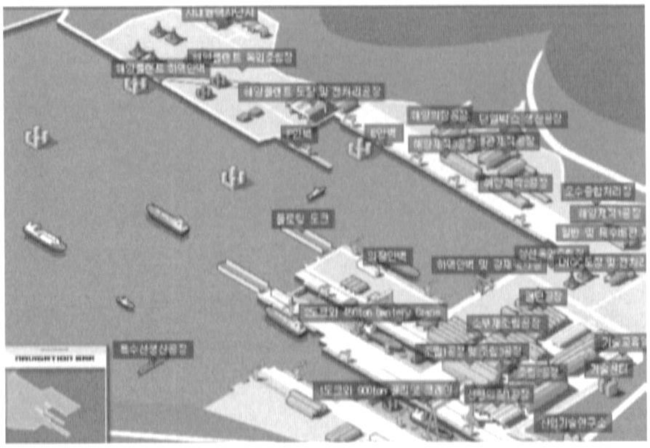

Figure 11-6 An example of the U-type yard layout

2.2 The Main Shops

2.2.1 Steel Stockyard

The steel stockyard is a place where steel plates and profiles are stocked for a while, as shown in **Figure 11-7**. It is equipped with magnetic cranes, jib cranes and roller conveyors to supply steel plates to the fabrication shop when required. The classification and systematic supply of the stocked steel plates and profiles as and when required by the fabrication process is regarded as the first stage of shipbuilding.

Figure 11-7 A steel stockyard

2.2.2 Fabrication Shop

In the fabrication shop the steel plates are surface treated and pre-coated before being cut automatically according to the design information. Some of them are then curve-formed as necessary. **Figure 11-8** shows such a fabrication shop equipped with automatic plasma and gas cutting machines. Both flat plates and profiles such as T-bars are cut and the edges prepared. Three-dimensional curved hull surfaces constituting bow and stern are also formed in the shop using a large hydraulic press or line heating technique.

Figure 11-8 A fabrication shop

2.2.3 Sub-Assembly Shops

The sub-assembly shops are located next to the fabrication shop in series. This arrangement allows sub-assembly parts to be transported by conveyors, enabling efficient control of the work flow between fabrication and sub-assembly processes and smooth running of the production process. Each sub-assembly shop must be furnished with a stock yard for detail pieces so that the assembly work can be specialized and simplified.

2.2.4 Indoor Block Assembly Shop

The location and direction is the most important factors to consider when planning the indoor block assembly shop. It is recommended to place the indoor block assembly shop in a location directly accessible from the fabrication shop and the sub-assembly shop, because its positional relationship with other indoor shops is more important than its relationship with the building docks. Sufficient area must be provided to ensure high transport efficiency and automation facilities.

2.2.5 Outdoor Block Assembly Shop

The outdoor block assembly shop should be located close to the indoor block assembly shop rather than the fabrication shop. Finishing work on blocks is carried out here in preparation for the inspection by ship-owners and classification societies.

2.2.6 Area of Work Bays

Work bays with adequate area are essential for accurate and easy assembly of three-dimensional curved blocks. The required area of work bays depends on various factors, including the location of the yard, annual building volume, in-dock building period, production capacity, monthly assembled steel weight, ratio of workloads for indoor and outdoor assembly shops, block sizes, assembly methods and turnover rate of the work bay.

2.3 Pre-Outfitting Shop

Various kinds of outfitting items and equipment including pipes, cables, electrical equipment, machineries and so forth are manufactured and installed in the blocks here prior to erection to reduce total ship production time and cost. **Figure 11-9** shows a pre-outfitting shop where outfitting items manufactured by subcontractors or branch factories in accordance with the design information are installed in blocks with the help of about five ton capacity overhead cranes.

Figure 11-9 A pre-outfitting shop

2.4 Painting Shop

Figure 11-10 shows painting shops where fabricated blocks are shot blasted and then spray or brush painted. The blasting facilities include high speed steel grit blasting machines, dust collectors and vacuum recovery systems.

The painting shops also need dust collectors in addition to humidity control machines (dehumidifiers and heaters) to create the best condition for painting and to prevent air pollution.

Figure 11-10 Painting shops

2.5 Building Berths and Docks

A building berth or dock is a place where blocks are erected to complete a hull as shown in **Figure 11-11**, and requires the services of cranes to lift the blocks.

Many shipbuilders are trying to make the blocks as large as possible, and, as a

Figure 11-11 Building docks

consequence, goliath cranes with lifting capacity of 1500 tons or more are used in some large modern shipyards.

2.6 Outfitting Quay

Figure 11-12 shows outfitting quays where outfitting and tests of a completely erected ship are carried out after launching. The quay must be equipped with mooring bitts, access ladders and jib cranes of about 30 to 60 tons capacity to support quay works.

Figure 11-12 Outfitting quays

3. Construction Processes

3.1 Fabrication Process

In the fabrication process plates and profiles are cut and formed into appropriate shapes as shown in **Figure 11-13**. The work is carried out in process groups in order to economize on steel materials and to improve the work efficiency.

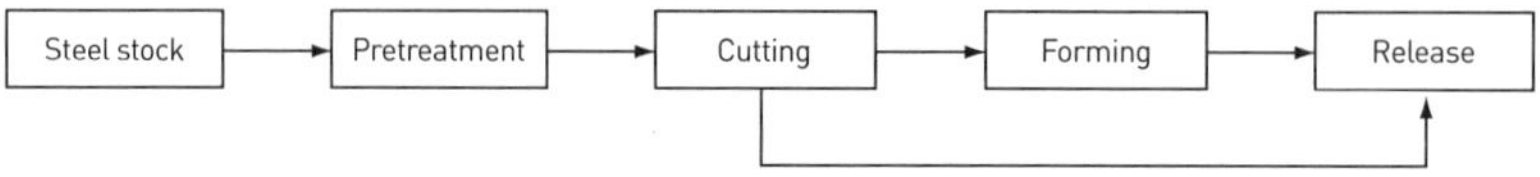

Figure 11-13 Fabrication process

3.1.1 Steel Stock

Steel materials approved by the relevant classification society are transported from steelworks and, on arrival at the pier, are unloaded with cranes. Plates and profiles are separately transferred to the designated stockyards.

3.1.2 Pre-Treatment

This is the first stage of processing the steel materials delivered from the stockyard. The mill scales and rust are removed from the materials, and then they are pre-heated for priming, shot blasted, painted and dried before being moved to the transfer bay. The materials deformed during transportation or unloading must be corrected with leveling machines before the pre-treatment.

3.1.3 Marking and Cutting

The materials are marked and then cut in batches of fabrication groups. The markings indicate the outlines of the pieces and attachment positions of small components onto them and are either drawn manually or automatically by the cutting machines. The cutting machines may be automatic or semi-automatic. The cut pieces are then transferred to a designated assembly shop.

3.1.4 Forming

The cut pieces are bent or folded into the intended curved shapes. The work can be undertaken either through cold working or line heating process. Cold working process inflicts plastic deformation on materials by applying mechanical forces at the ambient temperature whereas line heating bends a piece by introducing local shrinkages by means of applying heat on the material, usually in lines, and then cooling it quickly, for example, by spraying water on the area.

3.2 Assembly Processes

Small pieces prepared in the fabrication shop are then put through the sub-assembly processes. Stiffeners are welded onto flat plates in the sub-assembly process, while web frames are welded onto shell plates in the sub-block assembly process. Assembled components and sub-blocks are welded together to form a much larger unit in the block assembly process. **Figure 11-14** illustrates typical work in progress during these processes.

(a) Sub-assembly (b) Sub-block assembly (c) Block assembly

Figure 11-14 Sub-assembly and block assembly processes

3.2.1 Sub-Assembly and Sub-Block Assembly

In the sub-assembly process, individual small fabricated pieces like stiffeners, brackets, etc. are simply fitted on the plates, and welded together to form stiffened plates. Some of these assemblies are joined together to build up sub-blocks as shown in **Figure 11-15**.

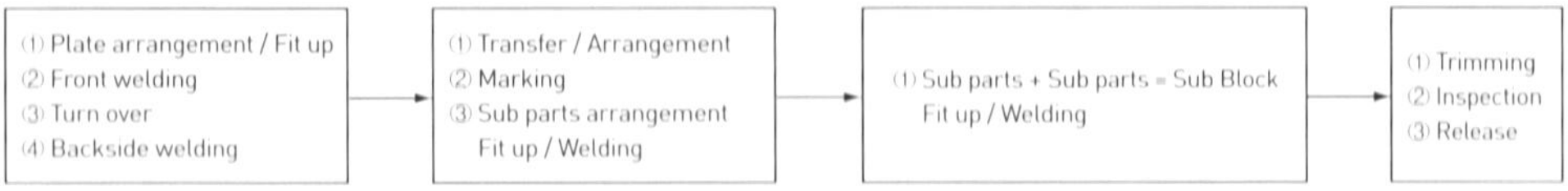

Figure 11-15 Sub assembly process

3.2.2 Block Assembly

In the block assembly process, sub-blocks are put together to make up large blocks appropriately sized for erection. In some cases, several blocks may be assembled into a super-block in the vicinity of the dock before erection, and this process is known as pre-erection. The assembly process of flat blocks amidships is different from the assembly of the curved ones around the bow, stern, engine room and so on. **Figure 11-16** shows a block assembly process, which is often known simply as assembly process.

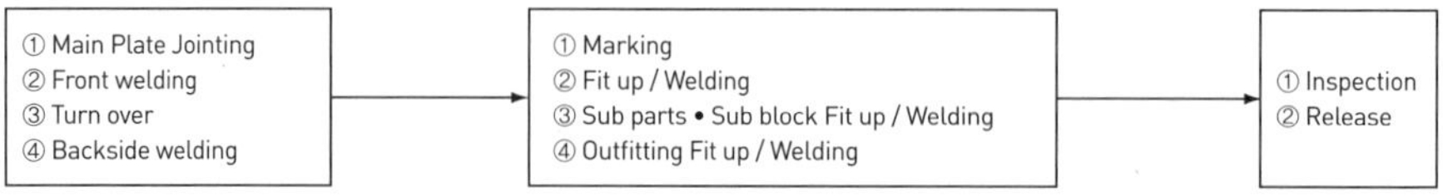

Figure 11-16 Unit assembly process

3.2.3 Production of Outfit Items

The main items produced are the piping and fittings and steel fittings for various systems on board.

- Pipe shop: small (150 A~200 A) and large (over 200 A) diameter pipes for hydraulic and water pressure systems.
- Steel outfitting shop: columns and steel outfits.
- Machinery shop: pipe fittings for the pipe shop and other items on request from the installers.
- Painting shop: painting work of fabricated pipes and steel fittings.
- Stock yard: smooth distribution of finished fittings through storing, sorting and supplying on demand.

3.2.4 Pre-Outfitting

Outfitting work used to be conducted midway through erection in a dock or on a building berth in the past, but much of the work is now done beforehand on land to improve work efficiency, and this process is known as pre-outfitting. Currently, items like large pipes, cable trays, machinery foundations, etc. are installed in the shops before the completion of block assembly. Such pre-outfitting process considerably reduces overall work period and volume of outfitting work required, thus improving the quality as well as productivity.

3.2.5 Pre-Painting

The pre-painting work is done before erection or pre-erection (PE) of blocks and consists of surface treatment process, mainly performed by blasting, shown in **Figure 11-17**, and painting process to form a film of protective coating as shown in **Figure 11-18**.

Figure 11-17 Surface treatment (blasting) process

Figure 11-18 Pre-painting process

3.3 Erection Process

The erection process is the final construction stage of a ship's hull. The pre-outfitted blocks are connected to each other in the proper order on the berth or in the dock to produce the designed hull form. It is important to check that its shape conforms to the design and the dimensions are accurate. Preparations for the completed ship to be launched or floated out safely must be made on the berth or in the dock before the erection process begins. The erection work can be carried out in succession or in tandem.

3.4 Outfitting Processes

The outfitting work is carried out after the erection and is usually divided into four groups: machinery outfitting, hull outfitting, electrical outfitting and accommodation outfitting.

- Machinery outfitting: installation of main engines, shaft systems linking the engines and propellers, rudder systems and auxiliary machinery like boilers and generators, steel outfitting, insulation and painting.
- Hull outfitting: piping and steel outfitting work in areas other than accommodation and engine room, including installation of mooring and anchoring systems, cargo handling systems, masts, hatch covers, etc.
- Electrical outfitting: installation of electrical and display equipment as well as all other equipment required for communication and navigation.
- Accommodation outfitting: installation of side scuttles, entrance doors, wall panels, ceilings, air conditioning, sanitary facilities, kitchen equipment, refrigeration equipment, recreation rooms, living rooms, bedrooms, offices and all other hotel facilities.

3.5 Final Finishing Work

3.5.1 Post-Erection Painting (Hull Painting)

The whole shell plates must be painted upon completion of erection. This process is important, because painting is directly related to the prevention of hull corrosion and the color scheme affects the final appearance of the ship.

3.5.2 Quayside Work

The launched ship is moored to a quay, and outfitting and painting work not done in the building dock is carried out. Various machines and equipment are tested and their performance examined, and the preparation for sea trials is made. The process typically takes two to three months.

3.5.3 Sea Trials

The trials are the final inspection process of a newly constructed ship. The trials consist of: quayside trials conducted while the ship is moored to a quay within the yard; and sea trials carried out on open sea to test the controllability and seaworthiness of the ship. The sea trials are divided again into preliminary trials done by the shipyard alone and an official one in the presence of the ship-owner, classification society and other competent authorities.

The items examined at the trials differ in the type and size of the ship concerned, but they must be confirmed by the relevant classification society and the owner. All items examined must comply with the safety regulations and the rules of the classification society. If the results of the trial do not meet, or exceed, the terms of the contract on fuel consumption rate, speed, tonnage, etc., it is customary to make commensurate compensation, or bonus payment. The ship can be delivered to the owner only when the trials are completed successfully.

3.5.4 Naming Ceremony

The naming ceremony is held right before the delivery of a ship, and traditionally a woman names the ship, because ships are regarded as female. A rope connecting the ship to the place where the ceremony is held is cut with a hatchet to signify the birth of the ship, reminiscent of the umbilical cord being cut when a baby is born.

3.5.5 Registration Inspection

The classification society and the relevant government authority issue a ship inspection certificate after checking that the ship is fully equipped with anchors, anchor chains, mooring ropes, boats, rescue equipment, lifesaving appliances and so on, and examining the results of the sea trials. The owner must request the relevant authority to carry out a tonnage measurement inspection for registration purposes. The ship then must be registered at a designated home port and the owner must request the flag authority to issue a Certified Copy of Ship's Registration. With the acquisition of a nationality, the ship finally becomes a legal entity.

3.5.6 Delivery

The ship is delivered to the owner when the outstanding outfitting work and the trials are completed. On the day of delivery owner's and yard's representatives gather together on board to hold a ceremony. At this time the ownership of the ship is transferred from the shipyard to the owner and the ship is ready for its maiden voyage. The delivery process is completed when the ship is towed to the outer port and various delivery documents are signed and exchanged at a place designated by the contract.

4. Outfitting Work

4.1 Introduction

The shipbuilding process is divided into hull construction and outfitting. The outfitting work installs various equipment and facilities necessary for transporting cargo and passengers, and therefore deals with a wide range of facilities and the materials are just as diverse, including copper, stainless steel, nonferrous metals, plastics, composite materials, ceramics, novel metallic materials, cement, tile, rubber and glass. Naturally there have been many changes in outfitting work over the years, and the relationships with other construction processes have become more complex as well. This has made it necessary to develop and implement production management techniques as well as efficient and effective equipment.

In the past outfitting work used to be done mostly when the hull construction is complete or block erection is over. However, in modern construction methods as much pre-outfitting work as possible is carried out in the assembly shops so as to shorten the overall time required for outfitting. This is clearly illustrated by the fact that the ratio of man-hours required for the same amount of work done in the assembly shops, in a dock and at a quay is about 1:3:5. The advantage of pre-outfitting is obvious. The outfitting work is largely divided into pre-outfitting, pre-painting and post-erection outfitting.

4.2 Pre-Outfitting

The pre-outfitting work is completed before the blocks are erected on a building berth or dock. It includes fitting out the blocks before they are painted with as much items as possible, such as pipes, steel equipment, electrical equipment and machinery control equipment, and fitting out super-blocks during pre-erection work after painting with group units and other equipment and parts. These are referred to as piping, steel outfitting, electrical outfitting and instrumentation, respectively.

The aim of introducing pre-outfitting process is to maximize the amount of outfitting work prior to dock erection to improve its efficiency and to save time and cost by carrying out the work concurrently with the block construction. There are many challenges in implementing effective pre-outfitting process and some of the problems can be alleviated by taking some measures as follows:

① Designate places devoted to the pre-outfitting process.
② Use exclusive facilities and instruments.
③ Carry out as much work as possible indoors.

The following will result as a consequence:
① Improvement in work efficiency owing to smooth workflow due to fixed locations.
② Improve the work environment through indoor work for improvement of painting quality.
③ Best use of block storage spaces by three-dimensionalizing them.

When an assembled block is moved into a pre-outfitting shop as scheduled, items to be fitted and various jigs are delivered. The fitting sequence of the items is determined according to their shapes and types as indicated in the drawings. They are then fitted up into the blocks with appropriate jigs and tack welded, where required, before being welded into place.

Those items requiring approval from the classification society or the owner are inspected when they are fitted up. Having been non-destructive tested (NDT), the block proceeds to the pre-painting stage. Typically pre-outfitting work takes about three months, and during that period piping as well as steel, electrical and wood outfitting must be completed to minimize post-erection outfitting work on board. Therefore, it is important to manage the pre-outfitting shop so as to allow workers of diverse trades to work side by side harmoniously with minimum interference to each other.

Dock turnover rate is an index for measuring the productivity of a shipyard. A higher dock turnover rate indicates that the shipyard can construct more ships with its given facilities. Mega blocks and pre outfitting are indispensable for improving the dock turnover rate, because they can shorten the in-dock period considerably. Thus, pre-outfitting has now become a center of effort to improve yard productivity and many shipyards are continuously striving to increase the pre-outfitting rate.

4.3 Pre-Painting

Before the fabrication stage hull materials like steel plates and profiles are surface treated to remove rust by shot blasting and primed indoors. The pre-painting is a painting work carried out on the blocks as soon as they are assembled, and it also consists of shot blasting and painting as shown in **Figure 11-19**. Shot blasting uses steel grits projected out of a nozzle at high speed to remove rust, mill scale, patina and general muck to improve the adhesion of paint to the surface. The blocks are then spray painted to form the film of a specified thickness. If the film thickness is insufficient at any area, it has to be touched up usually by brush to achieve the required thickness. Although it varies with building specifications, pre-painting is ordinarily done twice inside and four times outside.

(a) Blasting (b) Painting (c) Brush painting

Figure 11-19 Pre-painting

4.4 Post-Erection Outfitting

The post-erection outfitting is done on a berth, in a dock or afloat beside a quay when the block erection is complete, and consists of machinery outfitting, hull outfitting, electrical outfitting, accommodation outfitting and hull painting.

4.4.1 Machinery Outfitting

The machinery outfitting is a process of installing machines and equipment necessary for the operation of a ship as shown in **Figure 11-20**. The work comprises the installation of main engines, shaft systems linking the engines and propellers, rudder systems, auxiliary machinery like boilers and generators, steel outfits, etc. It also includes insulation and painting of these items where necessary.

(a) Main engine (head part) (b) Generators (c) Purifiers

Figure 11-20 Machinery outfitting

4.4.2 Hull Outfitting

It refers to piping and steel outfitting work in all areas except accommodation and engine room. The work includes installation of mooring and anchoring equipment, cargo handling gear, masts, hatch covers, etc. as shown in **Figure 11-21**.

(a) A windlass (b) Hatch covers on a container ship (c) A deck crane

Figure 11-21 Hull outfitting

4.4.3 Electrical Outfitting

Electrical outfitting is a work installing equipment and cables for supplying electricity to operate various equipment used for navigation, mooring, cargo handling and accommodation. Electrical outfitting consists of cable laying and installation of various electrical and electronic equipment, such as switching gear, transformers, distribution panels and lighting, just to name a few.

4.4.4 Accommodation Outfitting

An important element of accommodation outfitting is fitting out the wheel house and associated facilities as shown in **Figure 11-22**. Not only the rooms have to be fitted out with windows, doors, wall panels and ceilings, but also a variety of equipment required for monitoring and controlling the ship has to be installed. They include radars, communication equipment, navigational aids, computers and various display units apart from the control consoles. Crew's accommodation and public spaces (including cabins, living rooms, bedrooms, toilets, shower rooms, kitchens, pantries, cold storage, laundry, corridors, stairways, offices and recreation facilities) also need to be fitted out with windows/scuttles, doors, panels, ceilings, flooring, air-conditioning, lighting and so on. In other words, this outfitting work creates a living and working environment and facilities for full hotel services.

(a) Radio (b) Wheelhouse (c) Radar

Figure 11-22 Accommodation outfitting

4.5 Final Hull Painting

Upon completion of erection, the hull receives final painting. Any damage to pre-painting received during erection and welded joints during erection are repaired, and the shell plates, other hull parts, cargo holds and tanks are repainted as shown in **Figure 11-23**.

The shell plates are in direct contact with seawater and they are susceptible to damage, corrosion and fouling. Therefore, the shell plate coating is of utmost importance and must be carried out in an environment of properly controlled temperature and humidity. The painting is repeated at appropriate intervals to allow complete drying until a specified film thickness is obtained. The number of coats for ballast and potable water tanks depends on their usage. The Performance Standard for Protective Coatings (PSPC) came into effect for ships delivered on or after July 1, 2012, and it is a good reminder of the importance of painting. This regulation tries to enhance ship safety by preventing premature damages of paint film in ballast tanks. The post-erection painting is the most important painting process, because it provides the primary protection against hull corrosion and the color scheme largely decides the final appearance of a ship.

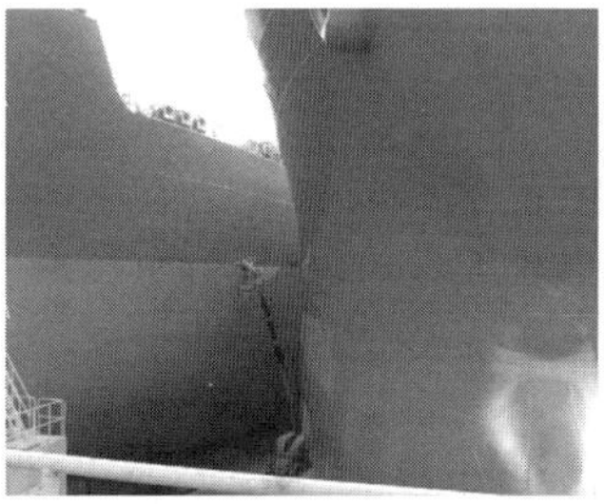

(a) Shell plate painting

(b) Cargo hold painting

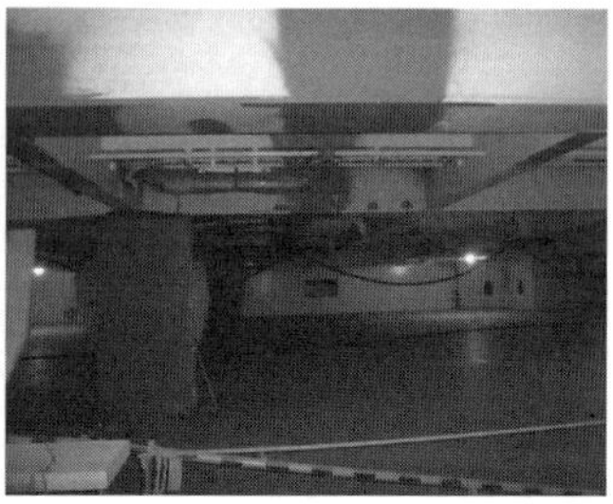

(c) Ceiling and deck painting

Figure 11-23 Hull painting

5. Erection

5.1 Outline

5.1.1 Pre-Erection

Pre-erection (PE) is a process of combining two or more assembled blocks to form a larger one. PE done before and after the pre-painting is called pre-PE (or 1st PE) and post-PE, respectively. Furthermore, PE blocks may be assembled again in a grand assembly process to make mega blocks within the limit of the capacity of erection cranes.

5.1.2 Erection

The erection is the final stage of hull construction process, in which the normal or pre-erected blocks are combined together in a dock to form a hull. The joints between the blocks are welded automatically or manually, and careful checks on the dimensional accuracy are required throughout the process. Ordinarily, the erection process begins with the engine room blocks and takes two to three months.

5.1.3 Berth Erection

In this method the hull is erected on an inclined building berth so that blocks are erected at an angle to the horizon as shown in **Figure 11-24**. Upon completion, the hull is allowed to slide off the berth on rollers or cradles greased with fat. This used to be the main method in the past, but these days its use is limited to the erection and launch of relatively small ships.

Figure 11-24 Berth erection

5.1.4 Dock Erection

In this method the ship is erected in a dry dock as shown in **Figure 11-25**. When the hull construction is finished, the dock is flooded and the hull is simply towed out of the dock. Most contemporary shipyards use this method, because it allows the erection to take place horizontally, making it easier to control the hull shape and there is virtually no risk in launching large ships from docks. Furthermore, a very large dock may allow more than one ship being built simultaneously even at different phases.

Figure 11-25 Dock erection

5.1.5 Floating Dock Erection

Floating docks have usually been used for repair work in the past but recently they have also been used for shipbuilding as well by some shipyards in their effort to increase the

Figure 11-26 Floating deck erection

ship production without having to expand the building berths or dry docks. Instead of large gantry cranes, large crane barges are used to service the floating docks as shown in **Figure 11-26**.

5.2 Erection Work

5.2.1 Arrangement of Keel Blocks

Keel and bilge blocks are placed on the floor of the dock to distribute the weight of a ship as evenly as possible and arranged as shown in **Figure 11-27** to prepare for hull erection. Such blocks are generally made of cast steel or concrete and topped with renewable wood block at the contact surfaces. The blocks are installed about 1.7 to 2 m above the floor and at least 3 m away from the side walls of the dock to give reasonable access to the bottom and side plates for painting and hull outfitting. Its arrangement differs depending on the weight of the ship and each block must be made to support about 40 to 50 tons.

Figure 11-27 Keel blocks

5.2.2 Block Erection

Blocks are lifted into their designated positions in a planned sequence by cranes, as shown in **Figure 11-28**. Erection of blocks weighing more than 1,000 tons is not rare these days, because ships are getting larger and pre-erection of blocks is becoming more common. The location of blocks must be decided after taking careful consideration of the ensuing work. Erection sequence and schedule of each block is made into an erection sequence network diagram and this becomes the basic data to work out the pre-processing (fabrication, assembly, pre-erection, pre-outfitting and pre-painting) plan. The erection usually begins with an engine room bottom block and proceeds towards the bow and stern until the hull is completed.

Figure 11-28 Block erection

5.2.3 Block Positioning

A block to be erected is placed in a preliminary position to check its dimensions (breadth, length and so on) against the drawings and then moved into the proper aligned position. The alignment is finely adjusted to make the joints match up precisely using block lifts and hydraulic cylinders, and this process is known as block setting. Measuring instruments used in block setting include automatic levels, three-dimensional total stations and theodolites; jigs and fixtures include strong backs, T-bars and ring pieces; and tools include 50-ton hydraulic cylinders, tape measures, ink pads for marking and plumb lines.

In particular, three-dimensional total stations can measure three-dimensional coordinates (X, Y, Z) of any point within a tolerance of one millimeter, and consequently contribute greatly to improving the accuracy of block setting and increasing the confidence of the owners and classification societies. During the setting process blocks of poor accuracy may have to be cut for correction and, in such events, special care must be taken to prevent fire.

5.2.4 Block Fitting-Up

Any excess on the edge is trimmed off precisely, the edges are prepared for welding by e.g. beveling, and then the block is fixed temporarily tack welded in position. The tack welding must be done by skilled welders according to the standards, and precautions must be taken to prevent interference with the welding work proper, especially at butt joints.

5.2.5 Block Welding

Prepared blocks are welded into their allotted positions of the hull. Welding can be done either automatically or semi-automatically. **Figure 11-29** shows typical semi-automatic flux cored arc welding (FCAW) machines. FCAW uses continuously-fed consumable tubular electrodes containing a flux core which, together with CO_2 shielding gas, protects the weld metal. It is highly efficient, easy to use in all welding positions and it provides stable arcs

Figure 11-29 FCAW (flux cored arc welding)

in various welding environment. For these reasons, it has now become a standard welding method for shipbuilding. It is good for welding of long butt joints of thick plates and vertical upward welding, and, therefore, it is often used for butt welding between blocks during erection and block assembly process as well.

A typical automatic welding method is electrogas welding (EGW), as shown in **Figure 11-30**. EGW is a continuous vertical position arc welding process applied to vertical butt welding and is also called simple electrogas arc welding (SEG-ARC). EGW is highly efficient, because it can continuously weld long joints at once and at high speed. EGW uses continuously-fed electrode wires (1.6 mm diameter) cored with flux to protect weld area along with a shielding gas, generally CO_2. Because EGW is practiced vertically, pair of water-cooled backing shoes made of copper alloys are attached on both sides of plate along the welding groove to prevent melting metal and slag from leaking down. Occasionally ceramics instead of copper alloys are used for the shoes, or the back beads formed by FCAW may replace the shoe on that side.

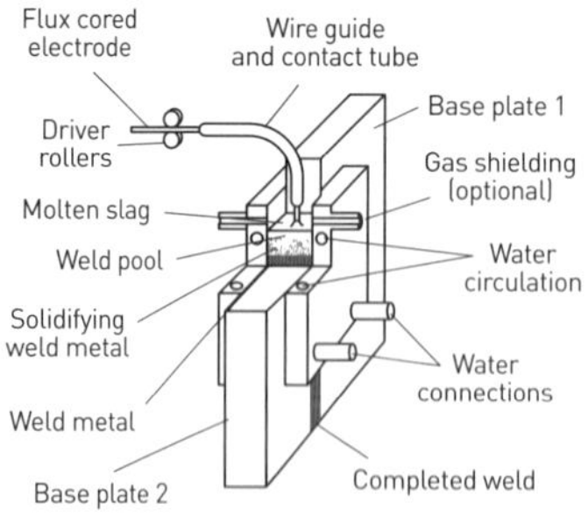

Figure 11-30 EGW (electrogas welding)

5.2.6 Inspection

Inspectors from the classification society and the owner's representatives examine completed sections of the hull. The inspection mainly focuses on the hull structure and welding, and the inspection includes nondestructive testing (NDT), leakage test with pressurized air or water and strength test.

6. Painting Techniques and Paints

6.1 Painting Techniques

Painting is the final process of ship construction and one of the core processes affecting the life span and performance of a ship. Painting has been done by conventional methods until now, but as the shipbuilding output per employee of Korean shipbuilding industry is continuously increasing, automation of painting is required to sustain competitiveness of the industry. Automated robotic painting technology is important not simply because it replaces workers but because it overcomes the human limitations in achieving high quality precision painting. Although using robots is common place in car manufacturing, it is expected to take considerable time to realize robotic painting in shipbuilding yet, because the painting work in shipbuilding is not as repetitive as in the automobile industry.

6.2 Anti-fouling Paints

When a ship floats in seawater for a period of time, marine organisms, such as slime, sea weed and barnacles, would attach onto the surface of the ship and grow rapidly. These organisms increase the ship's resistance and weight, leading ultimately to a speed loss and increase in fuel consumption. In order to reduce such detrimental effects of marine biofouling, diverse anti-fouling paints (AFPs) can be applied to the lower parts of the hull.

Recently low resistance anti-fouling paints have been developed in response to the rising oil prices and introduction of more stringent CO_2 emission regulations. Furthermore, demands for long-life anti-fouling paint products have been increasing due to the extended dry-docking program which allows the mandatory dry docking for regular inspection and maintenance to take place every 90 months. AFP product technologies have been evolved from the past as shown in **Table 11-1**.

Table 11-1 Development of anti-fouling paints

Period	Main measures
Age of sail	Pitch coating, copper cladding
~ 1700s	Lead cladding was used from ancient times but British Admiralty prohibited its use in 1682 due to the electrolytic corrosion of iron components of the ships
~ 1860s	Metallic salts based on copper sulfate
~ 1970s	Rosin mixed with copper oxide and mercury oxide
1970s ~	Mixture of copper sulfate and various natural or synthetic resins 1) Conventional type: uses insoluble resin compounds for sea water 2) CDP (controlled depletion polymer) type: uses soluble resin compounds 3) SPC (self-polishing copolymer) type: uses organic tin-based resin compounds
2000 ~	ISO regulations for protection of marine ecology coming into force 1) Tin-free SPC type: tin is substituted by copper, zinc and silica in self-polishing resins 2) Foul-release type: uses silicon with low surface tension to discourage adhesion of marine organisms IMO prohibits the use of TBT (tributyltin) for protection of marine ecology (2003) 1) Tin –Free SPC AFP : tin is substituted by copper, zinc or silyl acrylates 2) Foul-Release AFP : resin with low surface tension prevents adhesion of marine organism

Nowadays, tin-free self-polishing copolymer (SPC) AFPs are the most widely used and they release biocides by hydrolysis or ion exchange reaction of the acrylic polymer with seawater. The use of biocide-free eco-friendly foul release type AFPs is increasing in response to the rising oil prices and various environmental regulations. Conventional type and controlled depletion polymer (CDP) type AFPs are used together for some coastal and other small ships mainly for economic reasons. **Figure 11-31** summarizes the properties of AFPs.

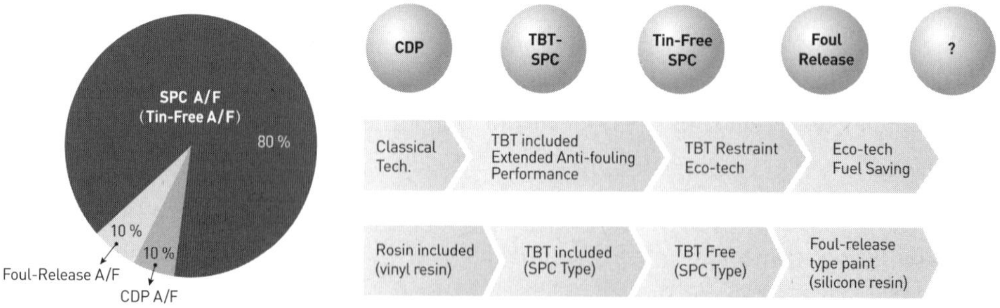

Figure 11-31 Market trend of anti-fouling paints and direction of their development

6.2.1 Conventional AFPs

The conventional AFPs are mixtures of biocides and insoluble materials such as polyvinyl resins or chlorinated rubber. Their anti-fouling performance degrades with release of the biocides and repainting is needed within two years. The price of the paint itself is not too high but substantial extra expenses occur in repainting because the remove the paint film flake left behind after the biocides leached away has to be removed first.

6.2.2 Controlled Depletion Polymers (CDPs)

The controlled depletion polymer AFPs are composed of biocides and slightly water-soluble binders, usually mixtures of rosin and polyvinyl methyl ether and vinyl chloride. They can provide anti-fouling protection for up to 30 months, but the dissolving rate of the rosin is non-uniform, leading to a rougher surface over time. Therefore, sealer painting or recoating after removal of the leached layer is needed during maintenance work to reduce the surface roughness.

6.2.3 Self-Polishing Copolymers (SPCs)

Self-polishing copolymers are non-soluble when dry, but undergo hydrolysis with sea water to become water-soluble and form a thin swelling layer of acid polymer. This layer gets minutely worn away by the friction with sea water when the ship is on the way. Although the SPCs are expensive, they provide anti-fouling protection up to 60 months and maintain smooth surface throughout their life, reducing fuel consumption and maintenance and repair costs. Furthermore, recoating requires less preparation than CDPs since leach layer is thinner and easier to remove.

The SPCs containing TBT-derivatives were used until 2002, but tin-free SPCs containing copper, zinc or silyl acrylates dominate the current market since IMO regulations no longer allow TBT-based paints.

6.2.4 Foul-Release Coatings (FRCs)

The foul-release coatings are biocide-free and the most eco-friendly. Foul-release paints are based on resins with very low surface tension, such as silicon elastomer or fluoropolymer, and provide a very smooth, slippery, low-friction surface onto which fouling organisms find it difficult to adhere. Even if any are successful to attach themselves to the surface, they are easily removed by friction with seawater when the ship is on the way. This type of paint is expensive, because it uses more expensive materials and the workability is poorer than other AFPs, but it is non-toxic. These paints have been gaining popularity, because they are eco-friendly and can provide smooth hull surface resulting in lower fuel cost.

6.3 Corrosion Protection

Corrosion protection is a system used to protect ships from corrosive environment, and the primary anti-corrosion measure is painting which produces a thin protective film isolating sea water from the hull surface. The fact that the marine environment is extremely corrosive and that controlling the painting environment outdoors is difficult has made the use of oil-based paints instead of water-based emulsion unavoidable.

However, due to the strict regulations on the emission of volatile organic compounds (VOCs), wider use of non-solvent paints or emulsion paints with comparable performance of oil-based paints is actively sought. Paints are basically synthetic chemicals which might contain substances harmful to the environment and humans, and consequently the production process and the materials used are strictly controlled worldwide. The paint industry recognizes this problem and is striving to develop environment-friendly paints. The users can also contribute to this process by not being simply content with what they have been using but actively looking for better products. This will encourage the paint manufacturers to develop and improve the environment-friendly paints continuously.

It is expected that the energy efficiency can be improved by enhancing the durability and effectiveness of the marine paints. Wider public awareness and continuous efforts are needed if we are to develop the ultimate eco-friendly paints which can eliminate environmental contamination from the hull coatings.

7. Special Ship Construction Techniques Developed in Korea

7.1. On-Land Construction Methods

In 2008 Korean shipbuilding industry had been experiencing a boom, and most shipyards had completely full order books with no spare dock capacity, forcing several shipyards to give up additional orders. To overcome this bottleneck, some forward thinking shipyards in Korea devised innovative construction methods which do not require docks. These methods broke the fixed concept that ships must be built in a dry dock or on a building berth. It heralded a new era in the advancement of shipbuilding technology in that it removed the constraints put on the building capacity of a shipyard by its dry dock capacity; they decided to build ships on land away from the docks.

It has been proved that the on-land construction methods work: various large marine structures, including drilling rigs for major oil companies, were successfully built on land and launched. Hyundai Heavy Industries was the first shipyard who applied the on-land construction method to building of large ships.

In the on-land construction process a completed ship is loaded on a launching barge moored to a quay, and the barge is towed by tugboats to a launching area at sea. Launching of the ship is then accomplished by submerging the barge so that the ship can safely float off.

Loading out the ship onto the launching barge is very delicate and technically demanding. The ship can be loaded out either longitudinally or transversely. Most shipyards use the longitudinal method in which ships are lifted with the so-called hydro-wheel system and then loaded longitudinally onto the launching barge with winches. On the other hand, Hyundai Heavy Industries use the transverse method in which the ship is lifted with air-pads and loaded onto the launching barge with skids as shown in **Figure 11-32**.

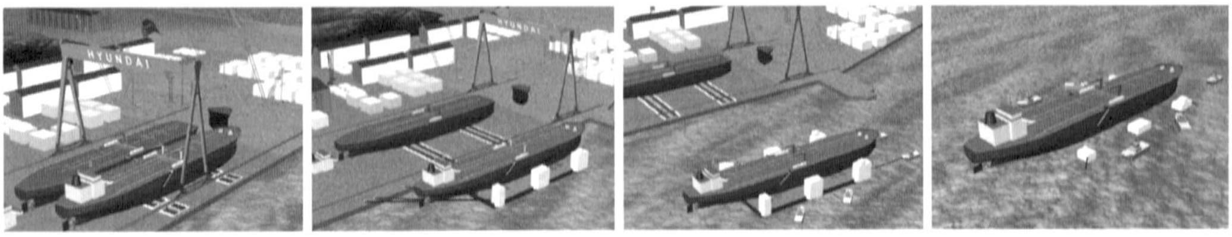

Figure 11-32 Transverse loading out and subsequent launching of a ship constructed on land (Source: HHI)

7.2 Floating Fore Body Erection Method (the 'Dam' Method)

Ordinarily, there is a limitation in the maximum size of ships a shipyard can build due to its facilities and dock sizes. Therefore, shipyards cannot respond positively to any invitation for bid for ships exceeding its facilities and capabilities.

Hanjin Heavy Industries has a dock the length of which is limited to 300 meters due to its geographical location. As a consequence, the company was unable to build ships longer than that length. To overcome this limitation, Hanjin developed a new erection method in which the fore body of a ship is welded onto the main hull afloat after it is launched. The method was successfully applied to the building of an 8,100 TEU container ship which is 25 meters longer than the dock length. It was hailed a remarkable new technology for ship erection and was admired by the shipbuilding industry all around the world.

In the Dam method, as large a portion of the hull as can be accommodated by the dock is assembled in it and the rest of the hull (typically fore body) is then erected separately in the dock at a place where a floating crane can access. The incomplete main hull is launched, and the fore body block is welded to it inside a watertight cofferdam (DAM) shown in **Figure 11-33**. A similar method was used by Scott Lithgow of Scotland in the 1970s and 1980s when they built a VLCC in two halves which were then welded together afloat using a cofferdam.

A schematic of the Dam method is shown in **Figures 11-34** to **11-37**.

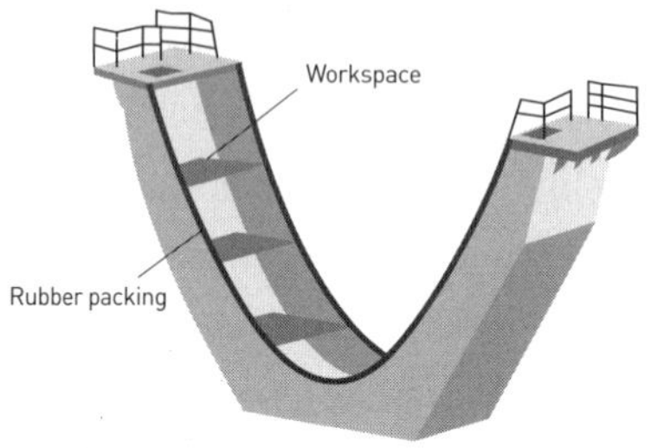

- Dimension : 3m × 14m × 8m
- Weight : 30tons
- Passageway : Inclined ladder 2ea.
  Emergency ladder 2ea.
- Ventilation : 2ea.
- Drain pump : 150tons × 4ea.
- Rubber packing : Watertight

Figure 11-33 The a cofferdam used in the floating fore body erection method
(Source: the Dam method, HHIC)

① 1st stage: Blocks which do not belong to the fore body are erected in the dock to form the incomplete main hull and, upon completion, a watertight cofferdam (Dam) is attached to the joint line with the fore body (**Figure 11-34**).

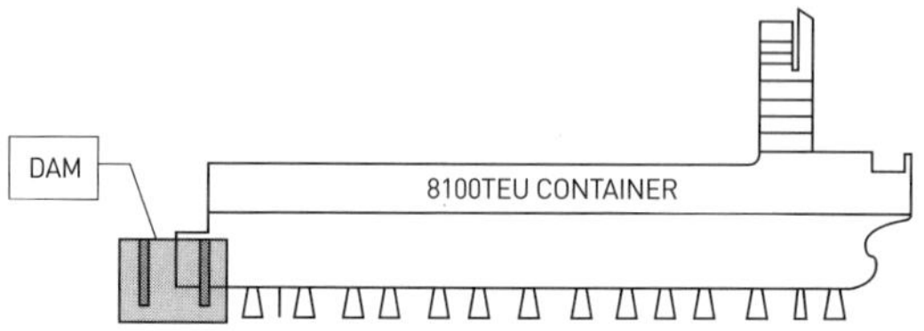

Figure 11-34 1st stage: In-dock erection of the main structure and installation of the cofferdam

② 2nd stage: The dock is filled with water and the dock gate opened to float out the incomplete main hull in preparation for receiving the separately erected fore body at sea.

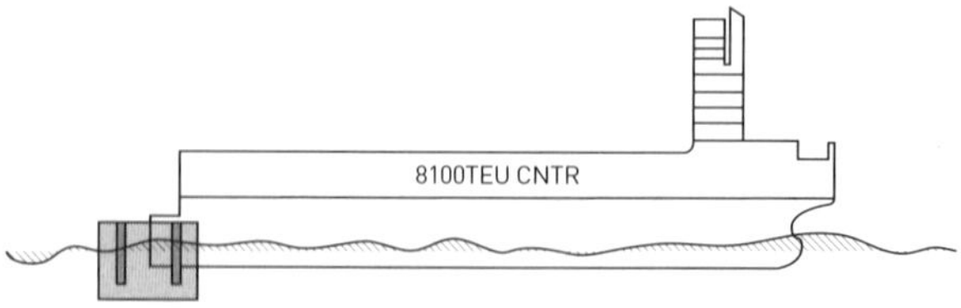

Figure 11-35 2nd stage: Floating of the main structure and launching

③ 3rd stage: The bottom block of the fore body is lifted with a floating crane and moved so that the two parts are aligned near the dock gate. The fore body block is tightly fit to the Dam already installed at the front of the main hull structure. The water in the Dam is drained to secure areas for welding work and the two parts are welded together.

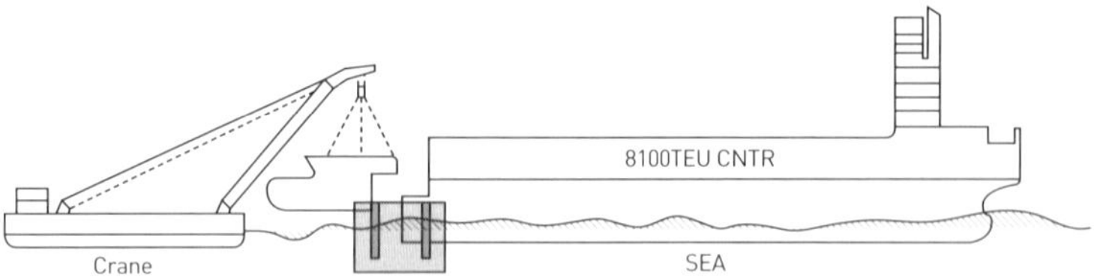

Figure 11-36 3rd stage: Connecting the first block of fore body to the main hull structure

④ 4th stage: The Dam is removed and erection of the remaining upper blocks of the fore body proceeds until the hull is complete.

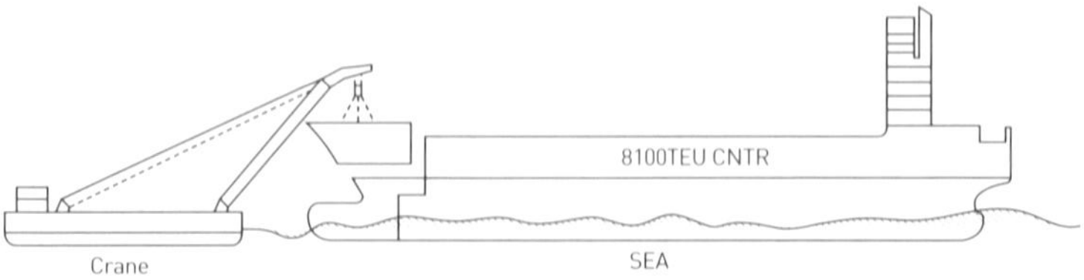

Figure 11-37 4th stage: Erection of the upper blocks after removal of the Dam

7.3 Extra-Large Block Method (the Mega-block Method)

In this method, several blocks are combined on land to form extra-large (mega) blocks which can be erected in a dock with the help of floating cranes. An entire hull can usually be subdivided into about ten mega blocks. Each of them is 'pre-erected' on land from several smaller blocks, and, when outfitted and painted, is erected in the dock. This method reduces the present average dock period of 2.5 months to 1.5 months and regarded as an innovative shipbuilding method greatly improving the productivity.

7.3.1 Evolution of the Mega-Block Method

Mega-blocks weighting up to 3,000 tons must be moved to the dock and lifted for erection. Therefore, some shipyards have decided to develop a suitable transporter of 3,000 tons capacity. They researched the heavy cargo transportation systems available around the world and carried out a comprehensive study of the problem through computer simulations. As a result a transporter was produced by combining three 1,000 ton-class transporters and equipping the whole with a micro-control system. It was successfully used in shipbuilding work in docks.

It must mean that the preconception of ships having to be built in dry docks has now become obsolete. However, the mega-block method had been limited by the capacity of dock cranes available. The essence of this technology is in that the mega-blocks weighing well over the dock crane capacity can be moved either on the ground or by the sea and erected in the building dock or at sea using alternative transporters. The methods have evolved now to the stage where much larger blocks are built and erected. Samsung Heavy Industries, for example, has recently developed the so-called giga-block method which uses two or more floating cranes to erect very large super-blocks weighing more than 5,500 tons, and even the much more ambitious tera-block method with ultra-large blocks weighing as much as 10,000 tons, as illustrated in **Figure 11-38**.

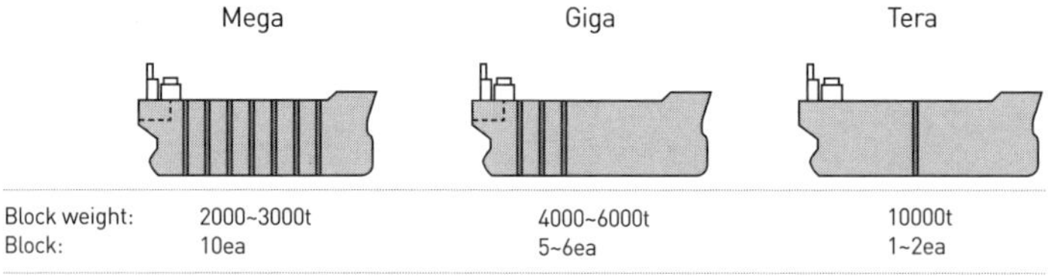

Figure 11-38 Mega-, giga- and tera-blocks (Source: SHI)

7.3.2 On-Land Skid Launching

A skid launching system (SLS) was developed and applied by STX Offshore and Shipbuilding Co. to build ships on land. After a ship is erected on land, the ship is made to slide along skid rails by a hydraulic transporter and is loaded onto a skid barge moored alongside a quay next to the building site. Then, the barge is towed to a launching location and submerged to float off the ship. Larger ships may be built in two separate parts at convenient locations. The two parts can then be joined together once they are moved onto the barge.

The method has now evolved into the stage where even large ships are built as a whole on land and transferred to the skid barge to launch upon completion, as illustrated in **Figure 11-39**. This skid launching method maximizes the overall productivity of the shipyard by building a hull in three months on land without occupying a dock, and at present time methods similar to the SLS is practiced by many shipyards.

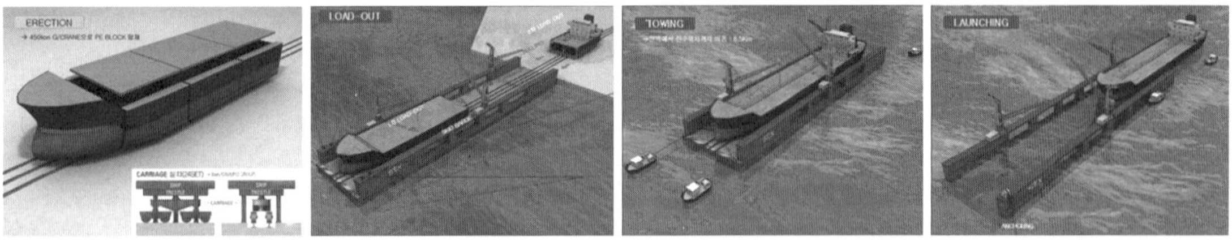

Figure 11-39 A skid launching system (Source: STX)

8. Launching

8.1 Definition of Launching

Launching is a process of floating a newly constructed ship in the water for the first time. When the ship is built on a building berth, it is known as berth launch, and from a dock, dock launch. In berth launching, the hull weight which has been supported by wooden blocks or pillars is transferred to a launching cradle or slip way which are heavily greased. When released, the ship then slides into the water under its own weight. During the process, the heavy ship structure is subjected to strong impulsive pressure and there is a danger of capsizing as well. Therefore, a berth launching must be planned meticulously after careful calculations and based on past records and experiences. The work must be conducted by skilled and experienced workers. The dock launching floats the hull by filling the dock with seawater. A precise ballast plan to correct trim and heel must be made to avoid damages to the ship's bottom when the ship is about to float up.

8.2 Type of Launching

Each shipyard must decide the best launching method suitable for its facilities, scale, ships being built, geographical layouts and other constraints.

8.2.1 Berth Launching (Gravitational Launching)

1) Longitudinal (stern first) launching

This type of launching system had been used by most shipyards throughout the world pre-

VLCC era. Even in modern times many smaller shipyards which do not have building docks still use it. The ships built on an inclined building berth with the bow pointing inland will clearly slide down stern-first when released, and hence the alternative name of stern-first launching. The process is shown in **Figure 11-40**. The reasons for launching stern-first are:

- Buoyancy of the ship increases faster than if launched bow-first.
- When the hull slides into the water, the buoyancy pushes the part in the water upwards, transferring the support of the ship's weight to the landward end of the ship. The stem has stronger structure than the stern and is therefore better suited to take the extra load.
- It is more convenient for the installation of stern frames, propellers, rudders, etc., if the stern faces the water.

Figure 11-40 Berth launching (longitudinal launching)

2) Side launching

In this method the ship is built so that it is parallel to the waterfront, and is allowed to slide sideways into the water when completed. It is used by small ship builders and shipyards facing a narrow waterway, such as rivers and canals. For example, many medium sized shipyards on the Rivers Clyde and Tyne in the UK were forced to use this method. The slope of the building berth is as high as 1/6 ~ 1/10 which is much steeper than the slope of longitudinal building berths. Therefore, the ship is likely to experience a high heel angle during launch. Thus, it is important to check carefully the height of the center of gravity and ensure water tightness of all the openings likely to be underwater during the process. Constructing a ship on this type of building berth is simpler than on a longitudinal launching berth, because the keel can be laid down horizontally and the frames and bulkheads can be erected vertically.

8.2.2 Dock Launching (Floating-out Launching)

Dock launching is the simplest method, but, of course, it requires the ship to be built in a dock. Since most large modern shipyards possess building docks, it is natural that they practice this form of launching. The completed ship is floated by flooding the dock as shown in **Figure 11-41**. Careful attention must be taken, however, to prevent excessive trim or heel of the heavy hull structure, when water is pumped into the dock and the hull begins to float.

Construction of a dry dock is very costly and takes a long time, and they are expensive to maintain too. Nevertheless, large shipyards keep a number of docks, because they are still the basic facilities for building large ships, despite the recent innovative shipbuilding methods not relying on the docks.

Figure 11-41 Dock launching (floating-out launching)

9. Inclining Experiments

When a ship is newly constructed or has just gone through an extensive repair/modification work, an inclining experiment must be performed on the ship to determine the vertical location of her center of gravity. The result is also useful in reaffirming or otherwise evaluating the estimations made during design. The result of the test forms the basis of determining the vertical center of gravity for all loading conditions. Since the height of the vertical center of gravity is a crucial factor in determining the ship's transverse stability, the inclining experiment is indispensable in ensuring ship safety. For this reason, an early part of the Trim and Stability Particulars (also known as Stability Booklet) which must be carried by all ships is dedicated to the procedures used for the inclining experiment and the results obtained therefrom.

9.1 The Principle of Inclining Experiment

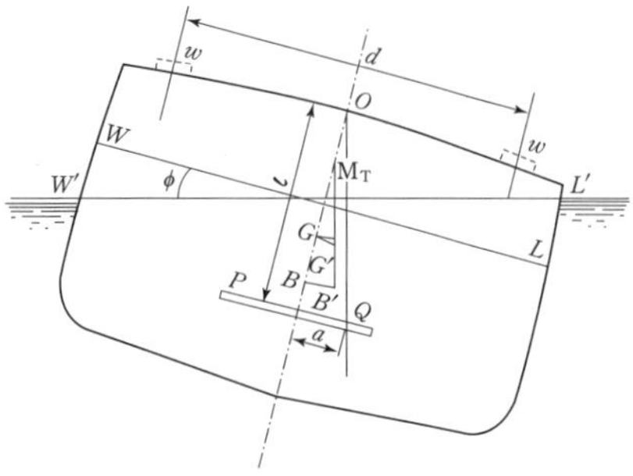

Figure 11-42 Inclining experiment

An inclining experiment of a ship is conducted by observing how much the ship heels when a known magnitude of heeling moment is applied. For small or medium sized ships the heeling moment can be provided by moving a few pairs of certified test weights, but large ships, or special type of ships which do not allow any additional weights, have to utilize their ballast tanks for this purpose.

The explanation of the inclining experiment below is given for the cases where test weights are used.

If a test weight with mass w is shifted transversely by a distance d on the deck of a ship of displacement Δ, the center of gravity G also moves transversely to a different position G' and, as a result, the ship will heel by an angle ϕ as shown in **Figure 11-42**. The distance GG' and the heeling moment have the following relationship:

$$w \times d = \Delta \times GG' \qquad (11\text{-}1)$$

If the heel angle ϕ is small, the distance GG' is approximately

$$GG' = GM_T \times tan\phi \qquad (11\text{-}2)$$

where M_T is the transverse initial metacenter.

From the **Equations 11-1** and **11-2**, we can find the transverse metacentric height GM_T as

$$GM_T = (w \times d) / (\Delta \times tan\phi) \qquad (11\text{-}3)$$

Then, KG, the height of the ship's center of gravity is given by

$$KG = KM_T - GM_T \qquad (11\text{-}4)$$

The value of KM_T can be found from the hydrostatic tables or curves.

The inclining experiment must be performed in accordance with a carefully prepared procedure. Before starting the experiment itself, the displacement of the ship as inclined must be determined. This is done by reading the drafts of the ship at various points. Based on the hydrostatics corresponding to the drafts, the lightship displacement can be determined by adding weights still to go on and taking off the weights of all items which are to come off the ship including the inclining weights. The contents of all tanks and their centers of gravity and specific gravity must also be recorded so that adjustments can be made later in producing the lightship displacement. The deadweight of the ship (cargo carrying capacity) can then be determined from the fully loaded displacement and the lightship displacement.

9.2 Preparations for Inclining Experiment

9.2.1 General Conditions of the Ship

① The weather conditions are to be as calm as possible, and in any case not any worse than Beaufort scale 3 (BF3) as defined in **Table 6-1**. The ship must be moored in a quiet, sheltered area so that the disturbance from wind, wave, tide, ships sailing nearby, etc. is minimal.

② Mooring ropes are loosened to minimize the influence of their tension on the ship.

③ The ship must be as upright as possible with the heel angle not more than 0.5° and trim not more than 1.5 percent of the length between perpendiculars.

④ All tanks must be completely emptied or filled, if at all possible, to minimize free surface effects. The tanks with free surface must be recorded carefully so that the allowances for free surface correction can be made later.

9.2.2 Provisions for Inclining Experiment

① Reference documents and drawings: inclining experiment procedure, hydrostatic curves, general arrangement plan, tank sounding table, etc.

② Test equipment: a dinghy (to provide access to draft marks), draft measuring instrument, seawater sampler, hydrometer, thermometer, plumb lines, communication equipment, flashlight, recording papers, writing instruments, etc.

9.3 Lightship Displacement Measurement

As stated above, the displacement of the ship as inclined is adjusted by adding the masses of items still to go on board and subtracting the masses of items which should be taken off. The following items are included in the lightship displacement in addition to the mass of the bare ship:

- water in boilers and condensers;
- main engine and boiler cooling seawater;
- hydraulic oil for steering gear;
- lubricating oil for main engine coolers; and
- all equipment required by shipping laws.

9.4 Inclining Experiment

The inclining weights are moved transversely in several distinct movements and the angles of heel corresponding to all the movements are measured. The mass and the center of gravity of the test weights must be certified by the classification society before use.

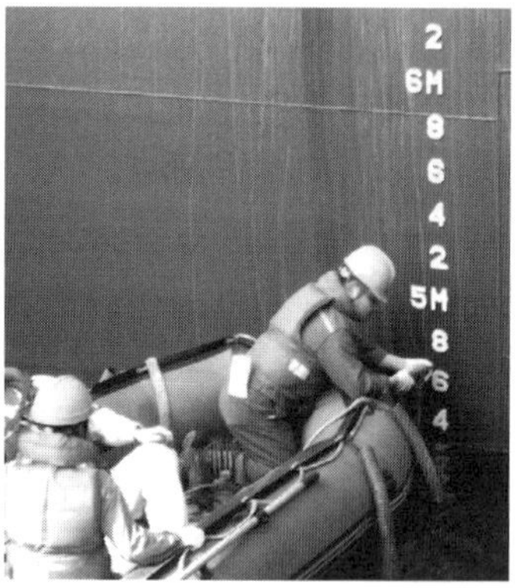

Figure 11-43 Draft measurement

9.4.1 Measurement of Drafts

Draft measuring instruments are used for reading the drafts which are used for estimating the displacement as shown in **Figure 11-43**. The readings are taken at the draft marks which are usually at bow, stern and midship, both port and starboard.

9.4.2 Installation of the Plumb Lines and Measurement of Heel Angle

At least two plumb lines are used so that bad readings can be identified and ignored. The plumb lines are set up as shown in **Figure 11-44**, and the deflection of the lines at the position of the wooden batten is read. The tangent of the heel angle is the deflection divided by the 'pendulum' length. Ordinarily, the inclining weights are moved eight times and the consequent tangent values are recorded as shown in **Figure 11-45**. From this GM_T and KG can be calculated using relationships **Equation 11-3** and **Equation 11-4**.

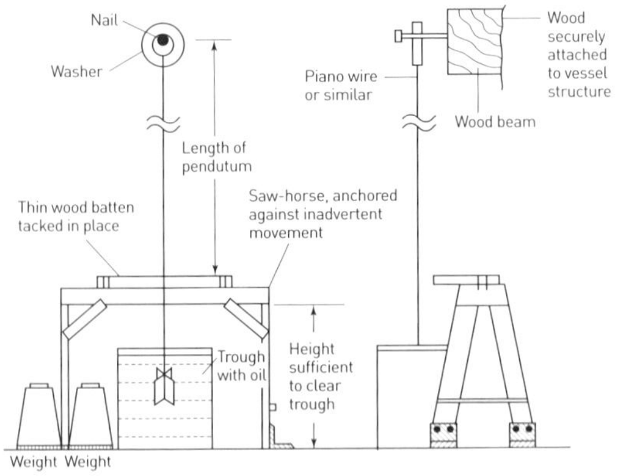

Figure 11-44 Installation of the plumb lines

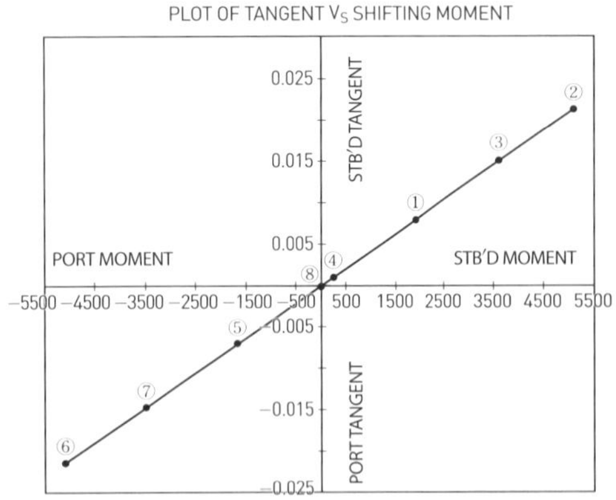

Figure 11-45 Results of inclining experiment

9.4.3 Care to Be Taken in Inclining Experiment

The inclining experiment is performed to find one of the items critical to a ship's safety and, therefore, it has to be conducted with utmost care with attention to details and accuracy of measurements. The following points must be noted:

- Personnel not essential for the experiment must come off the ship and those who remain on board must not move about, particularly while heel angles are read.
- Those who are reading the drafts must wear life jackets for safety and take care when the boat touches the ship.
- The plumb bobs must be kept in a bucket of water or oil to dampen high-frequency oscillations.

10. Sea Trials

Trials are performed to check the performance of the newly constructed ship is up to the standards as specified in the contract in every respect. A ship that has undergone a major repair must also be put through similar trials. There are two parts. Quayside trials which are conducted while the ship is moored to a quay to test the performance of machinery, equipment, pipework, electrical systems, instruments and various other systems which can be tested while stationary. Sea trials are performed to test the ship in its running condition, and must be conducted in conditions similar to the ones in which the ship is expected to operate.

10.1 Typical Tests in Sea Trials

1) Main engine starting test

Marine engines are started with the help of compressed air in the starting air tanks filled up by the main compressors. According to the regulations of classification societies, two starting air tanks must be installed, each capable of starting the main engine at least 12 times without being recharged by compressors.

During the test, the main engine is repeatedly started ahead and astern from the wheelhouse until the pressure in the engine starting air tanks falls below the blocking threshold, and then it is repeatedly started from the engine control room until the engine cannot be started any more.

2) Main engine load up & bunker change test

This test is the very first time that the main engine and the propulsion systems are operated and a load is applied to the system in open sea. The bunker change refers to the procedure of replacing diesel oil with less expensive heavier oils to save fuel costs. This process is tested during the sea trials.

3) Main engine safety test

Lubricants or coolants protect engines in their operation. Automatic slow down tests and automatic shutdown tests are conducted to check that these systems function normally.

4) Blackout restart test

A blackout can occur due to malfunctioning generator engines or electrical distribution systems. If a blackout occurs, emergency generators must be activated immediately and electrical supply must resume to numerous essential equipment and machinery, such as pumps, auxiliary machineries and navigational instruments, so that the ship can continue on its way with minimum disruption. This test is conducted to check that this arrangement functions correctly.

5) Dead ship start-up test

The dead ship start-up test is conducted to check emergency procedures and the recovery time required for a dead ship to return to the normal condition.

6) Anchoring test

The anchoring test is conducted to check anchoring and lifting performance of the ship's anchors. The anchors must descend smoothly to the sea bed when anchoring and be able to provide normal braking force when the ship tries to stop in emergency. Winches for lifting the anchors must have enough power to lift them faster than required by regulations.

- Minimum test length: three shots (about 82.5 meters).
- Lifting speed: 9 meters per minute or faster.

7) Crash astern test

Ships sometimes need to stop or go astern in an emergency situation. The crash-astern test is conducted to evaluate the ability of the ship's machinery to respond to full astern command from steady full ahead. Time taken for the engine and propeller to have developed a steady full astern rpm is recorded from the time the command is given. The stopping distance is also recorded.

8) Bridge control test

The bridge control system enables the engines to be controlled from the wheelhouse. The test checks if the bridge control system is capable of controlling the engine, and thus the propeller, accurately by testing it for a variety of scenarios, for example the engine or propeller rpm for entering and leaving ports.

9) Speed trial

The speed trial attracts the keenest interest from the owner, because it evaluates the factor which directly affects the cargo transport costs. Ship speed is usually measured these days by using Differential Global Positioning System (DGPS). Trials on a calm sea are desirable, as ship speed is highly sensitive to sea conditions, and in any case should not be conducted in seas rougher than sea state 3 Beaufort scale. If it is unavoidable to conduct the test at sea state between 3 and 6, sea state corrections must be made to the results in accordance with the terms of the relevant contract.

The speed is measured by making several consecutive runs in both directions over a prescribed course such as shown in **Figure 11-47**. The ship must approach the start of the measurement in a straight line at a steady propeller rpm for a sufficient distance to ensure zero acceleration before commencing each run.

Photographs of sea surface for the Beaufort scale 3 and 4 are shown in **Figure 11-46**. Experienced crew can tell the scale just from eye observation.

(a) Sea state 3 Beaufort scale

(b) Sea state 4 Beaufort scale

Figure 11-46 Feasible sea surfaces for speed trial

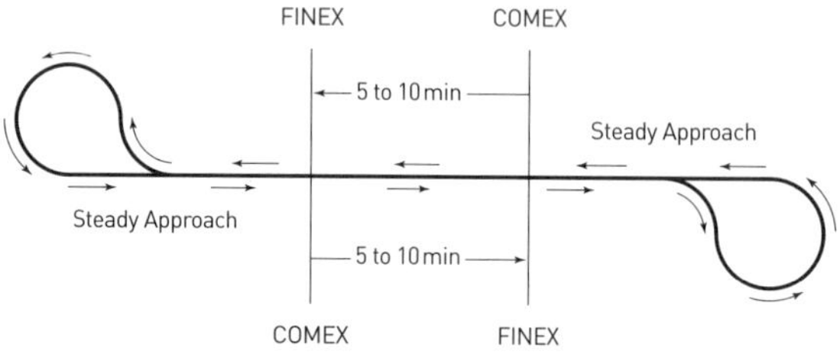

Figure 11-47 A typical path of ship during a speed trial

10) Maneuvering tests

In 2002 the International Maritime Organization (IMO) approved the Standards for Ship Maneuverability which specifies the type of standard maneuvers to be conducted and the associated criteria, and encouraged the application of these standards for vessels constructed on or after January 1, 2004. The maneuvering trials consist of:

- a turning circle maneuver to check the turning ability of a ship as shown in **Figure 11-48**;
- a Z-maneuver or zigzag test to measure the initial turning, course changing and yaw-checking abilities of the ship as shown in **Figure 11-49**;

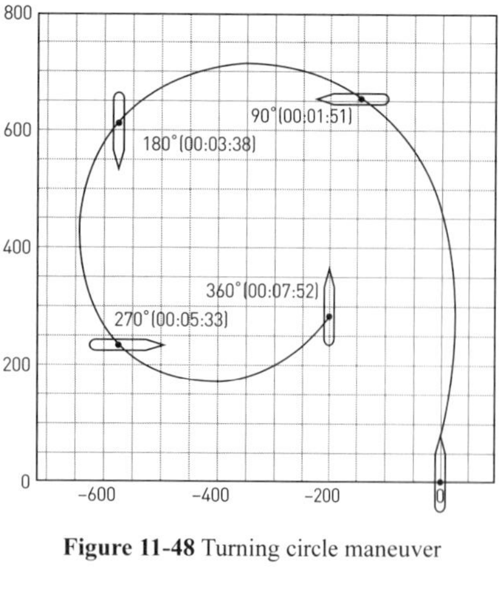

Figure 11-48 Turning circle maneuver

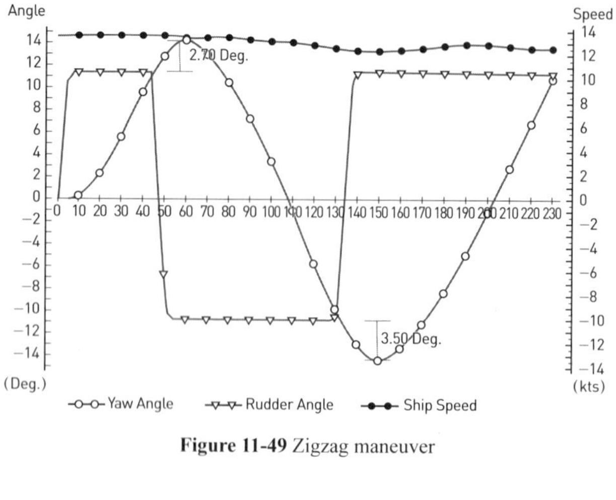

Figure 11-49 Zigzag maneuver

- a man-overboard maneuver to confirm the ability of the ship to go back to a point it previously passed through for the purpose of recovering a person-overboard; and
- a stopping inertia test to verify stopping ability of the ship through a crash astern maneuver performed after a steady approach at the full speed until the ship starts going backward.

More details on the maneuvering trials may be found in Chapter 6.

11) Steering gear test

Steering gear is one of the most important equipment controlling the ship's rudder and its failure can lead to the loss of the ability to control the ship. Under normal service speed, the time taken to move the rudder from hard over (usually 35°) on either side to 30° on the other side should not exceed 28 seconds when using the main steering gear. The auxiliary gear must be capable of moving the rudder from 15° to 15° within 60 seconds at one half of the service speed or 7 knots, whichever is greater.

12) Firefighting test

The firefighting test is conducted to confirm normal operation of the firefighting system installed in a ship. It must be demonstrated that the fire pumps and emergency fire pumps can provide enough sea water at sufficient pressure to all the fire stations including the engine room, when the pressure in the firefighting main drops below a threshold.

13) Cargo pump capacity test

The cargo pump capacity test is conducted to check the capacity of the loading and unloading pumps installed in oil tankers. Because it is not practicable to fill up cargo tanks with cargo oil just for the test, it is conducted with seawater instead. After pumping a certain amount of seawater into the tanks, the height of water surface is measured using the cargo monitoring system to check whether the actual volume transferred is the same as the measurement. The inert gas system prevents fire and explosion in the cargo tanks, and the function of this system is also tested here.

14) Unattended machinery space test

This test is conducted to check if it is possible to monitor, operate and control effectively the unattended engines remotely from the bridge and wheel house. The operation of various alarm systems in the wheelhouse is also tested.

15) Endurance test

The endurance test examines the operating conditions, maintenance of heat balance at various temperature settings and fuel consumption rates of equipment, including the main engine running at the maximum (MCR) and normal continuous rate (NCR).

- Engine performance measurement: the main engine output as a function of rpm.
- Fuel oil consumption (FOC) measurement.
- Fresh water generator capacity test.
- Onboard electrical load measurement, noise measurement, exhaust gas boiler (EGB) capacity test and EGB safety valve test.

16) Torsion meter calibration

A torsion meter is built of strain gauges and is attached to the main engine shaft to measure its shaft horsepower. The schematics of the torsion meter are illustrated in **Figure 11-50**. Zero shaft torque does not necessarily produce a zero reading on the meter, because the shaft rests under the influence of the friction in the stern tube, shaft bearings and engine. Consequently the torsion meter must be finely adjusted to read zero for zero shaft torque by the drag shaft tests or turning gear tests.

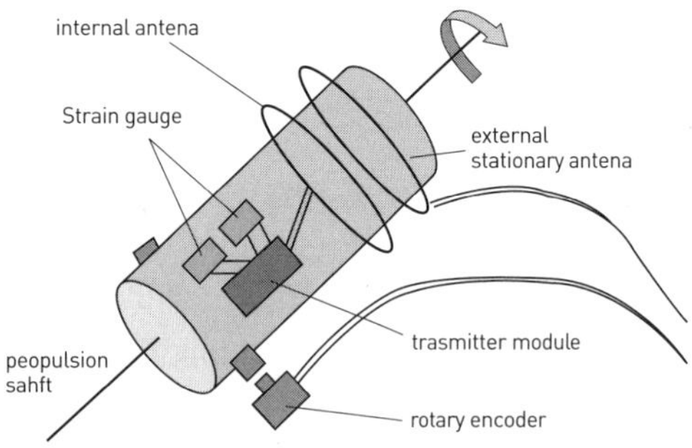

Figure 11-50 Torsion meter of main engine

17) Measurement of global vibration

The global vibration is measured to examine the level of vibration induced by the main engine running at various rpms. If the vibration level exceeds an allowable limit, corrective measures must be taken.

18) Measurement of local vibration

The local resonant vibrations induced by all equipment which can act as vibration sources, such as engines, propellers, generators, etc. are checked.

19) Measurement of torsional vibration

The torsional vibration of the propeller shaft arises from the discrepancies between the intermittent ignition and consequent fluctuation in torque of the engine and the reaction of the propeller. If the frequency of this vibration coincides with the natural frequency of the shaft, large amplitude vibration can occur. The shaft natural frequency is determined by observing the shaft vibration whilst slowly increasing the rpm. The resonant rpm is also called the whirling rpm or the whirling speed and the engine control system must be adjusted to avoid this rpm or speed of the engine.

Sea trials are important events happening once in a lifetime for the majority of the ships. The trials are conducted for the purposes of providing the ship's master with guidelines and data for operating the ship and fulfilling contractual obligations between the builder and the owner. Furthermore, the trials data constitute a set of full-scale performance data which are extremely valuable for determining model-ship correlations. The data will also add to the knowledge base of the shipyard which can be referenced in future designs.

References

1. Hyundai Heavy Industry, 2010. *Understandings of shipbuilding industry and its present status.* (Internal Report). Ulsan: Hyundai Heavy Industry.

2. Kim, H. et al., 2006. *Advanced Ships of Korea* (in Korean). Seoul: Jisungsa.

3. Production Technology Research Committee, the Society of Naval Architects of Korea, 2002. *Shipbuilding Handbook: Assembly work* (in Korean). Seoul: Korea Shipbuilders' Association.

4. Production Technology Research Committee, the Society of Naval Architects of Korea, 2003. *Shipbuilding Handbook: Erection and outfitting* (in Korean). Seoul: Korea Shipbuilders' Association.

5. Song, Y.J., et al., 2008. Shipyard layout design system by simulation (in Korean). *Journal of the Society of Naval Architects of Korea,* 45(4), pp.441-454.

6. The Society of Naval Architects of Korea, 1993. *Introduction to Naval Architecture* (in Korean). Seoul: Dongmyungsa.

7. The Society of Naval Architects of Korea, 2001. *Shipbuilding Engineering* (in Korean), 2nd ed. Seoul: Dongmyungsa.

CHAPTER 12

Production Management

1. Production Planning
2. Production Management System
3. Simulation of Production Process
4. Measuring Techniques
5. International Standards
6. Surveys

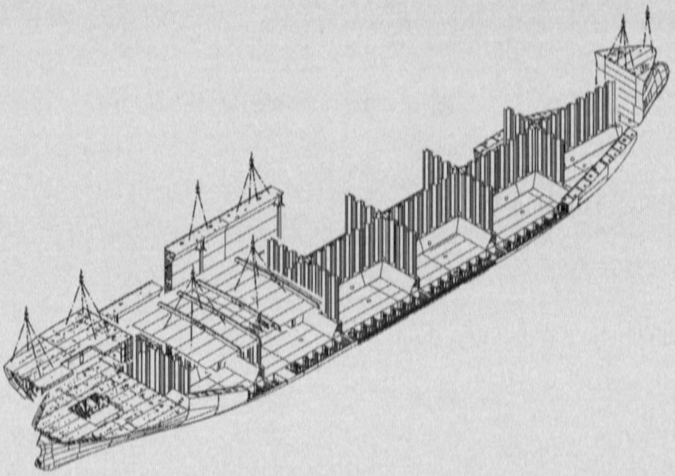

1. Production Planning

1.1 Block Division

1.1.1 Block Division Scheme

Block division has a strong influence on the whole production process (fabrication, assembly, outfitting and erection) and the production efficiency. However, determining the size, shape and weight of blocks is based on the capacity of facilities as well as the assembly practice of each shipyard.

The block division scheme should aim for maximizing the volume of assembly and pre-outfitting work and minimizing the dock workload and associated workload by taking into consideration the capacity of fabrication machinery and ability of the assembly shops, as shown in **Figure 12-1**. Through this process shipyards will try to minimize the total construction cost by using the optimum construction method and working environment leading to full utilization of the capacity of their facilities.

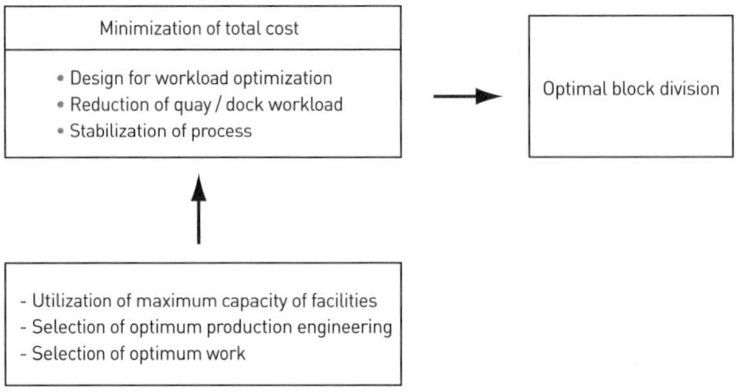

Figure 12-1 Basic concept of block division

1.1.2 Block Division Practice

1) Basic principles

The fundamental principles of optimum block division can be stated as follows:

- Minimization of erection work by using large blocks within the constraints of the capacity of the cranes.
- Normalization of blocks, if at all possible, so that they are of similar length and shape requiring similar workload and assembling time.
- Division of bow, stern and engine room blocks should be based on the outfitting work so

that unit outfitting can be contained within a block as much as possible.

- Ensuring the ease of assembly work and the minimization of work at elevated positions.
- Consideration of the workability of erection and the ease of accuracy control.

2) Block division policy

Blocks are either assembled units or pre-erected units, but the block division is carried out for the assembled unit blocks. The block division should be optimized for fabrication while satisfying the design conditions and meeting the requirements of the shop floor. Therefore, the division policy should be continually improved to suit the shipyard facilities and improving construction methods and technology.

As shown in **Figure 12-2**, a ship hull is split into a number of blocks satisfying, in principle, the conditions related to the following factors: keel laying blocks, crane capacity for assembly and erection, assembly process, assembly worktable rotation capacity, erection work and outfitting work. It should be noted, however, that it is not possible to satisfy all these conditions simultaneously, because some of them are contradictory. Therefore, each shipyard must decide how to divide the blocks based on the structural characteristics of the ship, and hence the difficulty in generic optimization of block division.

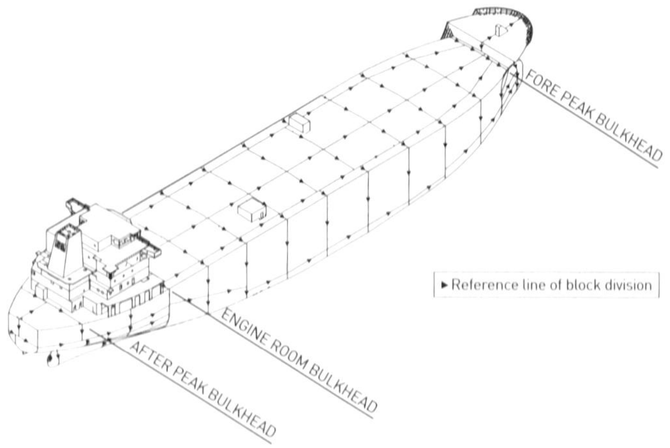

Figure 12-2 Example of block division

3) Considerations in block divisions

As stated earlier, the shape, size and weight of blocks are determined by taking into account the facilities of the shipyard. For example, the size of blocks must be calculated based on the size of assembly platforms, and the weight including the outfitting must be based on the crane capacity. If blocks are to be supplied by subcontractors, the capacity of their facilities should be taken into consideration. The block division must try to minimize erection work volume, work at high places, work in confined spaces and overhead welding during erection. For more effective design and production processes it is recommended to check the hull form information at various stages of erection in three-dimensional drawings as shown in **Figure 12-3**.

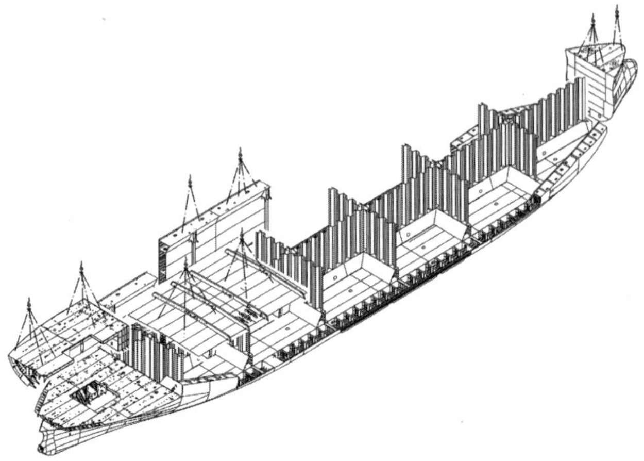

Figure 12-3 Three dimensional visualization of a partly erected hull

Long blocks are liable to twist and thin plate structures are apt to deform. Therefore, blocks must be divided with sufficient girders and stiffeners included to maintain accuracy during erection. In doing so, it is important to take into account the structural characteristics of compartments and make the most of the advantages of pre-outfitting. Especially the blocks of bow and stern regions require much work in confined spaces and complicated outfitting work. It is desirable to make these blocks as large as possible to maximize the work on land and pre-outfitting and minimize secondary workings.

1.2 Simultaneous Building

The workload involved in constructing a ship varies in time over the construction period. During the early stages the work volume and the shop floor areas required are relatively limited. However, as erection proceeds, the work volume increases until it reaches a maximum. This high workload level is then maintained until just before launching, when it decreases quite sharply. It implies that the required manpower fluctuates over this period too. Particularly when there are only a few building berths or docks, the total work volume can show wide fluctuations, resulting in some workers being left idle at certain times.

Furthermore, the erection work near the engine room must be completed as soon as possible to allow the main engine to be installed and the engine room to be fitted out. Given that the ships are getting larger and their construction tends to take longer, it is becoming more important than ever that the building docks are used as efficiently as possible with rapid turnover. To overcome these problems many shipyards have developed and implemented various advanced construction systems discussed below.

1.2.1 Semi-tandem System

The semi-tandem system can be used if the dock is much larger than the ships being built. In this system a ship is built in the dock next to the gate, while the stern part of another ship is built in the same dock beyond the bow end of the first ship. When the erection of the first ship is completed, the dock is flooded and the ship is floated out. The stern part of the next ship is moved towards the gate and the construction of this ship continues forward, while the stern of the next ship is built beyond its bow end, starting another cycle. In a large dock, an intermediate gate may be installed to facilitate the launching, and a transporter may be used to move the stern block.

Figure 12-4 depicts this process, when ships A, B, and C are built in that order. While ship A is being built, the stern part of ship B is pre-erected simultaneously in the dock at the location shown. When ship A is ready for launch, the dock is flooded and both ship A and the stern of ship B are floated. With the gate open, ship A is towed out and the stern of ship B is moved to the location shown. The gate is closed off and the dock drained. Construction of ship B resumes forward, while the pre-erection of the stern of ship C starts at the location indicated.

Using this system, the fluctuation of workload can be evened out, because the surplus workers around the time of ship A being launched can be reallocated to ships B and C. The dock turnover rate is also improved.

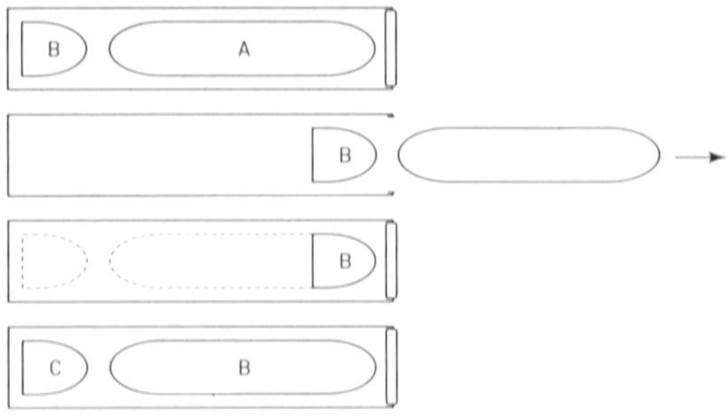

Figure 12-4 Semi-tandem building system

1.2.2 Canal Dock Building System

This system is similar to the semi-tandem system in that ship A and the stern of ship B are built simultaneously in the same dock. The key difference is the presence of the sea gates installed at both ends of the dock and a movable intermediate gate which allows the dock to

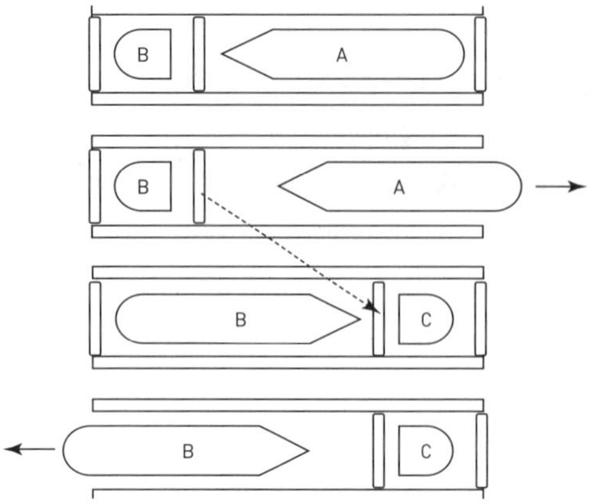

Figure 12-5 Canal dock building system

act as two separate ones. The gates at both ends make it possible for the construction of ship B to proceed independently of ship A's launching schedule. After launching ship A, the intermediate gate is moved further forward and the construction of ship B can continue without having to be moved at all. While ship B is being built, the stern of ship C can be pre-erected in the smaller part of the dock as portrayed in **Figure 12-5**. The canal dock building system, with the multiple gates dividing the dock into two, increases the efficiency of facilities and allows more even distribution of labor than the semi-tandem system.

1.2.3 Multi-Stage Building System

The multi-stage building system divides the whole working process into multiple stages to reduce the erection time in the dock. Upon completion of construction, ships are launched from the dock and immediately undergo the sea trials. There are two types of multi-stage building systems: the three-stage system and the five-stage system.

1) Three-stage building system

The traditional linear production system of erection in the dock followed by outfitting moored at a quay has been altered to finish the whole process within a huge building dock. The three stages are: stern erection; bow and parallel middle body erection; and outfitting, as shown in **Figure 12-6**. By controlling the work speed at various stages, the system facilitates seamless process management, workplace assignment, specialization of work, and use of various devices and facilities to reduce peak labor requirements.

Two types of dock can be used in this system: straight docks; and T-shaped docks with a side dock. Since sea trials can be conducted as soon as the ship is launched, the dockside outfitting work is minimized.

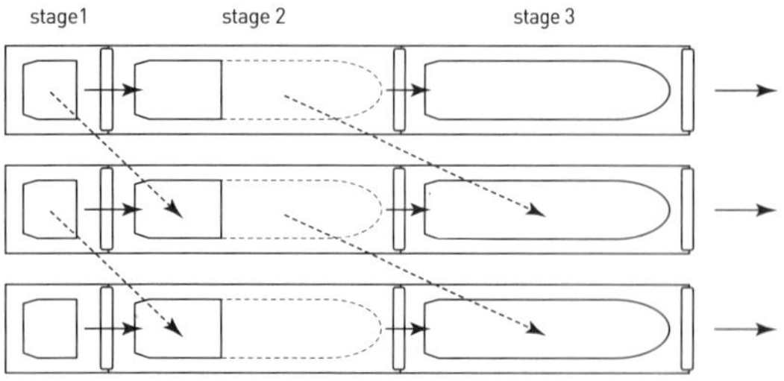

Figure 12-6 Three-stage building system

2) Five-stage building system

As depicted in **Figure 12-7**, the concept of five-stage building system, also known as the dual dock system, is similar to that of the three-stage building system. In contrast to the three-stage building system, however, the stern and the bow are built in different docks and then combined with the remaining body at the final stage, followed by launching. The stern and the parallel middle body are built in two separate production lines and the interrelationship between pre-erection, erection and dock work is improved, thereby more even distribution of workload is possible.

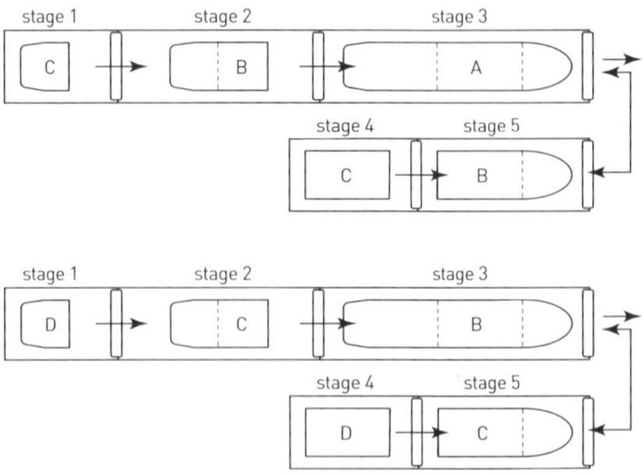

Figure 12-7 Five-stage building system

This building system consists of:

- Stage 1: construction of stern; installation of main engine and main pipes
- Stage 2: erection of upper structure; outfitting of engine and pump rooms
- Stage 3: hull construction; trial of main engine and auxiliary machinery
- Stage 4: construction of parallel middle body
- Stage 5: construction of bow.

1.2.4 Selection of Building System

The benefits and the drawbacks of each building system depend on the circumstances, including geographical location of the shipyard, ship type, technical capability and company policy. More specifically the factors to consider in selecting the building system are as follows:

(1) type of ships built;
(2) technical capability and production capacity of the shipyard;
(3) ease of even distribution of workload;
(4) minimizing dependency on heavy manual labor while maximizing mechanization and automation;
(5) good working environment avoiding up-and-down work and mixed materials work;
(6) avoiding simple repetitive work; providing variety to workers;
(7) possibility of workload levelling across grand-assembly, sub-assembly and fabrication processes; and
(8) possibility of pre-erection, pre-outfitting and integration of hull erection and outfitting.

2. Production Management System

2.1 Planning and Scheduling

Construction of a ship takes about 12 to 28 months from contract to delivery. The production process is complicated and the materials used are diverse. In order to execute such a complex operation smoothly and logically it is essential to plan its management meticulously. The management plan determines the work sequence and work schedule, and it also forms the basis of equipment usage, material supply, work force management and drawing management.

The production plan allows optimum usage of equipment and workforce through even distribution of work volume and workload, and cost reduction, productivity improvement and building time reduction by minimizing time loss through timely supply of materials. In other words, the production planning makes it possible for the shipbuilder to review the production process thoroughly to maximize its sales and maintain stable production activities.

2.2 Layers of Planning and Scheduling

Table 12-1 summarizes types, objects and functions of the planning and scheduling at each stage.

Table 12-1 Layers of planning and scheduling

	Schedule layer	Planning target	Output	Function
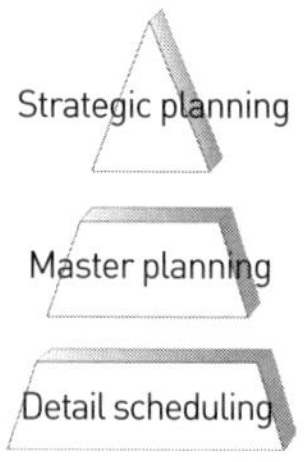 Strategic planning	Dock plan	Ship No. Dock Batch	Sales plan and building schedule Long term workload	Maximization of product sales Sales assessment (helps management-decision-making) Mid-and long-term business planning Event planning for each ship
Master planning	Midterm plan	Block Department Compartment Process	Erection plan Block schedule Drawing plan Material plan	Establishment and analysis of schedule Simulation for each stage and shop Load analysis and stock adjustment Allocation of work to subcontractors
Detail scheduling	Detail schedule	Production department Production team	Monthly work plan Weekly work plan Load analysis	Load analysis and adjustment Detailed schedule of workstages Budget management

2.3 Planning and Scheduling

2.3.1 Dock Plan (Strategic Plan)

A dock plan is the overall ship construction plan of the yard and is the basis of the company's corporate management. A basic dock plan shows the construction plan for all the building docks for the next 3 to 5 years, and the total sales for a year can be computed by summing the sale price of the ships, or part thereof, allocated to the docks in that year.

An example of production scheduling flow is depicted in **Figure 12-8**. When a contract is notified from the marketing department, a basic dock plan is drafted, taking into account the related sales, possible losses and the material schedule. Based on the standard workload distribution for the ship type, approximate workload, man-hours and workforce required are derived. Using the draft plan, more detailed workload is examined for fabrication, assembly and erection stages. If it is judged that the workload is feasible, the draft plan is confirmed as the dock plan.

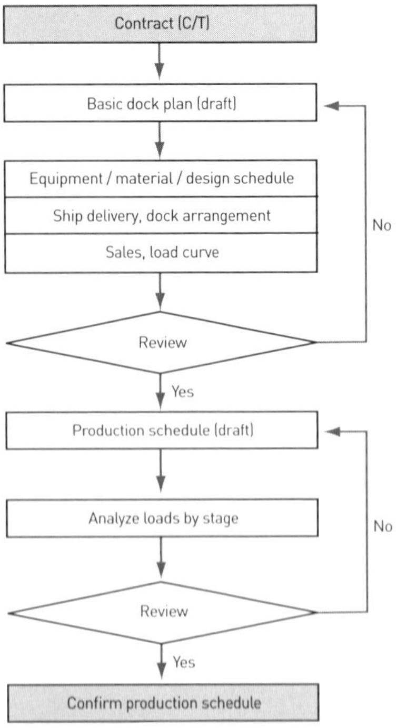

Figure 12-8 A typical production scheduling flow

2.3.2 Master Plan (Mid-Term Plan)

A master plan is made for preceding processes, dock or erection process and quayside process. The preceding processes include fabrication, assembly, pre-erection and painting and they usually take place indoors. The dock process refers to the erection of blocks produced by the assembly process. The quayside process encompasses painting, outfitting and sea trials.

The master planning deals with the detailed process plans, and the material procurement and production drawings are planned based on the master plan. Therefore, the master plan is the basic production schedule of a ship and it provides the outline framework of the production plan of the shipyard.

1) Erection network (dock process)

Once the dock period is determined from the dock plan, erection schedule can be established based on the past data for similar ships. If the dock period has to be modified for whatever reason, the intervals of the erection schedule thus produced are revised and adjusted.

The start time of erection for each block and the order of erection are adjusted to suit the block division and the erection sequence. Intervals of the erection schedule are decided from the erection time required for each block, based on the past experience. Using this data and the total erection time in the dock, the erection starting point can be established. The earliest start time (EST) is estimated by counting from keel-laying and the latest start time (LST) is calculated by counting backwards from launching. The line connecting two blocks with the EST of one and LST of the other at the same point in time is referred to as the critical path (CP). The schedule is examined with respect to the reach and availability of cranes, and if satisfactory, it is confirmed.

2) Preceding planning and quay planning

In **Figure 12-9** the process of preceding planning and quay planning is illustrated and their details are described here.

(1) Preceding planning

Preceding plans are derived by working backwards from the erection dates of the blocks specified in the erection network. The work duration and required processes are standardized by blocks and job lots and this data are used to generate the preceding plans. The work scope for each block for each stage is determined, and based on this the block fabrication and assembly schedule is produced. The preceding plan is then confirmed after adjusting the workload for more even distribution.

(2) Quay planning

The quay planning process produces the schedule of quayside or outfitting process. While the preceding planning is done around the blocks, the quay planning is done for each zone of

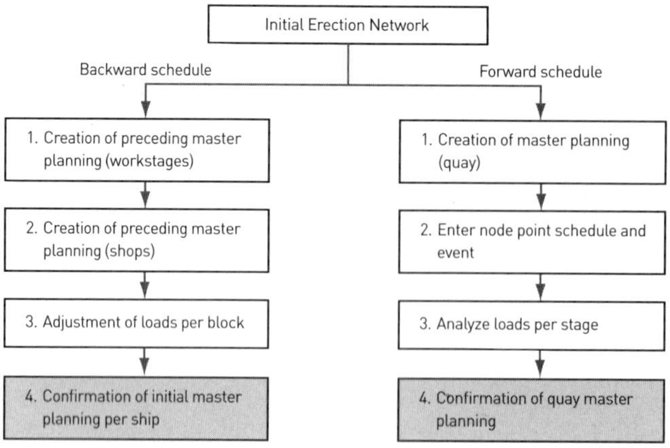

Figure 12-9 Flow of preceding and quay planning

a ship. In this planning, the key events of major outfitting equipment installed in each zone are determined and ultimately finalized by verification through load analysis.

2.3.3 Detail Scheduling

The detail schedule, as the smallest unit of planning and scheduling, is determined after the master plan is finalized through a process outlined in **Figure 12-10**. It is the detailed plan of work to keep to the master plan.

The detail schedule allocates various resources, such as facilities, equipment and work force, to each process on the basis of the master plan and coordinates the work schedule of various production departments. In this process current work progress, workload, weather condition and so on are taken into consideration. Projected completion dates are calculated by adding work durations to practically feasible start dates, determining whether or not the master plan requires adjustments. Based on this, the detailed schedules of the production departments and teams are produced.

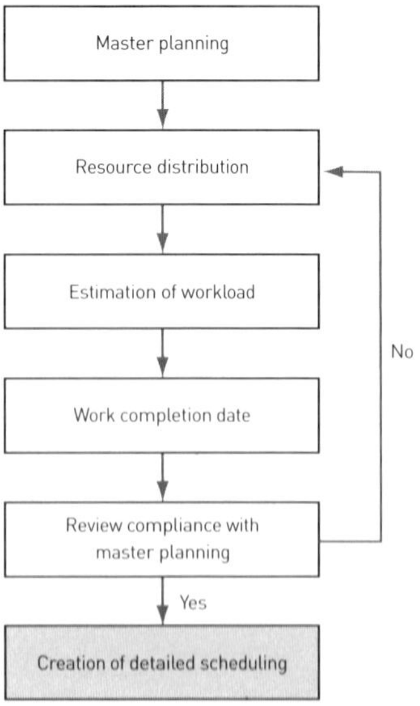

Figure 12-10 Flow of detailed scheduling

If the master plan is judged to be feasible, detail scheduling can proceed. However, if the master plan is judged to be impractical, an attempt must be made to make it feasible without modification by redistributing the resources. It may be necessary in some circumstances to revise the master plan.

3. Simulation of Production Process

3.1 Spatial Arrangement

3.1.1 Introduction

Shipbuilding is a large-scale industry and building merchant ships usually requires a yard occupying a wide area. It is crucial, therefore, to form a space schedule in order to use limited spaces most effectively and efficiently. Space scheduling prevents any bottlenecks in production process caused by the space constraints.

A process in a schedule cannot be undertaken, if an appropriate amount of space cannot be allocated to it. Therefore, all schedules must be generated to satisfy both spatial and workload constraints.

The primary aim of space scheduling can be said to be to make all processes in a schedule free from the spatial constraints. Spatial scheduling is particularly important in a situation where work space is limited and thus efficient use of space is critical. The traditional scheduling constraints are resources and delivery date, but dynamic spatial schedule must be included in them.

The conventional method is first to generate a schedule that satisfies the workload constraints using a searching method like the genetic algorithm, ignoring spatial requirements, and then, within the limitation of this fixed schedule, the required spaces are allocated almost as an afterthought. Even though this method does satisfy the workload constraints, its weakness is that some of the processes within the schedule may not be feasible due to shortage of space.

An alternative scheduling method may be to allocate the spaces first before considering the workload. There are three decision variables in space scheduling: the spatial allocation issue dealing with allocating positions of work within given spatial resources; the temporal allocation issue dealing with the starting and finishing times of work; and the resource allocation issue dealing with resource allocation based on workload.

If fixed worktables are used, a system of assembly using specific worktables is sometimes developed. In narrow work bays, blocks are arranged in a row. Some spatial allocation systems try to prevent work interference between polygonal blocks.

Space scheduling is used for block assembly, pre-outfitting, block painting and erection. It used to be carried out manually in the past, but even experienced personnel found it difficult to produce satisfactory schedule quickly due to its complexity. Therefore, computerization of space scheduling has emerged as a significant issue in improving productivity within the shipbuilding industry.

3.1.2 Block Assembly Process and Worktable Allocation

The assembly process assembles the fabricated plates and stiffeners, received from the fabrication shops, into blocks, and then delivers them to the subsequent process. The assembly can start from the ready date when the parts are received and must be finished before the due date. In order to produce many blocks required with limited resources it is important to ensure that high work volume is not concentrated in certain periods. **Figure 12-11** illustrates workload unevenly distributed in time and between various work bays. Such imbalance must

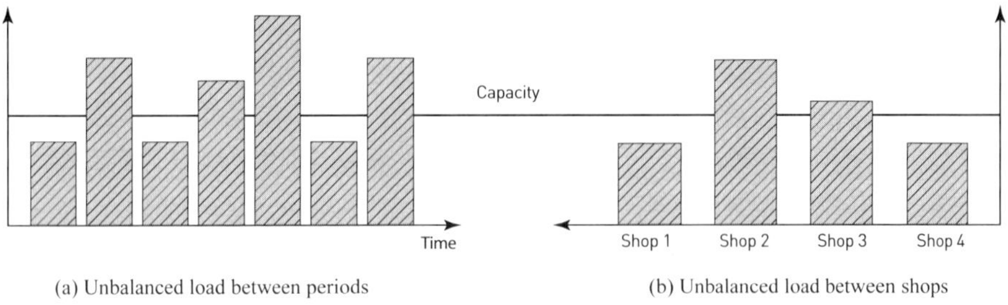

Figure 12-11 Unbalanced loads in periods and between work bays

be levelled out as much as possible.

The block assembly consists of several jobs, each of which requires corresponding work area and workers. The work areas are organized into three specialized shops: sub-assembly shop; unit-assembly shop; and grand-assembly shop. Each shop has a limited capacity which cannot be exceeded. It is important to level the work load in time and between the shops in order to maximize the productivity of these limited capacities.

There are two methods of levelling the load. The first of these adjusts the start date within the margin of work duration. This method can be used when the duration, the resource needed for the duration and the resource required in all works for the duration are fixed. The second method relies on adjusting the work intensity, which refers to the resources used for the work duration. Unlike the first method, it varies the duration (hence, the so-called rubber band duration) so as to make the work intensity evenly distributed.

1) Worktable allocation system

A block is the basic unit of shipbuilding and there are flat blocks with a flat bottom and curved blocks with a curved bottom or sides. Flat blocks are assembled as they move across roller conveyers, a process referred to as flow production method. Curved blocks are assembled on fixed worktables supported by adjustable pillars. Therefore, it is important to generate the worktable schedule based on the current monthly field schedule as shown in **Figure 12-12**. The worktable scheduling system aims to establish an efficient production schedule, based on the spatial constraints of the worktables and work capacity of the preceding

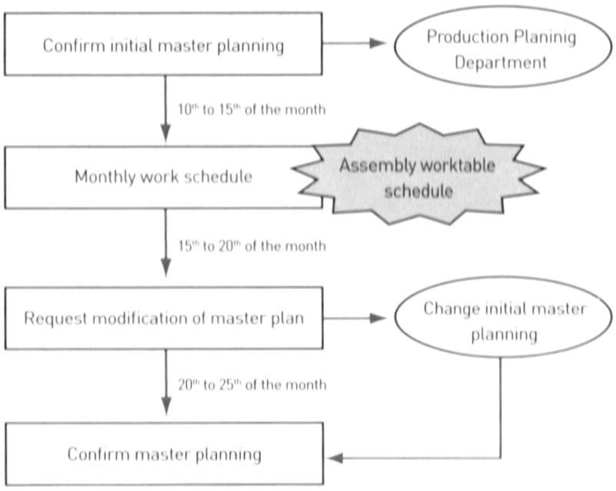

Figure 12-12 Flow of worktable scheduling

processes, so that the due completion dates of various blocks are met.

The development of such worktable allocation system has altered the shipbuilding workflow, as illustrated in **Figure 12-13**. Simply put, the data of all work processes are actively shared and updated between the stages for more efficient work planning.

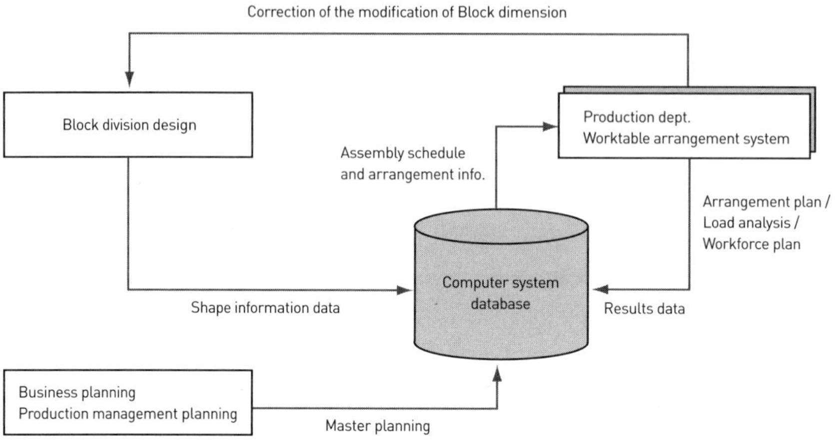

Figure 12-13 Modifications to the workflow due to the assembly worktable allocation system

2) Use of worktable allocation system

The worktable allocation system consists of four parts: information input; worktable schedule status; load analysis and simulation; and output.

(1) Information input

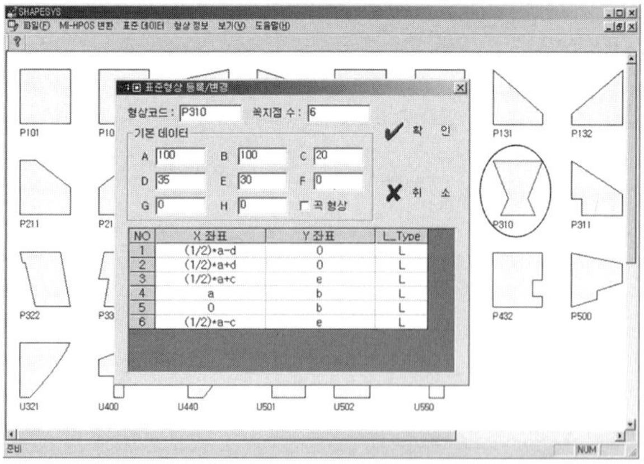

Figure 12-14 Registration and modification of standard block shapes

Input information consists of common information and block information. Common information includes holidays, work areas, and representative shape data for each block shape. Block information includes blocks already arranged and to be arranged. **Figure 12-14** displays the registration and modification process of the standard shapes of blocks.

(2) Status diagram of worktable allocation

As shown in **Figure 12-15**, the worktable allocation status diagram displays the current allocation status and it is possible to display, modify and generate various information through this facility.

(3) Load analysis and simulation

The capacity of assembly shops and the loading status are assessed through the load analysis. The changes in loading due to alteration in work schedule can be examined through simulation.

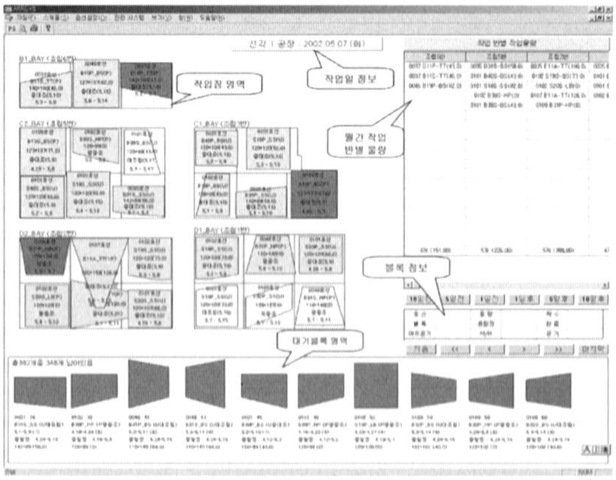

Figure 12-15 Arrangement information by shops

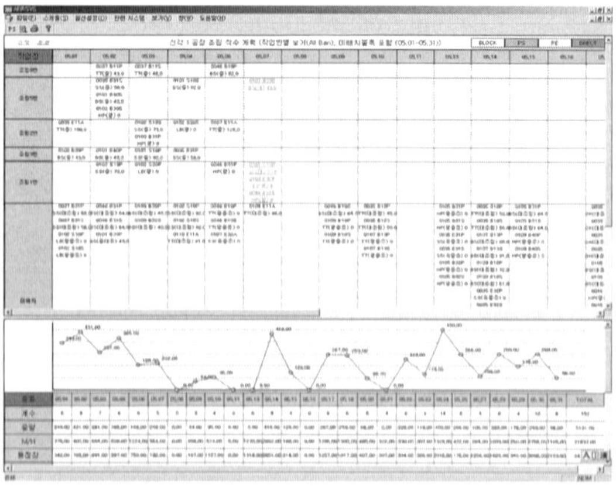

Figure 12-16 Worktable allocation information mosaic chart

(4) Output

The output displays or produces hard copies of results from automatic or manual planning, and the information includes block arrangement drawings or work schedule, as shown in **Figure 12-18**.

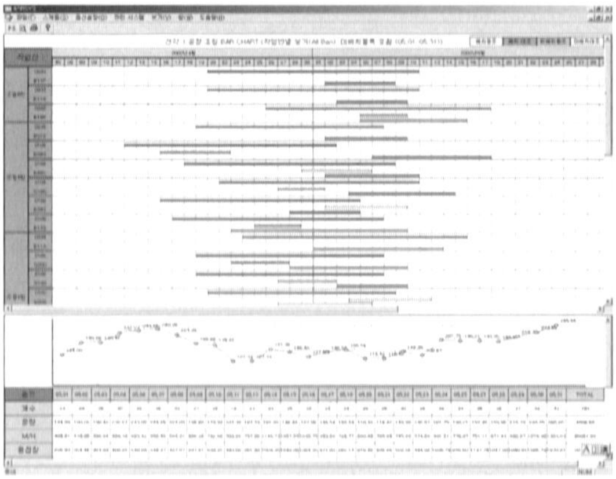

Figure 12-17 Worktable allocation Gantt chart

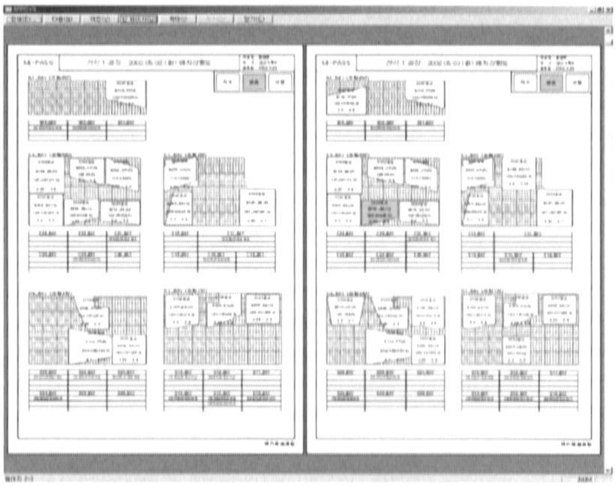

Figure 12-18 Worktable allocation information

3.2 Virtual Reality

3.2.1 Importance of Virtual Reality

VR (virtual reality) has become one of core technologies that could enhance the technological competitiveness of the shipbuilding industry.

3.2.2 Virtual Reality

VR is a technology which allows users to have immersive experience in, and interact with, artificial situations or environment generated by computers where a digital mock-up of a complicated model can be commonly accepted as an appropriate substitution. VR enables users to experience environment otherwise too large, expensive, dangerous or complex to do so in reality – such as shipbuilding – and perform challenging operations in a model created to mimic the real world. For example, VR is used for training pilots and drivers, rehearsals of clinical operations and games.

In the maritime industry VR is used for ship operational training, technical feasibility studies and design evaluation. Another example is shown in **Figure 12-19**, which illustrates a simulation of cargo loading at a design stage of a car carrier. VR is also applied to many other operations such as determining the accessibility of welding and painting to improve production work efficiency.

Figure 12-19 Car loading simulation for a car carrier

VR has now developed to a point in which the entire ship production process can be simulated by constructing a three-dimensional digital model of a shipyard (**Figure 12-20**). Utilizing VR offers many advantages to the shipbuilding industry. Firstly, improvement in accuracy by checking dynamic clash, avoidance of rework, design verification and mock-up test simulation are feasible. For example, interferences can be eliminated through design modifications, when indicated by simulation of equipment and facilities like cranes. Workload can be

Figure 12-20 Example of a VR shipyard

reviewed in advance through block erection simulation. Furthermore, optimal block erection sequence and safety of workers in their movement can be studied in this way.

Secondly, the accuracy of the process and production planning can be improved through simulating the entire process plan with the whole digital model of the shipyard. Optimizing the yard layout is possible through simulating the building process with a number of modified arrangements of facilities and shops which can lead to an improvement of the material flow and reduction of the work duration. A shipyard builds many ships at the same time, and the corporate management will benefit greatly if simultaneous construction can be simulated.

3.2.3 Applications

Determining where and how to use simulation is more important than the technical implementation of the simulation. Possible application areas are suggested in **Table 12-2**.

Table 12-2 Application of simulation in shipbuilding

Sales	Enhance sales effort through simulation and 3-D design visualization
Planning	Schedule/production planning
Fabrication	Curved parts fabrication
Assembly	Digital mock-ups, clash simulation
Construction	Block erection

These items in and of themselves can be regarded as the technological expertise of the yard and their practical application should be a core system of the company in their effort to become more competitive.

VR is a cutting-edge technology that has evolved rapidly in the shipbuilding industry in the last decade. As IT fusion and VR grows into a key sector, it is crucial that the Korean shipbuilding industry secure technical advantage through embracing the simulation technology whole heartedly.

4. Measuring Techniques

4.1 Overview

Ships are built through a number of fabrication and erection processes, and it is essential that the parts and blocks are made and constructed within the permissible tolerance limits, and therefore the ship production process requires strict quality control at every stage. The quality control is based on inspections and, to minimize the potential errors resulting from subjective judgment of the inspectors, a variety of quantitative measurement methods and techniques are used to produce more objective and scientific inspection data. In order to achieve more accurate and precise measurements more easily, each shipyard tries to develop its own jigs suitable for the purpose. As a result, new measuring instruments and techniques are developed based on novel ideas, and measuring techniques have become an increasingly important factor in determining production ability of the shipyard.

Accuracy control is a system of regulating the accuracy of the products at every production stage as faithful realization of the design within a range of allowable tolerance. Various instruments and devices are employed in shipbuilding for accuracy control: plumb lines, tape measures, automatic levels, theodolites and 3D optical measuring instruments.

The instruments themselves may become inaccurate due to temperature and weather condition, and the users may introduce measurement errors too. With the modern three-dimensional digital instruments, however, such errors can be reduced, thus making it easier to control the accuracy of hull fabrication. Furthermore, wider use of these state-of-the-art instruments has led to the construction of higher quality ships.

In order for the constructed ship to perform as detailed in the specifications the mechanical systems, such as main engines, auxiliary machinery, shafts and rudders, should be installed precisely in accordance with the drawings and instructions of the manufacturers. Therefore, the installation of these systems must be checked at every step to ensure its accuracy using specified measuring instruments. A log of such installations containing the accuracy achieved must be kept and the record maintained meticulously.

Shafts, rudders, main engines, generators and other auxiliary machines are installed usually in that order. The alignment of the main engine and the propeller shaft has to be checked and in this task measuring the shaft centerline requires great precision. For diesel engines, it is important to measure the deflection of crankshafts. There are also instruments that measure noise and vibration during sea trials.

Especially since the introduction of the more stringent performance standards for protective coating of seawater tanks, strict measurement and quality control is required at each stage of painting from pretreatment to final coating. Surface roughness, salt, dust, dry paint film thickness, temperature, humidity and adhesion of all painted surfaces must be measured and used to determine whether or not the quality of painting meets the standards. With the advancing technology, the use of instruments with automatic recording is becoming more widespread.

4.2 Introduction of Measuring Devices

4.2.1 Three-Dimensional Measuring Devices

1) Principles

Three-dimensional measuring devices measure a distance by transmitting light waves or radio waves and observing the phase difference of outgoing and returning waves. The refraction coefficient must be considered because the speed of radio waves of a fixed frequency can be changed by the physical properties of air.

2) Usage

Checking the errors in blocks and analyzing the accuracy is difficult, when one- or two-dimensional measuring methods are used. However, high quality can be obtained by accurately fabricating blocks based on three-dimensional digital measurement information (**Figure 12-21**).

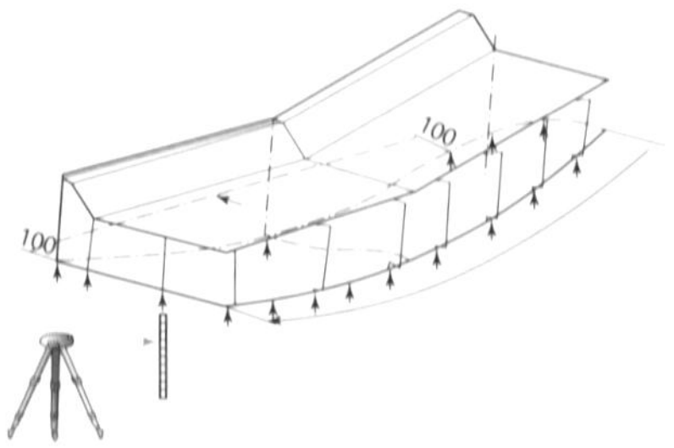

Measurement of block dimension by 2-D method

Measurement of block dimension by 3-D method

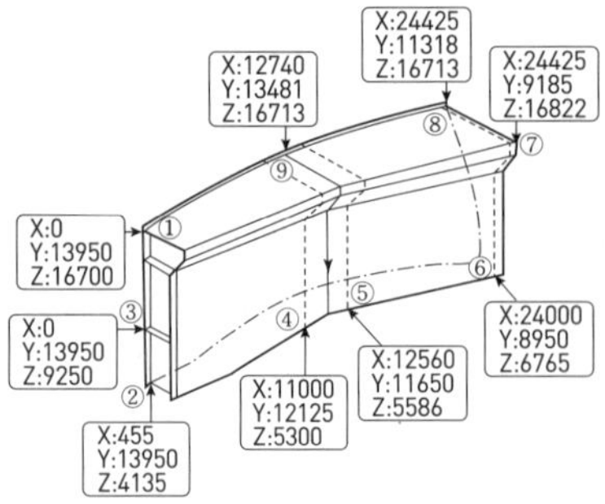

Figure 12-21 Block measurement and conversion of measured data into 3-D coordinates

3) Examples of usage

Detecting and solving potential problems likely to be encountered during the erection at the assembly stage can reduce the setbacks in the subsequent processes.

Recently, large shipyards are trying to improve the turnover rate of building docks by pre-erecting mega-blocks (from 2,500 to 3,000 tons) or giga-blocks (about 6,000 tons) so that the crane usage in the docks is minimized.

Because it is difficult to make fine adjustments in setting up such large blocks using a crane, the joints between blocks are checked and corrected for their verticality and dimensions to match up with the next block before setting them up for erection. Three-dimensional measuring devices are very useful for this purpose. Once the adjoining cross-section is measured accurately by such a device (**Figure 12-22** and **12-23**), the alignment data can be generated and the setting up becomes a much simpler task.

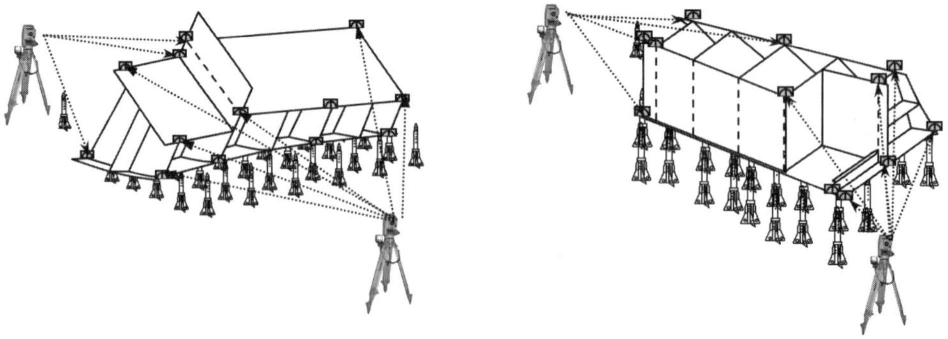

Figure 12-22 Block measurement at assembly stage using 3-D measuring devices

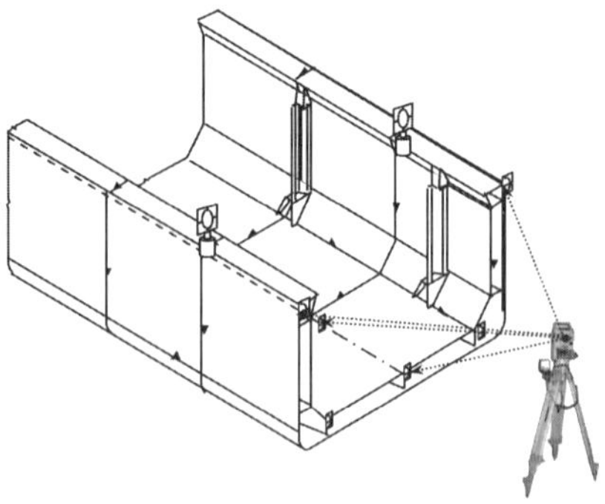

Figure 12-23 Example of measurement at the erection stage using a 3-D measuring device

4.2.2 Auto Levels

The height difference and short distances can be measured with the lens located at the top of the device. Errors made during production are minimized by verifying the accuracy of horizontality of blocks and sub-blocks. After verifying the accuracy, the unsatisfactory parts are modified, leading to improved work efficiency in the subsequent processes.

4.2.3 Non-Destructive Testing Equipment

A radiographic inspection device (**Figure 12-24**) uses short wavelength radiation to detect hidden flaws. The test piece is placed between the radiation emitter and a detector. When the radiation is emitted, the detector records the shading difference between sound and defective areas in two- or three-dimensions.

Figure 12-24 Non-destructive testing equipment

An ultrasonic detector (**Figure 12-25**) transmits ultrasonic waves into the test piece and they are reflected when they encounter internal discontinuity. The energy content of the reflected waves and the time taken for reflected waves to arrive back are analyzed to determine the location and size of the discontinuity. This device is used to detect internal defects and can also measure the thickness of plates. Magnetic particle inspection and liquid smear inspection can be used at various points during shipbuilding process as well.

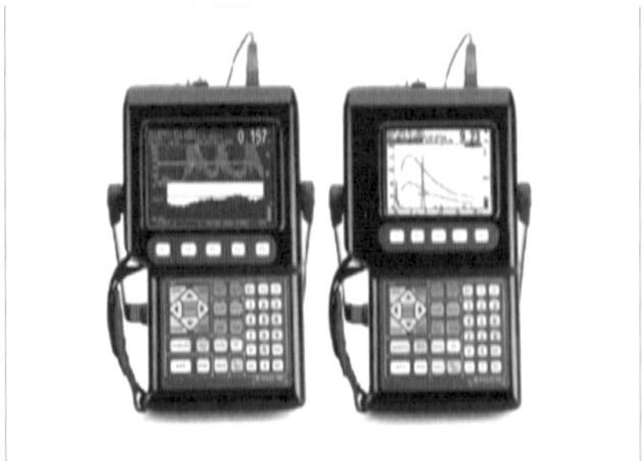

Figure 12-25 Ultrasonic flaw detectors

4.2.4 Dry Film Thickness Gauge

Figure 12-26 shows a dry film thickness gauge which measures the film thickness of the painted surface of steel material and usually a magnetic gauge is used. The end of the probe transmits a magnetic field and estimates the film thickness by measuring the returning speed of the magnetic field. Since the probe tip is wider than the surface peaks, there are bound to be many peaks in contact with the end of the magnetic probe. The film thickness measured, therefore, is their average value. Whether the measured film thickness is sufficient or not is determined after comparing it with the painting specification of each area.

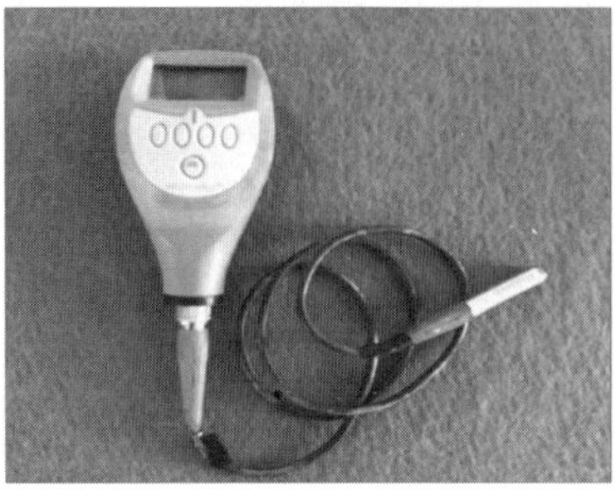

Figure 12-26 A dry film thickness gauge

4.2.5 Wet Film Thickness Gauge

A 'toothed' or 'comb' gauge can be used to measure the thickness of the paint coating immediately after it is applied, referred to as 'wet film.' The gauge is pressed firmly on the painted surface at right angles to the steel surface and then withdrawn after a moment whilst maintaining the right angles. The thickness of the wet film is a value between the largest wet tooth value and the smallest dry tooth value. In **Figure 12-27**, for example, the largest wet tooth value is 60 μm and the smallest dry tooth value is 70 μm, and therefore the wet film thickness is estimated to be around 65 μm. There is no specific rule about how many times the measurements must be taken, but it has to be done several times to ensure that the required dry film thickness is achieved.

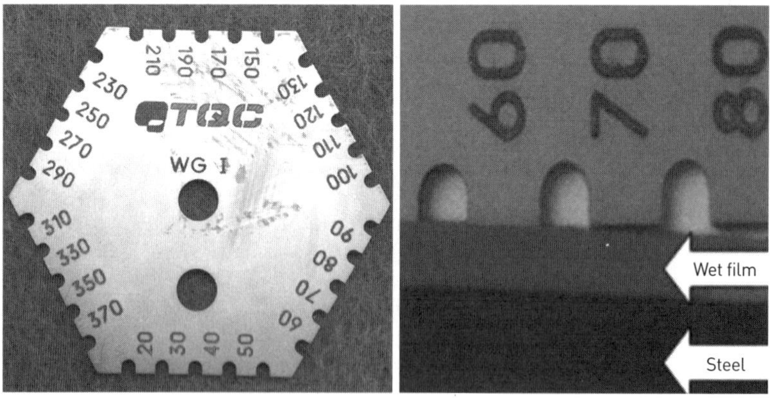

Figure 12-27 A wet film thickness gauge

4.2.6 Other Gauges

Other hull inspection devices include weld leg length gauges, undercut gauges, gap gauges and plate thickness gauges. Outside micrometers, dial gauges, cylinder gauges, inside micrometers, infrared radiation thermometers, multi-function testers, resistance testers, digital clamp-on testers, air pumps for adjusting pressure sensors, tachometers, hardness testers, insulation testers, wear gauges, displacement indicators, deflection measuring instruments and pump gauges are used for outfitting inspections. Wet and dry bulb thermometers and surface roughness sensors are used for inspecting painting and painting environment.

5. International Standards

5.1 International Maritime Organization (IMO)

Shipping and shipbuilding are truly international industries and in order to regulate their activities an international organization was established under UN. This organization, the IMO, deals with all maritime matters of international concerns including safety, security, pollution and litigation. It has so far adopted and revised 59 conventions and protocols, and over 2,000 codes. These conventions and protocols are initially proposed by committees and sub-committees. They are then discussed by the representatives of the member and non-member states, and the Assembly makes the decision to adopt them or otherwise. These conventions and recommendations address wide-ranging maritime issues as follows:

1) Safety and security

(1) International Convention for the Safety of Life at Sea (SOLAS)

(2) International Convention on Standards of Training, Certification and Watch-keeping for Seafarers (STCW)

(3) International Load Line Convention (ICLL)

(4) International Regulations for Preventing Collisions at Sea (COLREGS)

(5) Convention for the Suppression of Unlawful Acts against the Safety of Maritime Navigation (SUA)

2) Maritime pollution prevention

(1) International Convention for the Prevention of Marine Pollution from Ships (MARPOL)

(2) Convention on the Prevention of Maritime Pollution by Dumping of Wastes and Other Matter (LDC)

(3) International Convention Relating to Intervention on the High Seas in Cases of Oil Pollution Casualties (INTERVENTION)

(4) International Convention on the Control of Harmful Anti-fouling Systems on Ships (AFS)

(5) International Convention for the Control and Management of Ships' Ballast Water and Sediments (BWM)

3) Response and reaction

(1) International Convention on Maritime Search and Rescue (SAR)

(2) International Convention on Oil Pollution Preparedness, Response and Cooperation (OPRC)

(3) Protocol on Preparedness, Response and Cooperation to Pollution Incidents by Hazardous and Noxious Substances (HNS OPRC)

4) Liability and compensation

(1) International Convention on Civil Liability for Oil Pollution (CLC)

(2) International Oil Pollution Compensation Fund (FUND)

(3) Protocol on Preparedness, Response and Cooperation to Pollution Incidents by Hazardous and Noxious Substances (HNS OPRC)

5.2 SOLAS

5.2.1 International Convention for the Safety of Life at Sea

1) Purpose

The 1974 International Convention for the Safety of Life at Sea is the quintessential IMO preventive regulatory treaty for the safety of ships and marine pollution prevention. SOLAS regulates the design, construction (including structure, subdivision, stability, fire safety, main propulsion equipment, facilities, electrical equipment and maneuvering equipment), manning and operation of ships in order to ensure the safety of ships and protection of marine environment.

2) Main contents

The first form of modern SOLAS convention was adopted in 1974 but entered into force on May 25, 1980. It began with 13 articles and an annex containing technical provisions in nine chapters with three chapters being added in 1994 and another in 1997. Chapter XI was renumbered as chapter XI-1 and a new chapter XI-2 for ship security was added in 2002, bringing the total number of chapters in the Annex to 14.

The main articles of SOLAS 1974 As Amended outlines the responsibilities of 'Contracting Governments' and the procedures of becoming party to the Convention in addition to various regulations on the range of applications, amendment, entry into force and denunciation.

5.2.2 SOLAS Protocol 1978

1) Background

In response to the oil pollution from a spate of tanker accidents along the U.S. coastline in 1976–1977, the U.S. government released a statement in March 1977 on reinforcement of regulations for oil tankers entering American ports. As a result a conference was held in February 1978 in London. The conference adopted measures relating to tanker design and operation, and these were incorporated into 1978 SOLAS Protocol and 1973/78 MARPOL Protocol.

2) Main amendments introduced

① Adoption of inert gas system in tankers
② Additional installation of radars
③ Improved rudders
④ Ship inspection and certification issues.

5.2.3 SOLAS Protocol 1988

1) Background

An international conference on the harmonized system of survey and certification was held to resolve some practical problems arising from various conventions adopting different ship inspection systems. This has resulted in the adoption of SOLAS protocol 1988.

2) Main amendment introduced

Unified system of ship inspection and certification issues.

5.3 MARPOL 73/78

5.3.1 International Convention for the Prevention of Marine Pollution from Ships

1) Background

The International Convention for the Prevention of Marine Pollution by Oil (OILPOL 1954) was adopted as a regulatory measure in May 1954 and it entered into force on July 26, 1958. It went through many amendments until November 1973 when the International Conference on Marine Pollution adopted the 1973 International Convention for the Prevention of Marine Pollution (MARPOL 73). The purpose of the 1973 convention was to regulate not only oil pollution, but also the pollution caused by the discharge of other contaminants from ships. It was revised by the 1978 Protocol and renamed MARPOL 73/78 on November 4, 1980.

2) Composition of the Convention

(1) Main body: 20 articles
(2) Protocol I: Provisions concerning reports on incidents involving harmful substances
(3) Protocol II: Arbitration
(4) Annex I: Prevention of pollution by oil and oily water
(5) Annex II: Control of pollution by noxious liquid substances in bulk
(6) Annex III: Prevention of pollution by harmful substances carried by sea in packaged form
(7) Annex IV: Prevention of pollution by sewage from ships
(8) Annex V: Prevention of pollution by garbage from ships
(9) Annex VI: Prevention of air pollution from ships

5.4 ISM Code

5.4.1 Background and Application

Ships operate in harsh and often hostile environment and are frequently exposed to dangers. It is hardly surprising that the safety of ships has been a matter of great interest to mankind since time immemorial. In modern times it has become essential that all ships at sea have sufficient quality to withstand all the hazards that the environment and operation present them with during their relatively lengthy voyages so as to protect human lives and cargo. This not only requires sound hull form and structure but also various equipment and facilities such as fire-fighting equipment and lifesaving appliances.

Until recently IMO has mainly focused on the physical aspects of ships, including the improvement of ship design and equipment and crew training. When accidents, such as the capsizing of *Herald of Free Enterprise* in March 1987 and the stranding of *Exxon Valdez* in April 1989, kept occurring one after another due to human error, IMO decided to take action. **Table 12-3** summarizes some of these accidents.

Table 12-3 Marine accidents and related international conventions on marine safety

Ship Name	Year and place of accident	Accident description	Related international convention
Titanic	1912 North Atlantic	Sunk, along with 1,523 crew and passengers, after collision with an iceberg.	SOLAS
Torrey Canyon	1967 off southwest coast of UK	119,000 tons of oil spilled into the Strait of Dover after stranding; Under-qualified crew on ships flying flags of convenience became a hot issue.	MARPOL STCW
Herald of Free Enterprise	1987 near Zeebrugge, Belgium	Left port with the bow door open; capsized with 188 fatalities.	enforcement of SOLAS led to the creation of ISM Code
Exxon Valdez	1989 Prince Williams Sound, Alaska, US	Stranded, spilling 45,000 tons of oil and costing $3 billion for clean-up.	OPA 90 (The US Oil Pollution Act)
Scandinavian Star	1990 North Sea	Fire on board with 159 fatalities of passengers and crew.	expedited the development of the ISM Code

With the constant improvement in ship and equipment design, accidents caused by technical defects or malfunctions have been decreasing, while the proportions of accidents attributed to human error have been increasing. This has called for more systematic safety management of ships by the operating companies and this was how the International Safety Management Code (ISM Code) came into existence.

As summarized in **Table 12-4**, the ISM Code for some ship types first entered into force on July 1, 1998 and application was later expanded to all ships of more than 500 gross tons.

Table 12-4 Enforcement of ISM Code

Effective Date	Applicable Ships
1998.7.1	Passenger ships engaged on international voyages, oil tankers over 500 tons, chemical tankers, gas carriers, bulk carriers and high-speed container ships
2002.7.1	Other cargo ships and MODUs (mobile offshore drilling units) over 500 tons engaged in international voyages

5.4.2 Requirements

As previously discussed, marine accidents which occurred in the late 1980s, when the inadequate ship management was beginning to attract attention, were investigated and it was concluded that there were some serious shortcomings in ship safety management. Guidelines for management, at sea and ashore, of ships were developed and have been continually updated. ISM Code was amended a few times since its inception, for example by the IMO general assembly resolution A.1022 (26) which entered into force on July 1, 2010.

5.5 ISPS Code

5.5.1 Background and Application

The attack on the World Trade Center in New York City on September 11, 2001 reminded the world of the serious threat by terrorism once again. In response to such threats against port facilities and ships, particularly those carrying LNG and LPG, IMO adopted the International Ship and Port Facility Security Code (ISPS Code) as a new Chapter XI-2 of the Annex to the SOLAS Convention to strengthen the security of ships, personnel, cargo and ports at the Diplomatic Conference on Maritime Security in London in December 2002. ISPS Code entered into force on July 1, 2004, and its main contents are described in **Table 12-5**.

Table 12-5 Key points of the ISPS Code

No.	Item	Key points
1	General	Definitions, objectives, application and functional requirements of the company's security management system
2	Security policy	Establishment and implementation of security management system
3	The company	Obligations of the company
4	Ship security officer(s)	Duties and responsibilities of a ship's security officer(s)
5	Ship's master	Responsibility and authority of the master
6	Resources and personnel	Provision of human/material resources, information and qualifications
7	Ship operation	Ship operation and voyages
8	Emergency preparedness	Emergency measures and response
9	Control and compliance measures	Non-compliance, accident and risk reporting and analysis
10	Ship and facility maintenance	Maintaining ships in proper condition

11	Record	Document management
12	Internal verification, audits and inspection	Verification of the effectiveness of the internal audit and security management system
13	Certificates	Issuing, verification and management
14	Temporary certificates	
15	Verification	
16	Certificate format	

The ISPS Code applies not only to ships engaged in international voyages, but also to port facilities used by these ships. There was much confusion in the beginning of its application, but it is now accepted as the general security standard for ships and port facilities. This code is applied to all vessels of the following types engaged in international voyages: all passenger ships, including high-speed passenger craft; cargo ships, including high-speed craft, of 500 GT or larger; and mobile offshore units.

5.5.2 Main Content

ISPS Code consists of a mandatory part (Part A) and recommendations (Part B). Part A consists of definitions, objectives, obligations of governments and companies, security declaration, ship security appraisal, preparation and approval of ship security plans and people in charge of security. Part B consists of the appointment of an agency for security certification and the conditions of ship security plans.

5.6 International Organization for Standardization – Ships and Marine Technology Committee (ISO/TC8)

5.6.1 Organization

The technical committee on ships and marine technology of the ISO was formed in 1948, instigated by some European nations, and was designated ISO/TC8. It currently has ten specialist committees and many working groups covering all aspects of ship design, construction and operation which require standardization. The standardization work of ISO/TC8 includes "design, construction, structural elements, outfitting parts, equipment, methods and technology, and marine environmental matters, used in shipbuilding and the operation of ships, comprising sea-going ships, vessels for inland navigation, offshore structures, ship-to-shore interface and all other marine structures subject to IMO requirements."

ISO/TC8 works closely with IMO to establish international standards and consequently wields great influence on the shipbuilding and offshore industry.

5.6.2 Need for Standards

Many countries are currently engaged in a struggle over standardization, because the stakes in their standards winning global recognition are high. In particular, the United States and the European Union (EU) are fighting to have their own technology to be adopted as the international standard. Adding fuel to the fire are manufacturing companies seeking to adopt standards favorable to them and their technology.

Globalization and trade liberalization have gradually reduced and, even abolished, traditional trade barriers such as tariffs and import restrictions. However, non-tariff barriers (NTBs), such as technical codes, standards and conformity assessment procedures, have gradually come into prominence instead. These so called Technical Barriers to Trade (TBTs) have high potential of being misused by importing countries as undeclared obstacles to trade for protecting their own manufacturers. Consequently, the Agreement on TBTs, which entered into force in 1995 with the official launch of the World Trade Organization (WTO), gives rules for the use of such barriers. For example, the WTO/TBT Agreement makes it mandatory for any new technical regulations to adopt the relevant international standards. It is also trying to eliminate unequal trade environment and promote balanced development of all nations of the world through harmonizing national standards with the international standards.

6. Surveys

6.1 Introduction

6.1.1 Objectives of Ship Survey

All ships are properties of their owners, but are legally required to be surveyed and certified by competent authorities according to the relevant national laws, which, in Korea, are the Ships Act and the Ship Safety Act. Furthermore, if a ship is engaged in an international voyage, it is additionally subject to international laws and regulations, such as SOLAS, MARPOL and International Load Line Convention.

The purpose of ship survey is to secure the safety of ships by verifying whether or not the ship meets all relevant safety and environmental standards.

6.1.2 Statutory Survey and Classification Survey

Because ships and cargo are valuable properties, ship-owners will want to insure them against possible loss, while cargo owners will want to be certain of the safe carriage of their cargo. The marine insurance provider, on the other hand, needs technical information on the subject ship in order to determine the appropriate insurance premium for the ship and the cargo it is intending to carry. Given the differing interests of ship-owners, cargo owners and insurance companies, it is desirable that the survey and evaluation of the ship is carried out by an independent third party.

Historically, marine insurance businesses were conducted often in coffee shops in London and other English trade cities. One of these was a coffee shop run by Edward Lloyd in the late 17th Century. Shipping industry community – sailors, merchants, ship-owners and brokers – frequented the premise and often conducted their businesses of buying and selling ships, arranging carriage of cargo, discussing insurance matters and exchanging shipping news. Later, the members of the insurance arrangement set up a formal organization, the Society of Lloyd's, which is the very first classification society as we know it today. This organization, of course, later became Lloyd's Register of Shipping.

The nascent classification society established its own codes, awarded classification certificates to ships that passed its inspection, and added certified ships to the register of shipping. The register was published for all parties concerned, including insurance companies.

As shipping and shipbuilding technology developed, the classification societies also kept pace by developing their own technical rules on the design, construction and survey of ship structure, engines, electrical facilities and outfitting items. The classification societies have now evolved into organizations which approve design drawings of hull and equipment from initial design and oversee the whole process of construction until delivery. They continue conducting surveys on a regular basis after the ships are commissioned into service.

As classification societies have accumulated advanced technical expertise and use high quality surveyors to provide all kinds of technical services, their activities of evaluating ship structure and performance have now become critical in securing the safety of ships. Moreover, as the surveys carried out by classification societies are of similar kind to those carried out by government authorities, many countries have come to rely on the classification societies to carry out the inspections on their behalf as independent third parties. Prior to approval of the technical ability, trustworthiness and survey reliability of a classification society, the relevant government departments appraise the society's survey track records and global network.

Classification societies have been established in many nations other than the UK and there are currently over fifty classification societies all over the world. In 1969 the classification societies of UK (LR), U.S. (ABS), France (BV), Norway Germany (DnV GL), Japan (NK) and Russia (RS) recognized the need for mutual cooperation and formed the International Association of Classification Societies (IACS), which has been approved by IMO as a consultative body on matters of maritime technology. This has allowed the classification societies to make direct and indirect input into the international maritime affairs, such as formulation of international conventions. Ships registered with an IACS society receive favorable treatment in insurance premiums. The following twelve classification societies (listed in alphabetical order) are regular members of IACS as of 2014:

American Bureau of Shipping (ABS)
Bureau Veritas (BV)
China Classification Society (CCS)
Croatian Register of Shipping (CRS)
Det Norske Veritas Germanisher Lloyd (DNV GL)
Indian Register of Shipping (IRS)
Korean Register of Shipping (KR)
Lloyd's Register (LR)
Nippon Kaiji Kyokai (Class NK) (NK)
Polish Register of Shipping (PRS)
Registro Italiano Navale (RINA)
Russian Maritime Register of Shipping (RS)

Most ship-owners register their ships with a classification society, who will carry out surveys required by national and international regulations, and expect to secure favorable insurance premiums. However, it is the owners' prerogative to select a classification society to register their ships with.

6.1.3 Types of Survey

In Korea, the Ship Safety Act specifies the statutory surveys as: construction survey, special construction survey, intermediate survey, temporary survey, temporary voyage survey, international convention survey and special survey. Classification societies like KR carry out registration surveys during and after construction, regular post-construction surveys, intermediate surveys, annual surveys, docking surveys, propeller shaft surveys, boiler surveys, continuous surveys, temporary surveys, conversion surveys and so on.

There are other types of surveys depending on the circumstances and purpose, including Port State Control (PSC) surveys carried out by port state surveyors, cargo owner survey, insurance survey for damages, and field survey for selling ships.

6.2 Construction Survey

When a shipbuilder is contracted to construct a ship, the shipbuilder requests the services of a survey agency (commonly a classification society) in accordance with domestic laws, international conventions and the classification society's codes. The classification society approves the plans and thoroughly surveys every phase of the shipbuilding process: design, material, structure and fabrication of the hull, engine, outfitting and equipment. Many classification societies also provide technical consultative services as well.

6.2.1 Approval of Plans

In order for a design to be considered for approval, the shipbuilder is legally required to submit its plans, which must be prepared in accordance with domestic laws, international laws and the relevant classification society's codes. If necessary, plans must also follow the conditions defined in the construction specifications laid out in the shipbuilding contract.

Any reservations from field surveyors regarding the evaluated plans must be addressed and the revised plans must be reevaluated for approval. If the structure to be built is of a unique or novel character to which the existing regulations and codes cannot be applied, the validity of the design must be verified by direct application of theoretical methods such as structural and fatigue analysis.

6.2.2 Field Survey

Before proceeding with the survey, a survey plan has to be prepared and approved by a qualified surveyor, who must also be present at all times during the survey to make sure every phase is accounted for. The parties concerned – the shipbuilder, the surveyors from the classification society and the owner's representatives – hold a meeting prior to the survey. In the meeting, detailed procedures of the surveys and the tests (structural test, air-/water-tightness test, hose test, NDT, stability test, painting, trials, etc.) for each process are established and approved by the surveyor. Certain testing and measuring equipment specified in the relevant regulations must be used in surveys.

On-site surveys of hulls and engines must conform to relevant national and international laws, and codes of the classification society. After concluding field surveys, the survey agency issues certificates and survey reports, which must be kept on board at all times.

Field surveys assess the followings:

- Material and equipment: rolled steel, cast steel, forged steel, materials for fire prevention structures, anchors, anchor chains, ropes, side scuttles, equipment for unloading, hatch covers, watertight doors, steering gear, etc.
- Fabrication process: surface preparation of steel materials, marking, cutting, fairing, forming, etc.
- Assembly process: conformance of sub-assembly, unit-assembly, etc. with the approved drawings
- Hull block: quality of hull blocks conforming to the approved drawings
- Erection process: erection and launching capability of hull blocks in accordance with the approved plans
- Hull completion: the entire completed hull as specified by the approved plans
- Structure, air-/water-tightness, hose test: hull structure and closing equipment with water pressure or air pressure or water through hoses
- Hull painting
- Main engine installation: position of the installed main engine
- Shaft and propeller pressurized insert: installation of propellers in propeller shaft, and middle and stern shaft
- Rudders and rudder stocks
- Piping: joints of various pipes via pressure test and with its construction with respect to the specifications, both in the piping shop and the ship as installed
- Hull outfitting, machinery outfitting, electrical outfitting: entrance doors, manholes, hatches, hand rails, air pipes, sounding pipes, cables, ventilators and other outfitting in accordance with the approved plans
- Before launching: materials used before launching, principal dimension of hull, keel sighting, markings on external hull, anchors, anchor chains, completion of NDT on hull structure, etc.
- Fire prevention systems: fire prevention structures, ventilation equipment, escape equipment, etc.
- Onboard test: cargo handling equipment, hatch covers, ladders, generators, boilers, auxiliary machines, etc. after installation
- Stability: light ship weight and the position of center of gravity via inclining tests, light

ship weight estimation tests, and if necessary, oscillation tests, etc.

- Main engine dockside trials: (prior to sea trials) operation of main engine when the ship is moored to a quay
- Sea trials: (towards the end) operation and function of all facilities, machines, and electrical machines; speed test, crash astern test, emergency steering test, and turning circle test; operational state of engines and the state of ship during voyages
- National and international regulations: lifesaving equipment, fire control equipment, navigation equipment, equipment to prevent marine pollution, sanitary equipment in accommodation, etc.

6.3 Periodical Surveys

Upon the delivery of the ship, the owner has to file a request for survey to the relevant authority and the authority, or its agent, must carry out the necessary surveys in accordance with the national and international regulations and the classification society's codes. Furthermore, after the ship is put in service, the hull, engine, outfitting, and equipment are regularly surveyed.

Ship survey consists of regular surveys, which take place every five years; intermediate surveys, which take place between regular surveys; and annual surveys. Boilers and propeller shafts are also regularly surveyed. Certain ships, such as passenger ships and high-speed ships, may have different survey periods depending on national and international laws, and codes of classification societies.

6.3.1 Regular Surveys

These surveys are carried out to ascertain that a ship is in a satisfactory condition as to its hull, engine and outfitting, and, when maintained appropriately, can operate as originally intended until the next regular survey.

A regular survey is carried out with the ship docked or put on a birth, and the structural members are surveyed for their visual appearance, size and functions. Engines and machinery are overhauled and their operations are surveyed. The scope of survey is determined based on the age, size, type and state of the ship. A regular survey can be carried out in its entirety every five years or groups of surveys can be carried out annually over the course of five years.

6.3.2 Intermediate Surveys

The intermediate surveys take place between the regular surveys and the annual surveys, and are carried out to check whether or not the overall state of a ship satisfies the codes.

6.3.3 Annual Surveys

The annual surveys verify the overall state of a ship with the focus on examining the external appearance. A detailed survey may be required of certain items.

6.3.4 Docking Surveys

Docking surveys are usually carried out at least twice within the five years of the regular survey period. The bottom, side and bow of a ship, stern frames, rudders, sea chests, valves and propellers are surveyed by docking the ship.

6.3.5 Surveys of Propeller Shaft and Stern Tube

Propeller shaft is pulled out of the stern tube and surveyed. The propeller is usually separated from the shaft for this purpose and stern tube bearings and seals are surveyed.

6.3.6 Boiler Surveys

The boilers are surveyed usually at least twice within the five years of the regular survey period. The interior of boilers, the blow-off devices of safety valves, and the operational state of various safety equipment is surveyed.

References

1. Han, S.D. et al., 2008. Implementation and applications of simulation based digital shipyard(in Korean). *Transactions of the Society of CAD/CAM Engineers*, 13(1), pp.18-26.

2. Korean Register of Shipping, 2010. *50 years of Korean Register 1960-2010* (in Korean). Seoul: Korean Register of Shipping.

3. Korean Register of Shipping, 2011. *Rules for classification of steel ships, Part 1. Classification registry and survey* (in Korean). Seoul: Korean Register of Shipping.

4. Lee, C.J. et al., 2007. A study on discrete event simulation of shipyard outdoor block movement (in Korean). *Journal of the Society of Naval Architects of Korea*, 44(6), pp.647-656.

5. Lee, K.K. et al., 2006. Construction of scheduling support system for panel lines by digital manufacturing simulation (in Korean). *Journal of the Society of Naval Architects of Korea*, 43(2), pp.228-235.

6. Ministry of Industry and Resources, 1999. *Report on the development of the total shipyard production management system for the optimization of planning and production* (in Korean). 4th year progress report to Ministry of Industry and Resources (unpublished), pp.105-134.

7. Song, C.S. & Kang, Y.W., 2009. Simulation of block logistics at a big shipyard. *Transactions of the Society of CAD/CAM Engineers*, 14(6), pp.374-381.

8. Song, Y.J. et al., 2008. Development of a design framework for simulation based shipyard layout (in Korean). *Journal of the Society of Naval Architects of Korea*, 45(2), pp. 202-212.

9. The Society of Naval Architects of Korea, 2001. *Ship construction technology* (in Korean). Seoul: Dong-myungsa.

10. Woo, J.H. et al., 2005. A framework of plant simulation for a construction of a digital shipyard (in Korean). *Journal of the Society of Naval Architects of Korea*, 42(2), pp.165-174.

11. Woo, J.H. et al., 2005. Simulation modeling methodology and simulation framework for a construction of a digital shipyard (in Korean). *Journal of the Society of Naval Architects of Korea*, 42(4), pp.411-420.

CHAPTER 13

Naval Ships

1. Characteristics of Naval Ships
2. Platform Technologies
3. Naval Combat System Technologies
4. Survivability Technologies

CHAPTER 9
CHAPTER 10
CHAPTER 11
CHAPTER 12
CHAPTER 13
CHAPTER 14
CHAPTER 15
CHAPTER 16

1. Characteristics of Naval Ships

1.1 Definition

A naval ship is defined as a ship that is equipped with various mission systems, such as weapon systems, detection systems and electronic equipment, required for carrying out combat and support operations at sea. The term, 'warship,' is used for the same meaning.

A naval ship can also be defined as a weapon system that has the configuration of a ship. Accordingly, a naval ship has both characteristics of a weapon system and a ship system. This leads to significant differences between naval ships and commercial ships in terms of subsystems, equipment, applied technologies, design and construction procedures, among others.

1.2 Types

1.2.1 Naval Surface Ships

Modern naval ships are generally classified into the following categories according to mission, size and the sea area in which they operate.

- Battle ships
- Aircraft carriers
- Cruisers
- Destroyers
- Frigates
- Corvettes
- High speed craft
- Amphibious ships
- Mine warfare ships
- Auxiliary ships.

Battle ships were the dominant naval power up to World War II. But presently no battleship is in active service of any navy in the world. Ever since World War II, aircraft carriers have become the main naval power of the world's major fleets.

Cruisers, destroyers and frigates are the main surface combat units of modern blue water navies. They carry out multiple operations such as anti-surface, anti-submarine and anti-aircraft warfare. Land attack missions are another operational role that has been added recently, due to naval warfare moving to littoral areas. Corvettes and other high speed craft are the main surface combatant units for medium and small coastal navies. Amphibious ships, mine warfare ships and auxiliary ships carry out special missions, such as amphibious operations, mine detection and sweeping, and logistic support operations, respectively. **Figure 13-1** shows several types of modern naval surface ships.

Figure 13-1 A battle ship, an aircraft carrier, an amphibious ship and a destroyer (clockwise from top left)

1.2.2 Submarines

Modern submarines are generally categorized into two groups: nuclear powered submarines and conventional submarines. Based on the mission, propulsion plant and size, modern submarines can also be classified into six types as follows:

- Nuclear powered ballistic missile submarines (SSBN)
- Nuclear powered guided missile submarines (SSGN)
- Nuclear powered attack submarines (SSN)
- Diesel-electric submarines (SSK)
- Air-independent propulsion (AIP) submarines
- Midget submarines

Due to their independent underwater operations capability, submarines carry out a wide range of covert missions during peace time, including reconnaissance, surveillance and information gathering. In wartime, the main missions of submarines are protection of sea routes, defense of naval battle groups, and sea blockade of bases and sea routes of the enemy. Submarines also carry out strategic operations, such as precision strikes against high-payoff targets on the ground. **Figure 13-2** shows several types of modern submarines.

Figure 13-2 A nuclear powered ballistic missile submarine, a nuclear powered attack submarine, a diesel-electric submarine, an air-independent propulsion submarine (clockwise from top left)

1.3 Characteristics of Naval Ships as Weapon Systems

As mentioned earlier, a naval ship is a weapon system with the configuration of a ship. Therefore, naval ships have inherent unique characteristics as well as general characteristics of conventional weapons.

1.3.1 General Characteristics of Conventional Weapon Systems

General characteristics of conventional weapons systems that naval ships also possess are, in general, as follows:

- Diversity
- Complexity
- Accelerated obsolescence
- Covertness
- Limited demand
- High cost
- High probability of failure of new system development
- Ripple effect on technological and economic development.

1.3.2 Inherent Characteristics of Naval Ships

1) Complexity

Conventional weapon systems are composed of component units. A naval ship is also a composite system, but its scale is vast. It is a large complex system with many complicated subsystems, such as hull structure, propulsion system, auxiliary systems, electronic systems and weapon systems. Each subsystem is another complicated system in itself composed of a large number of components. Consequently, a naval ship is often referred to as a 'system of systems,' in recognition of the fact that a warship is the most complicated large weapon system.

2) Mission capability as tactical units

Conventional weapon systems are used continuously for several hours or days at a time but then are stored in designated places while not in use. On the other hand, a naval ship is in service all year round, except during overhaul.

The commander and crew reside in the ship and perform various tactical unit missions. Even while a naval ship is off-duty, shipboard personnel are engaged in administrative tasks or maintenance activities. Design of naval ships, therefore, must give serious considerations to the onboard habitability and working environment.

3) Employment of the prototype ship

A conventional weapon manufactured as a result of its development work is used only for the purpose of test and evaluation. Only the weapons mass-produced after its prototype was proved and accepted are deployed in actual battles.

It takes many years and enormous costs to design and construct a naval ship. Partly because of this, naval ships are designed in classes and a series of sister ships are then constructed, akin to sister ships in merchant fleets. In this process the first ship in the series (the lead ship) is used as a prototype for various tests and proving, but subsequently it will be deployed for active service. Therefore, it is crucial that all technical problems are resolved at the design stage before the construction of the first ship begins. This is one of the reasons why so much research and development effort is expended at all stages of design from the very formulation of operational requirements.

Despite what has been said above, a vessel of entirely novel concept may be tested for its operational and technical feasibility by building a smallest prototype practicable. In this case such a prototype is usually not deployed for active service.

4) Limitations in standardization

With conventional weapon systems the ultimate goal of producing a prototype and evaluating its performance is to obtain specifications for mass production.

A naval ship is built usually one at a time due to the constraints of defense budget and the need to keep the amount of work at the naval shipyards at a reasonable level. Consequently, the ships are built at different times, even though they may belong to a same class. This can introduce changes in operational requirements according to changing circumstances, and often causes sister ships being built with slightly different detail specifications. Standardization of naval ships, therefore, is quite limited.

5) Small quantity batch production

Conventional weapons are mass-produced based on standardized specifications. Navies, on the other hand, build a limited number of mission-specific ships as required so that they can carry out diverse naval missions effectively. As has already been said, a ship may be given slightly different specifications to its sister ships at the start of its construction. In this respect naval ships are very different from the conventional weapon systems. An exception to this was during World War II when the warring navies built large numbers of standardized ships without, or with very little, variation as fast as they could in large batches. However, this was an exceptional case which occurred in emergency war time circumstances.

1.4 System Constitution of Naval Ships

The main role of a naval ship is to carry out various operational missions, and sailing is but its basic function. For merchant ships and leisure vessels, voyage is their main function. This inevitably introduces significant differences between naval ships and commercial ships in their system constitution. A naval ship is a complex system consisting of a platform system and combat system, while a commercial ship is built for transporting passengers and cargo and consists of a platform system only.

Differences in the main mission make it inevitable that even their platform systems are different. Naval ships tend to have high speed and highly complex structure, while merchant ships have slow to moderate speed and their structure is simpler. The crew of a naval ship is composed of personnel for voyage and combat, whereas a commercial ship needs only personnel required for voyage.

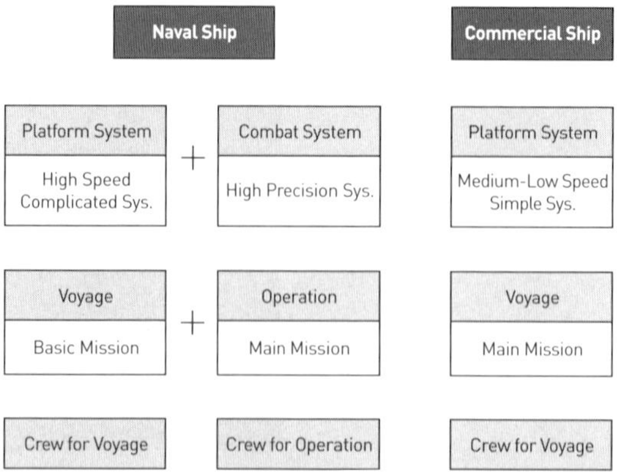

Figure 13-3 System constitution of naval ships and commercial ships

Figure 13-3 shows the major differences of system constitution between naval ships and commercial ships.

The simple addition of a combat system to a platform system does not create a naval ship, because the functions of the two systems are closely related to each other. Only when the two systems meet mutual requirements, they will be able to carry out the functions of a naval ship system successfully. Mutual reliance of the two systems can be illustrated more clearly by the following examples.

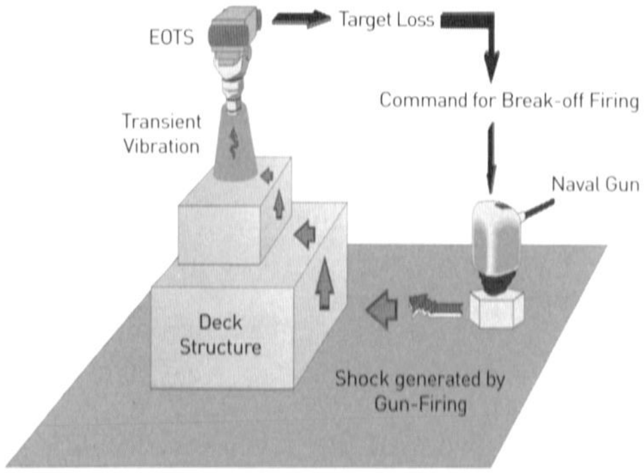

Figure 13-4 Effect of gun firing on a platform with weak structure

In modern naval warfare a target is detected by an electro optical tracking system (EOTS) before a gun can fire. When a gun fires, the shock is transmitted through the ship structure to the equipment on deck. If this happens on a merchant ship, the transient vibration will cause the EOTS to shake, and as a result the EOTS will lose the target making the gun unable to fire as represented in **Figure 13-4**. This means that continuous firing is not possible unless the platform structures is constructed strong and stiff enough to satisfy operational requirements.

The interdependence of the two systems can also be found in the effect of ship motions on operations. As shown in **Figure 13-5**, a ship-borne helicopter can only take off and land when the ship motions are within a certain operable limit. If the ship motions in sea state 4, for example, are less than the given limit, it can be said that the helicopter operations are

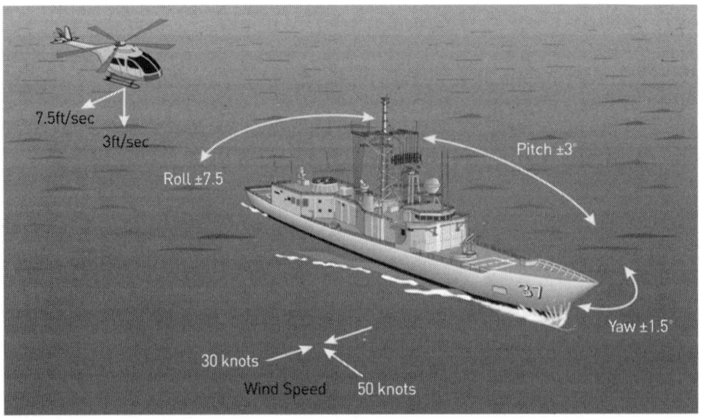

Figure 13-5 An example of ship motion requirements for ship-borne helicopter operations

possible up to sea state 4. This illustrates that the hydrodynamic performance characteristics of the ship platform decides the operability of helicopters, a core element of modern combat system of surface naval ships. Therefore, hull forms should be designed so that they satisfy the combat system requirements, as well as the performance requirements of the ship itself.

The inherent characteristics of naval vessels as a complex combination of platform and combat systems bring about several challenges in design, construction, operation and maintenance. When a developing country wishes to have a naval ship constructed, the order for the ship construction may be given to a domestic shipyard, but advanced combat systems of proven track record usually have to be imported from developed countries. However, the navies of the developed countries themselves, especially the U.S. Navy, must develop new combat systems to stay ahead of others, adding many years to design and construction of a ship. It can take well over 10 years to acquire a new naval ship system, and it can cause extra problems such as modification of the project and increased risk of project failure.

The complex nature of naval ship system can also bring about further problems in operation and maintenance. The ship itself has a lifetime of over 30 years, but the effective life of combat systems is very short, normally several years at most due to rapidly advancing military technologies. As a result, a newly built ship can carry out its operational missions effectively only for a short period before its combat system has to be modernized.

To solve the inherent problems mentioned above, the U.S. Navy has developed and applied new strategies and approaches for naval ship development, such as the evolutionary acquisition strategy, modular payload ship, advanced technology demonstration (ACTD) and integrated product and process development (IPPD).

2. Platform Technologies

2.1 Hull Form Technologies

2.1.1 Displacement Type Monohull

All but a small fraction of naval ships in service are of displacement monohull type, especially in the range of medium to large ships above the corvette class. This is mainly due to the desirable features of conventional monohulls, such as high payload capacity and effective availability of usable internal spaces. Furthermore, recent trends of war ships growing in size favor monohulls as well. The greatest weakness of conventional monohulls is that it is very difficult to increase the ship speed beyond 30 knots. However, this is not too much of an operational handicap, because effects of speed on operational capabilities have been reduced significantly in the last few decades. Various factors contribute to this, including enormous progress of reconnaissance and surveillance capabilities of satellites, high speed of military aircraft, and the increased range of advanced ship-borne weapon systems. Notwithstanding this, if operational requirements dictate higher speeds, alternative high speed hull forms should be selected. **Figure 13-6** shows a typical naval ship with a displacement type monohull.

Figure 13-6 A typical naval ship with displacement type monohull

2.1.2 Displacement Type Multihull

Some ships, such as catamarans, trimaran and SWATH (small waterplane area twin hull), have multiple hulls. Wave-piercing catamarans, which were developed by an Australian shipyard, have been approved by the U.S. Army and Navy for use as high speed logistic support ships. Recently, the Chinese Navy has selected missile boats with hull forms similar to wave-piercing catamarans. In view of this current trend, it is expected that wave-piercing catamaran hull forms will be widely used for naval ships, due to their large deck areas and high speed capabilities.

The practical use of modern trimaran, consisting of a slender monohull and two side hulls of low displacement, was pioneered by the United Kingdom. The Royal Navy developed the world's first ocean-going trimaran ship, the RV *Triton*, as a technology demonstrator. It

Figure 13-7 The littoral combat ship *Independence* of the U.S. Navy

displaces 1200 tons and has demonstrated its capability as a naval surface combatant. The U.S. Navy has also been interested in the ability of trimaran as evidenced by the construction and commission of a littoral combat ship (LCS) *Independence* in 2010 (**Figure 13-7**). Trimaran hull forms could offer a larger deck area, despite their small size, and achieve high speed. In this sense, trimarans are feasible candidates for future naval surface combatants.

The main advantage of the SWATH hull form is in its incomparable superiority of ride qualities. Much effort has been made to make the best use of its advantages for commercial and military vessels. As a result, some SWATH ships have been developed as passenger ships and leisure craft. In its military applications, the U.S. Navy constructed the first SWATH ship *Kaimalino* of 220 tons in the early 1970s. Since then acoustic surveillance ships have been deployed for operations in adverse sea states by the U.S. and Japanese navies. The concept of using SWATH as regular naval surface combatants, however, is still under review.

2.1.3 High Speed Hull Forms

The main high speed hull forms include planing hulls, hydrofoil ships, air cushion vehicles and surface effect ships.

Since the Egyptian *Komar* class missile boats sank the Israeli destroyer *Eilat* with *Styx* anti-ship missiles, high speed small craft with planing hulls became the most attractive naval combatant for smaller navies. Recently, however, the use of planing hulls has become less popular, although Scandinavian countries of Sweden, Denmark and Finland are developing and building high speed craft with planing hulls probably because of their geopolitical situations.

A typical example of military application of hydrofoil is the PHM *Pegasus* class hydrofoil craft of 250 tons that the U.S. Navy has deployed in the early 1980s, to counter the Soviet *Osa* and *Komar* class missile boats. Military hydrofoil craft have also been in service in Italy, Israel and Japan by the 1990s, but all of them have retired from active service. The main reason for their retirement is the complicated procedures required for maintaining the hydrofoil systems.

Due to their ability to operate on both land and water, air cushion vehicles (ACV) have been used mainly as amphibious assault craft, such as the U.S. Navy's LCACs and the Finnish Navy's would-be *Tuuli* class. Their shortcomings, the most prominent of which are high noise level and the size limitation, make them less attractive as amphibious craft.

Surface effect ships (SES) are air cushion vehicles having rigid side hulls unlike ACVs, and cannot, therefore, be amphibious. SESs have low air leakage and benefit from in-water propulsion system, and thus they can achieve high speed and the principle can be used for up to medium sized naval ships. SESs are currently used for high speed missile craft. **Figure 13-8** shows the Royal Norwegian Navy's *Skjold* class missile craft of 270 tons.

Figure 13-8 *Skjold* class missile craft of the Royal Norwegian Navy

2.2 Structure

2.2.1 Structural Design Loads

A ship's structure should be designed strong enough to withstand the loads that it is likely to encounter during its life. The structural loads are categorized into basic loads, sea-environmental loads, operational loads and combat loads, as shown in **Figure 13-9**. Of these, combat loads are not relevant for commercial ships, but for war ships they are classified as follows:

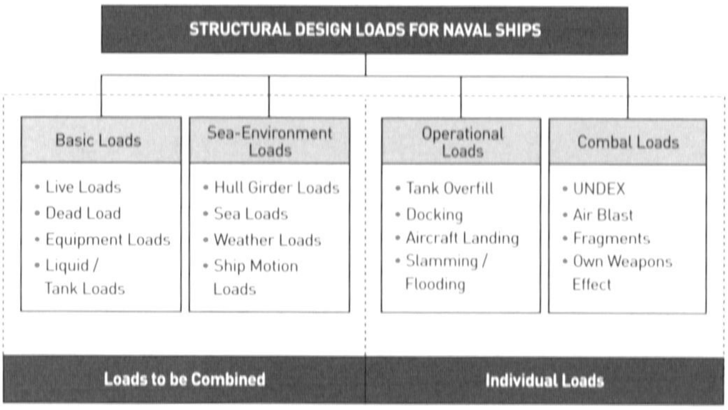

Figure 13-9 Structural design loads for naval ships

1) Shock loads from underwater explosions

Underwater explosions of certain weapons, including torpedoes and mines, lead to shock loads on vessels operating in the vicinity. The shock loads are divided into impact force caused by shock waves and vibration excitation force caused by expansion and contraction of gas bubbles. The impact force element can damage shipboard equipment and local structures, such as foundations, and must be taken into account in designing shock-resistance properties of the structure. The vibration excitation force on the other hand can cause whipping in longitudinal hull girder structures, and must be considered in the structural design for longitudinal strength.

2) Effects of own weapons

Gun blasts, gun recoil and missile blasts due to shipboard weapons being fired have considerable effects on local structure surrounding the guns and missile launchers.

3) Fragment loads

If a naval ship is hit by a bomb, shrapnel created by the explosion can cause serious damage to the ship structure and facilities. Therefore, fragment loads should be accounted for in design of armor for protection of key internal spaces.

4) Air blast loads

The explosion of a nuclear bomb generates a high-pressure air blast, which should be taken into account in the design of exposed equipment and local structures on deck.

2.2.2 Structures of Naval Surface Ships

The primary structural members of naval surface ships are outer plating, deck plating, bulkheads, and longitudinal and transverse frames. Unlike merchant ships for which the transverse frame system is normally used, naval ships traditionally adopt the longitudinal frame system, which is advantageous in structural weight and strength. The U.S. Navy is developing the advanced double hull (ADH) for application in the next generation warships). This structure gives higher survivability in the event of damages. The absence of transverse and longitudinal frames of ADH allows better utilization of internal compartments too. **Figure 13-10** shows the typical longitudinal frame system and ADH of the U.S. Navy.

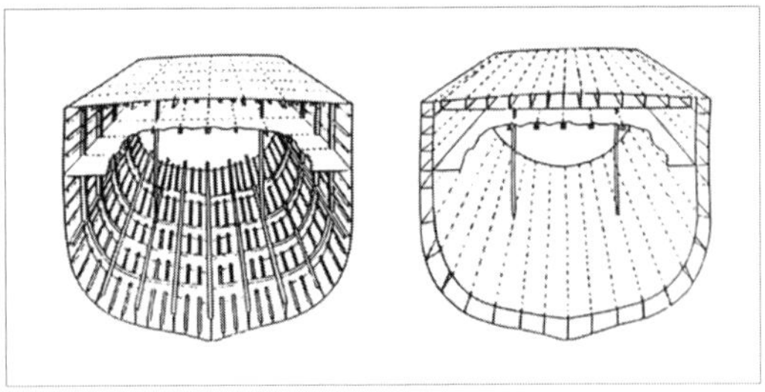

Figure 13-10 Typical frame system and ADH of the U.S. Navy

The weight of hull structure constitutes 40 to 50 percent of the light ship weight of an average naval surface ship with a conventional monohull. Therefore, structural weight should be reduced as much as possible to increase the effective payload.

The proportion of structural weight of U.S. and Russian naval surface ships is about 45%, whereas that of European naval ships is about 39% ~ 43%. The differences come from the applied structural design standards.

Figure 13-11, which is quoted from the structural design manual of the U. S. Navy, shows the influence of hull structure on the ship system in weight and cost.

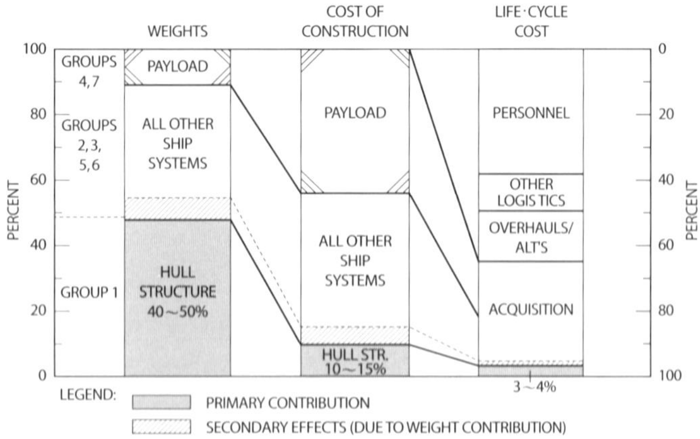

Figure 13-11 Structural weight and cost as percentage of light ship weight and total cost

2.2.3 Structures of Submarines

The primary structural members of submarines are the outer hull, pressure hull, bulkheads and ring stiffeners. In a single hull structure the outer hull acts as the pressure hull. If the outer hull and pressure hull are separate, the structure is called double hull. **Figure 13-12** shows various types of submarine hull structures, single hull, double hull and semi-double hull from the left.

The failure modes of a pressure hull caused by excessive water pressure are shown in **Figure 13-13**. The main failure modes are categorized as follows:

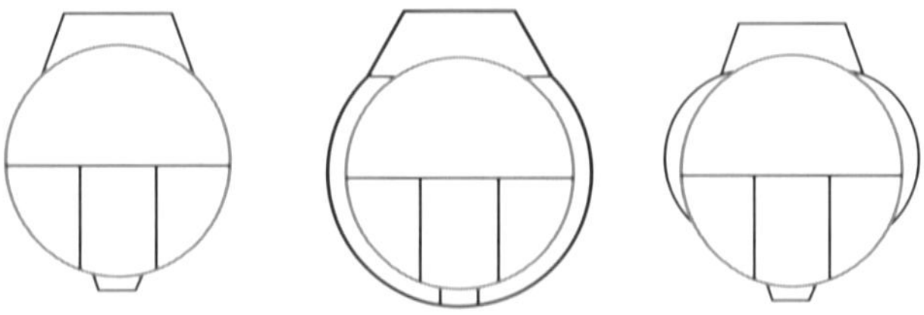

Figure 13-12 Types of submarine hull structure

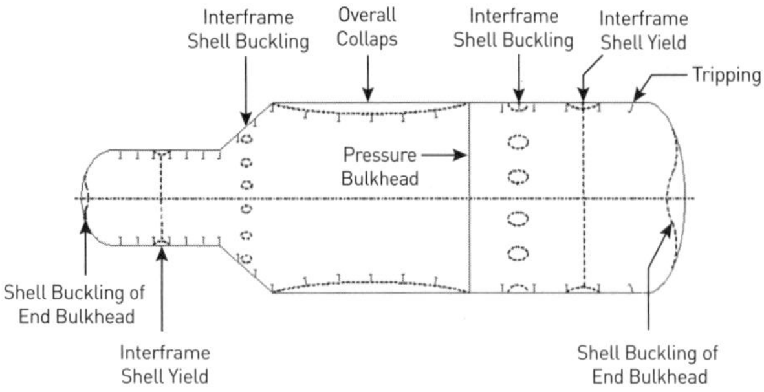

Figure 13-13 Failure modes of submarine pressure hull

- Inter-frame shell yield: a concertina-shaped yield or symmetrical buckling between ring stiffeners
- Inter-frame shell buckling: a non-symmetrical buckling failure mode between ring stiffeners
- Overall collapse: a total collapse failure mode of ring stiffened cylinders

The main materials for the pressure hulls include HY-80, HY-100 and NS-100 high tensile steel, and a titanium-steel alloy was also used to reduce structural weight in the former USSR.

2.3 Propulsion Systems

2.3.1 Propulsion Systems of Naval Surface Ships

The propulsion systems of naval surface ships consist of main engines, propeller shafts, reduction gears, generators, motors and propellers, and can be categorized into two groups: mechanical and electrical systems.

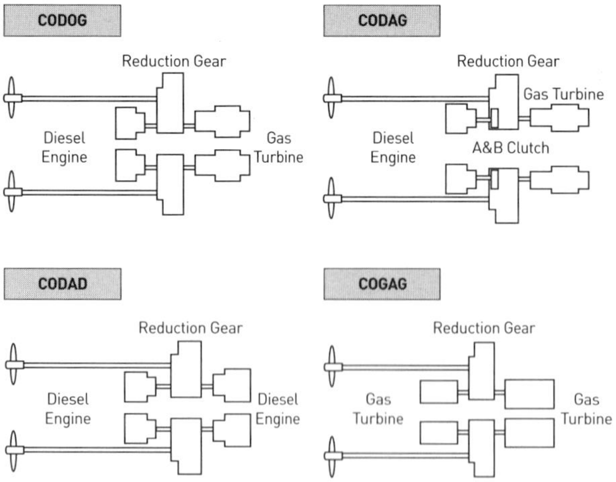

Figure 13-14 Examples of mechanical propulsion systems

1) Mechanical propulsion systems

Most naval surface ships have mechanical propulsion systems. There are various types, depending on the method of combining diesel engines and gas turbines, as shown in **Figure 13-14**.

2) Electrical propulsion systems

Some developed countries have been investing heavily in developing electrical propulsion systems for naval applications. The main reason for this is that the propulsion motors need to be connected to the generators only through cables

(**Figure 13-15**) and this allows free positioning of the generators and consequently the engine room layout can be optimized free from the constraints of direct mechanical connections. Furthermore, the length of the shafts can be minimized, reducing the weight and space required for the mechanical propulsion train considerably. Fuel cost reduction is also possible due to the superior controllability of electrical motors, such as starting and low speed operation, while the engines can be allowed to run at a constant optimum speed. Intensive research effort is being made to develop motors of smaller size but of higher output.

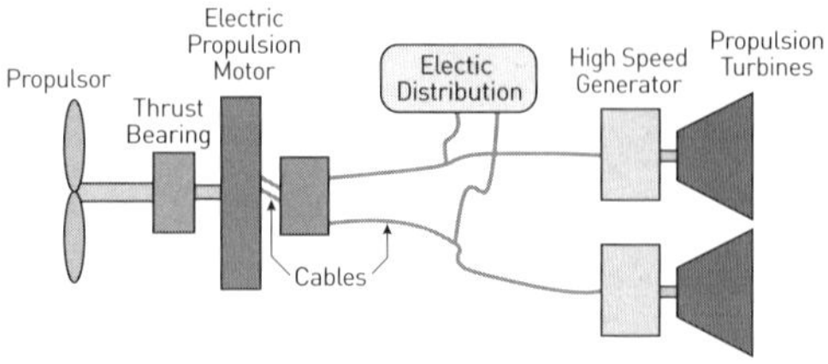

Figure 13-15 Concept of electric propulsion system

2.3.2 Propulsion Systems of Submarines

1) Nuclear power propulsion systems

Nautilus was the world's first nuclear powered submarine that the U. S. Navy commissioned in 1954. It was the beginning of enormously expanded submerged endurance capability of submarines, since such a submarine can stay dived virtually indefinitely and it needs charging with nuclear fuel only once or twice in its lifetime. As shown in **Figure 13-16**, the nuclear power plant of a submarine consists of a reactor plant, steam generator, steam turbine, reduction gear and propulsion motor. The underwater noise level from a nuclear powered submarine would have been high due to the noise of turbines and reduction gears, but it has been lowered substantially through the intense competition in submarine technology between the U.S. and the Soviet Union during the Cold War.

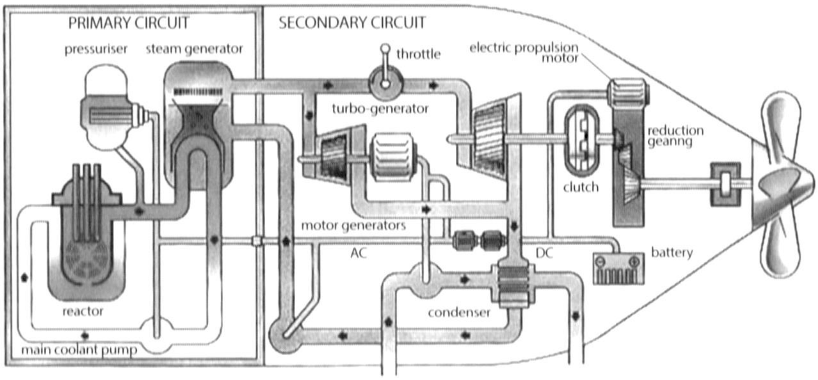

Figure13-16 Concept of a nuclear power propulsion plant for submarines

2) Diesel-electric propulsion systems

Conventional submarines are generally driven by a diesel-electric power plant, which consists of diesel generators, batteries and propulsion motors, as shown in **Figure 13-17**. Diesel-electric submarines have advantages of high reliability and comparatively low underwater noise level. However, they have a decisive handicap in the limited submerged endurance. A bank of fully charged batteries runs out of power in 4 or 5 days, in which event the submarine is forced to surface or snorkel. It goes without saying that the submarines are more easily detected and very vulnerable to attack while on the surface.

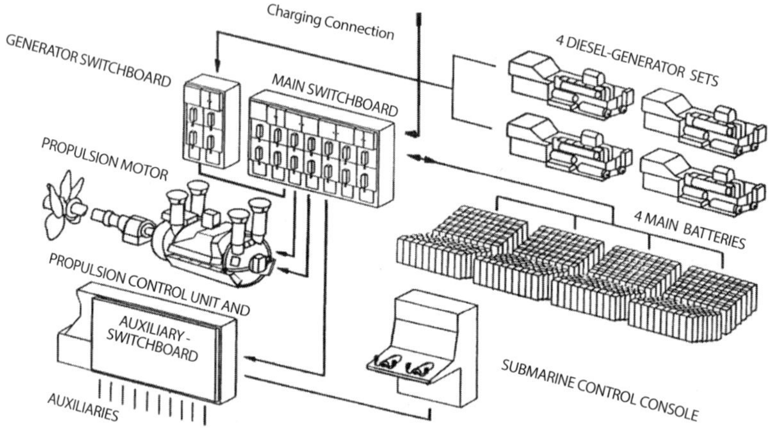

Figure13-17 Concept of diesel-electric propulsion system for submarines

3) Air-independent propulsion (AIP) systems

AIP systems do not need air intake to generate propulsive power and therefore can reduce or eliminate the need for the submarine to surface or snorkel, thereby overcoming the critical weakness of conventional diesel-electric submarines. This has prompted many countries around the world to conduct intensive R&D in this technology since the early 1980s. The currently available AIP systems are closed cycle diesel (CCD) engines, fuel cell systems, Sterling engines and MESMA (closed cycle steam turbines) system.

The CCD system uses standard commercial diesel engines that can be operated on a closed cycle by recycling the exhaust gas with some extra oxygen added from the store. The CCD system can be adopted relatively simply without relying on advanced technologies, but acoustic insulation is essential to reduce the engine noise.

The fuel cell system generates DC electricity from inputs of oxygen and hydrogen by an electrochemical process that is the reverse of electrolysis. It is therefore inherently silent and has high efficiency, but the problems to overcome include safety concerns in onboard storage of hydrogen and in logistics support.

The basic principle of the sterling engine system, developed in Sweden, is that it converts fuel into heat, then into mechanical work and then to electricity. The system uses an external combustion system as a heat source and helium as the working gas which is put through a series of thermodynamic processes between two cylinders. Due to the absence of explosion, the system is quite silent, but it is comparatively complicated and enlarging the capacity is difficult.

MESMA, developed in France, is a closed cycle steam turbine system. In the first part of the system, fuel mixed with stored liquid oxygen is burned and the resulting high temperature and high pressure gas generates saturated steam. In the second part, the steam drives turbines connected mechanically to an alternator. MESMA has an advantage in that small high output engines can be made, but oxygen consumption is excessive and thermal efficiency is low compared to other systems.

The AIP systems mentioned above can extend underwater endurance of conventional diesel-electric submarines to over 2 weeks without surfacing. Since Sweden's submarine *Gotland*, with the very first Sterling engine system in the world was commissioned in 1996, many navies have deployed AIP submarines. The Germans have built 212A class and 214 class submarines with fuel cell systems, while Italian, Greek and Korean navies also have AIP submarines. Japanese navy has AIP submarines with sterling engine systems.

2.3.3 Propellers

One of the main challenges in designing naval propellers is to deal with the cavitation phenomenon. Cavitation occurs when bubbles of vaporized water, generated by pressure drop on propeller blades, collapse during propeller rotation. The presence of cavitation causes blade surface erosion as well as reduction in efficiency, vibration and noise.

The primary method used for surface ship is to design the propellers so that cavitation does not occur in the cruising speed range by adjusting the cavitation inception speed. To eliminate cavitation that occurs at high speed, the PRAIRE (propeller air induced emission) system can be used. This system prevents water vapor collapsing by jetting high pressure air through air holes made on the blade edge. **Figure 13-18** shows the cavitation phenomenon and the effect of the PRAIRE system.

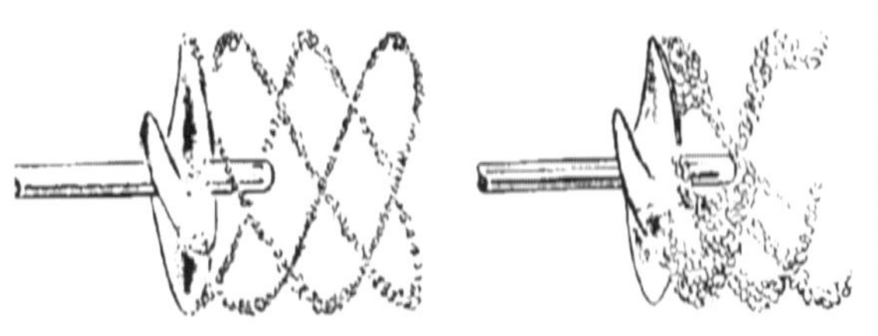

Figure 13-18 Cavitation and the effect of the PRAIRE system

Cavitation on submarine propellers is one of the main sources of underwater radiated noise, and much R&D effort has been directed at reducing it. As a result, highly skewed propellers have been developed and widely used. To reduce the underwater radiated noise further, pump jet type propulsors have been developed and installed on nuclear attack submarines of the U.S. Navy and the Royal Navy. Propeller types used in submarines are shown in **Figure 13-19**.

Figure 13-19 Types of submarine propellers

3. Naval Combat System Technologies

3.1 Introduction

In its broad sense, a naval combat system is defined as the whole system that is employed

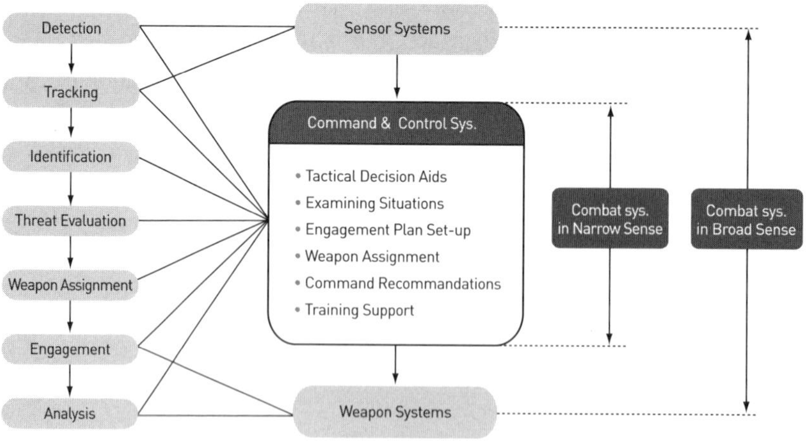

Figure 13-20 Definitions of naval combat system

for waging battles, such as command and control systems, detection systems and weapon systems. However, in its narrow sense it refers to the command and control system only. **Figure 13-20** shows the two definitions of naval combat systems. In this book, the term naval combat system is used in its broad sense.

3.2 Command and Control System

The main function of the command and control system is to provide a detailed tactical decision aid by examining tactics, setting up engagement plans and assigning weapons on the basis of sensor data obtained by the detection systems. Thus, the system could be referred to as an integrated information management system which connects detection equipment, armament, combat units and the commander (decision-maker) to each other.

Therefore, the system is software-based, and as a result, software development costs take up over 60% of total system development cost. Moreover, software updating takes up most of the costs required for system upgrade which is required several times during the ship's lifecycle in order to keep up with the changes in operational conditions including enemy threats and weapon systems.

The core function of the command and control system is to maximize the combat effectiveness of the vessel by processing a large amount of battle information in real time, using high performance computers.

Figure 13-21 Operation of shipboard AEGIS system

Currently, the principal threat to naval surface ships is multiple, simultaneous attacks of anti-ship missiles (ASMs). If an incoming missile is detected in favorable conditions some 10 to 20 nautical miles away, it leaves the ship two minutes or less to respond to such a threat. Even an ASM launched from an aircraft at high altitude and detected at a range of 100 nautical miles allows the defender only 10 minutes. For this reason, the defense capability against ASMs is of great concern to naval surface ships. Consequently, the U.S. Navy has developed the AEGIS system which can simultaneously identify, prioritize and track over 200 aerial targets and engage with a large number of them. It began to be deployed in cruisers and destroyers in the 1970s. **Figure 13-21** shows a scene of shipboard AEGIS system in operation.

The functions of command and control systems of submarines are basically identical to those of the surface ships. However, the command and control system of a submarine is mainly based on sonar information, whereas the system of a surface ship is mainly based on radar information. **Figure 13-22** shows the role of submarine command and control systems.

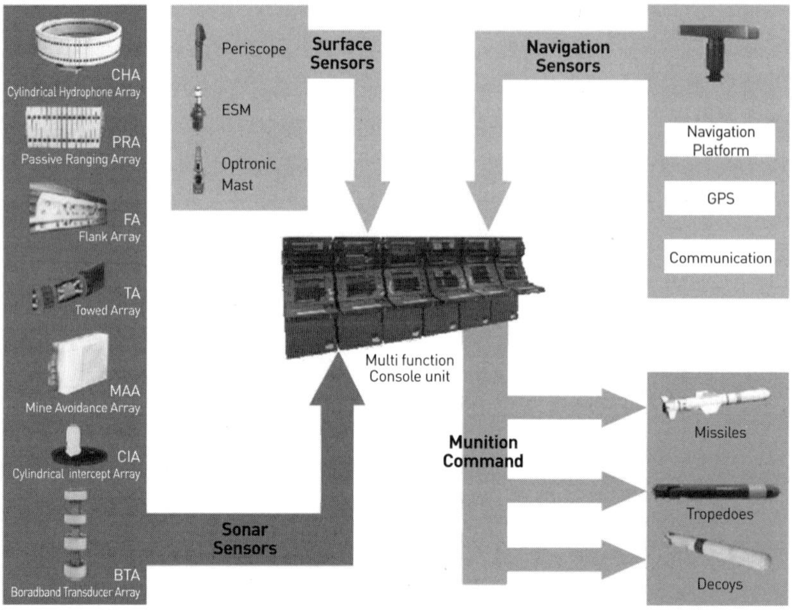

Figure 13-22 Role and function of command and control system in submarines

3.3 Detection Systems

3.3.1 Radars

Radar (radio detection and ranging) is a target detection system that uses radio waves to determine the range, altitude, direction and speed of a target. Radar emits radio waves through a transmitter and the waves reflected by any object are captured through the receiving antenna and processed by a signal processor. Worked on by researchers all over the world from the early 20th century and used to great effect during the Battle of Britain in 1940 and throughout the second World War by all warring parties, radars have advanced enormously since those early days. Naval radars, first used in the 1940s, developed into two dimensional and three dimensional systems and to the current phased array multifunction radar systems which have become the mainstream. A phased array multifunction radar system is able to perform simultaneous search and track of over 200 targets, and provide mid-course and final guidance updates to the interceptor missiles launched from a ship.

The phased array multifunction radar system was developed by the U.S. Navy in the late 1960s. It has since deployed AN/SPY-1 phased array radar systems, as part of the AEGIS combat system, on all the cruisers and destroyers in its fleet. Some European countries, including the UK, Germany, France and Italy, have also developed their own phased array multifunction radar systems jointly or on their own since the 1980s.

There are two categories of phased array radar systems: active and passive systems. AN/SPY-1 radar of the U.S. Navy is typical of the passive phased array radar, whereas APAR radar of Germany and the Netherlands, SAMPSON radar of the United Kingdom, EMPAR radar of France and Italy are active phased array radar. **Figure 13-23** shows phased array radar systems mentioned here.

AN / SPY-1D (U.S.A.) APAR (Germany) EMPAR (France / Italy) SAMPSON (U.K.)

Figure 13-23 Major phased array radar systems of the world

3.3.2 Sonar Systems

Sonar (sound navigation and ranging) is a detection system for targets on the surface or underwater using sound propagation and reflection to determine the direction and distance of a target. Currently, sonar system is the only means to detect, classify and track underwater targets. As shown in **Figure 13-24**, sound propagation underwater is influenced strongly by temperature and salinity of the water, and the seabed profile, creating acoustic refraction, reflection and reverberation. For these reasons, it is not easy to detect underwater targets by the use of sound waves. It is more difficult for surface ships to detect submarines than for submarines to detect surface ships, because the noise level of a surface ship is normally higher than that of a submarine.

If a submarine stays in the shadow zone, as illustrated in **Figure 13-24**, sound waves emitted by hull-mounted sonar equipment of surface ships cannot reach the target. Therefore, surface ships use additional sonar systems, such as towed array passive sonars and dipping sonars of ship-borne helicopters. For a submarine that operates mainly underwater, sonar systems are indispensible. **Figure 13-25** shows various kinds of sonar systems currently in use.

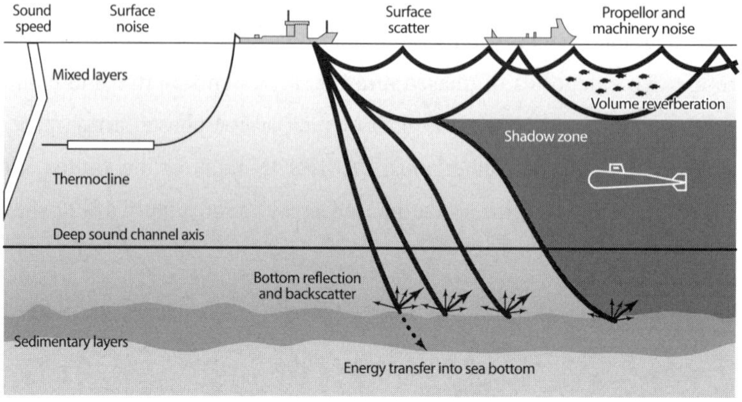

Figure 13-24 Sound propagation in oceanic acoustic environment

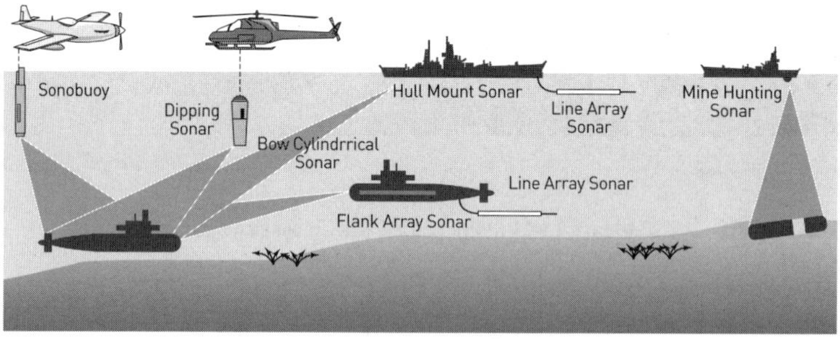

Figure 13-25 Sonar systems currently used

3.3.3 Periscopes

Periscopes are a passive sensor of submarines that are used to search and classify targets visually and are the very first detection equipment used when a submarine rises close to surface. Periscopes are generally categorized into attack and search periscopes.

Attack periscopes penetrate the pressure hull, and permit the submarine commander a direct view through an optical relay system. They are designed to be as small as possible, so that their wake on the surface is difficult to detect.

Except during attack situations search periscopes are more commonly used. Therefore, optical visual systems of search periscope are augmented with various optronic sensors needed for day and night operations, such as TV cameras, IR imaging systems and laser rangefinders. The enhanced search periscopes are known as optronic masts. More recently modular optronic masts have become popular which are non-pressure hull penetrating, since all sensors transmit their signals via cables, and the modularity makes it easy to replace and install new equipment. **Figure 13-26** shows a typical hull penetrating mast and non-penetrating modular hoisting mast systems.

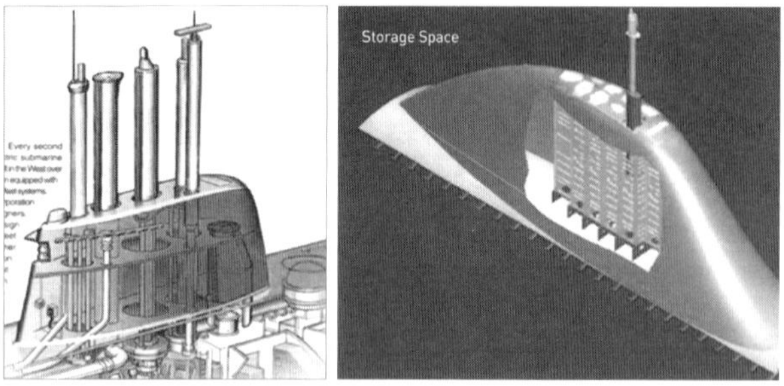

Figure 13-26 A typical hull penetrating mast and non-penetrating modular hoisting mast system

3.3.4 Electro-Optical Tracking Systems and Infrared Search and Tracking Systems

The electro-optical tracking system (EOTS) is a detection and tracking system that changes the original optical images to digital data to provide video information. The core function of EOTS is to increase security of the ship through rapid tracking. It acquires target data such as direction, altitude and distance and transmits this data to the command and control system. The EOTS system is usually composed of a daylight TV camera, IR imaging system and a laser rangefinder (LRF).

The infrared search and tracking system (IRST) is a self-defense sensor system. It detects, catches and tracks IR signals emitted by incoming anti-ship missiles (ASM) approaching at low altitude near the water surface. IRST is a useful piece of supplementary equipment, in the event of the radar and ESM systems of the ship not being able to detect ASMs approaching at low altitude.

3.4 Weapon Systems

3.4.1 Naval Guns

Battle ships, the largest surface combatant ships equipped with many large-caliber guns, were very good in attacking stationary targets at long range, but very vulnerable to simultaneous attacks from multiple aircraft. Naval battles, which used to be between ships before and during the First World War, changed to battles between aircraft and ships during the Second World War. To the dismay of many naval commanders of the second War, most battle ships were sunk or damaged severely by aircraft. Unsurprisingly, battle ships disappeared from fleet formation after this.

With the retirement of battle ships, their large 16 inch guns also disappeared from naval battles, leaving 100~130 mm guns as the main conventional armament. Naval guns are much inferior to guided missiles in their range and destructive power. However, guns are still widely used in naval surface ships because of their advantages, i.e. low acquisition and maintenance cost, high reliability, effective space utilization and ability to deliver large quantities of ordnance within a short period. The size of naval guns includes 76 mm, 57 mm, 35 mm and 27 mm, and diverse guns are favored by various navies of the world.

Since the ending of the Cold War naval warfare has shifted from the open oceans to littoral areas, and that is the reason for large surface ships to be given the additional mission of providing naval surface fire support (NSFS) to land battles. Consequently, the 127 mm principal guns have been improved for the U.S. Navy to use extended range guided munitions (ERGM) to secure range and accuracy required for NSFS operations. Since the incident of a suicide attack by a small boat on the U.S. Navy destroyer USS Cole in the port of Aden in 2000, the U.S. Navy's large surface ships have been equipped with additional small guns, such as the 25 mm machine guns, on deck to enhance capabilities against asymmetric threats such as surprise attacks by small fast boats and to assist in anti-piracy operations. **Figure 13-27** shows a 127 mm principal gun and a 25 mm machine gun mounted on a U.S. Navy's surface ship.

Figure13-27 A 127mm gun and a 25mm machine gun mounted on the deck of a U.S. Navy's surface ship

3.4.2 Missiles

1) Anti-ship missiles

As stated in Section 2, the Egyptian *Komar* class missile boats sank the Israeli destroyer *Eilat* with *Styx* anti-ship missiles during the Middle-East War in 1967. This was the first time that a ship was sunk by a missile launched from a boat, and ever since then anti-ship missiles

(ASMs) became the principal threat to surface ships. Currently, more than 75 countries deploy over 45 types of ASMs.

Diverse countermeasures against ASMs have also been developed, and operational environment for ASMs have become unfavorable due to geographic configurations in littoral areas that have become the main battle ground. For these reasons ACMs have also evolved continually in terms of extremely low altitude sea-skimming flight capability, weaving maneuvers during flight, pop-up maneuvers in the terminal phase, low radar cross-sections and low level signatures of seekers.

Figure 13-28 shows various paths of ASMs approaching a ship and illustrates the operational environment of ASMs in the littoral area.

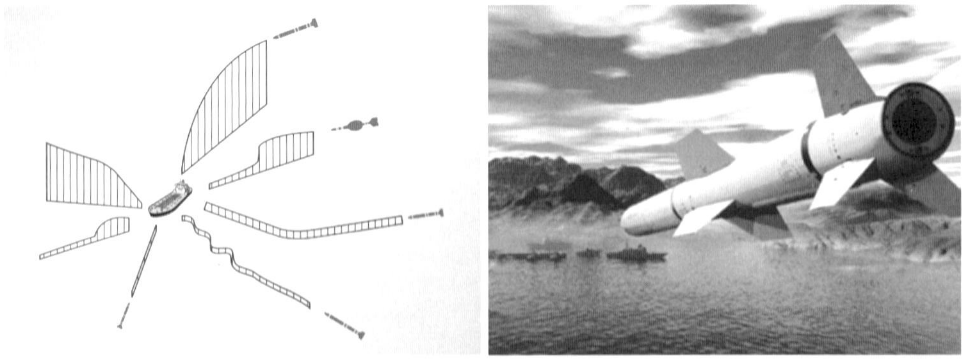

Figure 13-28 Approach paths of ASMs and typical operational environment in the littoral area

2) Anti-air missiles

Since the anti-ship missile *Exocet* launched from an Argentinian aircraft sank the Royal Navy's destroyer *Sheffield* during the Falklands War in 1983, aircraft equipped with ASMs have been recognized as an extremely effective long-range anti-ship weapon. For this reason, it has become essential for large surface ships to have anti-air missiles (AAMs), which can directly attack high altitude aircraft.

Typical AAMs are a family of Standard Missiles (SM) of the U.S. Navy, such as SM-1, SM-2 and SM-3. The SM-3 missile can attack even intercontinental ballistic missiles (ICBM) at their early and final flight phases. Therefore, AEGIS class cruisers and destroyers equipped with SM-3 missiles possess the capability of intercepting ICBMs. Since the vertical launching system (VLS) was developed, surface ships are capable of storing a large number of ASMs and AAMs and rapidly firing them. This is a significant improvement in combat capabilities against air attacks. **Figure 13-29** shows an SM-3 being launched and the VLS system of a U.S. Navy destroyer.

Figure13-29 Launching of an SM-3 missile and the VLS system of a U.S. Navy destroyer

3) Land attack cruise missiles

The *Tomahawk* cruise missile of the U.S. Navy is a typical land attack cruise missile and was developed originally as an anti-ship cruise missile in the 1970s and deployed as a precision guided munitions (PGMs) against vital land targets in the early 1990s.

Using cruise missiles like *Tomahawks*, submarines and surface ships can now precision attack targets over 1000 nautical miles away, while keeping themselves safe, with accuracy better than 80%. Since the Gulf War of 1991, and through Iraq War, Bosnia Campaign and Kosovo Campaign, long-range precision strikes using cruise missiles have been established as the most effective way of showing off their military might by the superpowers.

The range of *Tomahawk* land attack cruise missile is 900~1,300 km, and 2,500km if it is carrying a nuclear warhead. Tomahawk can get close to long range targets through the application of various navigation techniques, such as terrain contour matching (TERCOM), digital scene matching area correlation (DSMA), inertial navigation system (INS) and GPS. **Figure 13-30** shows a cruising *Tomahawk* missile and one being launched from a submarine.

Figure 13-30 A *Tomahawk* missile cruising and its launching from a submarine

4) Ship's defense system against ASMs

As mentioned already, anti-ship missiles (ASM) are the principal air threat to surface ships. The primary defense against ASMs is a pre-emptive strike on the hostile missile firing station, such as an enemy ship or aircraft, and destroying it. If that is not possible, ASMs penetrating the outer defense layer must be destroyed or evaded using the ship's self-defense system.

There are three defense layers against ASMs as shown in **Figure 13-31**.

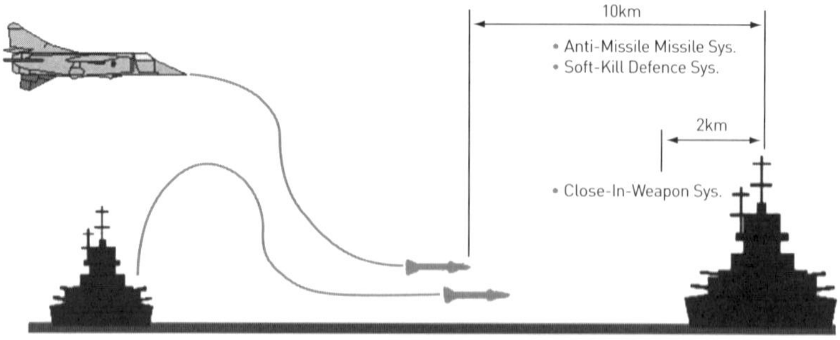

Figure13-31 Three defense layers against ASMs

The first layer is the anti-missile missiles, such as Rolling Airframe Missiles (RAM) or Evolved Sea Sparrow Missiles (ESSM), which can track and destroy air-borne ASMs directly at a greater range than the gun-based CIWS described below.

The second layer is the soft-kill defense systems, such as off-board expendable decoys and jammers to replay false signals, which are used to evade the incoming ASMs.

The third layer is the close-in-weapon systems (CIWSs) employing 20mm rapid-fire guns and are used in the range of 2km or less to intercept ASMs that penetrate the outer air-defense screens. CIWSs have a higher rate-of-fire of 3000 rpms. Currently, many navies have their surface ships equipped with CIWSs such as the *Phalanx* system of the U.S. Navy and the *Goalkeeper* system of the Netherlands.

3.4.3 Underwater Weapons

1) Torpedoes

Since invented by a British man Robert Whitehead for the Austrian navy in 1866, self-propelled torpedoes have become hugely popular. Originally used to attack surface ships, the weapon has been diversified and developed into powerful attack weapons of various kinds. Modern torpedoes may be classified into two groups: lightweight torpedoes used by surface ships and aircraft to attack submarines; and heavyweight torpedoes used by submarines to attack surface ships and other submarines.

As shown in **Figure 13-32**, lightweight torpedoes can be fired from a variety of platforms such as surface ships, aircraft and helicopters. On the other hand, heavyweight torpedoes are discharged from the torpedo tubes installed in the submarine bow. Submarines can also fire anti-ship missiles or land attack cruise missiles through the torpedo tubes. A submarine may be equipped with four to eight tubes, but usually carries no more than 20 torpedoes. A number of firing mechanisms are used, including swim-out, compressed air launch, hydraulic ram launch and air turbine pump (ATP) launch.

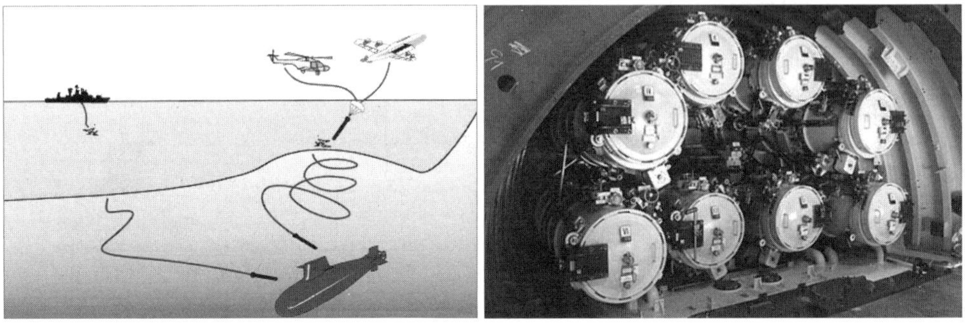

Figure13-32 Operational concept of lightweight torpedoes and torpedo tubes of submarines

2) Mines

Since the end of the World War II, mines have damaged fourteen U.S. Navy ships, whereas air and missile attacks have damaged three. This demonstrates what menacing weapons mines are to surface ships and submarines. Mines are inexpensive and can be used as a strategic weapon blocking and deterring enemy's shipping movements and are also effective as a psychological weapon. Therefore, mines are recognized as a very cost-effective underwater weapon. Since the 1990–1991Gulf War, the number of mines increased by 50% or more worldwide. Furthermore, new mines are emerging continuously that are intelligent and difficult to detect and classify.

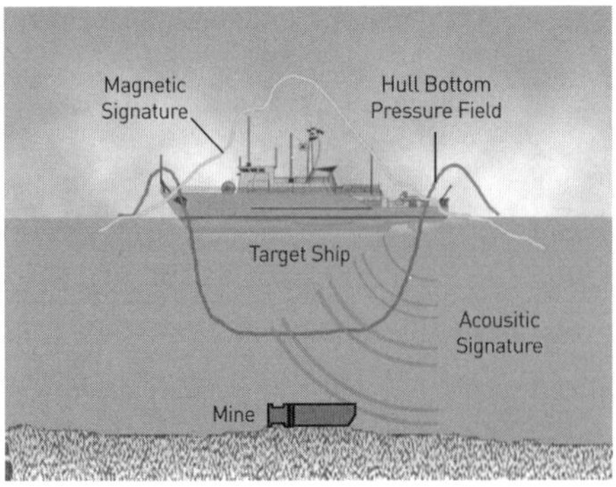

Figure13-33 Principles of detecting ship's signatures

Mines can be laid from a variety of platforms, but submarines are preferred due to its covert nature.

Figure 13-33 shows the principles of detecting a ship's signature by using magnetic, acoustic and pressure sensors built in mines.

4. Survivability Technologies

4.1 Survivability of Naval Ships

Modern naval warfare can be defined as multidimensional, complex warfare conducted on multiple battle fields: land, sea, air, space and cyber space. It is important to point out again that, following the ending of the Cold War, the main battlefield of naval warfare has shifted from the open ocean to the littoral areas. The complicated combat environment and the much strengthened asymmetrical combat capabilities of littoral countries have diminished the advantages of nations with traditionally powerful navies. Furthermore, rapid advancement of military technologies has diversified naval weapon systems and improved their performance.

Due to the drastically altered patterns of naval warfare, it is not possible to block or to evade an enemy's attack in advance completely. In other words, the probability of receiving preemptive strikes from an enemy is very high in modern warfare. Therefore, measures must be taken to minimize the risk of receiving preemptive strikes, but, even if a preemptive strike occurs, the ship must have ability to counterattack immediately.

The ship's survivability under attack is the basis for an effective counterattack, and that is why modern naval ships must have survivability features to enhance the ship's ability to survive in a combat environment as well as the attacking power. As shown in **Figure 13-34**, the survivability of naval ships depends on three elements: susceptibility, vulnerability and recoverability.

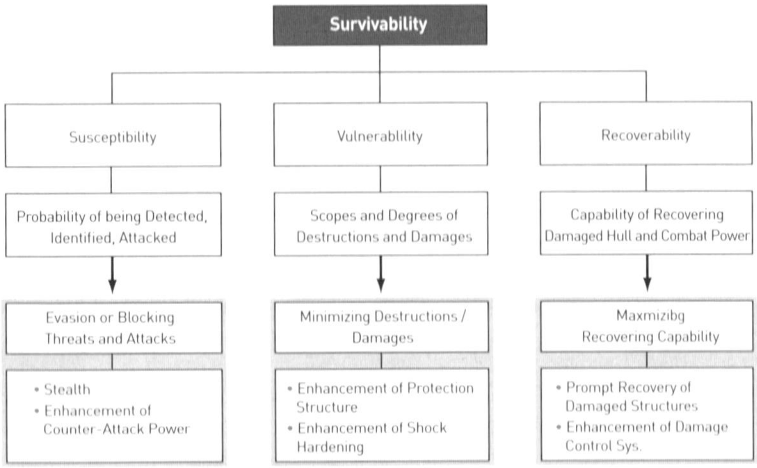

Figure13-34 Survivability elements of naval ships

4.2 Stealth Technologies

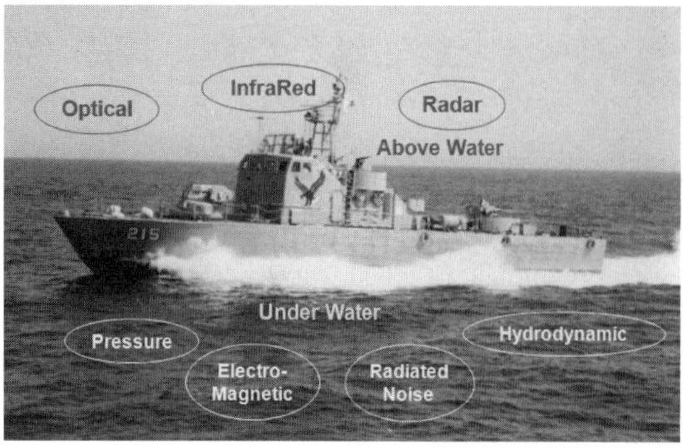

Figure 13-35 Signatures generated by naval ships

As described in **Figure 13-34**, the first step in enhancing the survivability of a naval ship is to minimize its susceptibility by minimizing the probability of being detected by an enemy. To achieve this objective the ship's signatures, which the opponent tries to catch, must be reduced by stealth technologies.

Naval ships generate a variety of signatures as summarized in **Figure 13-35**. Thus, the purpose of stealth technologies is to diminish primary signatures, such as underwater radiated noise, radar cross section, infrared (IR) signatures and electro-magnetic signatures.

4.2.1 Underwater Radiated Noise

The present day passive sonar systems can detect and classify the underwater noise radiated from target surface ships or submarines. Therefore, the underwater radiated noise level is the main factor determining whether one detects the opponent first or is detected by the opponent first. Especially the very survival of covert vessels like submarines literally depends on their acoustic stealth.

The underwater radiated noise of submarines is composed of sounds emitted by a variety of noise sources. These sources include vibration and noise generated by onboard machinery

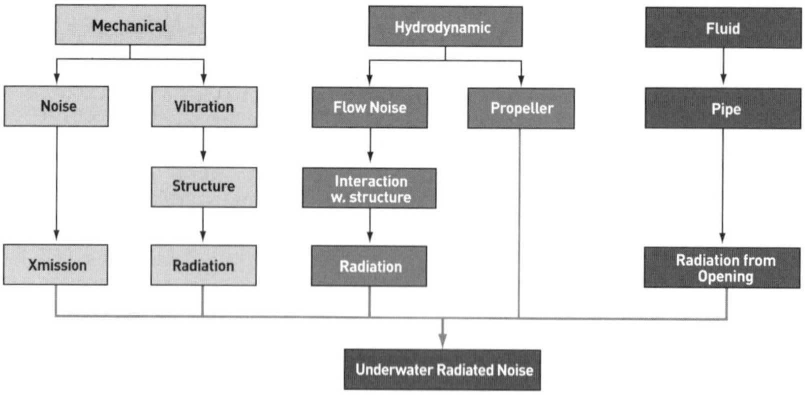

Figure 13-36 Noise sources and propagating paths of submarine underwater radiated noise

operations; propeller noise; hydrodynamic flow noise generated during its maneuvers; and noise generated by fluid movement in the piping systems.

Tonal noises of machinery are the main noise source of a submarine in the low and medium speed range, whereas broad band hydrodynamic flow noises are the main noise sources in the medium speed range; and broad band propeller noise is the main noise source in the high speed range. **Figure 13-36** shows noise sources and the propagating paths of submarines.

To reduce underwater radiated noise of naval ships, noise levels must be reduced at source. More specifically, machines and equipment should be selected based on their ability to operate quietly. Highly skewed propellers and pump jet propulsors on submarines make less noise, as already discussed in Section 2 of this chapter. To reduce hydrodynamic flow noise, the bow and the conning tower can be streamlined to minimize vortex.

In addition, the pathway of the inboard noise and vibrations being transmitted to the outside through the hull can be blocked. Various measures, such as isolation of vibration, application of dampening and absorbing material, acoustic insulation and underwater sound coating, have been used. **Figure 13-37** shows various measures to block the noise pathway.

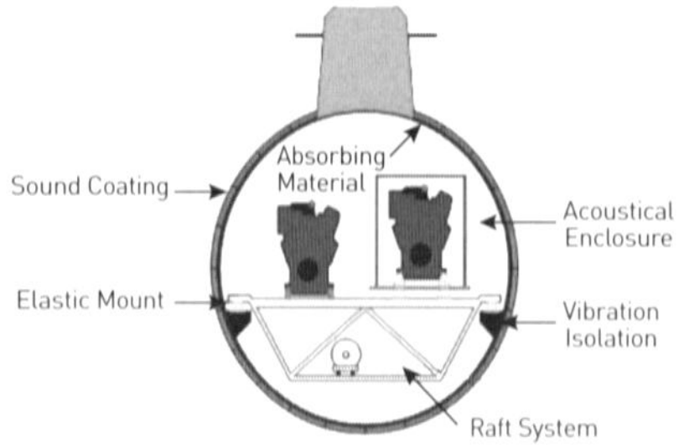

Figure 13-37 Measures to block the pathway of inboard noise and vibration to outside through the hull

4.2.2 Reduction of Radar Profile

Since World War II radars, the all-weather, long-range detection sensor system, have been utilized in various areas such as searching and tracking of aircraft and ships, battlefield surveillance, early warning and missile guidance. Naval surface ships are major targets of radar systems, and consequently, radar stealth has become a key element of naval surface ship design.

The detection range of radar depends on the strength of radio wave energy reflected by the target, and the index of the strength of the reflected waves is the radar cross section (RCS). The RCS of a ship is defined as the ratio of the strength of the radio waves emitted towards it to the strength of the waves reflected towards the emitter.

The pattern of radio waves scattered by the ship is complex and diverse, and depends on the shape of the ship's upper structures, as shown on the left side of **Figure 13-38**. The pattern and strength of radio waves thus radiated by the ship are shown on the right side of the Figure. The outside diagram shows the original reception strength of the ship, whereas

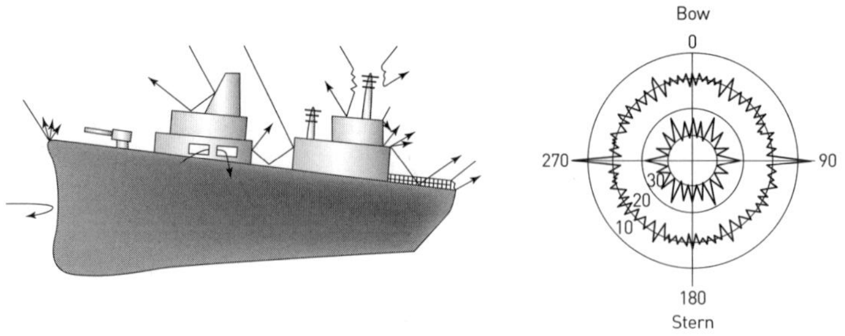

Figure 13-38 Radio wave radiation of a ship (left) and RCS diagram of the ship (right)

the inside diagram presents the reception strength of the ship decreased by the application of RCS reduction technologies.

In order to reduce the distance at which the ship is detected, therefore, it is clear that the ship's RCS must be reduced somehow and this is usually done through the following four methods:

- Optimization of the shape of ship's deck structure;
- Shielding;
- Application of radar absorbent material (RAM); and
- Optimization of topside micro-geometry.

Figure 13-39 shows the appearance of a new Italian frigate to which RCS reduction technologies were systematically applied in contrast to an old Russian cruiser which did not benefit from any reduction technology. There is no doubt as to which one will be more visible to radar.

Figure 13-39 Appearances of a new Italian frigate (right) and an old Russian cruiser (left)

4.2.3 Infrared Signature Reduction

Infrared (IR) is an electromagnetic wave, existing between visible lights and radio wave spectra, and can go through an opaque body which visible lights cannot penetrate. However, IR is absorbed or scattered by gases in the air or vapors in rain, snow, fog or smoke and thus its strength is decreased. Military IR imaging systems give better resolution, can be smaller and are cheaper compared to radar systems. But they cannot give distance information and are very sensitive to weather conditions, such as humidity and temperature. **Figure 13-40** shows various wave length bands and subdivision of infrared waves.

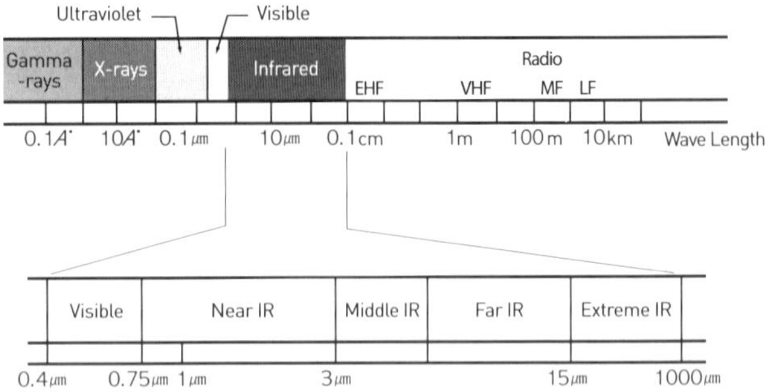

Figure 13-40 Various wavelength bands and subdivision of infrared waves

IR signatures of surface ships contain two main groups: middle infrared of exhaust gases from the funnel; and far-infrared radiated from the ship's deck and upper structure when heated by the sun. ASMs with an IR seeker detects and latches onto the mid-infrared signatures radiated by the funnel exhaust gases. A ship or satellite with an electro-optical tracking system (EOTS) relies on the target ship's far-infrared signatures to obtain infrared images.

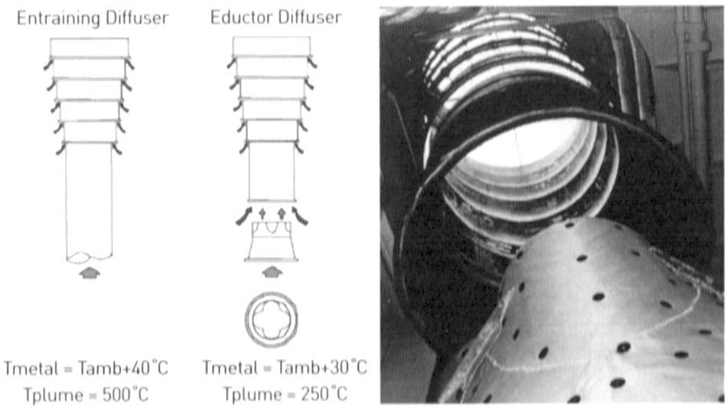

Figure 13-41 Funnel Cooling Systems of Modern Naval Surface Ships

The most important method of reducing the mid-IR signatures of a ship is to cool the exhaust gas by using ambient air or by mixing air with the exhaust gas. **Figure 13-41** shows the popular funnel cooling system applied in modern naval surface ships. Another method uses water sprinklers to cool the hot surfaces of the hull, deck and superstructure, and confuses the recognition ability of the IR imaging systems.

4.2.4 Electro-Magnetic Signature Reduction

Since their invention, mines have been used as a very potent and supremely cost-effective weapon for maritime operations for over 200 years. The earliest mines were mostly contact

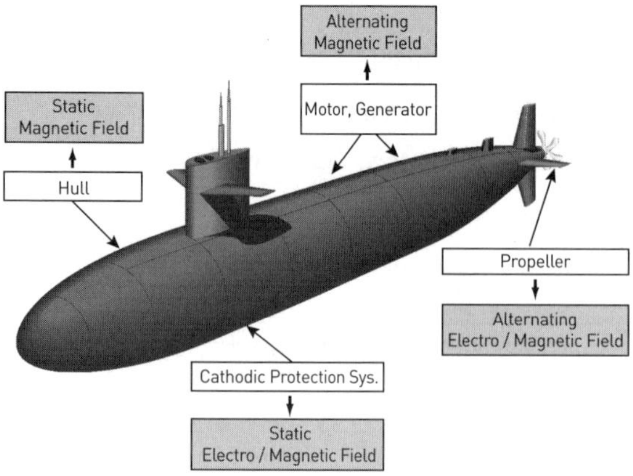

Figure 13-42 Electro-magnetic signature sources of submarines

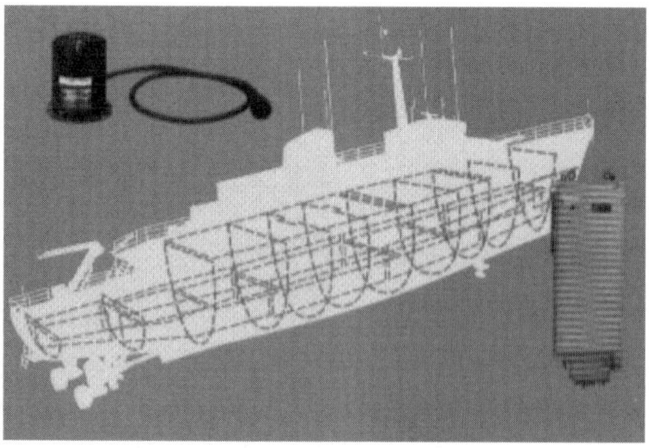
Figure 13-43 Degaussing coil system of a mine warfare ship

mines that needed to be touched by the target to be detonated, but modern influence mines are triggered by the ship's signatures, such as acoustic, electro-magnetic, water pressure and wake signatures. The fuses of influence mines incorporate usually one or more of the ship's signature sensors, but explosive timing is determined mostly by using an electro-magnetic sensor. Therefore, the first priority in overcoming the threat of mines is to reduce the electro-magnetic signatures of the ship.

The steel-hulled ships and submarines develop electro-magnetic signatures as they travel, due to the magneto-mechanical interaction with the earth's magnetic field that is generated by the rotation of the earth. In addition, electro-magnetic signatures of the ship are also generated by hull corrosion, cathodic protection systems, sacrificial anodes, propellers and electrical equipment such as motors and generators. **Figure 13-42** shows various sources of electro-magnetic signatures of submarines.

There are two main ways to reduce the electro-magnetic signatures of a ship: the permanent magnetic fields of the ship can be reduced or eliminated by using a specialized deperming facility; the permanent and induced magnetic fields of a ship's hull can be reduced or eliminated by using degaussing coils installed in the ship's hull. Both measures, deperming and degaussing, are applied to surface ships; whereas, in general, only deperming is used for submarines. The hulls of mine warfare ships are constructed of non-magnetic materials such as wood or GRP. Engines and other main equipment are also manufactured of non-magnetic materials to reduce electro-magnetic signatures. **Figure 13-43** shows the degaussing coil system installed in a mine warfare ship.

4.3 Ship Protection Technologies

4.3.1 Shock Hardening against Underwater Explosions

Even though naval vessels have become somewhat more transparent by using stealth technologies, it is not possible to remove the probability of being hit by an enemy weapon entirely. Therefore, to increase survivability of the ship, the vulnerability of the ship must be reduced as much as possible. One of the main measures used for reducing vulnerability is shock hardening technique to minimize damage to the hull and onboard equipment caused by underwater explosions.

If an underwater weapon, such as a mine or a torpedo, touches the hull or is detonated in close proximity of a ship, the ship can be sunk or receive irrecoverable damage. Even when a non-contact underwater explosion occurs at a fair distance away from the ship, shock waves and pulsating energy of gas bubbles can cause crippling damage to shipboard equipment and hull girders and local structures. This can lead to loss of ability to carry out the mission even if the ship survives. **Figure 13-44** illustrates the phenomenon of a non-contact underwater explosion and its effects on the ship.

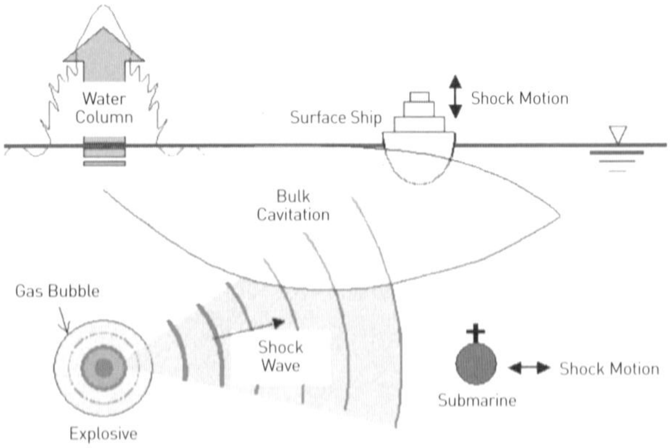

Figure13-44 Non-contact underwater explosion phenomenon and its effects on the ship.

A non-contact underwater explosion produces a powerful shock wave that strikes the exposed hull and causes shock response of the ship plating. The shock response is transmitted to equipment, foundations and other local structural elements. In this way, the entire ship structure and shipboard equipment are affected in a manner known as shock behavior. If no anti-shock measure has been taken, the shock behavior can damage equipment and local structures. Most ships have sufficient structural strength to resist the shock as long as the explosion is not too close by, but an explosion at close proximity is a different matter.

The gas bubbles generated by a non-contact underwater explosion expand and contract repeatedly while rising to the surface, and a pulsating energy is produced during this process. If the bubbles' pulsating period coincides with the natural period of the ship's hull girder, the hull girder can undergo low frequency vibrations, the so-called 'whipping response.' This

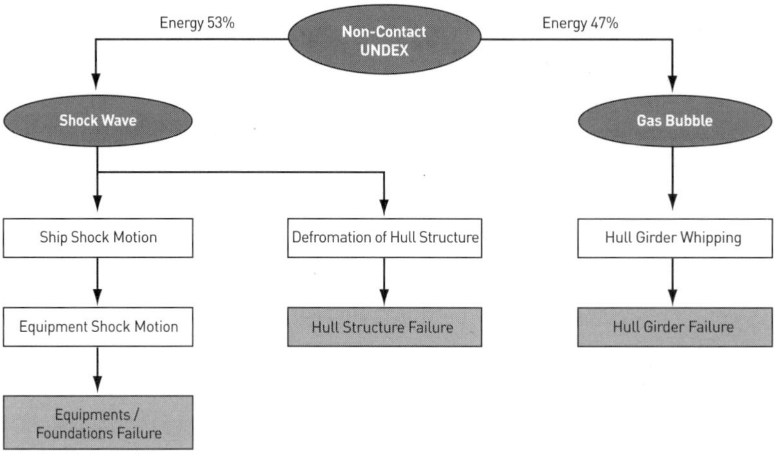

Figure 13-45 Types of damages to a ship caused by non-contact underwater explosions

can cause large deformations of the hull girder, and can even break the ship into two. **Figure 13-45** shows various possible damages to a ship due to a non-contact underwater explosion.

Shock hardening of naval vessels involves designing the hull and equipment so that they can withstand the shock of prescribed strength and verifying their shock resistance performance. There are four methods of shock hardening of naval ships as illustrated in **Figure 13-46**.

Empirical methods are based on experiential data and empirical formulae, whereas theoretical methods use analyses and calculations based on relevant theories. Both are used in the design of naval ships. When the equipment to be installed on board is manufactured,

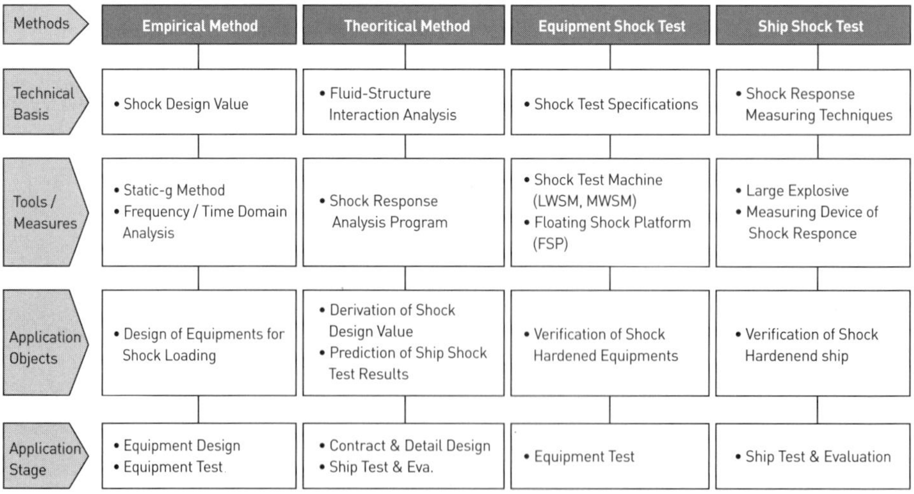

Figure 13-46 Shock hardening methods for naval ships

it can be tested using a shock test facility, provided the facility can accommodate the size and weight of the equipment concerned. A prototype ship is also shock tested by applying an explosion so that it receives about one third of the design shock strength.

4.3.2 Ship Protection Structures

There are examples of large surface ships being sunk or completely destroyed by single ASM, torpedo or mine in modern naval warfare, such as Middle East Wars and Falklands War. In 2000 a small fast boat laden with explosives carried out a suicide attack on a U.S. Navy destroyer and caused serious damage and many casualties. These real cases demonstrate that modern naval ships with all their sophisticated defense systems are still vulnerable today. Therefore, navies of the major powers of the world have developed and applied various new concepts of ship protection structures to increase their survivability.

Blast hardened bulkheads prevent longitudinal spreading of damages resulting from internal explosions. This can be achieved by either using double skinned bulkheads to minimize possibility of their destruction or making the top and bottom of bulkheads flexible to allow large deformation at the price of destruction of the bulkheads. The blast hardened bulkheads have been applied to DDG 51 class destroyers of the U.S. Navy and new German and Dutch frigates.

Box girders or thick plates can be installed at the main deck or hull bottom to give the ship improved residual longitudinal strength in the event of explosion damage to the hull. Developed in Germany, the concept was used in new German frigates. **Figure 13-47** illustrates the protection structures discussed here.

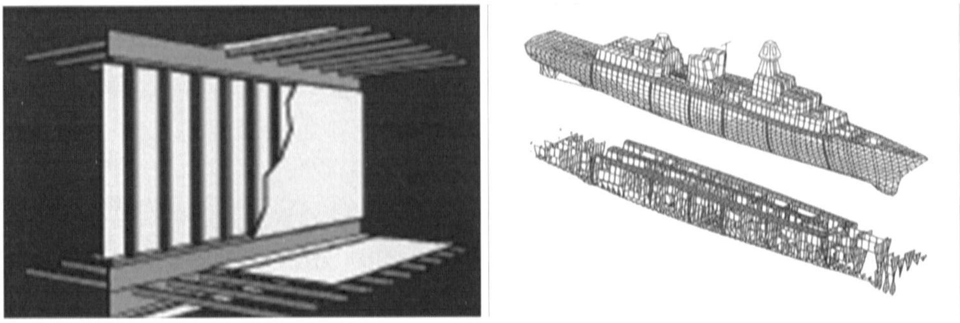

Figure 13-47 A blast hardened bulkhead and box girder type hull structure

The advanced double hull (ADH) consists of an internal and external skin structure (see **Figure 13-10**) and is being developed by the U.S Navy to improve the protection of the hull and internal space utilization.

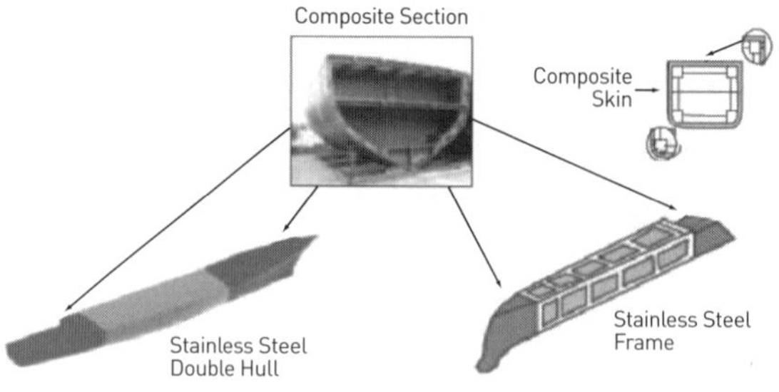

Figure 13-48 Advanced hybrid hull concept of the U.S. Navy

The inherent strength of the cellular structure between two skins increases the survivability of the ship, and the usability of internal compartments is increased due to the absence of transverse and longitudinal frames.

The advanced hybrid hull concept is also under development for future surface ships by the U.S. Navy. The middle section of the hybrid hull is a double hull type built of non-magnetic steel material, whereas the bow and stern are constructed of composite materials such as GRP. **Figure 13-48** shows the advanced hybrid hull concept under development by the U.S. Navy.

References

1. Annati, M., 2003. What if…? A ship's survivability depends on many factors. *Naval Forces*, 24(3), pp.110-112.

2. Jane's Naval Weapon Systems, 2006. *HDW weapon handling and launching system*. London: Jane's Information Group.

3. Longworth, B., 1996. Solution to the shallow water challenge. *Jane's Navy International*, 101(5), pp.10-18.

4. Naval Ship Weapons Engineering Station (NAVSEA), 1976. *Structural design manual for naval surface ship*. NAVSEA 0900-LP-097-4010, Port Hueneme: NAVSEA.

5. Pluemecke, G., 2005. *Introduction to Submarine Combat System*. [Seminar Note]. Atlas Electronik.

6. Podesta, C., 2003. The submarine sail: a platform within a platform. *Naval Forces*, Special Issue 2, pp.42-44.

7. Roshdy, G.S.B., 2002. Advanced ship protection and low signature hull. In: 3rd *Annual Navy-Industry R & D Partnership Conference*. Washington DC. August 2002. Washington DC: U.S. Navy.

8. Scott, R., 2005. Anti-ship weapons updated to target the shore. *Jane's Navy International*, [published 01 July 2005].

9. Song, C.T., 2000, *Trends of ship systems developments in future* (in Korean), Report No. NSRC-113-001340. Daejeon: Agency for Defense Development.

10. Song, C.T., 2010. *Naval ship systems engineering* (in Korean). [Lecture Note]. Ulsan: Ulsan University.

11. The Society of Naval Architects of Korea, 2012. *Naval Ships* (in Korean). Paju: Textbooks.

12. Truver, S.C., 2001. Aegis Evolves for the Future. *Jane's Navy International*, July/August 2001.

13. Yoo, Y.W., 2001. *Mr. Yoo's military world* (in Korean), (The Chosunilbo, Korea).
Available at: http://bemil.chosun.com/
[Accessed 20 June, 2015]

CHAPTER 14

Marine Leisure Craft

1. Forms of Marine Leisure
2. Types of Marine Leisure Craft
3. Design of Recreational Boats
4. Manufacturing Process of Recreational Boats
5. Berthing Facilities
6. Leisure Boat Industry in Korea

CHAPTER 9
CHAPTER 10
CHAPTER 11
CHAPTER 12
CHAPTER 13
CHAPTER 14
CHAPTER 15
CHAPTER 16

1. Forms of Marine Leisure

1.1 Definition of Marine Leisure

Marine leisure activities encompass all activities on or in water for enjoyment, entertainment, sports and education. A more formal definition of marine leisure activities may be 'all leisure activities away from daily living, directly or indirectly dependent on, or associated with, water at sea or inland waterways.' They may also be defined as all types of marine sports, either organized or purely recreational, possibly using various powered or non-powered equipment, at sea or inland waters.

Table 14-1 Types of major marine leisure activities

Items	Types	Key activities
Water activities	Sports	- Boarding (surfing, windsurfing) - Yachting & boating (sailing, canoeing, jet skiing, motor boating, ribbing, etc.) - Diving (snorkeling, scuba diving, etc.) - Parasailing, water skiing, etc.
	Hydrophilic form	- Bathing (swimming, dabbling, etc.) - Tideland experience - Sea fishing (boat fishing, shore fishing, etc.)
	Cruising form	- Ocean excursion (cruising, etc.) - Underwater sightseeing (submersibles, underwater tourism, etc.)
Associated activities		- Beach sports, playing with sand, sunbathing, relaxing on beach, etc. - Sightseeing, coastal walking, jogging, etc. - Visiting (aquaria, maritime monuments, etc.)

• Source: LEE, Soo Ho Current and Future of Marine Leisure Industry in Korea (2001)

1.2 Scope of Marine Leisure Equipment Industry

Manufacturing marine leisure equipment is an integrated industrial process requiring diverse materials, components and production processes, and therefore has much direct and indirect impact on a broad network of related industries. Although design and production of marine leisure craft is strongly related to shipbuilding, it is very different from merchant shipbuilding in materials, scale and usage, except perhaps luxury yacht building. The production system is more closely related to that of the automobile industry instead, with potential for high added value due to the fact that there are many types of leisure equipment produced in small batches. It is also easy for other related sectors, such as cars, electronic appliances, IT, textiles and toys, to diversify into the marine leisure industry.

The users of the equipment are varied, ranging from individuals to businesses and organizations, and they have to be designed and manufactured not only for good functionality but also aesthetics and ergonomic factors catering for diverse tastes. Production and sales are the two most important aspects of the marine leisure equipment industry and, therefore, it is essential to secure a sufficient number of users by popularizing the marine leisure activities. The most effective way of creating future demand is thought to be through cooperation with organizations which promote and host events, sports and marketing.

1.2.1 Industrial Aspects

Yacht and motor boat industry has a strong link with its upstream and downstream industries, such as shipbuilding, materials, sports, exhibition, finance and insurance

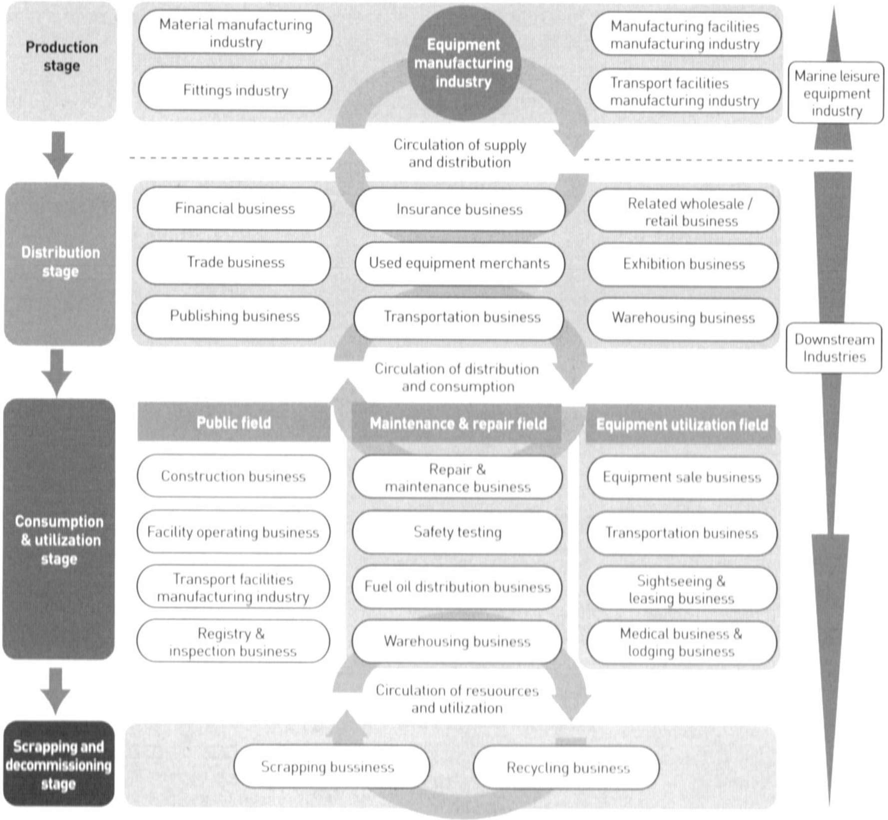

Figure 14-1 Relationships of various industries involved in marine leisure industry

businesses as illustrated in **Figure 14-1**. Small shipyards have suffered some hardships recently due to reduced demand for fishing boats as an aftermath of the Korea-Japan and Korea-China fishing quota agreements. It may be possible for them to find a new way out of this crisis by entering into the leisure market, if new demand for leisure craft can be created. In order to diversify from the traditional fishing boats into high value-added leisure equipment market they will need to improve productivity and acquire new production technical know-how. It may also be worthwhile for them to explore the leisure markets abroad.

With the bilateral Korea-US FTA entering into force, the Korean market for marine leisure equipment is now open to US companies. This has increased the concerns in some circles about the possibility of US products 'swamping' the Korean market. However, it is an opportunity as well as a threat, provided some serious effort is made to establish the foundation of this relatively new industry in Korea.

1.2.2 Technical Aspects

As shown in **Figure 14-2**, the marine leisure equipment industry is a complex state-of-the-art industry requiring integrated high technology in many areas. In addition to boatbuilding skills, it requires application of technologies originally developed in other industries, such as materials and fittings, automobile and mechatronics. For yachts and boats it is essential to apply up-to-date aesthetics, ergonomics and interior designs. It is worth studying the high marketing techniques used by manufacturers of consumer goods with strong brand loyalties like cars. It is quite clear that consumer analysis is equally as important as technology for them and therefore probably for the marine leisure equipment industry as well. Korea has

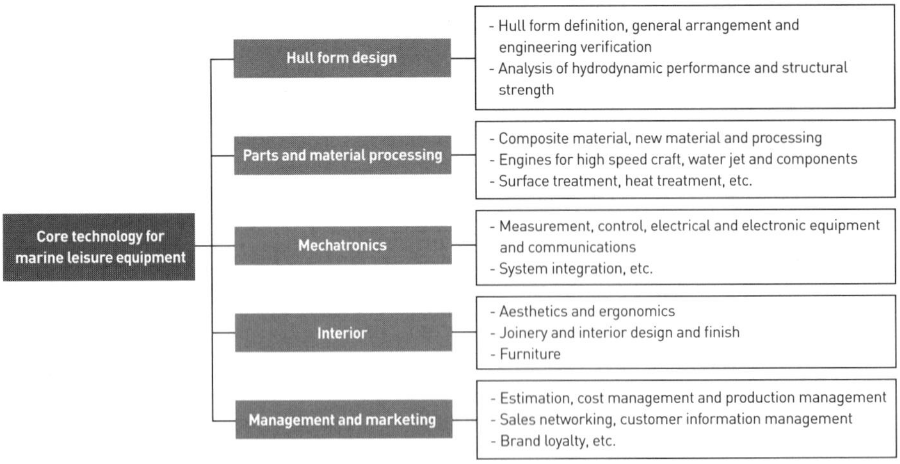

Figure 14-2 Technical aspects of marine leisure equipment industry

world-leading expertise in naval architecture and shipbuilding, and an excellent network of automotive parts suppliers. However, they need to be mobilized quickly to form a competitive industry in marine leisure equipment.

There are four key technological areas in marine leisure industry: craft, eco-friendly hybrid equipment, leisure facility and personal leisure equipment.

The technology area related to leisure craft can be divided into common technology, technology applicable to yachts, and technology applicable to motor boats. The key technologies applied to eco-friendly hybrid leisure equipment include renewable energy and environment-friendly systems. **Table 14-2** summarizes the knowledge, expertise and technology required for marine leisure equipment.

Table 14-2 Classification of knowledge, expertise and technology required for marine leisure equipment

Leisure equipment	Application	Core technology
Boats and yachts	Common	Analysis and evaluation of hull form and propulsion system Analysis of structural stability and strength Computer aided performance analysis and simulation (CFD, FEM) Model tests and seakeeping performance Production system, automated production, process management and standardization High quality gel coating and gloss painting of hull Eco-friendly lifecycle management system Interior design against boat motions Light material and finishing material Outfitting, equipment standardization Simple and small lifesaving appliances Processing and assembly of high performance, lightweight composite materials Automatic navigation systems and route monitoring systems Safety, salvage, search and rescue systems Non-destructive testing of composite materials Design and production of interior and exterior International inspection standards for export Design development of leisure craft High quality fairing and finishing of hull surface
	Sailing yachts	Design of hull forms and appendages Performance prediction and design of sails Stability Interior and fittings Design optimization and production of riggings Mast design Manufacture and assembly of hull and mast Development of fittings and riggings

Boats and yachts	Motor boats	Design of hull forms and appendages Manufacture and assembly of hull Stability Outfitting Development of sound-absorbing panel structures Emergency propulsion systems Redundancy design of communication systems Development of small, high performance inboard and outboard engines Waterjet and other propulsion systems
Eco-friendly hybrid	Renewable energy	Low carbon and eco-friendly renewable energy Charge and discharge control of eco-friendly renewable energy Hybrid propulsion systems using renewable energy Cabin powering with renewable energy
	Eco-friendly system	Hybrid high-efficiency propellers Eco-friendly outboard engines Gas emission reduction Reduction of waste water and purification Hull appendages design Scrapping and recycling
Leisure facilities and personal equipment	Mooring and anchoring equipment	Mooring and anchoring equipment Repair and maintenance and their management
	Miscellaneous	Personal watercraft development Windsurfing equipment, surf boards and fittings Inflatables Hovercraft Amphibious vehicles Electric river cruisers

2. Types of Marine Leisure Craft

2.1 Marine Leisure Equipment in Korea

Article 2, Paragraph 3 of the Marine Leisure Safety Act and Paragraph 2 of the Executive Order of the same law classify marine leisure equipment in Korea into 14 types. This classification is based on their forms and methods of use, but some of the equipment can be classified differently according to their manufacturing processes, and therefore clearer standards of classification are called for.

Furthermore, novel forms of marine leisure activities are appearing in almost bewildering rapidity these days bringing with them new types of equipment and jargons. Consequently the classifications given in **Table 14-3** may have become inadequate, and a new classification based on forms, propulsion equipment and production methods may be necessary.

Table 14-3 Types of marine leisure craft

No.	Type	No.	Type	No.	Type
1	Motor boats	6	Hovercraft	11	Canoes
2	Yachts	7	Water skis	12	Water sleds
3	Personal watercraft	8	Parasails	13	Aquatic bicycles
4	Rubber boats	9	Scull and sweep boats	14	Surf boards
5	Aqua scooters	10	Kayaks	15	Rowboats

• Source : Korea Coast Guard, Safety of Water Leisure Act (2009)

2.2 Types of Marine Leisure Equipment

2.2.1 Motor Boats

The world's first motor boat is known to have been created by Gottlieb Daimler of Germany in 1885 who designed a high speed engine of 1.5 hp and installed it on a boat. Both sailing yachts and motor boats are sometimes called yachts in the sense of leisure craft and they are often built of fiber reinforced plastics (FRP) and other composite materials because of their very low weight-to-strength ratio which is advantageous for high speed craft.

Figure 14-3 A motor boat

Motor boats (**Figure 14-3**) use internal combustion engines or electric engines as power generators, and are classified into inboard, inboard stern drive and outboard engines. They are often used for offshore and near-shore cruise, fishing and scuba diving.

2.2.2 Yachts

The word 'yacht' originates from a Dutch word 'jagen' meaning 'to hunt' or 'to pursue.' Generally, yachts are wind-powered craft normally with an engine for supplementary power and are used for sports and outdoor recreation.

Sailing dinghies are small yachts entirely powered by wind with no mechanical power devices. Dinghies may be equipped with the main sail (either Bermuda or gaff rig) and often a jib sail, and some boats also can carry a spinnaker in favorable conditions. They have a center board which can be lifted up out of the way when not in use. Because of their small size, the boats have to be balanced by the crew leaning out on the windward side when tacking. They are open boats with no covered accommodation on board and are used for racing and for fun, but some are designed for more sedate and longer cruising. They are usually less than six meters long, and the usual crew complement is two. Dinghies are immensely popular with youngsters in countries where sailing is a favorite water sport, and many people learn to sail on dinghies on inland or coastal waters.

Larger yachts usually have covered accommodation and are used for racing and cruising. There are a variety of sizes, and many are capable of crossing open seas. Indeed some adventurers have gone around the world on them and there are regular round-the-world races. Smaller yachts are single-masted with the mainsail and one or more jib sails with a spinnaker when conditions allow. Larger yachts may have up to three masts and there are a variety of types including yawls, ketches, sloops and cutters. Most yachts have an inboard engine and a feathering propeller for maneuvering in narrow waterways and harbors. Unlike dinghies, they have one or two fixed keels which have become extremely sophisticated in recent years particularly for racing yachts. A single-masted yacht, showing the main sail and a jib and towing an inflatable tender, is shown in **Figure 14-4**.

Figure 14-4 A yacht

2.2.3 Personal Watercraft

The first successful personal watercraft was invented in Japan by the motorcycle company Kawasaki who gave it a brand name of jet ski, which has since become a generic name. Since its launch in 1972 it has been spread to all corners of the world, gaining great popularity in the 1980s. Since its introduction in Korea, it has been called a 'water motorcycle' for its characteristic appearance of ride that resembles scooters on the road as in **Figure 14-5**. Although they are

Figure 14-5 Personal watercraft

sometimes called water scooters, they must not be confused with aqua scooters which are discussed later.

The main power source is an internal combustion engine, which drives a waterjet propulsor. Jet skis can be used on any water more than 30cm deep, unlike motor propeller boats which require full immersion of their screw propellers. Jet skis are also used to tow water skis and water sleds.

2.2.4 Rubber Boats

Rubber boats are inflatable boats made of PVC, hyperon and other rubbery substances glued together. Some are built to take an outboard motor while others are rowed. There are many sizes and forms, but the most sophisticated ones have the rubber buoyancy parts mounted on FRP hull and are called rigid inflatable boats (RIBs). Rubber row boats are mainly used in rafting and as tenders for larger vessels. Those equipped with high-horsepower outboard engines (usually RIBs) can tow parasails, water skis and multi-person water sleds.

The smaller inflatables can be inflated and deflated and are relatively light allowing easy handling and storing. They can be easily transported in vans or on car roofs, although RIBs may need a trailer. They are relatively cheap and therefore are popular for a number of water activities such as cruising and fishing. RIBs are often used for water safaris and as rescue boats for larger merchant vessels. A typical rubber boat with an outboard is shown in **Figure 14-6.**

Figure 14-6 A rubber boat

2.2.5 Aqua Scooters

Scooters are propulsion equipment used to help scuba divers move in the water. Originally invented by an East German man who used it to escape to West Germany during the Cold War era, scooters are now widely used by recreational divers.

They are designed to have neutral buoyancy in water and have plastic casing to reduce weight. Internal combustion engines may be used if equipped with an air-breathing snorkel, but often

Figure 14-7 An aqua scooter

batteries encased in a metal cylinder are used instead (shown in **Figure 14-7**) to drive a propeller at the end of the cylinder.

2.2.6 Hovercraft

Hovercrafts are ships partly supported by air cushion created by injecting high pressure air into a large air pocket underneath the hull. Because of the lift created by the air cushion, hovercraft can float above the running surface, allowing them to become amphibious, operable both on water and ground. The name hovercraft was a trademark of British Hovercraft Corporation (BHC) but has now become a generic name for the vessel type.

Hovercrafts are propelled by air propellers like a propeller airplane and are operated by pilots rather than captains. Although they are amphibious, ground operation is not easy because of noise and dust they generate. Some hovercrafts were used as fast car ferries across English Channel carrying about 250 passengers and 30 cars and were capable of travelling at 80 knots or more. Hovercrafts for leisure are of much smaller variety (**Figure 14-8**), often used in salt marshes and shallow inland waters.

Figure 14-8 A hovercraft

They are popular enough in certain countries to the extent that even assemble-it-yourself one-person hovercraft kits are available. Nevertheless, they are very rare in Korea as yet.

2.2.7 Water Skis

Water skis are aquatic leisure equipment that slide on the surface of water, towed by a fast power craft as illustrated in **Figure 14-9**. The first skis were created by Ralph Samuelson of America in 1922 and since then spread all over the world. They became very popular when the World Water Ski Union was established in Europe in 1946.

Figure 14-9 Water skiing

Water skiing has combined characters of surfing and skiing, and, apart from giving one great enjoyment, it can improve one's skill in control, predictive ability and balance. There are many types of waterskiing, including jump skiing, slalom skiing, trick skiing and wakeboarding. Since they cannot propel on their own, they must be towed by personal watercraft, motor boats or rubber boats.

2.2.8 Parasails

Parasailing is a recreational kiting activity in which a person is towed behind a power boat while attached to a specially designed parachute-like wing, known as a parasail (**Figure 14-10**). Developed in the 1950s for training airborne troops in France, it was introduced into the UK where it quickly developed into a leisure activity. It started gaining popularity in America in the 1960s, and by the 1980s has become widely spread.

For safety it must be used in quiet areas of open sea where there are no obstacles.

Figure 14-10 Parasails

2.2.9 Scull and Sweep Boats

It is said that the origin of scull and sweep boats was the boats used as a means of transport on the Thames in the mid-17th Century, which was far more convenient than roads at the time. Soon it developed into racing, and the first such competition was held in 1715. The first University Boat Race (between the Boat Clubs of Oxford and Cambridge Universities) was held in 1829, and then annually ever since 1856. The Fédération Internationale des Sociétés d'Aviron (FISA) was founded in 1892.

Figure 14-11 A sweep boat (eight)

Scull or sweep boats are long and narrow, and propulsion is achieved by rowing with oars over both sides of the boat. They are more often used in racing than for recreation, and such races require physical strength and stamina as well as rowing techniques and teamwork. The current FISA's eight official categories are: single scull, double scull, quad scull, coxless pair, coxless four, coxed pair, coxed four and eight (**Figure 14-11**). Since the speed is of the utmost importance for these boats, weight reduction is critical. This is achieved by using the state-of-the-art composite materials.

2.2.10 Kayaks

Kayaks have originated from leather boats used by Eskimos during summer hunting, often made for one person. The boats are light enough for one person to carry, and they are long and narrow for speed as illustrated in **Figure 14-12**. Such form gives them a low center of gravity so that they can be used in high waves. Because the boats are covered on top, they can be turned back up when capsized by the rower swinging or moving the paddles underwater. They may be built of wood, plastic or composite materials.

Kayaks (**Figure 14-12**) are very popular all over the world, particularly where there are many inland bodies of water or relatively calm coastal inlets. Kayak races are very popular as well.

Figure 14-12 Kayaks

2.2.11 Canoes

The word canoe originates from the Spanish word 'canoa,' meaning ship. They are also known as Canadian canoes, because many were used by the American Indians residing in Canada. Canoes, in their very primitive form, must have been used by mankind from the very beginning of our history for transport and hunting. As seen in **Figure 14-13**, a single-blade paddle is used for propulsion and control. Their size varies a great deal, anything from a one-man canoe to very long ones which can carry a score or so people.

Figure 14-13 A canoe

It appears that inflatable canoes are being preferred in Korea at present, maybe because they are more stable than kayaks while combining the characteristics of kayaks and rafts.

2.2.12 Water Sleds

Water sleds are rocket-shaped equipment made of PVC or synthetic rubber material, and are towed by power boats. They are also known as 'banana boats' from their shape, as seen in **Figure 14-14**.

Figure 14-14 A water sled

Sledding is similar to water skiing in that the sled is towed by a motor boat, and it requires good sense of balance on the water. It can be enjoyed by beginners since it does not require special skill, but teamwork and balance are required for longer sleds. Sleds are constructed with three segregated air chambers.

2.2.13 Aquatic Bicycles

Aquatic bicycles are a type of human-powered boats that are propelled by pedaling, and hence the alternate names pedalos and pedal boats. Duck boats in Korea are a form of pedal boats. The usual arrangement has the pedals on the deck of a twin hull boat, as seen in **Figure 14-15**. Many are one-person or two-person boats and they can often be seen on park lakes. Although most of them are human-powered, electrically driven versions are also available.

Figure 14-15 An aquatic bicycle

2.2.14 Surfboards

Boards used for both wave surfing and wind surfing are known as surfboards in Korea, but they are quite distinct not only in how they are used, but also in their construction.

Wave-riding on surfboards originated from the surfers of original Hawaiian natives. The boards are built of wood or polyurethane foam to provide buoyancy and rigidity. The surfer swims out to meet rollers coming in, and at the top of a swell stands up on the board and ride down the wave slope.

Figure 14-16 Wind surfing

Wind surfing, on the other hand, is somewhat similar to sailing. The board has a light sail hinged at the bottom which the surfer holds and controls while standing on the board as shown in **Figure 14-16**.

Both types of boards are shaped at the bottom to provide some lift while the wind surfing boards have a small skeg at the tail end to reduce leeway drift whilst assisting directional control.

2.2.15 Rowboats

Rowboats (**Figure 14-17**) are small boats built of FRP, plastic, metal or wood and propelled with oars. Their size varies but they can usually carry two to five persons. They are used for recreation

Figure 14-17 Rowboats

but some are used as tenders for larger vessels. Foldable boats and inflatable boats have recently appeared and they are attractive to some people because they are easy to transport and store.

2.3 Luxury Yachts

Luxury yachts are not mentioned in the Korean Marine Leisure Safety Act, but they have recently received much attention throughout the world for their high added value. The large super luxury class yachts are known as mega yachts while a somewhat smaller variety are known as super yachts, although clear distinction is not possible. **Table 14-4** summarizes the characteristics of luxury yachts.

Table 14-4 Characteristics of luxury yachts

Purpose	• High class leisure boats with luxurious passenger accommodation
Size	• Length 120ft (36m) or more : super yachts • length 200ft (60m) or more : mega yachts • Mega yachts of length 100m or more and costing 200 million US dollar have appeared
Performance	• Comfort : comfortable and luxurious interior • Appearance : sleek and classy hull form • Safety : safe and comfortable even for long voyages • Functionality : good maneuvering and seakeeping performance
Current status	• The world super yacht market is growing by over 10% every year. • Super and mega yachts are mostly built in Italy, USA and some other countries with track records in luxury yachts. • Korea has not explored this market yet.

Super and mega yachts are very luxurious compared to ordinary cruising motor boats as seen in **Figure 14-18**, and are built for or hired by very wealthy clients. They usually have excellent stability and maneuverability characteristics and aesthetically pleasing appearances. Moreover, their luxurious interior provides comfortable and pleasant living spaces. Consequently, these yachts are mostly custom-built of FRP, high tensile steel and aluminum alloys to order. Their cruising speed is usually about 30 knots, running high powered gas turbine engines. While the merchant shipbuilding industry has declined in the west, small and medium shipyards building leisure boats

Figure 14-18 Superyacht *Gyeonggi BADAHO*

including luxury yachts are steadily expanding in sales and employment.

It is believed that Korean shipbuilding industry will have sufficient competitive advantage in the international luxury yacht market, if its world-class infrastructure and technical know-how of merchant shipbuilding is mobilized in a systematic way.

3. Design of Recreational Boats

3.1 Introduction

Even small factors in the design and manufacturing processes of marine leisure boats can influence their overall performance. Their design must focus on achieving good stability, comfortable accommodation and low maintenance costs. There are many technical problems to solve, and research and development in leisure industry, unlike in merchant ships, is usually undertaken by a few highly skilled specialists.

Some researchers in the west have been developing three dimensional product life cycle management (PLM) systems for leisure boats to manage information on design, production, service, procurement and quality spanning over the whole lifecycle from design to decommissioning as illustrated in **Figure 14-19**. Such PLM systems are applied to the product development of leisure boats, cars and airplanes in Korea as well to secure systematic technical know-how in development, design and production.

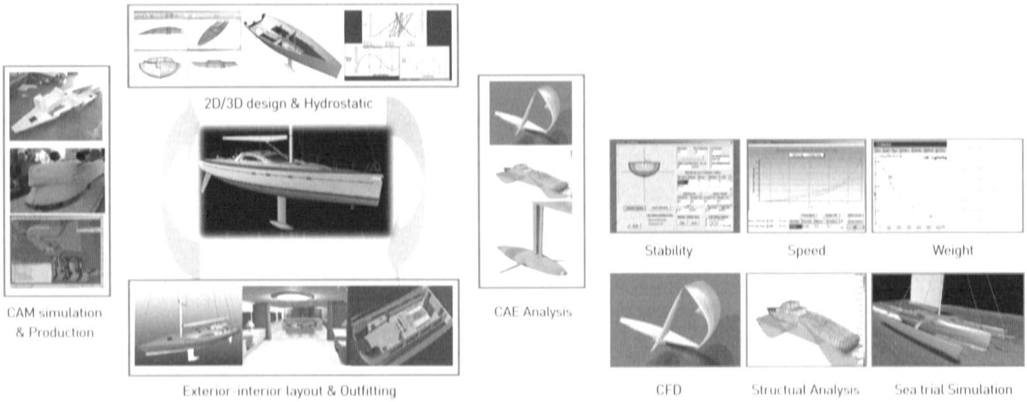

Figure14-19 Leisure boat design based on a PLM system

3.2 Design of Leisure Boats

3.2.1 Hull Form Design

Hull forms of marine leisure boats have been traditionally designed manually or using computer programs in a sequential manner. Having determined the principal particulars, the concept and basic plan are produced, and based on these, a preliminary hull form is derived and its hydrostatic properties are calculated.

Recent hull form design practice takes the computer-aided design (CAD) into the next level where the old separate sequential tasks have become systematically integrated. Today the surface model CAD systems are used to generate hull forms while the product information system relies on solid model CAD systems which have already been widely used in mechanical and car designs.

Design of a leisure boat should start with clearly defining its uses, hull type, size, operational weather conditions, target performance, general arrangement, style and equipment, as summarized in **Figure 14-20**. These factors are decided based on the owner's (or customers') requirements, and the designer can then set up the design targets and constraints based on them. An iterative method of determining various design parameters for optimum performance as well as meeting the design requirements, known as design spiral and illustrated in **Figure 14-21**, is often used. The design parameters to be determined in this manner include speed and seakeeping performance, safety, structural strength, weight, arrangements and costs of the vessels.

The performance goals of the boat are established and an approximate cost is estimated during the early planning stage before the design proper begins. The rough cost estimate is usually based on past experience of similar boats, but sometimes an estimate may be obtained from boat builders. There are many companies that specialize in leisure boat design, such as Dixon Yacht Design, 3IDEE and Ocke Mannerfelt just to name a few, and

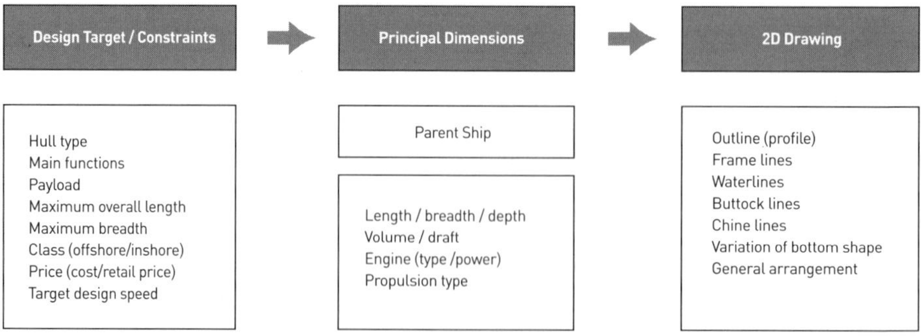

Figure 14-20 Design process of ordinary leisure boats

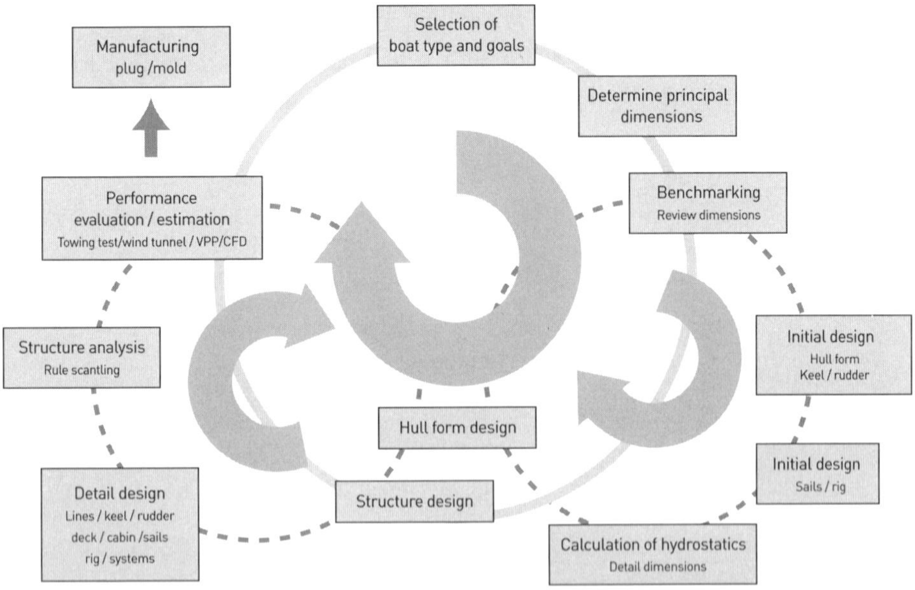

Figure 14-21 Design spiral for leisure boats

this has produced such proliferation of boat designs and production standards.

There are many different methods of expressing hull forms, but the most common is the method of three orthogonal projections of body plan, waterline plan and buttock plan (or profile), which collectively are known as a lines plan. The frame lines are the outline shapes of transverse sections and are shown in the body plan, while the waterlines are the horizontal section shapes of the boat. This has been the traditional method used for the design of all vessels including merchant ships, warships and submarines as well as boats. The only difference is that, while the baseline is the datum for most vessels, the datum line of yachts is usually set at the waterline, probably because it is difficult to define the baseline of a yacht.

These days three-dimensional modeling CAD systems are widely used and there are systems useful in yachts and boats, such as AutoShip, FastShip and MaxSurf. These systems are complete basic design systems, dealing with everything from the early sketches to definition of the basic hull form as well as calculating hydrostatic particulars and estimating resistance and propulsion.

In conventional design process of sailing yachts, the technical design is carried out sequentially after the form design is complete. If the form is modified in this process, the technical design has to be examined and the affected parts have to be redesigned. However, PLM-based design practice can proceed with technical details with an assumed three dimensional model. When the form design is complete, the technical design can be simply updated to suit the form.

If a minor modification is made to the form even when the detailed design is about to be finalized, simply updating the hull form is sufficient for a new three-dimensional model to be produced automatically, making tedious redesigning work unnecessary.

3.2.2 Design of Appendages

For recreational boats, the design of keels, rudders and sails are as important as the hull form design. Especially for sailing yachts, their performance is dependent on the performance of these 'appendages,' and their design has received attention from researchers for a long time. Keels play two very important roles. The first is to reduce leeway drift and the second is to lower the center of gravity of the vessel which leads to higher righting moment to withstand the heeling moment produced by the sails. Since many yachts are built for racing, the keel has to be designed in accordance with the race rules and taking into account the weight distribution and the hull form.

The keel takes up about 30-50% of the total weight of an ordinary yacht, and therefore it is essential to reduce the keel weight if the total weight of the boat is to be reduced. Special keels are not attached to motor cruisers, because they only cause additional resistance, while leeway drift and heeling moment is not a cause for concern. A keel will contribute much to the boat's stability and course-keeping ability, but a keel for such purpose does not need to be as large or prominent as the blade keel of sailing yachts.

Rudders are the most basic piece of equipment for controlling the direction of travel. In a sailing yacht, it also contributes to reducing leeway drift along with the keel, and small high-performance rudders are desirable for minimum resistance.

The rudder should be able to generate sufficient steering forces, but care must be taken as a yacht pitching wildly can lose steerage momentarily when the rudder comes out of water. The location of the rudder stock is the most important factor in rudder performance, and it has to be designed to provide a good balance of the torque while the steering gear must be able to provide a good feedback of the rudder force. The rudder should be given a high aspect ratio to maximize its performance, while its surface area should be minimized to reduce its frictional and induced resistance.

The appendages can be designed according to a PLM-based process presented in **Figure 14-22** using the lines plan acquired from the hull form design and the initial arrangement. Specific arrangements of major equipment are reflected in the three dimensional model. Appendages and equipment considered at this time are for maneuvering, propulsion and operation of the boat and includes the following:

- Rudder and steering gear;
- Keel and sails;

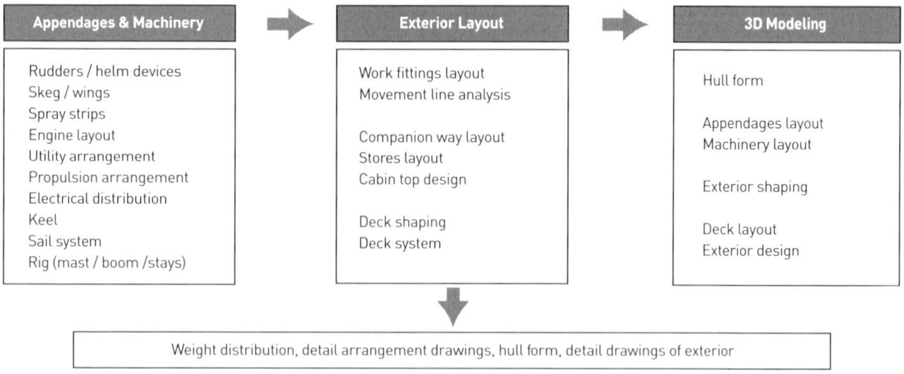

Figure 14-22 Design process of appendages based on PLM

- Energy-saving devices attached to the boat's surface;
- Arrangement of machinery, shafting and propulsion systems;
- Arrangement of tanks and piping (fresh water, sewerage, sanitary system, etc.);
- Supply of electricity and arrangement of various electrical devices;
- Mooring equipment;
- Rigging and sail operating equipment; and
- Radar, navigation and communication equipment.

The arrangement of deck fittings is decided concurrently with the exterior design of the upper part of the ship. The functions of deck equipment are analyzed and their optimum locations determined. The deck arrangement can then be finalized by analyzing the circulation paths and accessibility. Items determined at this stage are as follows:

- Arrangement of functional equipment;
- Analysis of circulation paths and accessibility, securing of working space;
- Arrangement of cargo and portable equipment stores;
- Exterior style design for cabin top;
- Design of deck appearance; and
- Arrangement of general equipment.

Based on the above, general arrangement and detail arrangement are determined.

The aforementioned series of processes are integrated through three-dimensional modeling which includes not only information of curved surfaces of the vessel but also of forms and properties. This makes visual models easy to generate enabling intuitive assessment, and facilitates rapid progress to production design.

4. Manufacturing Process of Recreational Boats

4.1 Introduction

The leisure boat market is led by specialist companies, such as Sunseeker, Blohm+Voss, Master Craft, Yamaha and Bayliner. These companies have secured a high international market share through integrated services from design to manufacturing. Most of the leisure boat builders in Korea are small companies such as GHI Yachts, Kwangdong FRP Industrial and Advanced Marine Tech. However, the industry is showing signs of growth with the recent introduction of advanced design and manufacturing technology and government support, and own brand names have begun to appear. Before manufacture of leisure boats can begin, a production design has to be produced. High-quality key manufacturing technologies, such as RTM (resin transfer molding), VaRTM and modular process using 5-axis CNC milling machine, are essential in developing and manufacturing world-leading products. Production, as illustrated by **Figure 14-23**, is a very influential factor in the cost and the manufacturing period of a boat.

Some high-cost equipment, such as five-axis CNC milling machines, vacuum lamination devices and infusion devices, are required in the manufacturing phase. The cost of equipment and facilities are much more substantial than those used in the design development stage. Therefore, it is essential to consider future plans and growth in designing the production processes and the factory layout.

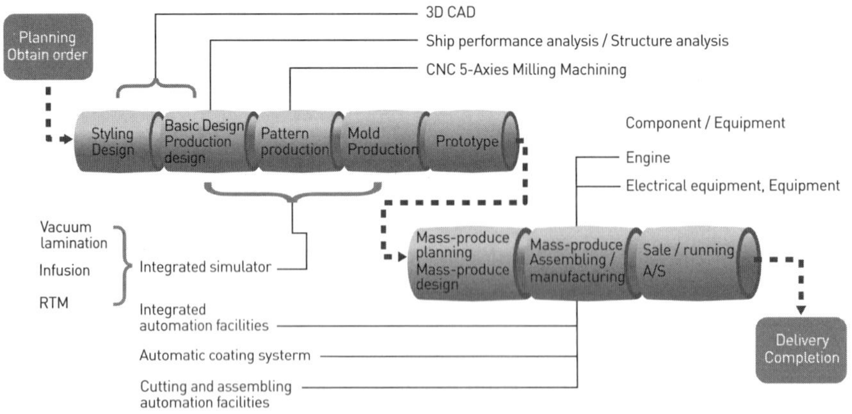

Figure 14-23 Production of leisure boats

4.2 Production Techniques of Leisure Boats

4.2.1 Manufacturing Process

Although there are some variations between the manufacturers and the products, the manufacturing process of recreational boats can be said to start from the production of the mold. Production design and construction scheduling is an important and lengthy business in merchant shipbuilding. However, such steps are often left out by most Korean boat builders or outsourced to a specialist design offices. For this reason, production process of leisure boats is as illustrated in **Figure 14-24**. Usually making the mold and applying gel coats are considered to be the beginning of the production process. The process is largely composed of five stages: mold production; hull production; hull assembly; outfitting assembly; finishing; and inspection. Detailed processes are presented as a diagram in **Figure 14-25**.

FRP materials are made of synthetic resins and reinforcing material. The major functions of the resin are to transfer the pressure between the reinforcing fibers, to bind the fibers and to protect the fibers from mechanical and environmental damages. There are two types of resins: thermosetting and thermoplastic.

Thermoplastic resin softens when heated, and can be molded or formed in its heated semi-fluid state. It hardens and sets once it cools down. Typical thermoplastic resins are polyethylene (PE), polyvinylchloride (PVC) and polypropylene (PP). Thermosetting resins are initially in a liquid state until solidified by using a catalyst, heat or a combination of both. Such resins include unsaturated polyester (UP), epoxy and polyurethanes (PURs). The majority of resins used in the manufacturing of leisure boats are unsaturated polyester resins.

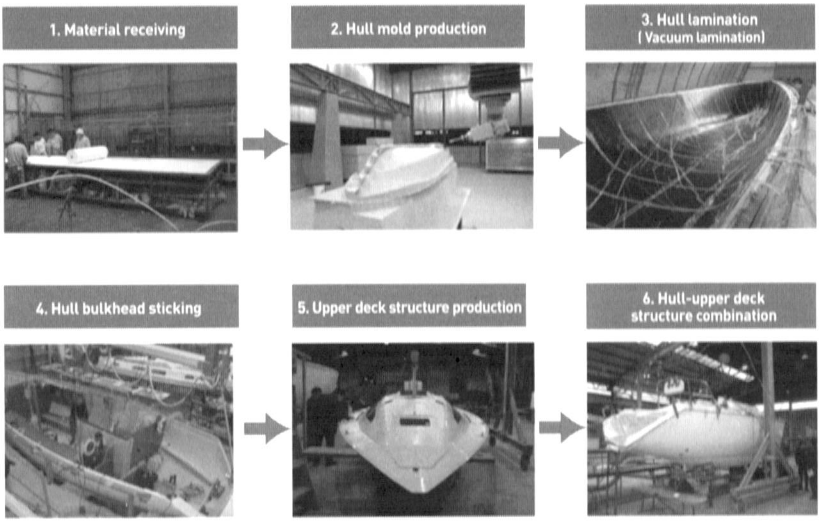

Figure 14-24 Manufacturing process of leisure boats

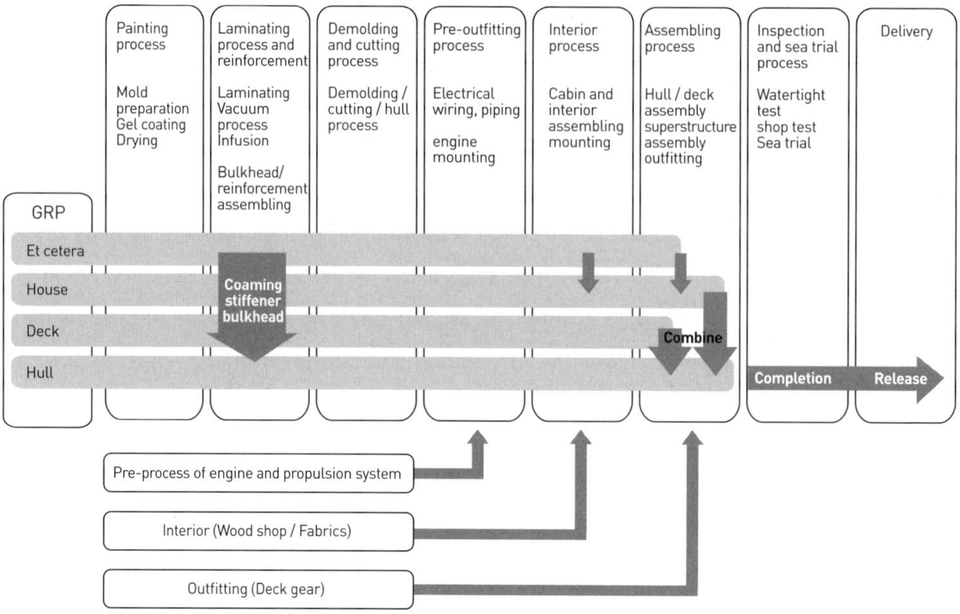

Figure 14-25 Detailed manufacturing process of leisure boats

They take up approximately 75% of total resins in use, but various other resins are also used according to application and purpose.

Reinforcing fibers are used to give the structure tensile strength by taking up the load in the lengthwise direction of the fibers, and areas where loads are concentrated can be specially reinforced by adding more fibers.

The reinforcing materials used most often are glass fibers, but carbon, aramid, UHMW (ultra-high molecular weight) polyethylene, polypropylene, polyester and nylon fibers are also used. Reinforcements are classified into five types according to their physical forms: multi-end and single-end roving, mats, cloth and 3D fibers, non-directional reinforcements, and pre-impregnated composite fibers.

Such reinforcements are possible in any form as long as they satisfy the requirements of the final product and can be produced by the wide range of available processes. Various other fiber structures are possible and reinforcements can sometimes be pre-formed according to requirements and as the production processes allow.

4.2.2 Five-Axis Milling Machine in Producing Plugs

Hull molds had been traditionally manufactured manually. This practice was apt to introduce dimensional errors causing imperfect assembly and consequent degradation of quality in the final products. One of the automatic equipment which can overcome this problem is five-axis milling machines shown in **Figure 14-26**. They are becoming globally accepted as the machine that can answer the requirement of small quantity batch production, automation and high productivity. This technology is actively researched and developed for application in rapid and precision manufacture of products with complex surfaces, such as leisure boats, in Europe and Japan.

Figure 14-26 Manufacturing a hull form plug using a five-axis milling machine

4.2.3 RTM/Infusion Vacuum Forming Technique

RTM (resin transfer molding) and infusion vacuum forming illustrated in **Figure 14-27** are new FRP construction methods. They are receiving much attention by the recreational boat industry in many countries due to their many merits, including uniform strength achievable, high fiber-to-resin ratio, higher strength-to-weight ratio, and relatively low initial investment, ease of using all kinds of fibers, minimum finishing work and reduction of odor.

The conventional hand layup method often used currently in Korea is incapable of giving the necessary quality and strength required for special vessels and recreational boats. The key point in using the RTM/infusion vacuum forming technique is to use only the appropriate amount of resin by using additional materials such as sandwich foams. Once the production is set up, a failure in any aspect can lead to much financial and time loss. Thus, it is essential to simulate, review and inspect the process meticulously before and after setting it up.

Figure 14-27 RTM/infusion vacuum forming

5. Berthing Facilities

5.1 Overview

The word marina comes from a Spanish or Italian word meaning 'a promenade by the sea' which in turn was derived from Latin marinus. The modern meaning of marina was started by NAEBM (National Association of Engine and Boat Manufacturers) of America in the early 1900s. A marina is a special harbor with moorings for recreational boats and yachts, and has ancillary convenience facilities, such as bunkering facilities, repair facilities, boat chandlers, shops, restaurants and accommodation.

Marinas also have parking facilities for cars and trailers needed for overland transport of boats, and slipways through which boats can be launched or recovered using trailers. Boat hoists are also usually provided to lift boats in and out of the water. Boats are sometimes stored ashore to protect them when they are not going to be used for an extended period of time. Examples of marinas are shown in **Figure 14-28**.

Figure 14-28 Marinas

5.2 Marina Facilities

As seen in **Table 14-5**, Marinas must have 11 essential basic facilities as summarized in **Table 14-5**. Users may find it more convenient if there are additional facilities, such as for entertainment, accommodation, equipment rentals and banks. In this section the said 11 facilities will be discussed in more detail to aid understanding of marinas.

Table 14-5 Required facilities of marinas

<table>
<tr><th>No.</th><th>Facility</th><th>Remark</th><th>No.</th><th>Facility</th><th>Remark</th></tr>
<tr><td>1</td><td>Berths</td><td rowspan="6">Seaward facilities</td><td>7</td><td>Restroom facilities</td><td rowspan="5">Landward facilities</td></tr>
<tr><td>2</td><td>Potable water service</td><td>8</td><td>Vehicle parking</td></tr>
<tr><td>3</td><td>Fire protection systems</td><td>9</td><td>Boat launching facilities</td></tr>
<tr><td>4</td><td>Electrical power supply</td><td>10</td><td>Boat chandlery</td></tr>
<tr><td>5</td><td>Gangways</td><td>11</td><td>Repair facilities</td></tr>
<tr><td>6</td><td>Bunkering service</td><td colspan="3"></td></tr>
</table>

5.2.1 Berthing Facilities

Berthing is the most fundamental facility that a marina provides. Marinas are often located on the coast, and are affected by tidal differences, and, as a consequence, floating berthing systems are usually used. However, fixed berthing systems are used in places where there is no tide, such as rivers and lakes.

The most important factor in designing the berthing facilities is the size and number of boats to accommodate. The floating pontoon berthing system must be given sufficient buoyancy to support the weight of equipment, facilities and people on the system. The fixed berthing facilities must also have sufficient structural strength to support the weight of equipment and people on them.

5.2.2 Potable Water Service

The potable water service is a facility similar to the clean water supplies which we take for granted in our daily lives. It provides fresh water of drinking quality to the berthed boats. When preparing for a voyage, fresh water is taken aboard in the boat's fresh water tanks through a potable water supply unit shown in **Figure 14-29**.

Figure 14-29 A fresh water supply unit

Hose bibs can provide connecting points for fresh water hoses. They should be located at convenient spots for the users while not causing obstruction to human traffic.

5.2.3 Fire Protection Facilities

A fire protection system consists of passive and active fire suppression systems, and must be installed at effective locations for suppressing possible fires from various sources, such as boats, berths, buildings, bunkering facilities and other ancillary facilities. Fire-fighting equipment must be installed so that sufficient amount of fire extinguishing material can be supplied through the system as shown in **Figure 14-30**.

Fire extinguishing systems are generally divided into two types:

- Low-volume low pressure emergency fire extinguisher: equipment to suppress fire soon after its breakout and installed at various locations of the berthing facility.
- High-volume high pressure main fire extinguisher: fire-fighting facility, usually fire hydrant, to be used when the fire has spread and a small fire extinguisher is insufficient to suppress the fire.

Emergency first aid capability

A main fire hydrant

Figure 14-30 Fire-fighting equipment

5.2.4 Electrical Power Supply

Figure 14-31 Electrical power hookup unit

When a boat is berthed, it can be hooked up to the shore electrical power supply system to run various onboard appliances, such as televisions, air conditioners, refrigerators and electric heaters. The shore power hook up units (**Figure 14-31**) can supply electricity of 110～220V, depending on the country and sometimes regions. It is important to match the supply voltage with the onboard system voltage.

5.2.5 Passageways

A network of passageways interconnects various parts of the berths and shore. As mentioned earlier, the berthing systems are liable to change their vertical positions due to tide. Therefore, access gangways and passageways are hinged at all connecting points as illustrated in **Figure 14-32**.

Figure. 14-32 Gangways

The gangways must be designed to take the weight of people and equipment on them. To prepare for emergencies, it is sensible to have redundant bridges about 0.9m wide sufficient for two adults to pass each other freely. Installation of guardrails to prevent falling is essential. Access for disabled persons must be provided and this requires planning at a very early stage of marina layout design.

5.2.6 Bunkering Facilities

Leisure boats use gasoline or diesel engines, and consequently the refueling facilities in marinas are very similar to those usually found at fuelling stations on land. Such facilities are vulnerable to fire hazards, and must have effective fire-fighting facilities near at hand. Measures to prevent marine pollution must also be taken in preparation for any unfortunate oil leakage.

5.2.7 Public Washrooms

Washroom facilities at marinas vary widely from unisex to gender-segregated washrooms and these days it is mandatory in many countries to provide washrooms for wheelchair users and other disabled persons. They should be equipped with toilet bowls, urinals, sinks, mirrors, shower facilities, lights and drinking water fonts. These facilities must be placed at convenient locations throughout the property close to the berths and car parks.

5.2.8 Parking Facilities

Parking facilities at marinas are slightly different from the usual car parks in that they must provide spaces for cars, recreational vehicles, vehicle/trailer combinations and recreational vehicle/trailer combinations.

Each marina often specializes in catering for certain types of boats, for example, sailing yachts, power boats or fishing boats. Since they have different number of people and equipment on board, it is important to determine the size of parking facility and number of spaces to be made available for various vehicle types in its planning .

5.2.9 Boat Launching Facilities

Boat launching facilities are used to put boats into and out of the water. **Figure 14-33** shows a slipway through which usually small boats are launched on trailers which are moved manually or towed by a vehicle. The boats can be retrieved in the reverse process. **Figure 14-34** shows a boat hoist in action. The hoist lifts the boats, large and small, bodily in and out of the water. Unlike the slipways, marinas charge a fee for the use of the hoists. Sailing yachts normally moored in season may be lifted out by a hoist for storage ashore during the off-season.

Figure 14-33 A slipway

Figure 14-34 A boat hoist

5.2.10 Repair Facilities and Boat chandlery

Repair facilities are used to carry out repairs and routine maintenance of boats and their function is similar to automobile garages. Boat chandlers sell equipment and spare parts/consumables needed for boats and their repairs. Boatyards will have cranes as shown in **Figure 14-35** and covered sheds shown in **Figure 14-36**.

Figure 14-35 Cranes in a boatyard

Figure 14-36 A covered repair facility

6. Leisure Boat Industry in Korea

6.1 Yachts

When the shipbuilding industry in Korea started developing in the 1970s, a few companies recognized the importance of recreational boat industry and launched into yacht building business. Kyung Il Yachts Industries, a subsidiary of Hyundai Group, and Miwon Group tried to export yachts. In the early 1980s, systematic yacht production facilities were established and an attempt was made to produce a 35ft-class yacht *Blue Bird* as the original equipment manufacturers (OEMs). However, the attempt ended in failure due to the quality of the produced boats not meeting the clients' expectations. An effort to distribute them in Korea also met with cool response because there was no proper Korean market for yachts at the time.

Figure 14-37 The *Blue Bird*

Although the *Blue Bird* (see **Figure 14-37**) could not be sold abroad, its performance as a yacht was outstanding. Two Korean men (Ro Hyung-Moon and Lee Jae-Woong) crossed the Pacific Ocean on one in 1980, arriving in Los Angeles after 76 days at sea, shortening the then record of 94 days held by a Japanese team by 18 days. There are still a few boats of the *Blue Bird* class produced by Kyungil Yachts remaining in Suyeong Marina in Busan and Choongmu Marina.

After the *Blue Bird*, Kangnam Corporation also started to build a few yachts for export but soon stopped. Since the end of the 1980s, small FRP boatyards also developed or copied some foreign yachts but were not very successful due to lack of technical knowhow.

Figure 14-38 *KORDY 30*

The Korea Research Institute of Ships & Ocean Engineering developed a 30ft sailing yacht named *KORDY 30* (**Figure 14-38**) supported by the Ministry of Oceans and Fisheries in an effort to popularize marine leisure activities in Korea. During the development process a systematic assessment of its design and performance was made from the viewpoints of hydrodynamics and structural dynamics. For the hydrodynamic assessment, experiments were carried out on sail operation, resistance, maneuvering and seakeeping performance. Methods of quantifying these performances through simulation were developed

Figure 14-39 36ft. racing yachts developed by Gyeonggido

and these can now be used for designing hull forms, appendages and the arrangement of appendages.

Various realistic cruising conditions were simplified into simple external loading conditions to examine structural strength and stability of the hull, deck, mast and the connections of various appendages. A systematic research of the complete process, covering interior design, hull production, manufacture of outfitting and appendages, assembly of the mast and rigging, was carried out on *KORDY30* over a period stretching nearly a year from the plug and mold manufacture in December 2004 to sea trials in Suyeong Bay in Busan in October 2005. The yacht was considered to have functionalities and performance comparable to reputable foreign yachts by the specialists including the national yacht team, who undertook realistic sea trials and performance assessment work. This is marked as the first step toward developing high-performance sailing yachts and establishing a yacht industry.

In 2008 Gyeonggi Province hosted the World Match Racing Tour, one of the top three yacht festivals in the world, in an effort to stimulate the marine industry. This provided an opportunity for Korean companies to produce approved racing yachts. A prestigious designer from New Zealand designed the 36ft-class sailing yacht *Beyond 36* and Advanced Marine Tech of Korea built the boats for this event. Outstanding sailors in the top 10 world rankings who competed in the races in 2008 praised highly on the quality and technology of the official racing yachts. Together with the Gyeonggi International Boat Show, the event became a show case for Korean small boat building expertise. As a result, companies like Woonam Craft Co. and Blue Galaxy Co. decided to enter the high-tech business of racing yacht manufacture.

6.2 Power Boats

Motorboat races which the Korea Sports Promotion Foundation started at Misari in 2002 were the catharsis for the Korean industry to start manufacturing the race equipment, such as racing boats, outboard engines and race adjudication equipment. A racing motorboat displaces about 170kg with a one-man crew weighing 50kg aboard. The boat must produce a speed of at least 40knots out of a 32hp 2-stroke engine.

As shown in **Figure 14-40**, motorboat racing requires all the contending racers to use identical boats, and therefore new manufacturing technology became necessary to mass produce boats of uniform quality and performance. A part of the proceeds from the races is used for developing small craft and leisure equipment industry. The Ministry of Oceans and Fisheries started a family boat development program to promote safe leisure activities in 2001. In response to this, the Korea Research Institute of Ships & Ocean Engineering produced the *Marine Family* (**Figure 14-41**) in November 2003. This development project provided an opportunity to examine all the technological factors involved throughout the whole process, including hull form design, model making, high-speed towing tests, plug and mold manufacturing, prototype construction and sea trials. It also provided an opportunity to learn interior design and fittings of family-sized boats, including tables, portable toilets and showers. Excellent operational and maneuvering performance was obtained through designing a hull form appropriate for the wave and tidal conditions of Korea's coastal environment. The *Marine Family* has 6.53m length overall, 2.18m breadth and 1.3 GT. The boat draws 0.27m of sea water when the full complement of 8 persons are on board, and is capable of 25 knots (46.3km per hour) cruising speed. It has been received favorably by marine leisure specialists and is expected to contribute to popularizing marine leisure activities.

Figure 14-40 Korean-made racing motor boats

Figure 14-41 The power boat *Marine Family*

Figure 14-42 Planing boat with good rolling behavior

In 2006, the Research Institute of Small to Medium Ship Building carried out a project to develop a boat form with good roll motion characteristics, supported by the fund created with the profits from the motorboat races. Sea trials were carried out with a prototype boat thus developed, and it was shown to be capable of a maximum speed of 51 knots and the roll amplitude at this speed was found to be at most 3 degrees. It also exhibited a stable cruising attitude with a constant trim angle of 3 ~ 4 degrees.

Another project developed a stepped planing hull form for high-speed rescue craft capable of being deployed even in bad weather conditions, and it achieved a speed of 35 knots at 90% MCR. This work has been credited to have laid the foundation for small high-speed craft design capabilities in Korea.

References

1. Hideaki, M. et al., 1998. *Science of yachts.* Translated to Korean by Hong, S.W. et al., 2006. Seoul: Jisungsa.
2. Lee, H.B., Rhee, S.H. & Yoo, J.H., 2011. Analysis of a two-dimensional section of deforming yacht sails (in Korean). *Journal of the Society of Naval Architects of Korea*, 48(4), pp. 308-317.
3. Kim, S.-H. et al., 2010. Strategy for securing the core leisure boat technology (in Korean). *Journal of the Society of Naval Architects of Korea*, 47(2), pp.36-44.
4. Sim, S.M. et al., 2006. A Study on the techniques for sailing yacht model tests with sailing condition (in Korean). *Journal of the Society of Naval Architects of Korea*, 43(1), pp.32-42.
5. Shin, J. G. et al., 2006. A Study on the structural design and structural analysis for small yachts (in Korean). *Journal of the Society of Naval Architects of Korea*, 43(1), pp. 75-86.

CHAPTER 15

Offshore Structures

1. Principal Platforms for the Development of Marine Resources

As the world is becoming industrialized at an increasingly faster pace, the demand for hydrocarbon fuel steadily increases, while production finds it difficult to keep up with it. Renewable energy sources have received much attention in recent years, but they cannot yet replace hydrocarbon fuel because of their inefficiency and low availability. The situation of world energy supply is more precarious and unpredictable because many oil producing countries try to use the oil as a weapon in their international politics by adjusting the quantity of oil and gas supplied to control the prices and exploiting its economic power to their advantage. Many countries around the world are preoccupied with the problem of securing energy resources, and it is hardly surprising that conflicts can arise sometimes between neighboring countries over natural resources. In this context search for hydrocarbons in immensely challenging regions of the globe is only natural. Successful production of oil and gas in offshore locations has been going on for over half a century, not counting the early near-shore activities, and it is certain to be intensified as the known reserves are exhausted.

The development procedure for offshore oil and gas, shown in **Figure 15-1**, appears to be similar to that of onshore production, but offshore production has to contend with many more problems and much harsher environment not usually encountered on land, making it considerably more expensive. Despite this, major oil companies have made huge investments in the exploration and production (E & P) of oil and gas, as their prices will tend to rise with the continuing increase in energy consumption.

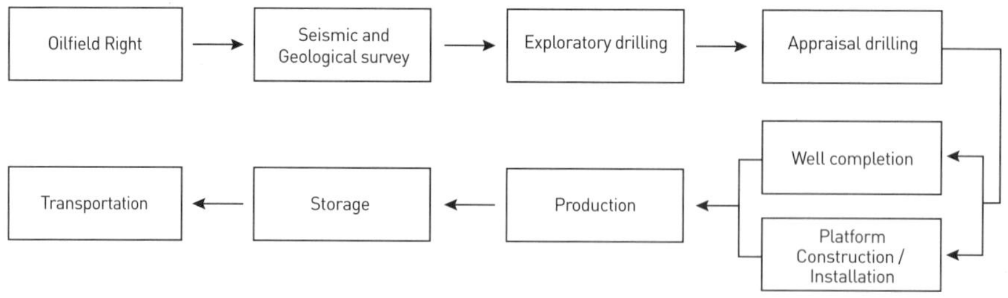

Figure 15-1 Process of offshore oil and gas development

Figure15-2 shows the history of oil and gas development and the distribution of production platforms in various water depths in the Gulf of Mexico (GOM).

Offshore oil and gas development began at the end of the 19th century in Summerland,

Summerland, California
(Source: http://en.wikipedia.org/wiki File:Oil_wells_just_offshore_at_Summerland,_California)

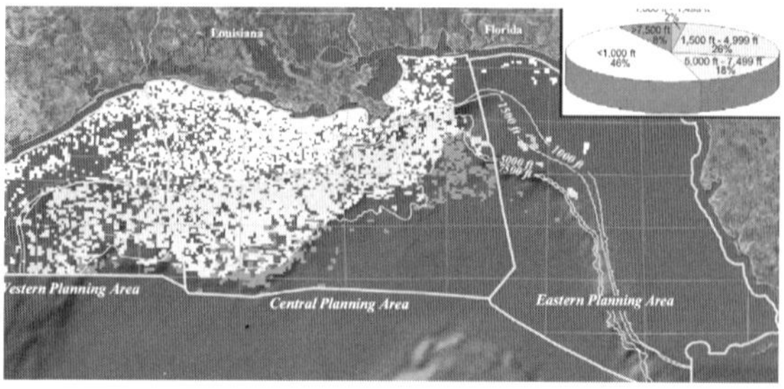

Gulf of Mexico
(Source: Deepwater Gulf of Mexico 2009, U.S. Minerals Management Service)

Figure 15-2 History of oil and gas development and depth-wise distribution of production platforms in Gulf of Mexico

California, USA in very shallow waters, but now giant oil companies are drilling and producing oil in waters up to 3,000m deep.

Table 15-1 shows the number of production platforms for bands of water depths in the GOM as of 2009. Altogether over 7,000 platforms are in operation, and they are expected to move into ultra-deep waters in time.

Offshore oil and gas development requires more specialized facilities compared with the onshore development, and the key facilities can be categorized as shown in **Table 15-2**.

Table 15-1 The number of platforms for various water depths in the GOM:

Water depth		Number of platforms in operation
ft	m	
<1,000	<305	3,096 (42.4%)
1,000–1,499	305–457	152 (2.1%)
1,500–4,999	458–1,524	2,066 (28.3%)
5,000–7,499	1,525–2,286	1,398 (19.1%)
>7,500	>2,286	598 (8.1%)

Table 15-2 Key facilities for offshore oil and gas development

Development Step	Category of facilities	Facilities
Survey	Survey ships	Satellite, seismic survey ships, semi-submersibles, airplanes
Drilling	Bottom-supported drilling rigs	Jack-ups
	Floating offshore drilling units	Drillships, semi-submersibles

Production	Fixed platforms	Fixed jacket production platforms, jack-ups, gravity based structures (GBS), guyed towers
	Floating production units	FPSOs (floating production, storage and off-loading), TLPs (tension leg platform), spars, semi-submersible production units
Storage	Storage tank	FPSOs, FSOs (floating storage and off-loading)
Export	Transportation by ship, subsea pipelines	Shuttle tankers, export pipelines, flow lines, risers

2. Drilling Platforms

The process of making a hole in the seabed for exploration, appraisal and production is called drilling and requires a rig adapted for offshore operation and mounted on a special structure which can be a jack-up, drillship or semi-submersible as shown in **Table 15-2** and **Figure 15-3**.

A jack-up rig　　A drillship　　A semi-submersible rig

Figure 15-3 Mobile offshore drilling units
(Courtesy of Samsung Heavy Industries)

2.1 Floating Platforms

1) Drillships

The first drillship in the world appeared in 1953, but at that time the industry preferred submersibles, jack-ups and semi-submersibles for operations in relatively shallower waters up to about 500 m deep until the offshore activities started moving into much deeper waters for which drillships can be very attractive. Outwardly, a drillship is very much like any other ship but with a drilling derrick installed on board as shown in **Figure 15-4**.

Figure15-4 World's first drillship Submarex (1953)
(Source: The offshore Industry, Mike Utt, Society of Petroleum Engineers)

A multi-leg mooring (known as spread mooring) system and/or a dynamic positioning system (DPS) is used for station keeping. The ship-shaped drillship has a few advantages over other units:

① It has high mobility due to low forward motion resistance and can move quickly to another location using its own propulsion system.
② Less frequent supply is required because of its large payload capacity.
③ Intact stability is good and it has good structural stability.

In addition to the low resistance, they also have lower air resistance and submerged surface areas compared to semi-submersibles. It should be noted, however, that even though their large waterplane area gives good intact stability, it also causes higher motion responses in waves.

Traditionally, drilling has been labor-intensive (typically a crew of 150 working in two shifts), but it is expected that more automation will reduce the workforce required in the future. Some units are equipped for dual drilling (**Figure 15-6**). The dual drilling equipment can also be used on semi-submersibles.

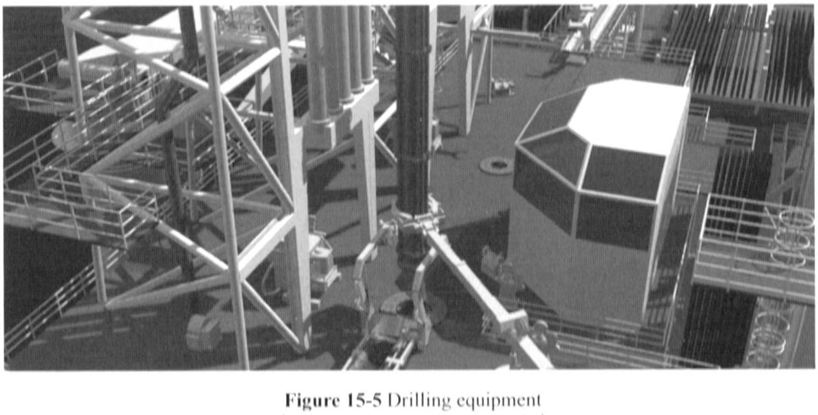

Figure 15-5 Drilling equipment
(Courtesy of Samsung Heavy Industries)

Figure 15-6 Drillships (single drilling type and dual drilling type)
(Courtesy of Samsung Heavy Industries)

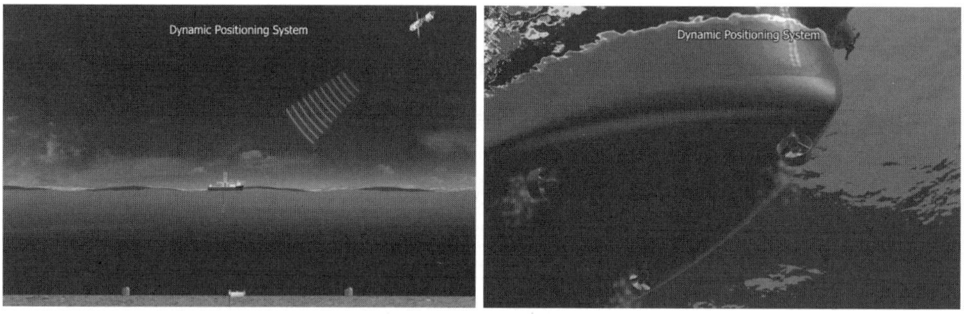

Figure 15-7 Dynamic positioning system
(Courtesy of Samsung Heavy Industries)

A dynamic positioning system (DPS) maintains the position and heading of the vessel by automatically controlling the vessel's own azimuthal thrusters using feedback control. The current position of the vessel is detected by a number of position reference systems including global positioning system (GPS), underwater acoustic systems and gyro compasses. Wind sensors and motion sensors are usually added to the sensing system.

2) Semi-submersible drilling units

Semi-submersible drilling units usually rely on DPS for station keeping and can be operated in all water depths. The current record stands at about 3,400m (12,000ft).

The hull of a semi-submersible consists of buoyancy-providing pontoons and columns which support the deck structure, hence the alternate name of 'column-stabilized drilling units.' The deck house, the main power generators, production and/or drilling packages and stores are arranged on the deck.

Semi-submersibles are de-ballasted for transit so that the top of the pontoons is just above the water surface. Some have self-propulsion capability, but those that do not have their own

Figure 15-8 A semi-submersible drilling unit
(Courtesy of Samsung Heavy Industries)

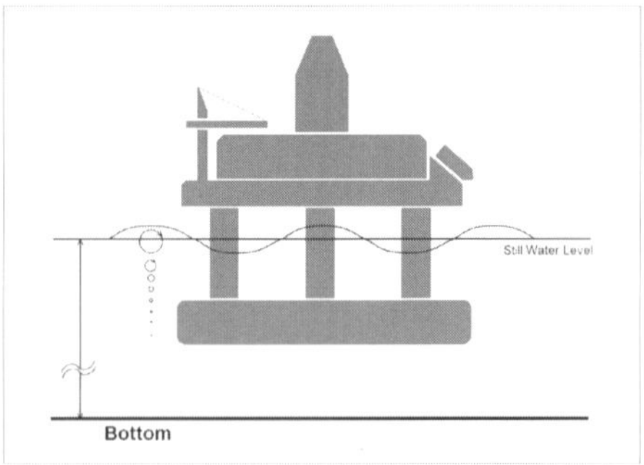

Figure 15-9 Effects of surface waves for different depths

propulsion system rely on tug boats or heavy transport barges for heavy offshore structures to move.

Upon arrival at the drilling site, the unit is ballasted by filling the tanks in the pontoons and columns to the normal operational draft leaving sufficient air gap from the water surface to the underside of the deck to avoid wave impact. In the operational draft the wave load on the pontoons is very small, since the wave forces decrease exponentially with water depth. Furthermore, the waterplane area is very small compared to the total mass displacement, and this makes the natural frequencies of heave, pitch and roll very low, virtually decoupling the motions of the unit from the waves. Because of these factors semi-submersibles provide steady platforms for drilling operations.

Their small waterplane area makes the wave load very small, but it also makes the tons per cm immersion very small, which in turn gives much lower deadweight capacity compared to drillships. Consequently they require regular supply of consumables such as drilling pipes, mud, cement, food, fresh water and so on.

Figure 15-10 shows offshore sites appropriate for the two different floating drilling units

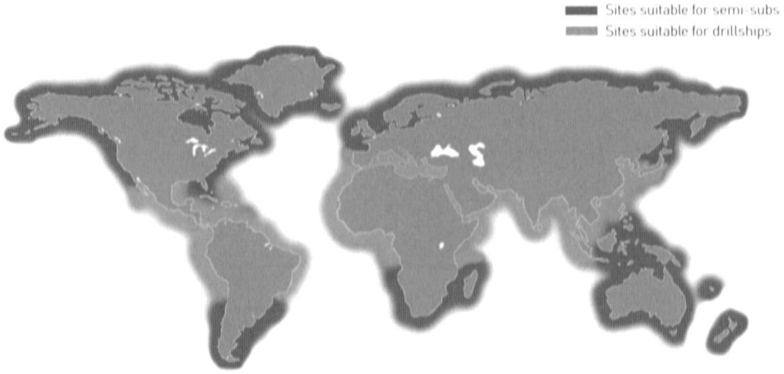

Figure 15-10 Offshore sites appropriate for drillships and semi-submersible drilling units

discussed above. The areas marked with darker shade have rougher environment and therefore call for semi-submersibles while the lighter areas are more benign and drillships can be used.

Table 15-3 Comparison of principal particulars and performance of drilling units

Items		Drillships		Semi-submers-iblerigs		Remarks
Size		Length	227.8m	Hull	118m	-
		Breadth	42m	Topsides	83×73m	
Draft		19m		22m		-
Displacement		96,000t		33,000t		About 3 times
Transit Speed		12knots		9knots		Semi-submersibles are less mobile.
Drilling operation limit state	Hs	6m		9.6m		The superior seakeeping performance of semi-submersibles results in more working days than drillships.
	Wind speed	27m/s		32m/s		
	Tz	7.0s		12s		

3) Jack-up drilling units

A jack-up has a barge-like main hull and three or four self-elevating legs. In transit the legs are carried fully raised and the main hull provides the necessary buoyancy to carry the whole structure and all the machinery. Arriving on site, the legs are lowered until they touch the seabed. The main hull takes on extra ballast water to increase its weight and jacking down continues. The legs begin to take on the weight of the structure and they dig into the seabed soil until the soil resistance is too great: a process known as spudding in. Each leg has a 'spud can' at the bottom end which is designed to allow easy penetration in the beginning but offer high resistance once penetrated. When the legs are securely in place, the extra ballast water is pumped out and the main hull is jacked up into the air with sufficient air gap to put it out of the reach of waves. The jack-up unit now offers a stable working platform supported at the bottom. The legs are often constructed from a lattice of K-shaped trusses, but smaller units may have tubular legs. Alternatively the legs may be attached to a bottom mat in which case the legs to not penetrate the soil but rest on the mat. Normal hull shape is triangular having three legs one in each corner but there are rectangular jack-ups too.

Jack-up rigs have been immensely popular, because of the ease in installation and the stable platform they provide once installed. However, their critical weakness is that they cannot operate in deep waters because of the very nature of being bottom-supported. Despite some elaborate structural modifications, they can operate in up to only about 200m water. Furthermore, they have suffered quite a number of casualties from toppling while in operation and capsizing during transit (high center of gravity and wind heeling moment) over the years.

3. Production Platforms

Most offshore areas around the world with known hydrocarbon reserves, both near shore and continental shelves, are already actively exploited with many different types of production platforms as illustrated by the case of Gulf of Mexico shown in **Figure 15-2**. Nevertheless, the search for new fields is continuing and it is inevitable that the offshore activities move into deeper and more hostile environment. As the water depth increases, the platforms become larger – and the cost of construction, transportation and operation increases and becomes less efficient. Fixed platforms are impractical for deep waters and, therefore, floating production platforms like FPSOs are required.

Key equipment of floating production platforms is outlined in **Table 15-4**. Production platforms have facilities to separate water, oil and gas from the multi-phase fluid from the

Table 15-4 Key equipment of production platforms

Equipment	Functions
Oil processing plant	Separation of water, oil and gas from the multi-phase fluid and chemical sta bilization for oil storage
Gas processing plant	Compression for gas transportation and/or reinjection and gas liquefaction
Produced water treatment plant	Reinjection and/or discharge of separated water after treatment
Water injection equipment	Filtering and clearing air and sediments from seawater and injecting
Off-loading system	Transportation of oil from storage tank to shuttle tanker
Turret and mooring system	Connection of production risers to hull; and mooring
Flare system	Flaring off waste gas produced in production process
Safety and firefighting system	Monitoring and prevention of fire and blowout
Utility system	Supply of electricity, cooling water, compressed gas, etc. necessary for self-contained offshore production
Hull & marine system	Operation of ballast, navigation, communication, etc.
Living quarters	Living space for crew and workers and shelter in emergencies like fire and explosion

well, to chemically stabilize oil and gas, to liquefy gas if necessary, and, in some cases, to store and offload oil and gas. Special designs for the separators are required for floating platforms, since they are subjected to motions due to wind, waves and current. Many of these facilities are arranged on the upper deck and the storage tanks are in the lower hull. Floaters also have off-loading equipment and a mooring system. If the platform has no storage tanks, a separate storage vessel with shuttle tankers or subsea pipelines can be used for exporting products.

3.1 Floating Offshore Production Platforms

1) FPSOs

The FPSO (floating production, storage and off-loading) shown in **Figure 15-11** is a self-

Figure 15-11 Floating production, storage and offloading (FPSO) unit
(Courtesy of Samsung Heavy Industries)

contained production platform suitable for deep water oil and gas production, and is often deployed far offshore. In addition to the normal production facilities, it has its own storage space for oil and gas. Hydrocarbon thus produced and stored in FPSO is then transported to shore facilities via subsea pipelines or a fleet of shuttle tankers.

In comparison to fixed production platforms, floating production platforms suffer from the motions which can lead to less efficient operations of gravity-based separators and flare systems.

The production process on FPSOs is illustrated diagrammatically in **Figure 15-12.**

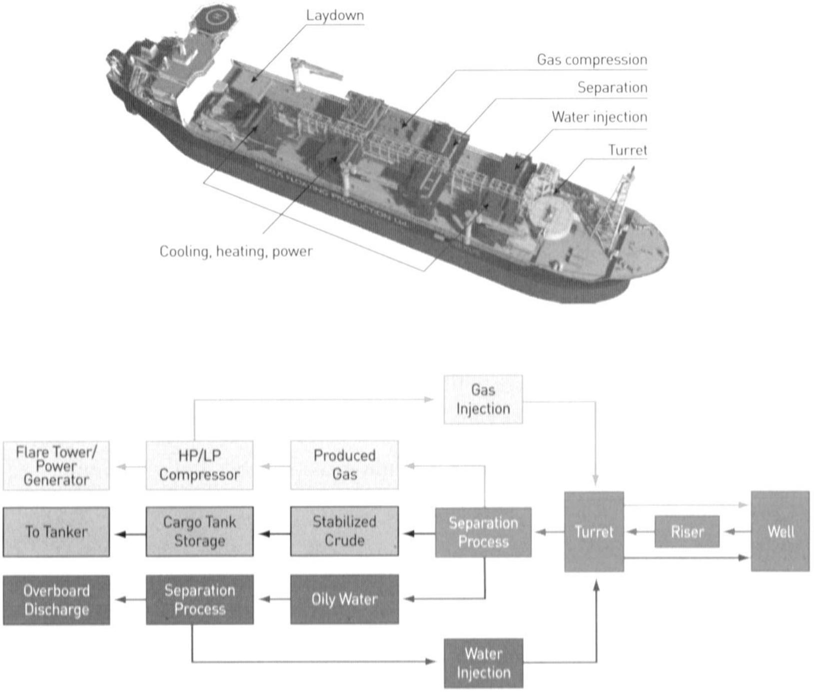

Figure 15-12 Production process on FPSOs
(Courtesy of Samsung Heavy Industries)

The multi-phase fluid pumped up from the well is separated into water, gas and oil by production separators. The crude oil is then filtered and stored in tanks before being offloaded from the FPSO to a shuttle tanker moored to it in tandem or side-by-side. The gas collected through a high pressure (HP) and low pressure (LP) production separators is de-moisturized and compressed to be used as fuel gas or to be re-injected into the reservoir to maintain the reservoir pressure. Unused gas is burned off safely from a flare tower. The produced water is also filtered and re-injected or discharged into the sea.

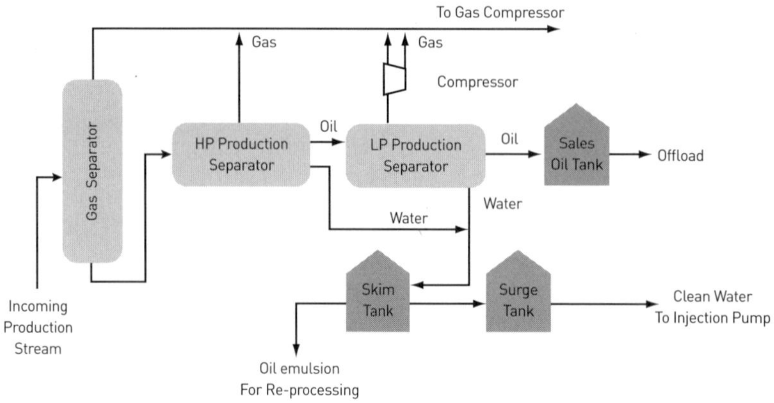

Figure 15-13 Oil production process

2) LNG FPSOs

Natural gas is much cleaner compared to other fossil fuels, and its supply and demand is expected to grow rapidly. The total world demand for natural gas has been increasing at the rate of about 8% per year, due to increased consumption in developed countries and many developing countries turning to gas for power generation as a cleaner form of fuel than coal. Gas produced in shallow waters is sent via pipelines to shore liquefaction plants, from where it is transported to the consumers in the form of LNG, or liquefied natural gas (**Figure 15-14**). In deep waters subsea pipeline is often not available, and the only practical way to transport the gas to a shore plant is to compress it as much as possible and use compressed gas carriers. That, however, is much less efficient, as the energy density of the maximum compressed gas is about 2.5 times less than LNG. That is one reason why deep sea gas fields have been rare.

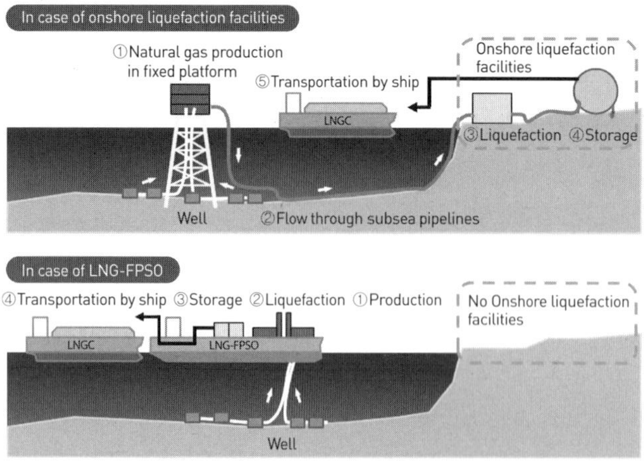

Figure 15-14 Gas production process of LNG-FPSO

In order to meet the growing demand for gas, however, deep sea fields have to be exploited, and for this purpose a variety of technologies for gas production and cleaning have emerged. One of the most promising novel concepts is LNG-FPSO, the LNG equivalent of FPSO shown in **Figure 15-15**. It has a gas liquefaction plant on board and can produce natural gas, filter, liquefy and store it until it can be offloaded by LNG carriers. Therefore, it can reduce the

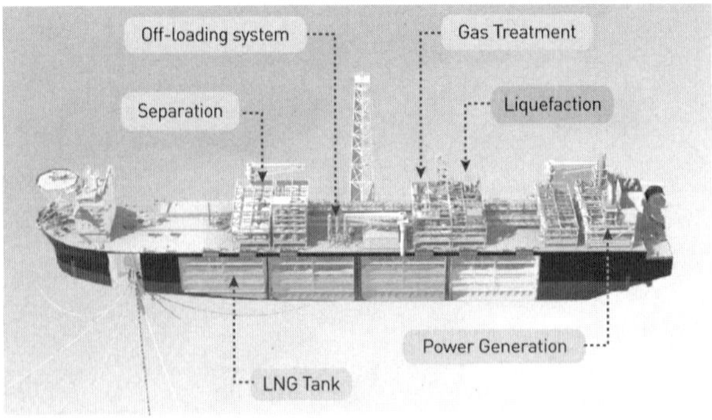

Figure15-15 Arrangement of LNG-FPSO
(Courtesy of Samsung Heavy Industries)

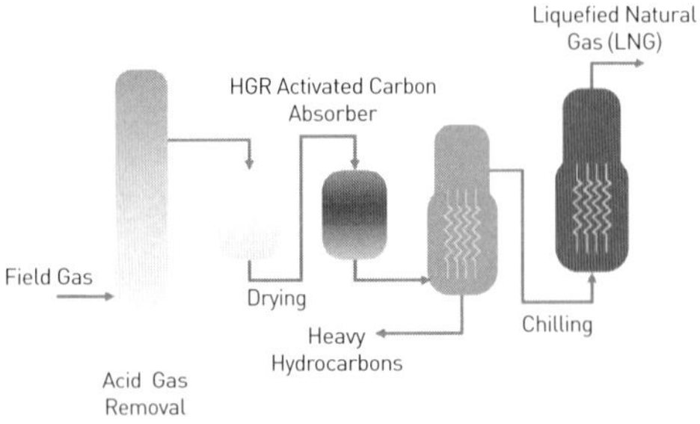

Figure 15-16 Production process of LNG

infrastructure required for gas production and thus save on development costs. Although no LNG-FPSO has been put into service as yet, development of the core technologies required for its realization has been nearly completed and LNG FPSOs are close to field deployment. The core technologies include small high efficiency appliances for liquefaction plants, large liquefied gas storage tanks and reduction of sloshing in them, liquefied gas off-loading technology, mooring systems, gas treatment while undergoing motions, low-temperature piping equipment, and corrosion-protection of pipes against CO_2 and H_2S.

Figure 15-15 shows the general arrangement of key modules of production facilities and the gas production process of LNG-FPSO is shown in **Figure 15-16**.

3) Semi-submersible production units

A semi-submersible drilling rig was explained earlier in this chapter. A semi-submersible production unit shown in **Figure 15-17** is a floating platform like FPSOs, but the unit is based on semi-submersible forms to provide good motion characteristics. The topsides components and functions are similar to those of FPSOs, although the square or rectangular deck space makes the arrangement of facilities much easier than FPSOs. However, it is unable to provide a storage tank and therefore must be connected to a subsea pipeline or in conjunction with an FPSO or FSO at a nearby location.

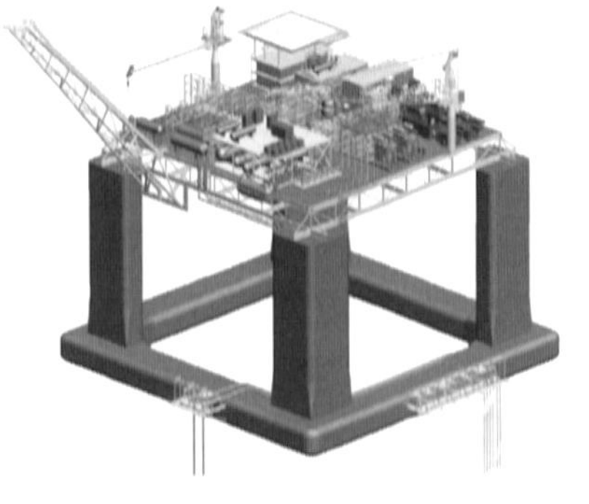

Figure 15-17 A semi-submersible production unit
(Courtesy of Samsung Heavy Industries)

4) TLPs, spars and other structures

(1) Tension leg platforms

Tension leg platforms (TLP) (**Figure 15-18**) were developed as an alternative to fixed and floating production platforms for intermediate water depths, taking the obvious advantages of fixed production platforms and relatively low initial cost of floating production units. Strings of steel pipes in tension, called tendons, connect the bottom of the platform to the anchor template on the seabed, making its natural period of vertical motions very short due to the high stiffness of the tendons.

Figure 15-18 TLP
(Source: Samsung Heavy Industries)

The natural heave period of TLPs is well outside the ocean wave period range as shown in **Figure 15-38**, making its motion response in waves small. The outward appearance of TLPs is quite similar to semi-submersibles, but their heave motion response in waves is much better than semi-submersibles. The natural heave periods of semi-submersibles are also outside the normal wave periods, but by being longer.

TLPs have their shortcomings too in that they cannot have storage tanks and thus can only be deployed at sites with subsea pipelines or in conjunction with an FPSO or FSO at a nearby location. Its small heave motion allows dry Christmas tree to be used with oil flow regulation valves and safety valves to be installed on the platform above the water surface allowing easy access for maintenance & repair and rapid response to emergency situations.

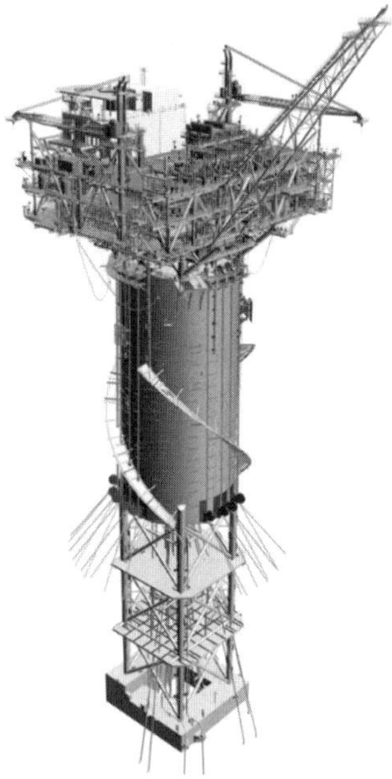

Figure 15-19 A spar (Samsung Heavy Industries)
(Source: http://www.pagiticas.com/cronus/tahitiflowlines.php)

(2) Spars

A spar is a floating platform and is in the form of a large diameter, single and long vertical cylinder supporting a deck structure and is anchored to seabed as shown in **Figure 15-19**. Its waterplane area is relatively small compared to the mass and therefore has good heave motion response in the normal short period waves.

However, when it is excited by a swell with a long period, it may experience high resonant response due to its low natural frequency and small damping. In order to overcome this problem, damping plates are installed in the lower part of the spar and these also cause the heave added mass to increase. Vortex suppression devices such as helical strakes are attached to the surface of the spar to prevent vortex induced vibratory motions.

Production spars have small storage tanks and the deck spaces are also very limited so that they can be equipped with only small to medium size facilities. The low heave motion of spars allows the use of dry trees with all their advantages.

4. Offshore Support Vessels and Structures

4.1 Offshore Support Vessels

1) Pipe layers

Increasing offshore activities and growing demand for energy inevitably increases subsea pipeline installations. Subsea pipelines are used for transporting crude oil and gas, but they can also be used for moving all kinds of fluid necessary for offshore operations. It is also expected that the demand for removal or replacement of aging pipelines will continue to rise in the future and this will add the demand for the services of underwater cable-laying and pipe-laying vessels.

On continental shelves or in coastal areas, a network of subsea pipelines can be more efficient than storage tanks on floating platforms. Subsea pipe-laying vessels are special-purpose vessels necessary for installing such infrastructure and they are also used for installing high capacity power cables and optical fiber cables.

The method of laying pipes is determined according to the characteristics of installation area, budget and time available, and a pipe-laying vessel is selected to suit the water depth. Pipe-laying methods are summarized in **Table 15-5**.

The maximum water depth for pipe-layers is determined by the following factors:

- capacity of mooring system of layers;

- size of stinger[1];
- capacity of tensioner[2];
- diameter and thickness of pipe; and
- thickness of pipe coating.

A reel-lay type vessel can lay also subsea umbilical cables.

Table 15-5 Subsea pipe-laying methods

Laying method		Characteristics
S-Lay	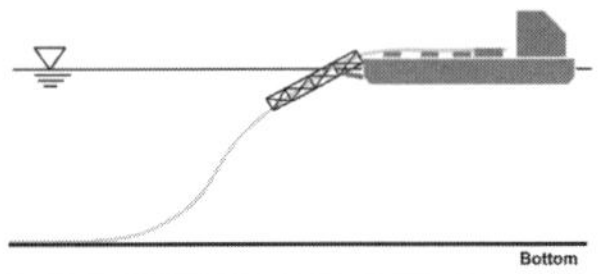	A traditional laying method. Maximum water depth and laying speed dependent on the layer's capacity. A stinger is installed at stern to keep the pipe in a gentle S-shape. Suitable for shallow water.
J-Lay	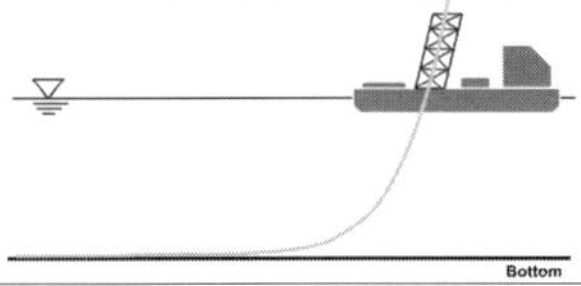	Developed to overcome the depth limitation of S-lay method. A near-vertical tower on the vessel allows the pipeline to be lowered forming a J-shape.
Reel-Lay	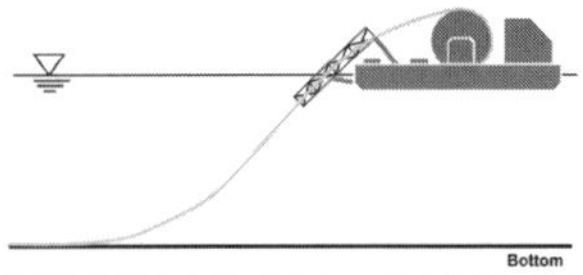	The pipes are welded into a pipeline and wound on a reel on shore. The line is reeled-off on site as installed, and therefore the laying is relatively fast. However, concrete coating is not possible and pipe diameter that can be handled is limited.

2) Decommissioning units

With the increasing awareness of the importance of oceanic environmental protection, what to do with the defunct offshore installations has been receiving much attention. This problem has led to the development of decommissioning vessels.

Exploitation of offshore resources started in the 1950s, and therefore there are many offshore installations aged 20 to 30 years or more by now. The structures which have served their purpose and are no longer needed have to be abandoned. These abandoned structures, if left where they are, are not only unsightly but also quite dangerous as they can explode or collapse without warning and there is always the risk of additional pollution. Therefore, they have to be removed or reused in some way.

In order to solve such problems a new decommissioning market worth billions of dollars was born, and new regulations on decommissioning have been introduced to regulate this

1 A device attached to the pipe-layer to support the pipe being laid to prevent excessive bending of the pipe.

2 A device used to pull on the pipe as it is lowered to maintain appropriate tension.

new sector of the industry. Installing offshore structures is difficult, but removing them can also have a few problems as follows (**Figure 15-20**):

- collapse of platforms due to corrosion;
- explosion and sea pollution due to remaining oil and gas;
- difficulties due to motions of the decommissioning vessel and crane barge;
- heavy lift vessels and cranes needed; and
- cutting required to separate topsides from the substructure.

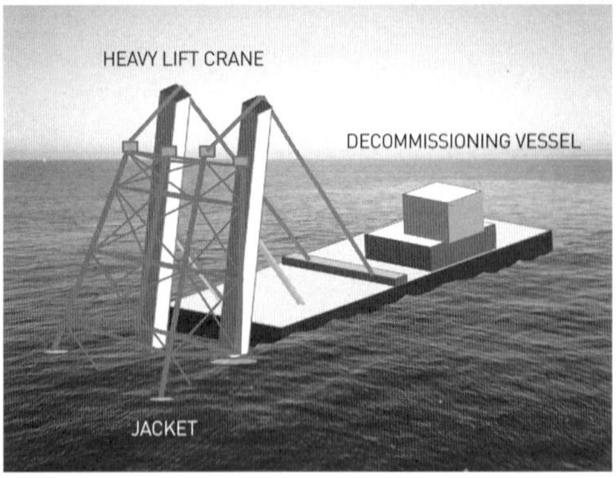

Figure 15-20 Decommissioning platforms

Figure 15-21 shows the number of newly installed platforms since the 1950s and the record of decommissioned old platforms in the Gulf of Mexico since the 1970s.

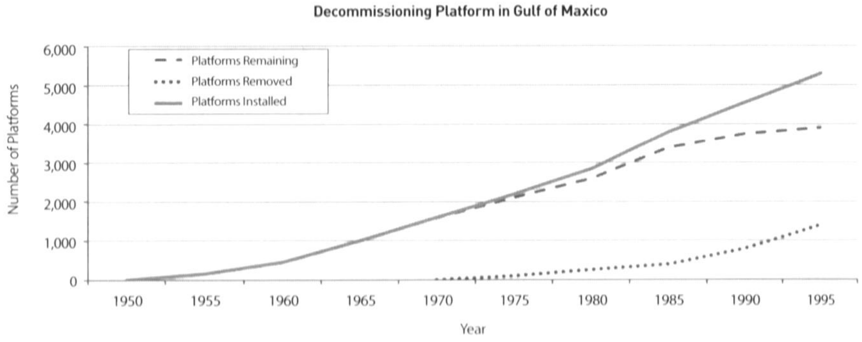

Figure 15-21 The number of platforms installed and decommissioned in GOM since the 1950s

3) Heavy lift vessels

Heavy lift crane vessels are often employed for construction and installation of offshore structures. Using very large capacity heavy lift vessels improves the efficiency of fabrication and installation since they are capable of handling very large modules which minimizes the time and expenses required for labor-intensive module-to-module connections. For small to medium size production platforms the whole topsides can be lifted and placed over the jacket or lower hull in one operation, saving much time and effort. Lifting capacity ranges from 3,000 to 14,000 tons and they are categorized into two types depending on shape and operational method: revolving cranes and sheerleg cranes with wide booms for lifting large structures.

A floating crane barge (8,000 tons)

A 14,200-ton semi-submersible crane vessel

A heavy lift barge with a revolving crane (7,450 tons)

Installation of platform using a heavy lift vessel

Figure 15-22 Heavy lift crane vessels
(Courtesy of Samsung Heavy Industries)

Land-based large cranes are sometimes limited in their movements due to ground instability such as subsidence, but floating cranes have no such problems. Depending on the operating environment, a suitable vessel can be selected from available semi-submersible crane vessels or crane barges (see **Figure 15-22**).

4.2 Offshore Structures and Facilities

1) Offshore wind turbines

Renewable energy sources are receiving much attention these days as one of the solutions for the urgent need to replace fossil fuels. Wind turbines started in North America and Europe, and their commercial use has already been well established on land and at sea, making wind farms common sights in several countries there. Exploitation of wind power has now spread to all corners of the globe and Asian countries are no exceptions. For a variety of reasons, including higher efficiency, freedom from the limitations in turbine size, aesthetics and objections from pressure groups, the wind farm industry has been moving out to offshore locations.

With the development of wind power generation technology, the turbines have grown to 3MW and are becoming larger. Presently, 7MW land-based or offshore wind turbines are under development, and over 10MW wind turbines are expected to appear in the near future. Onshore wind farms with large wind turbines are noisy, take up much space, and can be accused of spoiling the landscape, while at many offshore locations required space is much easier to secure, wind is more constant and its speed is usually higher.

Figure 15-24 shows various types of offshore wind turbines, and a schematic representation of a typical offshore wind farm is shown in **Figure 15-25**. So far offshore wind farms have been installed mostly within 10km from shore, but there have been complaints that they interfere with ship navigation and fishing activities. Consequently, offshore wind industry has begun to move into deeper waters far offshore and for this purpose several novel concepts of floating offshore wind turbines have been proposed and developed. The subsea power cables from such offshore wind farms are connected to subsea power transmission stations which send the collected power to shore transformer stations as shown in **Figure 15-25**.

Figure 15-23 A typical offshore wind farm

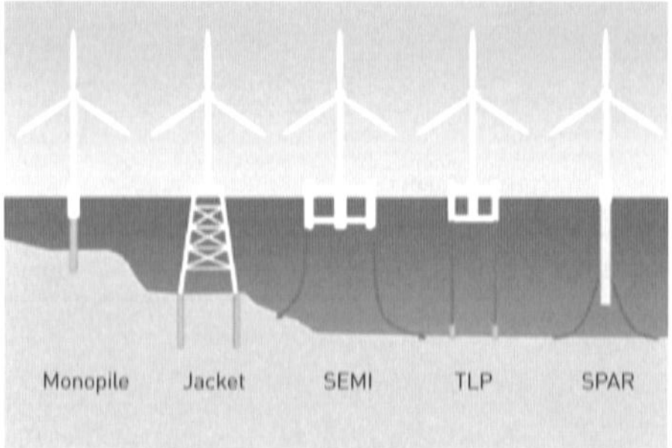

Figure 15-24 Types of offshore wind turbines

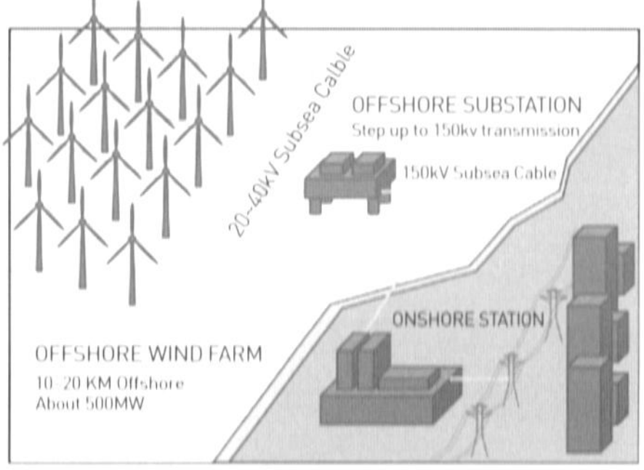

Figure 15-25 Schematic of an offshore wind farm

2) Deep-sea oil and gas development

In deep waters the main drilling and production packages can be placed above sea surface, while some of the crucial equipment have to be installed on the seabed. As stated above, floating structures are required in deep sea hydrocarbon production, but they suffer from motions. The most effective solution to such problems is doing all the work underwater, while only those items which can be sent through flexible pipelines are transferred between the surface structure and the seabed. Subsea systems are therefore essential in deep sea oil and gas development.

The main function of subsea systems is to maintain production smoothly and safely by controlling and monitoring subsea production facilities remotely from floating offshore platforms. They can be used for both deep sea drilling and deep sea production, and these are shown in **Figure 15-26**. Usually there are risers connecting the drilling or production facilities

Figure 15-26 Riser and subsea pipelines
(Courtesy of Samsung Heavy Industries)

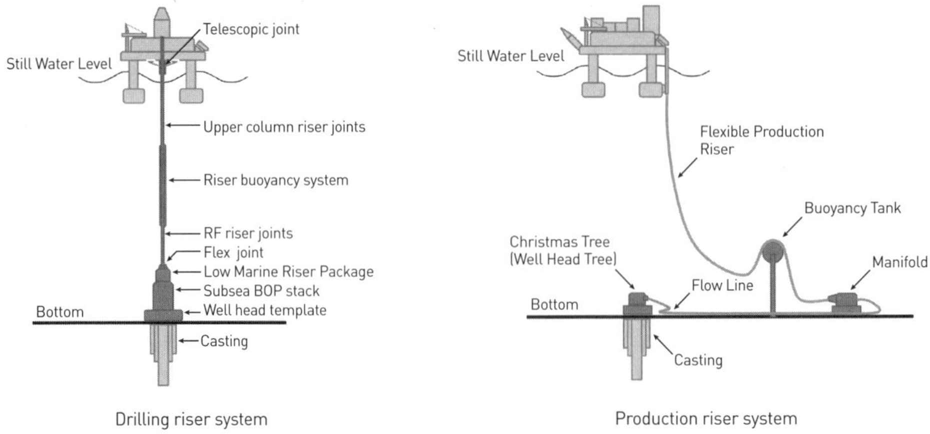

Figure 15-27 Drilling risers and production risers

on the surface to the subsea equipment and facilities, and they serve various purposes.

Subsea completion has now been widely accepted because the production lead time can be much reduced compared to using traditional platforms.

The drilling riser in **Figure 15-27** is a conduit for circulating mud and protecting drill pipes, and the production riser is a pipe that transfers multiphase flow from a subsea wellhead or manifold to a floating production platform. Flexible pipes are used for production risers to overcome the problem of floating platform motions. In deep waters, buoyancy tanks (mid-water arch, or MWA) are installed to reduce weight, fatigue load and bending moment acting on flexible risers. Key subsea system equipment and their functions are shown in **Figure 15-28** and summarized in **Table 15-6**.

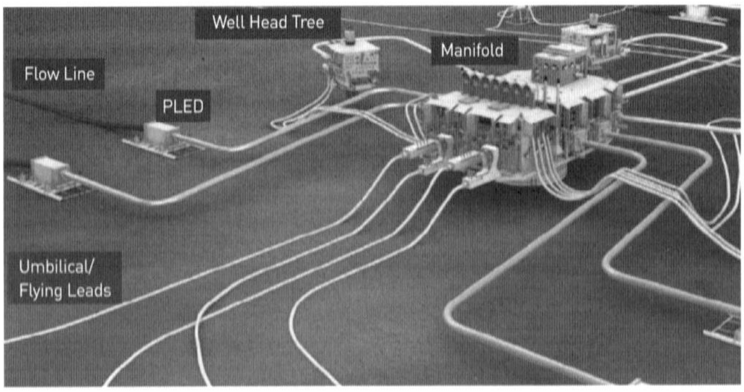

Figure 15-28 Subsea systems

Table 15-6 Principal subsea system components

Items	Functions
Manifold	A pipe junction used to collect multiphase flows from several neighboring subsea wells and transfer it to production risers.
Well head tree	Wellheads and trees are the main equipment for oil production, water injection, and down-hole operation. They are installed on the casing head to seal the annular space between casing and tubing, to control wellhead pressure, to adjust well flow rate and to transfer multiphase flow to flow line.
Flow line	A pipe that connects well head trees to a manifold for transporting multiphase flow on the seabed.
Umbilical cable	Cables used to supply required utilities to an apparatus. An umbilical can, for example, supply hydraulic power, electrical power and fiber optic command signals to subsea equipment (to provide chemical/hydraulic and electrical connections between umbilical termination assemblies (UTAs), subsea distribution units (SDUs), trees and manifolds, etc.)
BOP (Blow-out preventer)	Equipment used to prevent uncontrolled flow of liquids and gases and extreme erratic pressures (formation kicks) during well drilling operations.
Production riser	A pipe that connects the sub-sea system to a floating unit for production, injection and export.
Drilling riser	A conduit that provides a temporary extension of a subsea oil well to a surface drilling rig, enclosing drill pipe to form an annular space for mud circulation and bringing up cuttings.

3) Deep-sea remotely operated vehicles (ROV)

In shallow waters underwater work used to be often undertaken by divers, but the offshore hydrocarbon activities have moved into much deeper waters where human divers cannot intervene directly. Special equipment is necessary for installation and operation of subsea systems such as shown in **Figure 15-28**. One typical piece of such equipment is remotely operated vehicles (ROVs) and one has been developed for exploration and research at a depth of up to 10,000m. Some recent floating drilling rigs are designed for water depth of 3,000m, and they usually have one or two ROVs on board for underwater work.

An ROV (see **Figure 15-29**) is equipped with six to ten thrusters, cameras, lights, various sensors, underwater robotic arms, ballast tanks and batteries. An ROV is operated and controlled through an umbilical cable which supplies power and transmits command signals from the surface and transmits video and other sensor data to the operator. The

An ROV (remotely operated vehicle)

BOP stack installation process monitored by ROV

Figure 15-29 Remotely operated vehicles

ROV operator can monitor the deep water conditions in real time and use the robotic arms for repair and maintenance of subsea facilities. The weight of an ROV varies depending on its size and usage, but is usually from 0.3 tons to 2 tons. The ROV shown in **Figure 15-29** is monitoring the installation process of a BOP stack onto the well template.

Key functions of ROVs are:

- subsea wellhead monitoring;
- support for inserting drill pipes into wellhead;
- installation, maintenance and repair of subsea systems;
- manipulation of simple devices like valves;
- pipe and cable cutting and high pressure jetting;
- exploratory survey of seabed and investigation of subsea systems;
- replacement of subsea system components; and
- video recording of underwater work.

In the *Deepwater Horizon* accident of 2010, ROVs were employed for leakage monitoring, and recovery and clean-up operations of crude oil as in **Figure 15-30**.

Figure 15-30 ROV operation in Deepwater Horizon accident in GOM, 2010
(Source: http://en.wikipedia.org/wiki/File:ROV_Deepwater_Horizon_BOP.jpg)

4) Mooring systems

FPSOs and other floating structures require a mooring system for station keeping against environmental and operational loads. Types of mooring systems vary depending on water depth as shown in **Figure 15-31**.

Wire type	Chain type	Wire & chain combined type	Intermediate sinker type
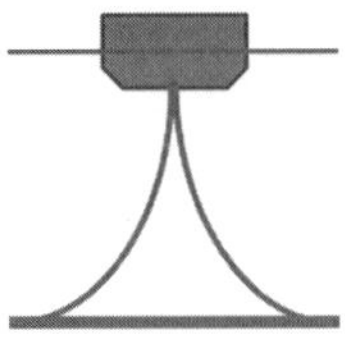	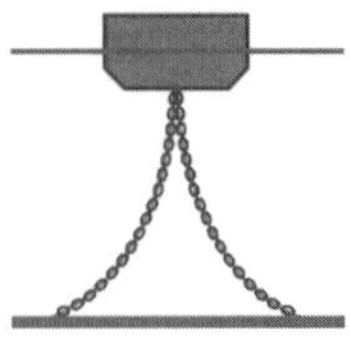	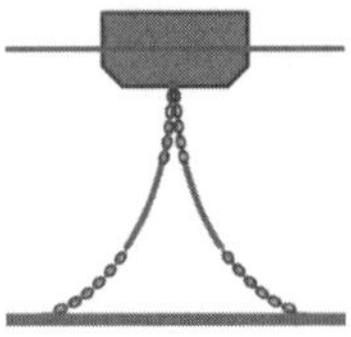	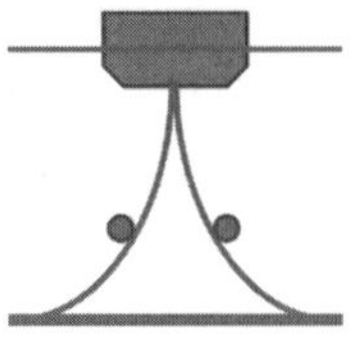
50~200m	20~500m	400~1,500m	20~1,500m

Figure 15-31 Types of mooring systems for various water depths
(Source: Lee. J, "Overview of Design of Offshore Structures", Kwangmoon Publishing Company, 1989)

There are two methods of mooring FPSOs depending on the presence of a turret as illustrated in **Figure 15-32**.

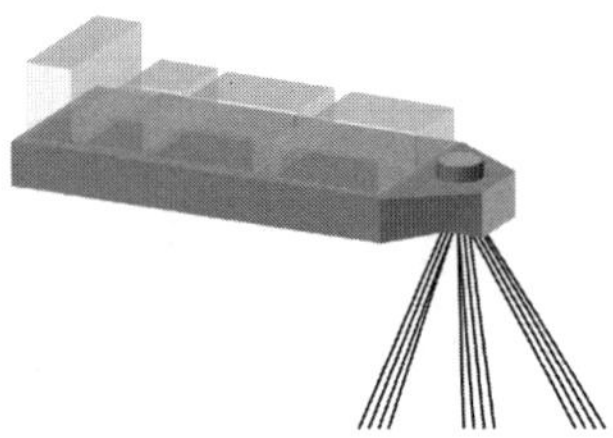	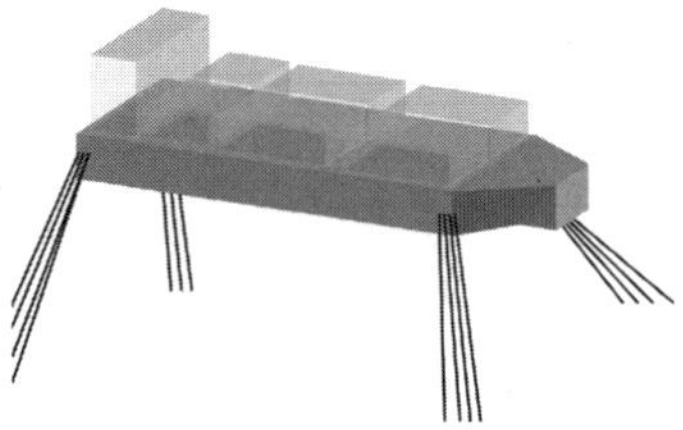
Turret mooring system	Spread mooring system
Can be used only when there is a turret. Good for strongly directional environmental loads as the vessel can weathervane.	Suitable for weak and non-directional environmental loads, and thus there is no need for weathervaning.

Figure 15-32 Mooring systems with/without a turret

Figure 15-33 shows the arrangement of mooring lines and production risers in a turret mooring system. Mooring lines are spread out so that they do not interfere with each other and subsea flow lines. Turret size depends on the size of the FPSO and external loads. The turret in **Figure 15-34** is of a vertical cylinder type with a diameter of 10 to 30 m, which is integrated within the hull and transfers external loads to the mooring system. The turret has bearings allowing it free 360° rotation to weathervane.

Some FPSOs have turret installed in a yoke built in front of the bow and connected rigidly to the ship (external turret system).

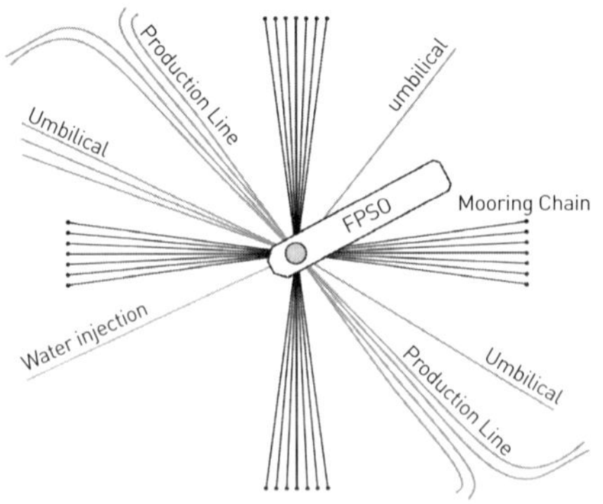

Figure 15-33 Arrangement of mooring lines and production risers in a turret mooring system

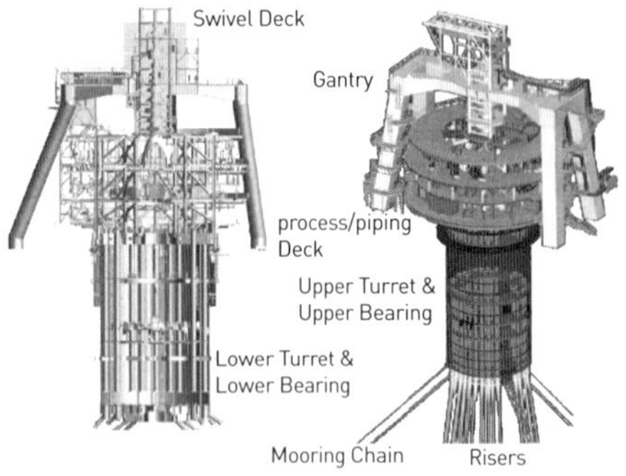

Figure 15-34 Internal turret system

5. Structural and Hydrodynamic Characteristics of Offshore Structures

5.1 Modules and Links

Land structures are usually designed strong enough to withstand earthquakes and wind. However, offshore structures are placed in moving ships or in a narrow and confined space above a substructure, and therefore there are more constraints and factors to consider in designing offshore structures than land-based ones. Most land-based steel structures are built with nuts and bolts, but offshore structures are welded so that they can withstand severe conditions for a long time.

The welded production platform shown in **Figure 15-35** consists of four to twenty modules containing various equipment and outfittings necessary for oil and gas production. The topsides are modularized for safety and efficient construction, and the modules are defined according to their structures, functions, size and weight of facilities. Through modular construction it is possible to maximize use of the construction yard spaces, reduce construction time, maximize pre-painting and pre-outfitting, and minimize work at elevated positions. Sequential construction method, where the topsides are constructed after the vessel or the base structure is complete, is more likely to experience delays. However, concurrent construction of the hull and the topside modules can have definite advantages and can lead

Figure 15-35 Topside modules
(Courtesy of Samsung Heavy Industries)

to early completion, especially since the modules can be lifted onto the base structure as soon as they are ready.

Work on board is both dangerous and inefficient, due to the confined working environment and the height of the work site. By doing welding, pre-outfitting and pre-painting on land as much as possible where it is much safer and access is much easier, on-board work can be minimized. However, a heavy lift vessel is necessary to lift the completed modules onto the substructure. Thus, the size of modules can be limited by the available crane capacity (**Figure 15-36**). In planning the lifting operations, thorough structural analyses on deflection and bending should be carried out in order to avoid damaging the modules being lifted.

Module fabrication on land

A module being lifted by a floating crane

Figure 15-36 Construction and installation of modules
(Courtesy of Samsung Heavy Industries)

Sagging or hogging deformation of ship-shape floating platforms due to waves can affect the topsides design greatly. In order to minimize the effect of base deformation on the topsides structure, a bearing support system can be used. The bearing support system is installed between the topsides and the base, and it can help mitigate the impact loads on the topsides while transferring and distributing the loads to the base.

In order to enhance the effect of the bearings, elastomers shown in **Table 15-7** are inserted into the bearing support system. Elastomer layers have an effect on vertical stiffness and load transfer, and the overall horizontal displacement depends on the thickness of the elastomer layers.

Each module is not solidly fixed to the base, but is made to have the connection boundary condition as shown in **Figure 15-37**. In this way the modules are isolated from the base and the neighboring modules so that each has its own resonant period. This prevents energy transfer and resonance. The boundary conditions in **Figure 15-37** show that the first column is fixed, column 2 has fore and aft freedom, column 3 has transverse freedom and column 4 can move both fore and aft and transversely.

Table 15-7 Types of Module support

Type	Appearance	Characteristics
Welded support	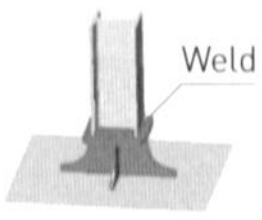	Used for flexible module structures or small platforms. The hull deflection greatly affects the topsides.
Sliding pad type	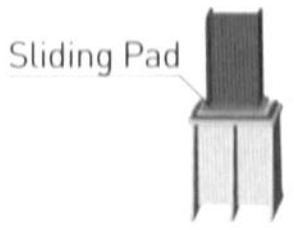	Used in small and medium size modules together with welded support, since there is no vertical deformation. Little effect of the base deformation on the upper structure.
Elastomeric type support		Used in large modules or FPSO modules. Little effect of hull deflection on topsides. Expensive due to complexity. Strong against dynamic loadings. Vertical and rotational displacement is easily absorbed.
Pot/spherical type support		Used in FPSO modules. Hardly any effect of hull deflection on topsides. Precision quality management is necessary due to directionality.

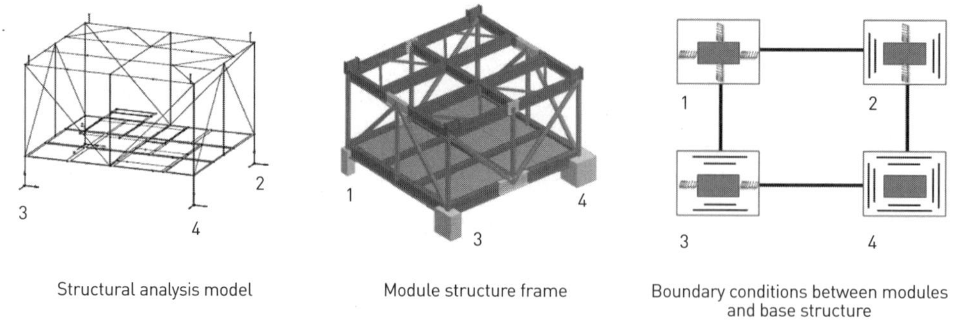

Figure 15-37 Module configurations and support conditions

5.2 Seakeeping performance

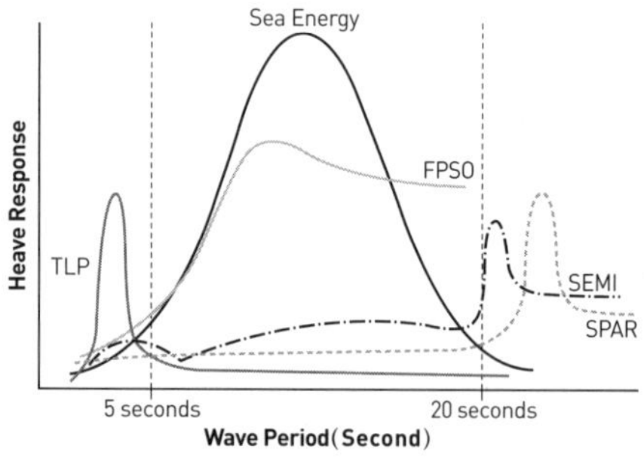

Figure 15-38 Heave responses of floating platforms

Motion response of floating bodies in waves is dependent on their size, mass and geometry. The main factors determining the motion response of a body are the wave period, natural period of the body and the natural period of the mooring system. As shown in **Figure 15-38**, FPSOs show the highest heave response in waves of 4 to 16 seconds, while TLPs have very short natural period of around 2 to 4 seconds, which is the natural vibration period of the tension legs, taking it well outside the predominant wave periods. Motion response affects the operation of the production facilities directly, and motions beyond certain operable limits cause downtime. For example, a three-phase separator may not be able to operate when the angle of inclination is more than 4° and drilling operations may have to stop when heave motion double amplitude is more than 5m.

A merchant ship can take shelter in a harbor or change course when rough weather is expected. Offshore platforms, on the other hand, cannot take such evasive action easily and therefore must be designed to be able to stay on site even in severe storms without being damaged. Offshore platforms have daily oil or gas production target and therefore downtime should be kept to an absolute minimum. Structures with good motion response characteristics are bound to have less downtime and consequently will be more competitive. Therefore, it is important to select the optimum type of structure for the operating environment, production capacity required and water depth.

6. General Arrangement of Offshore Platforms

6.1 Overview of Drillships

The main feature in the outward appearance of a drillship is the drilling derrick tower amidships as shown in **Figure 15-39**. The derrick lifts, lowers and positions the drill string over the well bore. The drill bit at the end of the drill pipe is rotated by a top drive or kelly drive at the drill deck level. This configuration of drilling rig has not changed much since it was developed for land-based drilling.

Drillships have higher loading capacity and mobility compared to semi-submersibles, but they are more susceptible to wind, waves and current, making them more suitable for sites with relatively mild weather conditions (significant wave height less than 7 ~ 8m). In order to restrain drillships within a certain horizontal radius (allowable footprint) from the well bore, a suitable station-keeping system is necessary. Spread mooring can be used in shallow waters, but in deep water drilling the only practical solution is dynamic positioning system.

Traditionally the oil and gas industry has employed rotary drilling rigs with a circulation system for liquid mud. All drilling rigs, regardless of whether for land or offshore drilling, also have pipe stand where pipes are stored vertically in readiness for being added into the drill string. Some of the work done manually have been gradually automated in modern drilling rigs, especially the pipe handling systems.

All offshore drilling units have a riser system which is a conduit between the wellhead and the drilling platform (see **Table 15-6**).

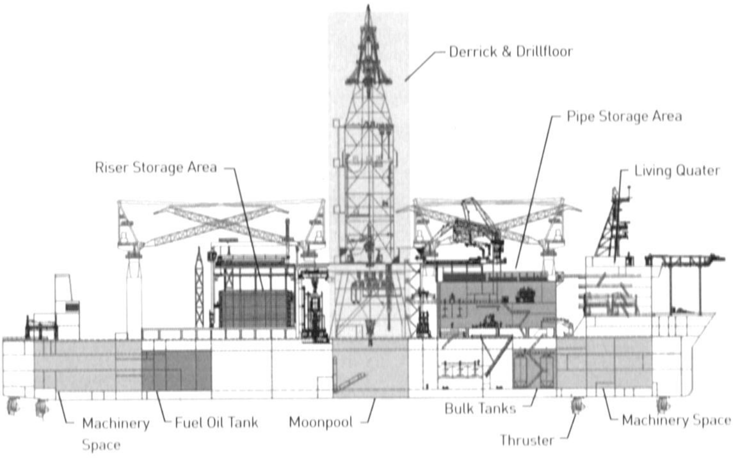

Figure 15-39 General arrangement of a drillship

The general arrangement of a drillship can be divided into two groups: hull interior arrangement and topside arrangement. Engine room, switchboard room, ballast tanks, sea water tanks, oil tanks, pump room, living quarters, drilling moon pool, fuel oil tanks, fresh water tanks, mud storage tanks and bulk stores are arranged in the hull. The topsides contain drill floor, derrick tower, riser storage, BOP stack storage, Christmas tree storage, drill pipes and casing storage, shaker, mud pit, etc.

6.2 Overview of Semi-Submersible Drilling Units

The major strength of semi-submersibles as mobile drilling units is their superior motion characteristics in operating draft compared to ship-shaped units. Owing to this feature, semi-submersibles are widely used as floating production units (FPU), mobile drilling units, heavy lift units and floating hotels. However, they are never used as FPSOs because of their small deadweight capacity compared to ship-shaped units.

Geometrically, a semi-submersible unit consists of three parts: pontoons, columns and deck. Modern semi-submersibles have two parallel pontoons, each of which is shaped like a narrow ship or an enclosed barge. They provide most of buoyancy and constitute the lowest members. Each pontoon supports 2 to 4 columns and the columns support the deck.

Figure 15-40 A semi-submersible drilling units
(Courtesy of Samsung Heavy Industries)

Although the outward appearance of semi-submersible drilling units is somewhat different from drillships, the drilling equipment is identical. The term 'semi-submersible' originates from the bottom supported 'submersible' drilling units, and has nothing to do with drilling operations. **Figure 15-40** shows a semi-submersible drilling unit in transit. Once at the required site the units are ballasted down to their operating draft before drilling operation can begin.

Pontoons contain ballast tanks, oil tanks, sea water tanks, liquid mud tanks, fuel oil tanks, pump room and fresh water tanks. In columns, piping and cabling, access to the pontoons, bulk stores and ballast tanks are arranged. Generator room, mud pump room, machinery shops, switchboard room, mud sack storage/mix room, drill floor, derrick, riser store, BOP stack store, Christmas tree store, drill pipes and casing rack, living quarters and mud circulating system are placed on deck. In short, below deck facilities in drillships are placed in pontoons and columns. Topside compartments and equipment are similarly arranged on open deck in both types.

6.3 Overview of FPSOs

The floating production, storage and off-loading (FPSO) unit in **Figure 15-41** has an integrated facility for oil production, storage and off-loading, and is suitable for deep sea oil production. With the industry moving into ever deeper waters in search of new hydrocarbon resources, the traditional fixed production platforms can no longer serve the purpose.

Figure 15-41 A FPSO
(Courtesy of Samsung Heavy Industries)

An FPSO consists of three parts: topsides for oil production; a hull to support the topsides and to store oil; and accommodation block for crew cabins and overall system control. Information on the wells, subsea production systems, subsea manifolds and risers, subsea flow lines and subsea control systems is required for designing an FPSO as well as the metocean data of the operational site and water depth.

The topsides are designed appropriate for the processing capacity of multi-phase fluid, production capacity, water treatment capacity, gas treatment capacity, water injection capacity and so on. It is common to arrange the following equipment and facilities away from the accommodation block in this order: generators, water treatment equipment, water injection equipment, oil and water separators, gas treatment facility, compressors and flaring facility. For the safety of the crew, gas flare tower and modules with high risk of explosion or fire are arranged farthest from the accommodation block.

The topsides are divided into several modules for efficient production and arrangement in the limited space. The module size is usually 20 to 30 m long and wide, depending on the lifting capacity and optimum arrangement. A module is installed on 4 to 6 supports, depending on its size and shape.

The principal particulars of the hull are determined based upon the capacity of cargo tanks required, deck area required for topsides, and the sizes of engine room, ballast tanks, and tanks for various consumables, such as fuel oil, lubricating oil and potable water.

Large FPSOs with a storage capacity of about 2 million barrels have two longitudinal bulkheads forming three cargo tanks (port, center and starboard), while smaller FPSOs have a central bulkhead forming two cargo tanks one each on port and starboard. The size and number of cargo tanks should be determined to allow optimum quantity and frequency of loading for a given daily production. The number of cargo tanks should be determined after careful studies, because an increase in the number of tanks allows more flexible operation, but it increases the amount of equipment and facilities as well as structural weight.

Depending on the composition of the multi-phase fluid from the well and the production process on the topsides, some tanks required for oil processing may have to be installed in the hull. This should be carefully examined as early as possible and reflected in the hull design if required. FPSOs, which are deployed at offshore locations with mild sea conditions, for example off western Africa, tend to have large cargo storage capacity and have barge-shaped hull with spread mooring systems. On the other hand, FPSOs used in such hostile waters as the North Sea tend to have cargo tank capacity of less than 1 million barrels, and are ship-shaped in order to minimize the external loadings with turret mooring system assisted by thrusters.

FPSOs must be able to operate without docking or major repair for about 20 years at a time, and therefore safety and reliability is the top priority and all the equipment and facilities used must be highly reliable. It is also important to design FPSOs to allow easy and smooth maintenance and repair.

7. Topsides Design of Floating Offshore Plants

7.1 Offshore Plants

Offshore plants must be designed taking into account various factors including environment, space, construction and safety. Robust and impeccable design is essential to mitigate the high risk of accidents and to avoid huge potential loss due to operational failures.

The design, construction and operation of offshore structures are strictly regulated by many standards, guidance, regulations and codes as summarized in **Table 15-15**. It is also required to follow various procedures and produce several complicated documents as shown in **Table 15-8**. Some engineering and consulting companies in Europe and North America, who are familiar with such regulations, often undertake offshore plant design. Recently, however, some shipyards have started basic and detailed designs based on the accumulated experience and reflecting new construction techniques they have developed.

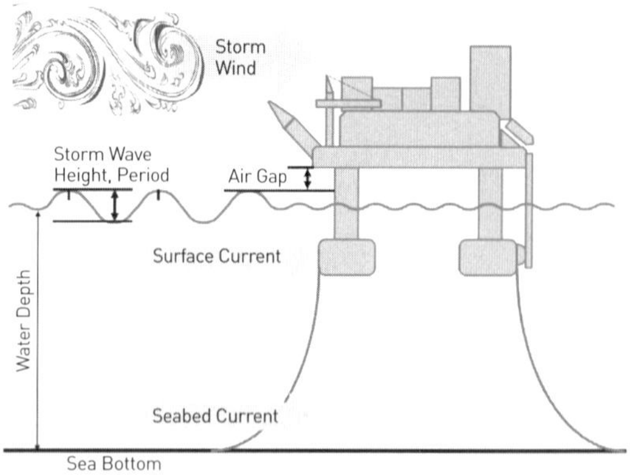

Figure 15-42 Environmental condition of offshore platform

Table 15-9 summarizes the major differences between onshore and offshore plants. Design and construction of offshore plants can be regarded as simpler in some ways than land-based plants, but there are substantial difficulties due to the motions of the floating units and the space limitations.

A comparison between offshore plants and ships is given in **Table 15-10**.

Table 15-8 Procedures and documents for design and construction of offshore plants

Items	Related documents and procedures
Management of design drawings	Design philosophy
	Develop design basis
	HAZOP HAZID study and review
	Document control
	Design verification/3rd party assessment
	Calculation/Analysis
	Engineering deliverables and report
	Design approval (IFD, AFC)
Management of material and purchasing	Request for quotation
	Technical bid evaluation
	Procurement specification
	Place order
	Material handling procedure
	Vender data/drawing approval
Management of construction and quality	Material preservation procedure
	Material tracing procedure
	Inspection and test procedure
	Weight control report
	Test and commissioning procedure

Table 15-9 Comparison between offshore and onshore plants

Type	Offshore plants		Onshore plants	
Process	• Diverse reservoir compositions • Multi-phase fluid → crude oil	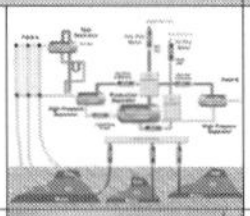	• Typical process optimization • Crude oil → refining	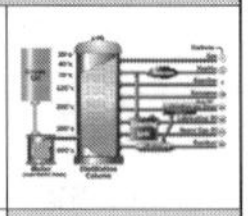
Environmental Condition	• Control of floater motions in waves • Sea water corrosion		• Coastal and undeveloped areas • Strengthening foundation and quay construction needed	
Spatial arrangement	• Limited space → multi-layer structure • Optimization of module arrangement		• Level arrangement • Large scale	
Construction and installation	• Construction in shipyard → transportation • Increasing size of structures		• Construction & installation on site	

Table 15-10 Comparison between offshore plants and ships

Factors	Offshore plants	Ships
Purpose	Drilling, production and storage of oil & gas	Transport of cargo
Design objectives	System performance and operation (Drilling, production & storage) Mooring or positioning stability	Maximum deadweight capacity Speed and stability Economical efficiency in transportation
In common	Floater & steel structure	Floater & steel structure
Structure	Beam structure Beam structure + Plate structure Plate structure	Plate structure
Main equipment	Multi-phase separator Processing equipment	Propulsion system Cargo handling system Navigation system
Trends in Technology	Size becoming larger Deeper waters New types replacing existing ones	Size becoming larger Incremental improvement of performance Research of novel concept
Leading institute	Oil majors Engineering companies	Shipyards & classification societies
Participation	EPCI[3]	Design & building
Participants	Many companies (engineering, construction, transportation, installation, commissioning)	Single party (shipbuilder)

3 Type of turn-key order encompassing engineering, procurement, construction and installation.

7.2 Safety Check

1) Formal safety assessment

Protection from danger is an essential consideration from the beginning of design for all gas and oil production facilities. Emergency response plans must be prepared and well-rehearsed in advance so that all personnel can be evacuated safely in the event of an emergency situation. All foreseeable accidents must be analyzed to quantify their statistical frequency and consequence. If the risk (product of frequency and consequence) is higher than what regulations require or pre-determined criteria, corrective measures must be taken or design modifications must be made.

2) Hazardous area

Any space where the air can contain flammable, explosive and toxic gases should be marked clearly as a hazardous area on both the manual and the general arrangement drawing as shown in **Table 15-11**.

Table 15-11 Hazardous area marks

Safe area		A space where there is no hydrocarbon mixtures and other dangerous gases.
Hazardous areas	**ZONE 0**	Spaces where explosive gas-air mixture exists constantly or for a long time.
	ZONE 1	Spaces where explosive gas-air mixture can occur during normal operation.
	ZONE 2	Spaces where explosive gas-air mixture is unlikely to occur or if it does, remains for a relatively short time. Spherical boundary of 3 m radius with hazardous source at center.

3) Hazard identification

Hazard identification (HAZID) is carried out by dividing the topsides into several sectors and analyzing the likelihood of dangerous situations arising and any problems in working environment in each sector. HAZID provides information so that measures for preventing accidents and emergency response procedures can be devised for possibly hazardous areas, such as process deck and moonpool areas where dangerous and explosive materials are treated, areas with the risk of water or oil/gas leakage, areas with the risk of helicopter crashes, storage areas of noxious chemicals and areas with the risk of heavy objects falling.

4) Hazard operability

Areas with a risk of interrupting or shutting down the production process are identified and the hazard operability (HAZOP) procedures are examined. Failure mode effect analysis (FMEA) must be performed and the emergency shutdown (ESD) procedures devised for major equipment and facilities.

5) Safety equipment

In hydrocarbon production facilities a fire is usually expected to be accompanied by explosions and the risk is very high. In order to protect life and facilities in this condition some active and passive fire protection systems must be available (**Table 15-12**).

Table 15-12 Fire protection system

Active fire protection (AFP)	AFP limits fire spreading and assists firefighting. Hazardous areas should be identified and fire extinguishers and sprinklers must be installed so that damage and loss of structure or key facilities due to fire can be minimized.
Passive fire protection (PFP)	Fire isolation areas are designated to prevent fire spreading. Apply protective coating against the failure of main structures, secure evacuation routes, and install firefighting equipment such as fire hydrants.

Sensors are installed so that the facilities can be monitored 24 hours a day from the control room in order to prevent accidents. The usual sensors used are as shown in **Table 15-13**. Various national and international rules and regulations require stringent specifications and quality management for all safety equipment to be able to operate in bad weather and while in violent motions.

Table 15-13 Monitoring and measuring sensors

Monitoring and measuring sensors	Photographs	Monitoring and measuring sensors	Photographs
Open path flame/gas detector		Visual strobos	
Smoke detectors		Platform status indicators	
Heat detectors		Audible speaker alarms	
Gas detectors		Manual break glass stations	

• Source: http://www.directindustry.com/

A consistent safety guideline must be used from the beginning of design in deciding quantity, locations and operation of equipment so that the safety systems can act as an organic whole. Apart from the monitoring equipment, hydrant piping, sprinklers and fire extinguishers must also be installed.

Design for safety applies to not only equipment but also entrances and exits, passage ways, stairways and handrails that personnel use in their work or movement. It is especially important to secure sufficient space for emergency evacuation, as shown in an example in **Figure 15-43**.

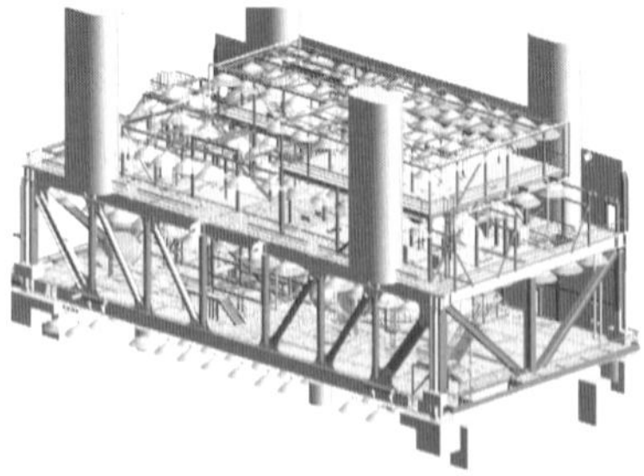

Hydrant system

Evacuation route

Figure 15-43 Safety systems of offshore plants

7.3 Material Selection

1) Structural steel

Steel used for offshore structures is custom produced in accordance with the owner's specifications or the order specifications in addition to the normal requirements of national standards (API, NORSOK, etc) and classification rules (LR, DNV GL, ABS, BV, KR, NK, etc). **Table 15-14** shows additional requirements for various areas of operation.

Steel at low temperature is susceptible to brittle fracture due to fatigue or impact. Therefore, only steel with validated Charpy V-notch number (CVN) in design temperature must be used for offshore structures.

2) Painting

Paints are chemical products which, when applied on the surface of an object, form a dry film coating over the surface and thus provide protection from corrosion, damp-penetration and fouling. Recent advancement in paint technology has produced paints which can control conductivity, suppress marine growth and change color according to temperature variation.

Table 15-14 Additional requirements for various major operational regions

Items		GOM	North Sea	Arctic
Grades		API 2W50T-M	NV E420/F36 NV E36/D36/A36	RS FH36/EH36 RS DH36/AH36
Additional requirements	SR	S1, S4, S11	UT, Z-testing, stain aging test	UT, Z-testing
	CVN[4]	41J@-40℃, T	34J@each grade condition, L	34J@each grade condition, L 41J@-60℃, T
	CE	Report	Report	Report
	CTOD[5]	-10℃ as per API RP2Z	N/A	-10℃ as per API RP2Z
Third party inspection		N/A	DNVGL	RMRS

The durability of paints is greatly affected by surface treatment as well as specifications of paints. The work procedure of painting offshore structures is shown in **Figure 15-44** and the painting properties for the three zones shown in **Figure 15-45** are as follows:

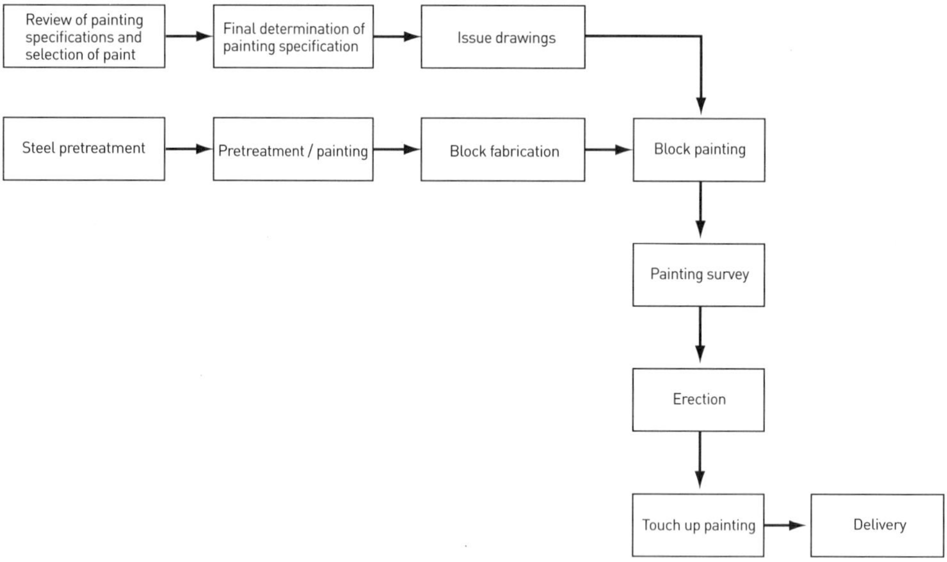

Figure 15-44 Painting work procedure

4 Standards for Charpy V-notch number. Charpy V-notch test is a standardized high strain-rate test which determines the amount of energy absorbed by a material during fracture.

5 Stands for crack tip opening displacement and is a fracture toughness test.

Figure 15-45 Painting properties in three zones
(Courtesy of Samsung Heavy Industries)

- Atmospheric zone: primarily anti-corrosion and aesthetics.
  Organic or inorganic zinc + epoxy + urethane paint
- Splash zone: mainly anti-corrosion and anti-abrasion.
  Epoxy primer + epoxy glass flakes
- Immersion zone: mainly anti-corrosion and anti-fouling.
  Epoxy primer + epoxy glass flakes + anti-fouling paint

7.4 Construction Method

1) Load-out/mating

Topsides and substructure of offshore structures are built separately for efficiency and convenience of fabrication and are assembled together on site or before transportation to the site. Topsides of a production platform can be constructed in one piece on the ground and in this case its weight can be anything between 10,000 and 40,000 tons. Since it will be difficult to lift it by cranes, skid rails are laid on the ground and the heavy topsides are loaded-out using hydraulic jacks known as strand jacks (**Figure 15-46**) to a transport barge.

Figure 15-47 shows the topsides and lower hull of a semi-submersible being loaded out to the transport barge. This work requires careful preparation considering quay height, quayside water depth, wave height, tidal difference and ballast capacity of the barge.

Strand jack　　　　Skid rail

Figure 15-46 Load-out facilities
(Courtesy of Samsung Heavy Industries)

Load-out of topsides module　　　　Load-out of semi-submersible

Figure 15-47 Load-out of offshore structures
(Courtesy of Samsung Heavy Industries)

The positioning of the topsides on the lower hull at the site is performed by lifting the topsides with a crane vessel and/or by floating it over the substructure on the barge submerged to an appropriate depth. Guide cones (**Figure 15-48**) are used to help mating the two structures.

Load-out and erection work can only be carried out in calm weather, but relative motions between the topsides and the lower hull are inevitable, however small. When the two structures contact because of the relative motions, the mating cones (**Figure 15-48**) stabbing into the female cones can mitigate the positioning difficulties to some extent. During the mating process, the female cones may be filled up with sand or hydraulic shock absorbers may be employed for impact absorption.

Mating guide cones

Mating cones

Figure 15-48 Mating guide cones

8. Maritime Laws and Safety Rules

8.1 Maritime Laws and the Responsible Authorities

Offshore production platforms are independent facilities producing oil and gas, and are limited in mobility and accessibility. They are constantly exposed to hazards of fire and explosion due to the nature of the substances being produced. Any accident can become catastrophic resulting in loss of many lives and irrecoverable environmental pollution. A halt in their normal operations due to any cause can be hugely expensive because of the long-term high investment involved in petrochemical industry.

In order to prevent such accidents nations and institutions have produced stringent regulations and standards appropriate for local conditions and some of these organizations are shown in **Table 15-15**. Major oil companies including Exxon, Shell, BP, etc. also have their own high technical specifications and standards. Unlike merchant ships, offshore facilities must comply with the rules and regulations of the country to which the hydrocarbon field belongs rather than the country of the operating company's nationality.

Table 15-15 Some organizations which have rules, regulations and codes relevant to offshore structures

AISC	American Institute of Steel Construction
API	American Petroleum Institute
AWS	American Welding Society
BS	British Standards
DEn	Department of Energy (United Kingdom)
DNV GL	Det Norske Veritas Germanischer Lloyd
ASTM	American Society for Testing and Materials
NMD	Norwegian Maritime Directorate
HSE	(UK) Health and Safety Executive
Major Oil Companies	Owners
NACE	National Association of Corrosion Engineering

8.2 Accidents of Offshore Structures

Figures 15-49 ~ 15-53 highlight major offshore accidents.

- Capsized in Newfoundland, death toll 84
- Design and construction fault
- Non-compliance of regulations and careless management
- Ballast control system malfunction
- Shortage of life saving appliances and insufficient emergency drills
- Led to improvement of Canadian offshore regulations
- Reflected in IMO MODU code
- Led to revision of rules and requirements concerning ballast control and damage stability
- Led to improvement of evacuation facilities and tightening of crew training requirements

Figure 15-49 *Ocean Ranger*, a semi-submersible drilling unit, 1982
(Source: http://en.wikipedia.org/wiki/Ocean_Ranger)

- Explosion due to gas leakage, North Sea, UK, death toll 167
- Loss of 3,500 million U.S. dollars
- No provision for explosion at design stage
- No consideration of crew's evacuation
- A turning point in the UK laws on offshore structures
- Led to installation of fireproof walls and safety zones in offshore structures
- Revision of requirements and rules on ballast control and damage stability
- Intensification of crew's training and escape equipment

Figure 15-50 *Piper Alpha*, a fixed production platform, 1988
(Source: http://en.wikipedia.org/wiki/Piper_Alpha)

- Capsizal accident off Brazil, death toll 11
- Flooded due to ballast tank damage
- Sank due to explosion and flooding
- Loss of US$ 350 million
- Not followed proper procedure in repairing ballast tank
- Insufficient training and knowledge of emergency procedures
- Led to strengthened requirements for provisions against malfunctions
- Led to requirements for risk evaluation for abnormal events
- Led to establishment of safety procedures
- Led to call for overall risk analysis

Figure 15-51 *Petrobras 36*, a semi-submersible drilling unit, 2001
(Source: http://en.wikipedia.org/wiki/Petrobras_36)

- An accident off Timor
- Oil leakage from the well
- Fire due to continuous processing
- Crew's insensitivity to safety
- Loss of US$ 180 million
- Operational safety guide required
- Led to safety procedures being established

Figure 15-52 *West Atlas*, a jack-up rig, 2009
(Source: http://en.wikipedia.org/wiki/West_Atlas)

- Explosion and sinking in GOM, 11 people missing
- Non-compliance of safety regulations, BOP malfunction
- Catastrophic pollution due to 4 million barrels of crude oil spilt
- Fishing near the site banned
- Scale of loss still being determined
- Enormous cost of cleaning-up
- Led to offshore safety requirements being revamped
- Led to tougher drilling license conditions and safety regulations
- Led to petroleum industry crisis management funds being established

Figure 15-53 *Deepwater Horizon*, a semi-submersible drilling unit, 2010
(Source: http://en.wikipedia.org/wiki/Deepwater_Horizon)

References

1. Faltinsen, O.M., 1990. *Sea Loads on Ships and Offshore Structures*. Cambridge: Cambridge University Press.
2. Jo, C.H., 2001. *Introduction to Subsea Pipeline* (in Korean). Seoul: Daesun Publisher.
3. Lee, J., 1989. *Introduction to Offshore Structural Design* (in Korean). Busan: Kwangmun Publisher.
4. Mather, A., 1995. *Offshore Engineering: An Introduction*. London: Witherby & Company Limited.
5. Oilfield Publications Ltd. (OPL), 2002. *Offshore drilling & production concepts of the world*. 5Rev Ed edition: Oilfield Publications Ltd.
6. Pulsipher, A.G. et al., 2001. *Forecasting the number of offshore platforms on the Gulf of Mexico OCS to the year 2023*, OCS Study MMS 2001-013. U.S. Department of the Interior, Minerals Management Service (MMS): New Orleans.

CHAPTER 16

Life of Ships and Prospects of Ship Technology

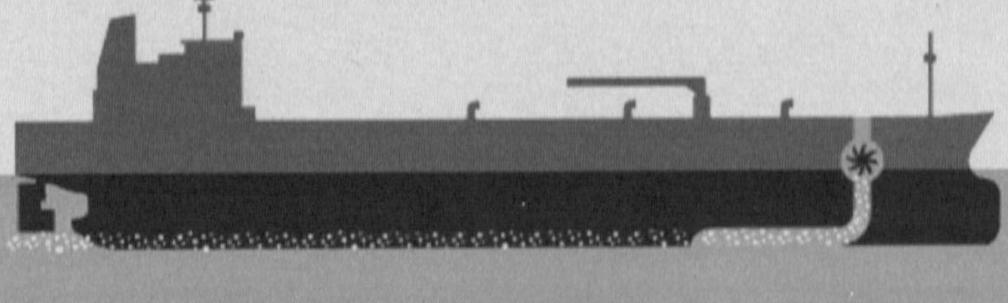

1. Life of Ships

A ship's life starts with the processes of its birth, including the decision to have it built, its design, construction and commissioning. It is then deployed in active service for economic and other purpose, and its life ends with the terminal processes of retirement from service and usually dismantling and recycling. In broader sense, the lifecycle of a ship must include the reasons for the investors to invest in a ship and the decision-making to withdraw the ship too. By analyzing the market environment, where diverse investors come together, in addition to the simple supply and demand relationship between 'owners' and 'builders,' the lifecycle of a ship can be more deeply appreciated.

International shipping is a typical high-risk high-investment industry, and before launching into it a thorough understanding of the external and internal environment surrounding the industry is required. An investment is made when the expected return from it in uncertain future exceeds its definite value at present. The purpose and form of the investment should therefore be clearly established for investing in ships. Public and private sectors have different purpose and form of investment. Commercial shipping investment can be categorized into several different forms: replacement investment; expansion investment; new entry investment; and others such as modification and improvement.

When new ships are built, the total fleet size may appear to increase, but the effect of decommissioning and accidental losses must be taken into consideration. An investment into new ships for expansion implies an increase in shipping service supply, but the same investment into second-hand ships to replace retiring tonnage results in a decrease of the total shipping supply. Therefore, if second-hand tonnage is used, neither new entry nor expansion makes any difference to the total supply of transport capacity.

A ship's life span is also influenced by the requirements and mandatory regulations of various bodies such as IMO on environmental protection, maritime safety and recycling.

1.1 Shipping Market Cycles

Approximately 90% of the world's merchandise trade depends on international shipping. Since the volume of seaborne trade is cyclic, the investment cycle of the shipping business can be understood as the market participants' response to the cyclic changes of the cargo volume. In other words, the demand cycle of ships is connected to the pattern of how the major stake holders of shipping (owners, financiers, government, shipbuilders, charterers, ship breakers, etc.) decide freight rates, new shipbuilding orders, scale of scrapping, employment, governmental regulations and financing conditions in response to changing market condition.

The market response mechanism is composed of a time-invariant long cycle, short business cycles of decline, depression, recovery and boom, and seasonal cycles. The earliest study of shipping demand cycle was carried out in the 17th century when the world seaborne trade was increasing rapidly and, as a conclusion, a theory of 7-year cycles was put forward. However, it was later shown that the maritime business had a close relationship with the world economy, although a long-term trend existed independent of the general industrial cycles. Such a long-term trend has been known to manifest itself more than 100 years after the events. It is mainly due to revolutionary technological advances, dramatic economic changes and regional changes, but it is not easy to identify clearly even when such data as shown in **Figure 16-1** is available.

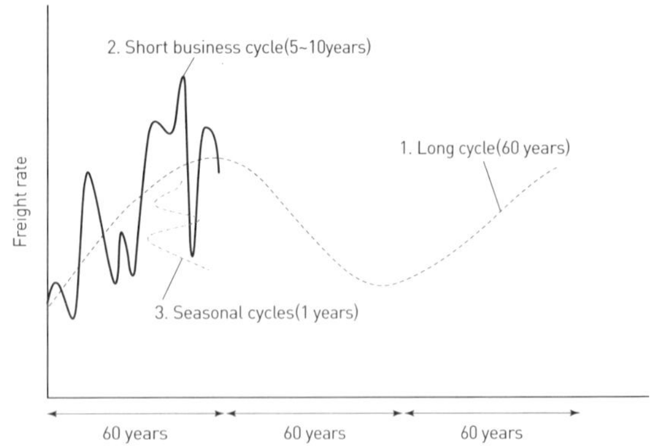

Figure 16-1 Market cycles of shipping

The short business cycles are determined by the fluctuations of cargo volume according to the state of world economy, and the consequent variations of the value of ships (in terms of freight rate, price of new and used ships and price of ships to be sold for demolishing) and capital investment environment (yield, financing costs, etc.). Ultimately, short business cycles, when combined with the long cycle, reach the high and low marks in 3–12 year cycles. All of these factors are mainly determined by the conditions of the shipping business market. Lastly, there are the seasonal cycles within each year, which are less influenced by the course of time. For example, the freight rates of tramp bulk carriers decrease around July and August of each year, which is the off season for them. Although this is accompanied by the short business cycle as well, this fluctuation manifests independent traits as a separate component.

The key component of the shipping cycles is the short business cycles. This can be construed as a group behavior of the shipping market participants, and this leads to a collective expansion when an expansion occurs because of optimistic expectations of the market movement. The collective expansion is bound to bring about over-supply by the time the peak expansion period is over, and this inevitably results in a transitional period of adjustment. If the adjustment process is prolonged, a short recovery period can occur. When this state of over-supply is resolved for whatever reason, the expansion period starts once again.

1.2 Composition of the Market

The market that a ship-owner can participate in can be divided into four sectors as illustrated in **Figure 16-2**:

- New-building market
- Freight market to charter and/or earn freight
- Sale & purchase market to buy or sell existing ships
- Demolition market for final disposal of ships

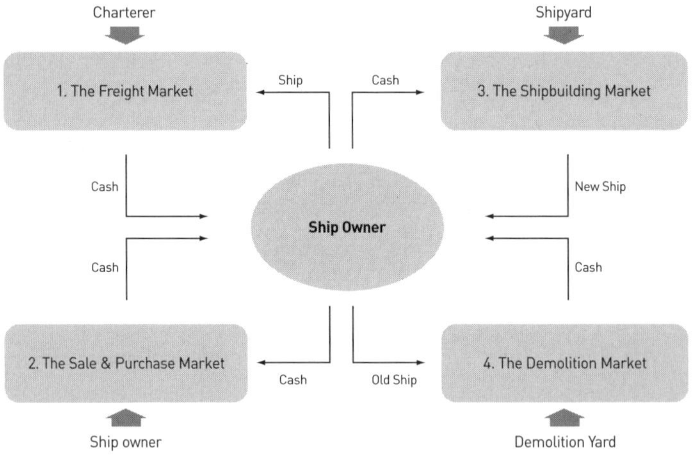

Figure 16-2 Composition of shipping market

The party that owns a ship is an individual owner, a company or a group of them, and they maintain a very close relationship with all four market sectors. For example, when freight rate increases rapidly in the freight market, it affects the sale & purchase market briefly before the effect is spread to the new building market and eventually the demolition market. In other words, the commodity known as income is exchanged between participants of different markets using a ship as the medium. Ultimately, the forms and quantity of the income from the four markets change the demand cycles of the marine transport market.

Recently, more specialized markets have started to appear in response to the demands from cargo owners and more complicated and diversified financial transfers such as futures transactions. Although these markets may appear to be different from the existing markets, the market mechanism based on securing cargo for the ship is the same.

2. Lifespan of Ships

2.1 Required Lifespan of Ships (Physical Lifespan)

For the past few decades market investments have been expanding due to the increase in tonnage necessary to satisfy the demand of seaborne trade. As a result, shipyards started to pay particular attention to prolong the economic lifespan of ships they build. Ships are designed to possess sufficient structural strength and stiffness to withstand all loads expected during its life time. The sufficiency of such structural provisions determines the durability and physical lifespan of a ship.

The basic analysis method consists of calculating the tensile and compressive stresses of the longitudinal structural elements due to bending of the ship in a standard wave and comparing the resulting stresses with permissible values for the material used. The use of beam theory is under the assumption that the hull girders can be treated as a simple beam. A corrosion allowance must be added to the scantlings needed for the required longitudinal strength calculated in this manner. In other words, the final thickness of the structural elements is the sum of the required values obtained from the general beam theory analysis and an allowance for corrosion expected throughout the life of the ship.

As ships became larger to benefit from economy of scale and new ship types appeared to carry new types of cargo more efficiently, the empirical methods used to determine structural stiffness and strength based on existing data of conventional vessels have reached their limit. For large ships, for instance, torsional strength, transverse strength and local strength as well as longitudinal strength need to be analyzed and provided for in structural design. Container ships have regions where the decks are largely open, so analyses of torsional, transverse and longitudinal strength must be carried out. Naturally the designers turned to direct strength calculation methods based on the statistical theories of ocean waves for the structural analysis and design.

The International Maritime Organization (IMO) came to recognize the need to address the fundamental causes of several recent marine accidents due to structural failures. At the 90th IMO Council (C90, December 2003), it was resolved to start work on goal based standards (GBS) for new ship construction to set the safety goals and functional requirements throughout the lifecycle of ships. The resulting regulations were finally approved and adopted by the 87th Maritime Safety Committee (MSC87, May 2010) initially applicable to all bulk carriers and tankers of 150m or more in length for which the building contract is placed on or after July 1, 2016.

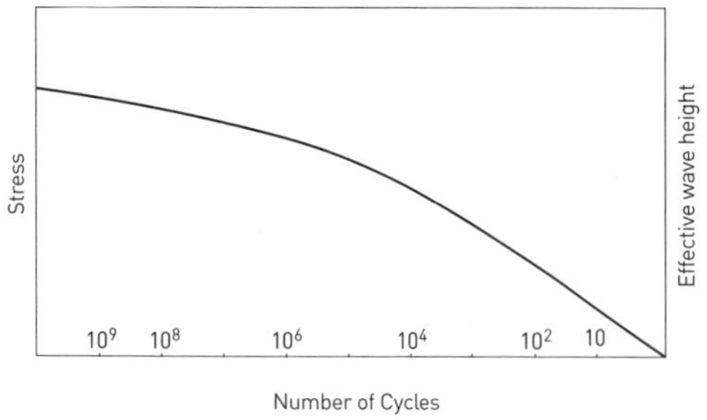

Figure 16-3 Probability level approach for determining ship design life

The high-level objectives of these new standards consider specific design lifespan of a ship to be a major factor in determining the operational and environmental conditions and corrosion. The functional requirements to meet the objectives have specified a ship's lifespan as 25 years and the corresponding conditions are as given in **Figure 16-3** based on the environment of the North Atlantic. This means that the specified design lifespan of 25 years is going to be used for determining maximum expected load in the ultimate strength calculation, expected fatigue life and material life (affected by corrosion). Of course, the actual service life of a ship may be shorter or longer than this, depending upon the operational conditions and repair and maintenance practice.

2.2 Expected Lifespan of Ships (Economic Lifespan)

The economic lifespan of a ship is the period from its delivery to the time when the ship can no longer render economic benefit to its owner or charterer, now or in the foreseeable future, through transportation or other trade activities. Important factors that determine the economic lifespan of a ship include repair and maintenance, market situation, charter policies, technical advances and obsolescence, residual value at scrapping and regulations.

The average economic lifespan of oil tankers and bulk carriers in the 1970s was 15 years and 20 years respectively. However, as construction and maintenance technology improved in the 1980s, respective economic lifespan was extended to 25 years and 30 years by the end of the decade. It goes without saying that the remaining lifespan of used ships is shorter than that of new ships and consequently their profit-making periods are shorter too. Also, the period during which the ship's price can increase, resulting in capital gains, or the residual value at scrapping can increase is also shorter. Although a new ship may be expected to have a longer lifespan, a well-maintained used ship can have a longer lifespan than a poorly constructed new ship.

A large scale initial investment is necessary to have a new ship built, and therefore it requires much more effort to analyze and understand the shipping market thoroughly compared to when buying used ships. In other words, the new building contract negotiation process, quality management during construction until delivery, waiting period for delivery and complex financing mechanisms can become a barrier for a new entrant to the market.

Therefore, ship-owners usually order a new ship at the time when the market is expanding, and they expect to run the ship until the end of its economic life or sell or decommission after a certain period of time as custom dictates.

Exceptionally there are some speculative orders too, particularly when the shipping market is depressed. If the market improves at the time of delivery of the ship, the owner may sell it at a higher price than its build cost and take the profit. On the other hand, when used ships are purchased, the degree of immersion into the shipping business is low in comparison to new ships.

Of course, used ships are often purchased to meet the long-term needs of existing market movers, but other investment capital may also be attracted to them as they are relatively easier to buy. In other words, speculative purchases of ships, which are motivated by the expectations of steep rise in prices in a short time, often target second-hand ships. For new ships the building plans have to be examined thoroughly. However, purchase of used ships only requires a review of the number of ships to buy, and thus decision-making is relatively easy.

2.3 Actual Lifespan of Ships

An examination of the age of ships in service in the past few years shows that the average age of the world fleet decreased during 2009 as new tonnage was delivered and more ships were scrapped during the international financial crisis. The average age per deadweight ton also decreased, indicating that the newly delivered ships tended to be larger than most of those in the then existing fleet. Indeed, vessels built during the last four years are, on average, six times larger than those built before 1990.

Table 16-1 summarizes the average age of vessels by major types as of January 2010. It can be seen that container ships have the lowest age with 10.6 years, followed by bulk carriers (16.6 years), oil tankers (17.0 years), general cargo ships (24.6 years) and other vessels (25.3 years).

When the same data are examined for various different country groups (advanced economies, developing economies, transitional economies and open registry), it is easy to see that the major open and international registries have the youngest fleet among the country groups with the average of less than 16 years. Furthermore, about 25% of these ships have been delivered within the last five years, compared to the 15% in developing economies, 10% in advanced economies and 8% in transitional economies.

Table 16-1 Age distribution of vessels by ship type (as of Jan. 2010)

Classification		0~4yrs	5~9yrs	10~14yrs	15~19yrs	20yrs~	Average		Varia-tion
							2010yr	2009yr	
Bulk carriers	Ships	19.0	16.0	14.2	10.8	40.1	16.58	17.22	-0.64
	dwt	25.2	19.4	15.7	12.4	27.4	13.77	14.27	-0.50
Average vessel size(dwt)		74,809	68,046	62,375	64,563	38,537			
Container ships	Ships	31.3	21.7	20.9	12.8	13.3	10.56	10.92	-0.37
	dwt	38.9	26.0	17.2	9.5	8.4	8.72	9.01	-0.29
Average vessel size(dwt)		44,701	43,151	29,644	26,579	22,653			
General cargo	Ships	9.6	8.0	9.1	11.1	62.3	24.63	24.44	0.18
	dwt	16.1	9.8	13.5	9.8	50.8	21.40	22.12	-0.72
Average vessel size(dwt)		8,260	6,083	7,372	4,391	4,043			
Oil tankers	Ships	24.2	16.0	10.7	12.0	37.1	17.03	17.55	-0.52
	dwt	31.8	28.2	16.7	13.0	10.2	10.13	10.72	-0.59
Average vessel size(dwt)		55,138	74,066	65,636	45,454	11,514			
Other types	Ships	9.2	9.3	9.1	8.7	63.8	25.33	25.26	0.07
	dwt	28.3	14.1	11.3	8.4	37.9	17.47	18.24	-0.77
Average vessel size(dwt)		4,923	2,444	1,980	1,548	953			
All ships	Ships	12.7	10.8	10.2	9.9	56.4	22.93	23.00	-0.07
	dwt	28.8	22.2	15.8	11.7	21.5	13.35	13.97	-0.62
Average vessel size(dwt)		28,401	25,665	19,266	14,799	4,764			

• Source: Review of Maritime Transport 2010 (UNCTAD), Dec. 2010

As at the beginning of 2009, there were 99,741 merchant ships of 100 GT and above. In the same year 3,658 new vessels were delivered, while 1,205 ships were withdrawn and/or demolished, making the total fleet in January 2010 amount to 102,194 ships, representing a 2.5% increase from January 2009.

The demolition and recycling market is much more changeable compared to the new building market. This suggests that the decisions are made by the final owners of ships somewhat unpredictably but according to market conditions. In other words, changes in seaborne trade volume and freight rate cause fluctuations in the demolition market and they in turn are manifested as the price of recycled steel. In 2008 and 2009 the shipping demand decreased sharply and the price of recycled steel was $200/ton, a huge drop from $650/ton only a few years ago. However, with the recovery of the market since 2010, the price has maintained \$400/ton level.

Ships of 30 – 35 years old were dismantled the most often, but almost no ship under 18 years of age is dismantled, while some ships 40 years or older are still in service. Age of scrapped ships is steadily increasing as shown in **Figure 16-4** due to development in construction and maintenance technology.

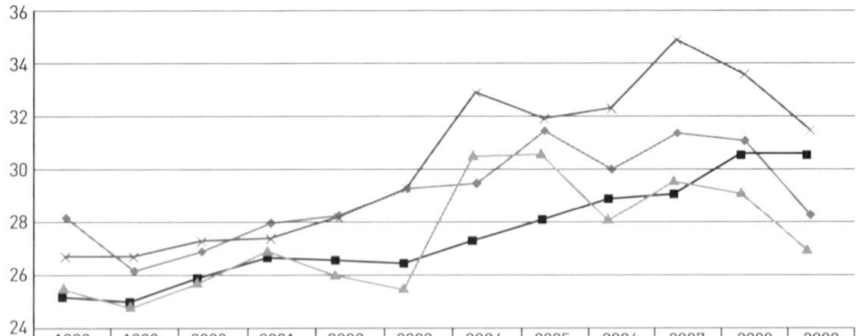

	1998	1999	2000	2001	2002	2003	2004	2005	2006	2007	2008	2009
Tankers	28.2	26.2	26.9	28	28.3	29.3	29.5	31.5	30	31.4	31.1	28.3
Dry bulk carriers	25.2	25	25.9	26.7	26.6	26.5	27.3	28.1	28.9	29.1	30.6	30.6
Container ships	25.5	24.8	25.7	26.9	26	25.5	30.5	30.6	28.1	29.6	29.1	27
General cargo ships	26.7	26.7	27.3	27.4	28.2	29.3	32.9	31.9	32.3	34.9	33.6	31.5

Figure 16-4 Age distribution of scrapped ships by ship type

In 2009 bulk carriers and container ships each took up 23% of the total volume scrapped, followed by car carriers (15%) and tankers (13%). Tankers and bulk carriers are showing a close relationship with the increase in the dismantling rate of ships built in the 1970s as well. The number of container ships scrapped increased sharply in 2009, but their proportion in the total container ship fleet is relatively small as most of the container ships appeared after the 1980s. Scrapping rate of general cargo ships is lower than other vessel types and there are some built in the 1960s in the current fleet.

2.4 Life Cycle Assessment of Ships

The environmental problems have become a global issue, and ship design also started moving towards cradle to cradle design (C2C design), in which process life cycle assessment (LCA) is essential, covering from design and construction of the ship to operation, dismantling and recycling.

The life cycle assessment of ships is an environmental assessment process currently being studied in Europe and Japan, and there is a movement to establish international standards for it. The assessment of construction is based on the information of vessel type, structure, functions, main engines and vessel weight. The assessment of operation is based on the information on cargo weight, years in service, maintenance planning (including frequency of hull painting and number of main engine overhauls and replacements) and voyage planning (including routes, sailing duration, speed and distance). The assessment of dismantling uses the information on materials recycled and energy and equipment needed in the dismantling process.

The environmental loads of ships in the form of resources consumed include coal, crude oil, LNG, wood, iron ore, glass materials and bauxite. Their waste products include styrene monomer, solid wastes, steel scraps and sludge produced during operation.

1) LCA at construction stage

The flow of the construction process for each vessel is analyzed, and the categories and amount of energy, material and parts used and waste materials and scraps produced during the construction are presented in a tabular form for each of the steps. Based on this data, a quantitative inventory of environmental impact in terms of energy and materials for building the ship and manufacturing the parts can be obtained. During this process, various support databases for life cycle assessments are referenced. Since not all consumable resources and waste materials can be accounted for, usually only the data of accountable resources and solid wastes are calculated. Through such data analysis the industrial inputs and outputs (process volume) and environmental inputs and outputs (environmental loads) can be interpreted for each stage of the construction process.

2) LCA at operation stage

As at the construction stage the flow of operational process is analyzed for each ship type. The data required includes schedule and route of the voyage and maintenance plans, such as the frequency of hull painting and main engine replacement. Based on such data, a quantitative inventory of environmental impact for each of the detailed operational processes is obtained. Through the analysis of this data the industrial inputs and outputs (process volume) and environmental inputs and outputs (environmental loads) can be interpreted for each stage of the operation process.

3) LCA at dismantling stage

In order to analyze the inventory at the dismantling stage, input and output items must be listed first. Waste products that can be recycled, such as steel plates or other steel scraps, are analyzed. The inventory of solid waste is also analyzed at this time. Then, the work undertaken during the breaking process and its flow is studied.

Finally, problems and challenges discovered through the LCA exercise are identified. In the future lifespan of ships will be directly affected by their economics, efficiency and inventory analysis and assessment required by the rules and regulations.

3. Future Prospects of Ship Technology

3.1 Direction of the Development of Future Shipbuilding Technology

Future ship technology in mid- to long-term will be focused on environment-friendly technologies and will aim for reduction of greenhouse gases, exploitation of deep-sea energy resources, safety and marine alternative energy. In addition, technology required for using the Arctic passage for shipping routes is also likely to receive attention.

IMO has strongly influenced international shipping through ever more stringent environmental and safety regulations on ships and offshore structures, and there is no doubt that it will continue to exercise great influence on future ships. The Marine Environment Protection Committee (MEPC), a sub-committee of IMO, is developing market-based and operational measures for reducing greenhouse gases produced by ships. Moreover, the Maritime Safety Committee (MSC), another sub-committee of IMO, is working to establish generic GBS (goal based standards) based on SLA (safety level approach) applicable to all vessels.

Sustainable technological development is currently pursued throughout the world in a number of fields, foremost of which includes environment, biology, information and safety, and it is thought that the ship technology will follow their lead. Next generation ships, green ships, environment-friendly logistics and global supply chain management system (SCM), especially shipping safety and security, are all areas that are expected to receive research attention in the future.

The technology of next generation ships consists of some core technologies that must be developed as presented in **Table 16-2**: LNG or CNG carrier core technology; high-performance vessel technology; ship design and construction system technology; and complex offshore plant technology. Some key projects in each category are also shown in the table.

Table 16-2 Major projects in the technology for next generation ships

Classification	Major projects
LNG carrier core technology	• Transportation of LNG and use of LNG for propulsion fuel • LNG storage control
CNG carrier core technology	• High speed CNG charging and loading/unloading systems • Design of large cargo tanks and CNG containers
High performance vessel technology	• Mega container vessels • Cruise vessels • Ice breaking vessels • Short sea shipping vessels
Ship design and construction system technology	• PLM (product lifecycle management) for shipbuilding • Eco-friendly construction technology • Automation for productivity improvement • Shipbuilding for GBS(goal based standards) • Special structure design for shipping safety
Complex offshore plant technology	• FPSO (floating, production, storage & off-loading) technology • LNG FSRU (floating, storage & re-gasification unit) technology

The core technologies for green ships include emission reduction technology and alternative energy technology to meet the new IMO requirements on the reduction of emissions of pollutants, especially greenhouse gases, from fossil fuel. Environment-friendly logistics is a systematic approach to minimize input resources and waste products in logistics. In shipping this translates into minimization of all pollutants, including engine emissions, paint, solid waste and bilge.

Maritime logistics technology consists of safety, information and security management systems for the whole supply chain, and includes biometrics and container security and sealing technology. Logistics information is a key technology area for establishing integrated logistics systems and includes EDI (electronic data interchange), SWC (single window concept) for logistics information, cargo identification, classification and marking and cargo tracking technology.

The scope of threats dealt with by safety management technology ranges from piracy to natural/man-made disasters as shown in **Table 16-3**. It is expected that innovations in marine technology will continue in the future through steady high-level investment in IT-based research and development. It is also crucial to invest generously in strategic education and training of the next generation talents who will lead the marine industry of tomorrow.

Table 16-3 Applicability of safety management technology

Threats	Safe management technology				
	Security	Safety	D&H	Environment	Spec. Proj.
Terrorism	●				
Crime	●				
Natural Disaster		●			
Accidents		●	●		
Pollution			●	●	
Epidemic					●

• D&H : dangerous & hazardous cargo
• Spec. Proj. (special projects) : epidemics such as SARS, swine influenza, foot-and-mouth

3.2 Green Ship Technologies

3.2.1 Trends of IMO Regulations on Greenhouse Gas Emissions

According to an IMO study report, annual carbon dioxide emissions by ships worldwide reached one billion tons, about 3.3% of the world's total carbon dioxide emissions of 30 billion tons, in 2007. IMO classified the carbon dioxide emission by industries as shown in **Figure 16-5**.

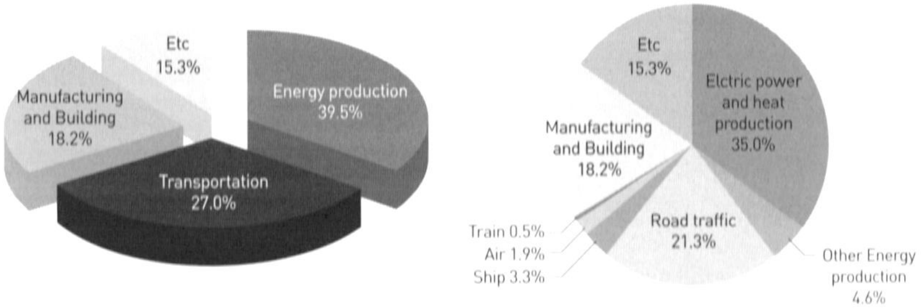

Figure 16-5 Quantity of carbon dioxide emission by industries
(Source: IMO/MPEC 59th session, "Prevention of Air Pollution from Ships", April, 2009)

In accordance with Article 2.2 of the Kyoto Protocol adopted in 1997, the 57th meeting of the Marine Environment Protection Committee (MEPC) discussed the issue of greenhouse gas emission by ships, which was delegated to the IMO to control. A draft agreement was reviewed and reduction goals were negotiated in the 60th Committee meeting. In 2011 IMO adopted mandatory measures to address energy efficiency of international shipping, which entered into force on 1 January 2013 under Chapter 4 of MARPOL Annex VI. These regulations on energy efficiency for ships made mandatory the Energy Efficiency Design Index (EEDI) for new ships and the Ship Energy Efficiency Management Plan (SEEMP) for all ships.

1) Regulatory measures of IMO

IMO is developing various measures across technology, operation and market for reduction of greenhouse gas emissions as summarized in **Table 16-4**.

Table 16-4 IMO's measures for reduction of greenhouse gas emission

Instruments	Application techniques	Target ships	Status
Technical measures Energy Efficiency Design Index (EEDI)	Energy efficiency improvement technology; Engine and exhaust gas treatment; New renewable energy and next generation power source.	New ships	Trying to make mandatory, MARPOL Annex 6 revision (revised in MEPC 62)
Operational measures Ship Energy Efficiency Management Plan (SEEMP)	Energy efficiency management technology (best practices)	New/existing ships	Voluntary participation, MARPOL Annex 6 revision
Operational measures Energy Efficiency Operational Indicator (EEOI)	Efficient operation technology and index (operational indicator)	New/existing ships	Voluntary participation, MARPOL Annex 6 revision
Market Based Measures (MBM)	Carbon emission trading (ETS); Carbon tax-fund (bunker levy)	New/existing ships	pursuing separate agreement

The EEDI of a ship is calculated by dividing its carbon dioxide emission by its carrying capacity (e.g. DWT multiplied by design speed), and it indicates the energy efficiency of the ship. If the obtained EEDI exceeds the required EEDI, the ship will not be allowed to operate. This measure was made mandatory for all new ships by IMO MEPC in 2013 and the basic idea is presented in **Figure 16-6**.

The operational measure of Ship Energy Efficiency Management Plan (SEEMP) was also made mandatory for all ships in 2013, while Energy Efficiency Operational Indicator (EEOI) and Market Based Measures (MBM) may follow suit in the foreseeable future.

- SEEMP: Documented tool for self-managing the energy efficiency of a ship.
- EEOI: Method of calculating and managing actual emissions of a ship.
- MBM: Discussion in progress on carbon taxes, emission right trading system, differential carbon taxes, etc.

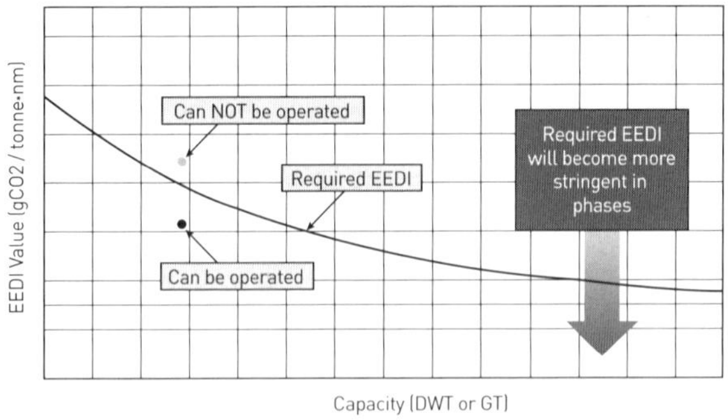

Figure 16-6 Conceptual diagram of the IMO EEDI requirements

3.2.2 Environmental Impact of Shipping and Green Technology

IMO's study has already shown that the amount of carbon dioxide from shipping is much less than road or rail transportation as presented in **Figures 3-5** and **16-5**. Nevertheless, an IMO study report on emissions of greenhouse gas (GHG) and other air pollutants shows that carbon dioxide emission from shipping still amounts to 3 to 5% of the world total. It also reports that sulfur dioxide emission is 4 to 8% and nitrogen oxide emission is 15% of world total.

Shipping companies and shipyards are under international pressure to meet the high environmental standards required by IMO. The environmental pollutants produced by ships include fuel and lubricants for engines and generators and waste water from ballast, bilge and ships' sanitation systems.

In order to minimize such environmental pollution, the green ship technology is focused on energy-saving and high efficiency propulsion systems. Efforts are also made in finding alternative and/or sustainable energy sources including biodiesel fuel, wind and solar power and liquefied natural gas, as well as in designing hull forms and developing devices to reduce energy consumption.

1) Energy-saving and improvement in propulsion systems

Reduction of energy consumption reduces not only environmental pollutants but also operating costs. Therefore, ceaseless improvement of energy-saving technology is an important future task for the international shipping community. Current research in this area includes slow steaming, filters for exhausts, and use of tidal current and biodiesel fuel. Recent research in propulsion systems includes use of solar photovoltaic panels and wind power, and resistance reduction through air lubrication.

A 2010 report by Det Norske Veritas predicts the annual global emissions from shipping to reach 1.53 billion tons in 2030. **Figure 16-7** taken from the report presents possible amount of reduction by technological areas on the horizontal axis and cost per ton of reduction on the y-axis, with the goal of annual reduction set at one million tons. This study excludes bio-fuel and hydrogen fuel on the basis that they are unlikely to be commercialized before 2030. The report expects that the most effective methods of reducing emissions include slow steaming, use of contra-rotating propellers, use of wind power through sails or kites, and use of clean gas fuels and heat from waste incineration. It concludes that, in order to maintain the 2009 level of CO_2 emission of 925 million tons until 2030, the emission must be reduced by 40% annually and the required cost is likely to be around $35 per ton of reduction.

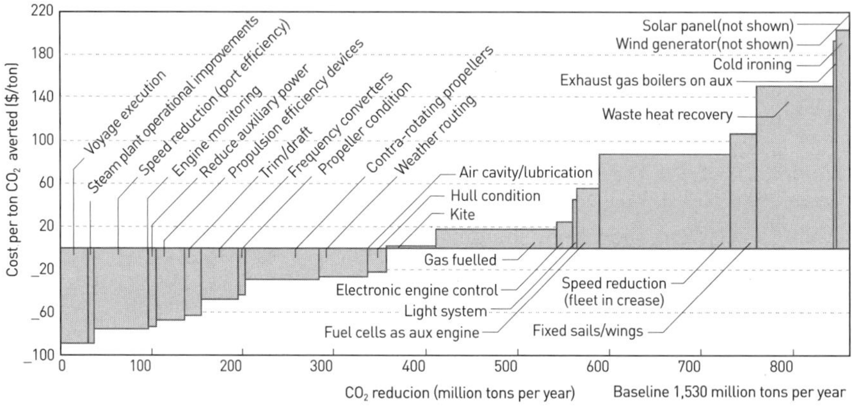

Figure 16-7 Prospects of energy saving technologies in shipping - cost and benefit in 2030

2) Biodiesel

Biodiesel is a form of bio-fuel which can be used in diesel engines, and is typically produced from vegetable oil or animal fat. The use of biodiesel has been tested extensively by automobile industry and was shown to emit lower greenhouse gas and sulfur dioxide but produce more NOx gases. Take-up of biodiesel is likely to be for economic reasons, and it is expected that sudden increase in oil price or increase in production of raw material plants may make it attractive as a bunker replacement.

3) Wind power and solar power

Sail-powered ships have been around for centuries, but new technologies in harnessing wind power may be giving sails a revival. The idea of using giant sails or kites, not unlike para-gliders, to propel large commercial ships was proposed some time ago, but it has been treated as just that – an idea. However, projects that tested the viability of the technology have demonstrated significant potential for reducing fuel consumption even on larger ships. *Beluga SkySails*, a German container carrier, had kites controlled by computers installed in

2008, and this is thought to have reduced average annual fuel consumption by as much as 35%. Solar power is another new energy source for ships. Although it has been used in some small ships, commercialization for large cargo ships is some time away yet. There are some experiments carried out with photovoltaic panels on large cargo ships, but more work is required for practical exploitation of the technology.

4) Liquefied natural gas

Use of liquefied natural gas effectively reduces emission of pollutants by short sea shipping and shipping on inland waterways. LNG as an oil-based fuel replacement can eliminate all SOx emissions, and cut emissions of CO_2 and NOx by as much as 20% and 80%, respectively. However, one of the main problems in using natural gas as the fuel for ships is its storage on board. LNG is obtained by compressing natural gas in very low temperature, reducing its volume to 1/600 of its gaseous state. Even then, the volume of tanks required to store LNG is more than double the volume of tanks needed for storing equivalent amount of conventional oil fuel.

5) Air lubrication system

Air lubrication reduces frictional resistance of a ship by forcefully blowing air out into the flat bottom, which takes up 80% of a ship's bottom in conventional cargo ships, to form bubble flows or air film as illustrated in **Figure 16-8**. The concept has already been developed to the commercialization stage, and it is expected to reduce a ship's fuel consumption by about 15%.

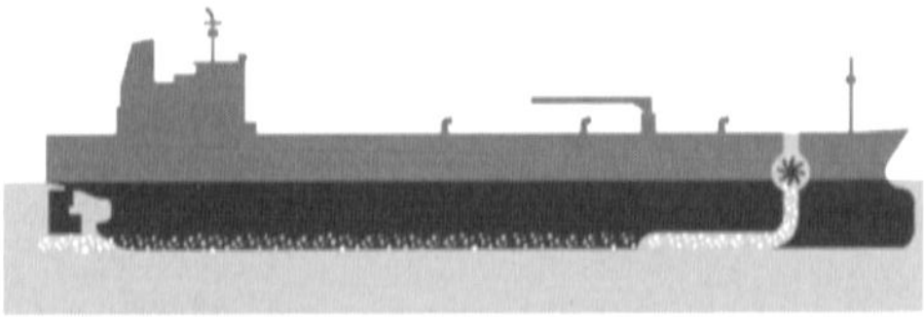

Figure 16-8 Air lubrication system

6) Hull form design

In parallel with the development of alternative fuels and power sources, shipyards are developing hull forms for higher energy efficiency. The main direction of effort is in optimizing the hull form for minimum frictional and wave resistance. Numerical techniques, such as computational fluid dynamics (CFD), are expected to contribute much in this effort. Application of computer simulation will be very useful in qualitative assessment of various hull forms for their resistance performance in diverse operational sea states. Computer programs will also be

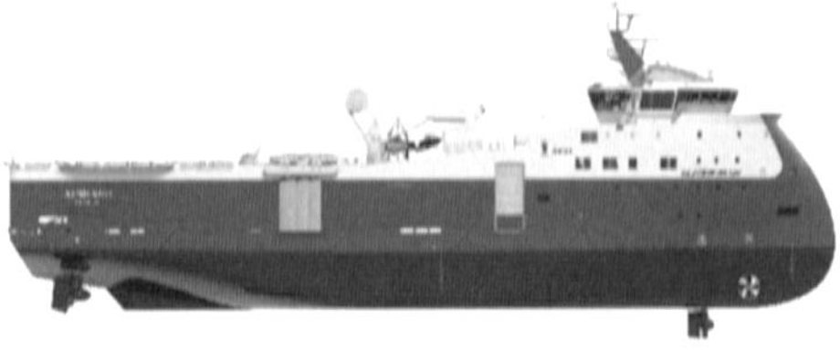

Figure 16-9 An 'X-bow' ship

used in ship optimization for stability, safety, noise and vibration.

For example, X-bow ships, as shown in **Figure 16-9**, are expected to have much lower wave-making resistance and consequently reduce fuel consumption by approximately 7-16%.

3.2.3 Low carbon green ship technology

Low carbon green ship technology can be divided into three areas: concept design; equipment and components; and exploitation of new power sources. These are summarized in **Table 16-5**.

Table 16-5 Classification of low carbon green ship technology

Concept design of ships	Equipment and components	Exploitation of next generation power sources
• Electric propulsion • New concept hull form, propulsion system • Automated navigation and optimization	• Superconducting generators/ motors/inverters and intelligent power control • Exhaust gas treatment (WHRS, SCR) • Eco-friendly engines (LNG, etc)	• Solar and wind power • Li-Ion Batteries • Fuel cells

Optimizing ships for minimum greenhouse gas emission from the earliest concept design stage can bring about changes of various scales in all aspects of ships: for example, ships utilizing new technologies such as electric propulsion systems or new power sources. It also includes system integration techniques, which integrate various green ship equipment and components to achieve an overall reduction in greenhouse gas emission while maintaining optimum sailing, and platform design techniques for realizing new concept hull forms and installing novel propulsion systems.

Equipment development can also contribute to emission reduction by using low carbon propulsion power sources or reducing exhaust gas emissions, and includes the following:

- High-efficiency superconducting propulsion systems, such as high temperature superconducting generators (HTSG), high temperature superconducting propulsion motors (HTSM) and superconducting inverters;
- Integrated/intelligent onboard electric power control system (smart grid);
- Novel state-of-the-art propulsion systems (pod propulsion systems, contra-rotating propellers, etc.);
- Engine exhaust gas processing technology (waste heat recovery systems, selective catalytic reduction, carbon capture and storage, etc.); and
- Environment-friendly engine systems including LNG engines, dual fuel engines, smart

gas turbines, etc.

New power sources include photovoltaic systems, wind power, fuel cells or secondary cell technology (fuel cells, lithium-ion batteries). Current technical development being undertaken in various parts of the world is as follows:

1) Europe

Europe is leading the green ship technology development through continuous research on ship safety and reducing greenhouse gases.

The Norwegian classification society Det Norske Veritas Germanisher Lloyd (DNV GL) has developed a new concept for container ships (Quantum: A container ship concept for the future, DNV GL). A concept ship was designed for minimizing greenhouse gases and for rapidly changing future shipping environment.

- Length 273m, cargo capacity 6,210 TEU (including 1,200 TEU refrigerated containers)
- Propulsion system of electric motors, four dual fuel engines of 33MW each, two propulsion pods of 11.5MW each.
- Fuel: LNG 5,000m^3, diesel 3,000m^3

2) United States

The United States is tightening regulations concerning ship safety and emissions from ships anchored near its coasts or in US ports.

- Relevant regulations will be strengthened in the aftermath of the BP oil rig explosion in the Gulf of Mexico on April 20, 2010.
- Naval construction capability will be maintained to defend the position of world's leading naval power.

The United States plans to change more than 50% of its navy's total energy consumption to alternative energy sources by 2020 in order to aid the world-wide effort of reducing greenhouse gas emissions.

3) Japan

The Super Eco Ship Project was the showcase project in Japan and took place from 2001 to 2008. Funded by the Ministry of Land, Infrastructure, Transport and Tourism, and led by the Japanese National Maritime Research Institute (NMRI), the main aim of the project was to reduce greenhouse gas emissions from shipping.

In 2009, Japan's NYK Line, one of top ten shipping companies in the world, published a blueprint for an environment-friendly ship, 'Super Eco Ship 2030' illustrated in **Figure 16-10**. The blueprint claims that, if all the technologies are developed as suggested, a ship's carbon

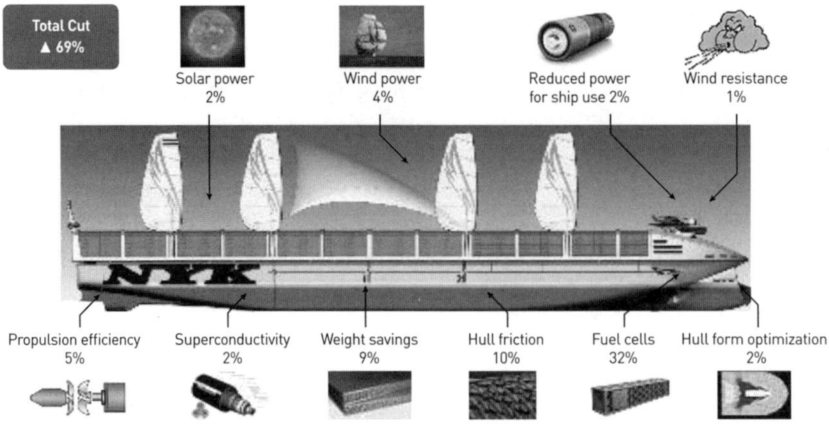

Figure 16-10 Outline of Super Eco Ship 2030 (NYK Line, Japan)

dioxide emission can be reduced by 69%. This involves integrated application of various technologies such as hull form optimization, wind power, electric motors and fuel cells.

Ship-owners' future requirements of ships are quite plain, and in order to realize the technologies to meet these requirements as soon as possible, a research association was set up to carry out feasibility studies, key technology development, manufacturing of test products and verification sea trials. Presently, models are being selected for each type of vessel for application.

4) Korea

Examples of technical development in progress in Korea at present time are as follows:

(1) Hyundai Heavy Industries

The company is developing a dual fuel diesel electric (DFDE) propulsion system, resistance reduction wings, hybrid engines and eco-ballast systems. Their most successful research achievement is the development of energy-saving rudders with winglets. The addition of the winglets on the rudder maximized the propeller thrust and it is expected to save fuel consumption by about 4 to 6%. This translates to savings of about $2.4 million a year when applied to a large container ship. The company claims that it will cost about $500,000 to install them, compared to the total expected life-time saving of $60 million, assuming a lifespan of 25 years.

(2) Daewoo Shipbuilding & Marine Engineering

The company has been developing pre-swirl stators, LNG cargo containment without loss through boil-off gas (BOG), waste heat recovery systems (WHRS) and LNG fuel engines. The most successful of these has been pre-swirl stators. Four stator wings installed in front of the propeller reduces the rotational loss of the propeller and enhances the wake inflow to the propeller. This reduces propeller-induced pressure fluctuation on the hull surface as well as reducing fuel consumption by about 5%. This has been installed on 320,000 ton VLCC *Astro Capricorn* for the Greek owner, Kristen Navigation Inc., and there is a plan to install it on 12 additional VLCCs and 12 container carriers.

(3) Samsung Heavy Industries

Since they declared Green Management in January 2010, the company has been working towards developing green ships with up to 30% less greenhouse gas emission, achieving environment-friendly work place and green network, and establishing the list of pollutant sources within the yard. The company's showcase achievement is the development of saver fins. Saver fins are fish fin-like single steel plates 2.5m by 0.5m attached to the outside of large ships, and can improve the flow around a ship. They were shown to reduce fuel consumption by 3 to 5% and vibration by more than 50%. The fins have been installed on about 20 ships, and through these simple devices fuel cost savings of $450,000 to $900,000 per VLCC have been achieved.

3.3 Green Shipbuilding

1) Ship construction processes that can cause environmental problems

Some shipbuilding and ship repair processes have direct impact on environment.

Such tasks in shipbuilding include the following:

- Handling of raw materials, surface treatment and processing of steel;
- Fabrication and assembly of blocks;
- Outfitting and welding during erection;
- Assembly of electrical and electronic equipment; and
- Preparation and installation of various fabricated parts that are not of a structural nature.

Ship repair tasks that can impact environment include the following:

- Surface cleaning and treatment of hull surface, cargo holds and tanks;
- Oil transfer and treatment of waste oil; and
- Maintenance of machinery and other equipment.

2) Metal processing and preventive measures against pollution

Most merchant ships are built of steel, although special ships like high speed craft and some naval vessels are built of aluminum, wood and FRP as well as steel. Metal processing includes many tasks, such as cleaning of the surface, cutting, pressing, boring, milling and grinding. Metal cleaners, cutting oil, lubricants and solvents used in these tasks and metal chips produced during cutting are sources of environmental pollution.

The most common methods of cutting metals are oxyfuel gas and plasma cutting, and the particulate matters (PM) and gas produced during oxyfuel gas cutting cause environmental pollution. Pollutants produced during welding include greenhouse gases, chemical toxic materials, ozone, PM, carbon monoxide, nitrogen oxides (NOx), sulfur dioxide (SO_2) and lead.

In order to minimize such pollutants, welding partitions, hoods, torch smoke reflectors and transportable ducts are most widely used to collect and process them.

3) Surface treatment and painting and preventive measures against pollution

Surface treatment is a process of cleaning the surface of a ship's hull in preparation for painting during ship construction and repair. Environmental pollutants occurring during surface treatment and painting are heavy metals, such as copper, lead, mercury and zinc, and highly volatile organic compounds such as solvents and thinners. **Table 16-6** shows pollutants produced during surface treatment and painting work.

Table 16-6 Pollutants from surface treatment and painting work

Pollution sources	Pollutants
Raw material(metal) steel, stainless steel, galvanized steel, aluminum, copper-nickel and other copper alloys	aluminum, cadmium, chrome, copper, iron, lead, manganese, nickel, zinc
Paint primers, AC(anti-corrosive) paints, AF(anti-fouling) paints, etc	copper, barium, cadmium, chrome, lead, tin compound, TBT(Tributyltin)
Surface Treatment Agent metal surface preparation agent (steel grid, steel shot, etc), slag surface treatment agent (coal slag, copper slag, nickel slag), synthetic compound (aluminum oxide compound, silicon carbide), natural oxides (silica sand)	arsenic, beryllium, silica, amorphous silica, cadmium, chrome, cobalt, lead, manganese, nickel, titanium, vanadium

In order to prevent pollution during surface treatment and painting, protective barriers must be installed and the waste water must be processed before releasing. Sand and metal slags used in surface cleaning must be collected and recycled.

4) Ship repair

Ship repair work includes regular maintenance, modifications, overhauls and damage repair of ships and their equipment. While many tasks of maintenance and repair are direct causes of environmental damage, they may not necessarily take place in shipyards. The processes which can cause problems in this regard are cleaning, painting and other surface treatment and disposal of waste oil, bilge, cargo waste, ballast water and sanitary waste. The provision of facilities for disposal of such pollutants is part of the MARPOL protocol, but many ports still lack adequate reception facilities, implying that ships sometimes have to resort to discharging their waste directly into the ocean in violation of the protocol. The IMO consequently devised an action plan to tackle the inadequacy of port reception facilities.

3.4 International Maritime Logistics and Supply Chain Security

As a result of many large container ships of over 10,000 TEU entering the international liner market in recent years, the 'hub & spoke' maritime logistics system has been formed. This system consists of main trunk lines connecting global hub ports like Busan, Shanghai and Rotterdam and feeder lines which connect local ports to the hub ports.

Logistics is an act of effectively planning, organizing, managing and executing transportation of cargo. Logistics integrates various modes of transport to form a door-to-door system of global supply chain. In this respect, recent IMO, 2010, Second IMO Greenhouse Gas Study 2009, IMO aspires to establish compound transport networks based on standardized logistical information networks and automatic port processing systems.

The EU is trying to establish an effective green, freely flowing, integrated logistics network by forming a seamless interface between maritime and land transport networks including EU-wide short sea shipping (SSS). The logistics information technology developed recently to achieve this goal includes 'electronic navigation' and 'one-stop logistics information,' and Europe is pursuing the international standardization of these matters.

The short sea shipping is a new concept proposed by Europe in the 1990s and is an integrated transport system with maritime transport at its center. This system is based on new technology and is different from the traditional coastal transport in that it produces high added value through close links between ground, sea and air transport modes. Furthermore, it can be said to be a greener transport system with lower traffic congestion and social costs achieved by shifting cargo from ground transport to shipping.

Waterborne transport is the most important transport mode, taking up 98% of total cargo volume (ton-km) around the world. In the future marine transport must be developed into eco-friendly transport system which can help alleviate the problems of traffic congestion and greenhouse gas emission. It is expected that such a development will require integrated logistics and green shipping technology based on information technology.

3.4.1 Green Logistics

1) Shipping and green logistics

Green logistics minimizes environmental pollution from conventional logistics and reverse logistics while simultaneously maximizing recycling as presented in **Figure 16-11**.

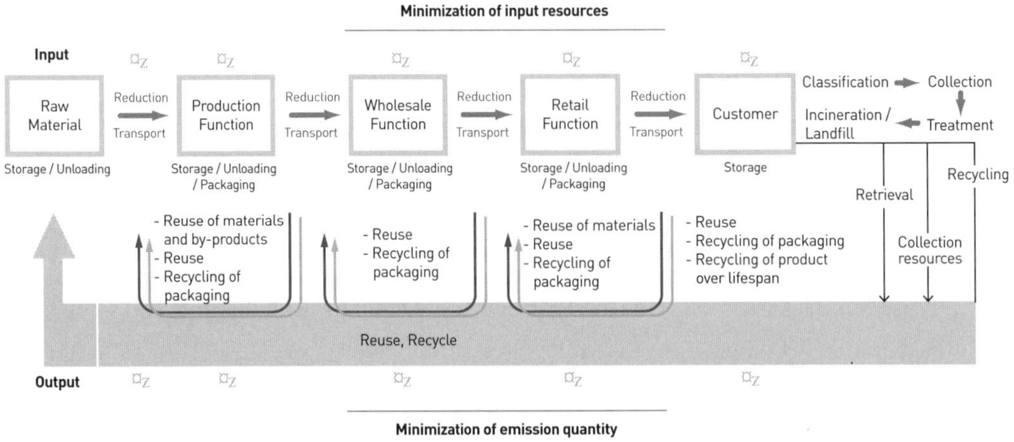

Figure 16-11 Concept of green logistics

General logistics activities are related to packing, transportation, loading and storage of goods. On the other hand, reverse logistics activities are related to the collection of disposed products and materials from the end consumer, and thence classification, secondary processing such as reusing, recycling or remanufacturing and transportation or redistribution for the final disposal of these products.

Compared to air and ground transport, shipping produces considerably less CO_2 emissions per ton-km. However, it makes up about 3.3% of total world CO_2 emissions (in 2007), and hence the more stringent regulations on emissions by IMO of 2013.

Materials disposed of from the ship during its operation directly affect the ocean and atmospheric environment. Major pollutants produced by shipping are presented in **Figure 16-12**.

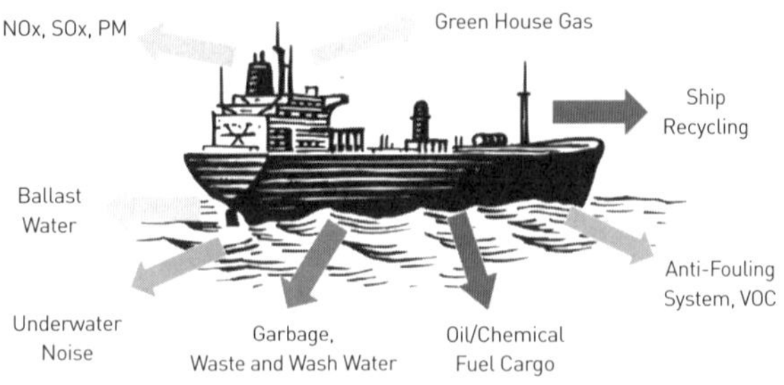

Figure 16-12 Sources of environmental pollution in shipping

2) IMO's green passport system for ship recycling

The green passport system was introduced by IMO's MEPC. This system requires shipyards to record all dangerous and hazardous material present on board a ship it is building and hand it over to the owner at the time of delivery of the ship. The ship operator maintains this record by recording all modifications and additions to the hazardous material. The certificate is then handed over to the ship breaker, when the ship is sold for demolition, so that these materials can be safely disposed of and accounted for. This policy came into effect at the beginning of May 2009, applicable to ocean-going ships of GT 5,000 tons and over and ship breaking yards.

3) US policy for reduction of ship's emission

The US introduced a green ship program on January 1, 2005 which set a speed limit of 12 knots for all ships sailing within 20 miles from any American port. The boundary was extended to 40 miles in 2009. This speed limit was not compulsory, but the complying ships were allowed to raise a green flag and were awarded a reduction on berthing fees.

On the other hand the use of fuels with low sulfur contents is compulsory in the US waters. In November 2006 the use of marine gas oil with sulfur content greater than 0.2% was banned within 40 miles of any American port. Furthermore, the use of marine diesel oil with sulfur content of 0.5% or more and marine gas oil with 0.15% or higher sulfur content was also banned within 24 miles of American coasts in July 2008. The allowed sulfur content was reduced to 0.1% in 2010.

The use of generators on all berthed ships is banned and ships are required to make use of ground power.

4) EU policies for reduction of ships' emission

The EU has banned the use of ship fuels with sulfur content exceeding 0.5% in inland water ways since July 2005, and even lower 0.1% since January 2010. The use of fuels with sulfur content exceeding 1.5% in the Baltic Sea has been prohibited since 2006. This was expanded to the North Sea and the English Channel in 2007.

The EU is also strongly demanding policies for regulating gas emissions from the international shipping community. At COP15 (the 15th Conference of the Parties to the United Nations) in January 2009, the EU proposed a goal of decreasing total shipping emissions to the 2005 level by 2020, and the 1900 level by 2050. It is expected that the EU policy on reduction of ships' emissions will be reflected in the IMO's mandatory greenhouse gas emission regulations effective from 2013.

5) Green logistics and ship design concepts

Pollutants from ships include engine exhaust gas, ship paints, cargos, wastes, waste water, bilge, ballast water and operational noise. Major design concepts for the reduction of such pollutants are summarized in **Table 16-7**.

Table 16-7 Major design concepts for realizing green logistics

Ship types	Major design concepts
All ships	Fuel : cleaner fuels, gas fuels, hydrogen fuel Bilge : clean engine room - integrated bilge system Recycling : design for recycling
Container ships	Diesel generators at bottom of ship Electric propulsion (front drive) + rear side-thrusters Forecastle navigation bridge and accommodation
Bulk carriers	Non-ballast ships
Oil tankers	Non-ballast ships Semi-submerged tankers

3.4.2 Supply Chain Security Regulations

1) Supply chain security programs

Supply chain security deals with a wide variety of situations, including protection from unlawful tampering, terrorism acts, international or national conflicts and natural disasters. The logistics security has become a major international issue especially since the 9/11 terrorist attack. The International Organization for Standardization (ISO) is working toward establishing international standards of logistics security under the slogan 'Standards for a Safer World.'

Figure 16-13 shows supply chain security management programs (SCSM), illustrating worldwide supply chain security initiatives and related management programs. The security management programs are composed of six areas: facility management, cargo management, human resources management, information and communication management, business networks and company management systems, and crisis management and disaster recovery.

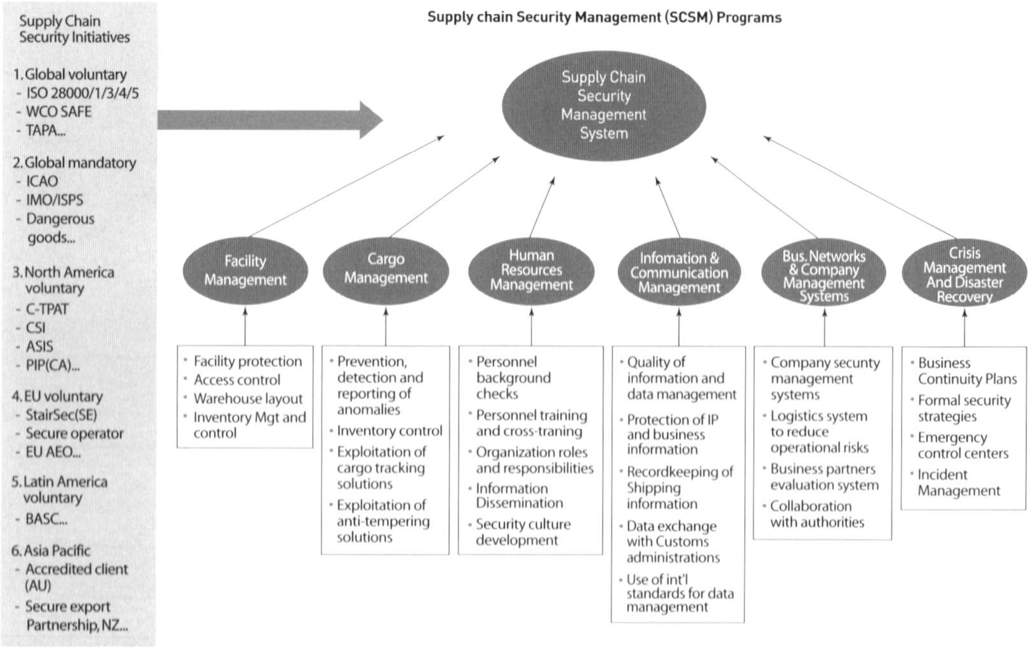

Figure 16-13 Supply chain security management (SCSM) programs

Security regulations on global and regional information collection for the above 6 areas are as follows:

- Global voluntary provisions
  - ISO: 28000/28001/28002/28003/28004/28005, etc.
  - WCO: SAFE Framework
  - TAPA, etc.
- Global mandatory provisions: ICAO, IMO/ISPS, Harmful Materials, etc.
- North American regional voluntary provisions: C-TPAT, CSI, ASIS, PIP (CA), etc.
- EU voluntary provisions: EU/AEO, StairSec (SE), Secure Operator, etc.
- South American voluntary provisions: BASC, etc.
- Asia Pacific voluntary provisions: Faithful Client (Australia), Safe Export Partners (New Zealand), etc.

Figure 16-14 shows the range of applications of security information collection systems for major global security provisions. For example, the ISO 28000 Series and the WCO SAFE framework are for all processes of security information collection, while C-TPAT and IMO/ISPS provisions of the United States are only applicable from the loading port to the unloading port, and the SAFE Port Act of the United States is applicable from the loading points of each container to the final destination, whereas the safe transportation program for the American Department of Homeland Security is applicable from the loading port to the final destination.

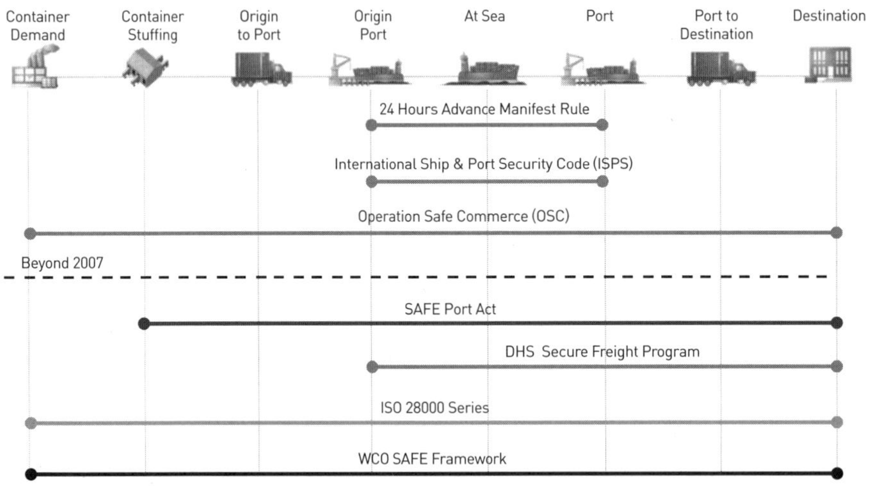

Figure 16-14 Application range of security information collection systems for major security provisions

2) Facility management

Facility management is the security system for all major facilities, ranging from the manufacturing facilities to storage facilities for loading and transshipment. The security facilities include physical barriers such as fences, logistics security equipment for access control, facilities monitoring and warehouse layout. The layout should be determined by considering the convenience of special cargo staking, protection of facilities, prevention of infiltration and security of computer on related drawings.

3) Cargo management

This aims to prevent illegal access to cargo and intrusion to cargo areas, and uses equipment and software solutions for real time monitoring. Related systems include cargo search systems, cargo tracking and monitoring systems and intrusion prevention systems.

4) Human resources management

This refers to management of personnel involved in supply chains and logistics processes. Main tasks are background checks on employees who have direct or indirect access to the cargo and cargo-related facilities, and education and training of employees on the roles and responsibilities of logistics security within the organization. Establishing basic principles on distribution and sharing of security-related information and organizational culture of heightened awareness on security is also within its remit.

5) Information and communication management

Information and communication management is a security system that is focused on blocking access to internal business information by unauthorized personnel, and includes qualitative management of information and data. Management of compatibility with security management data system used by clients and supply chain partners belongs to this area as well as application of international standards for the management of security-related data.

6) Business network and business management system

Logistics security is a shared responsibility that cannot be managed by a single company alone. Therefore, effective security calls for security management throughout the business network. It includes comprehensive security management system throughout the logistics chain and security assessment systems of supply chain partners.

7) Crisis management and disaster recovery

Most current logistics-related legal and systemic regulations put emphasis on early detection and prevention of security risks. However, if a security breach occurs, crisis management and recovery must follow. In order to do this, establishment of plans for continuing with the business and an official security strategy is crucial, and this must include the manuals for operating a crisis management center and event management.

3.4.3 Logistics Security Equipment

1) Container sealing equipment

Modern container sealing equipment is a combination of conventional mechanical sealing equipment and various electronic sealing equipments, such as RFID. Such sealing equipment can receive the records and monitoring signals of containers in real-time allowing them to prevent various risks in advance. **Figure 16-15** shows sealing equipment used on containers.

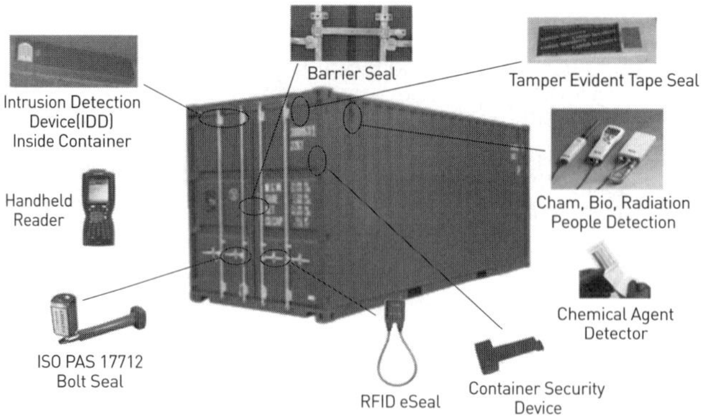

Figure 16-15 Sealing equipment of containers

2) Smart containers

Smart containers are ordinary containers equipped with the next generation intelligent container security equipment. A smart container device has the following main functions:

- Container tracking: short and long distance.
- Communication: multi-frequency channels, multiple communication protocols, data processing codes, broad-band communication, low speed data ratio, satellite communication, mobile phones and local communications.
- Six walls instruction sensors: UV ray detectors, surrounding detection sensors, sound sensors, air detecting sensors.
- Front door monitoring sensors: optical fiber sensors, light sensors, sealed cables, entrance micro switches.
- People detection: sensors, sound sensors, UV ray sensors.
- Weapons of mass destruction and explosives: external motion tracking devices, gamma ray sensors, biochemical sensors.

3) Container search devices

Container search devices search the interior of containers by using imaging scanners or other methods (**Figure 16-16**). An imaging search apparatus may use the neutron activation method, X-rays or gamma rays.

4) Security management solutions

Most of the current security management solutions have been developed by hardware companies around their own hardware. GE and SAIC are examples of such companies.

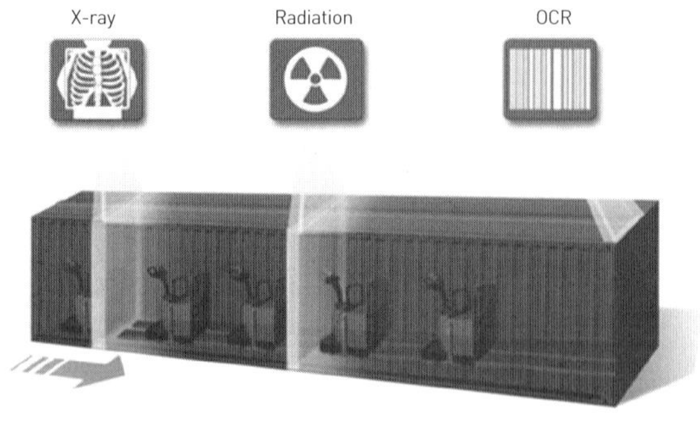

Figure 16-16 Container search apparatus

The majority of security management solution companies construct the interface system between the early systems and the existing systems and the logistics security work is carried out by the certified integrated information system which have been supplied from the authorized organization in information technology outsourcing.

References

1. Branch, A.E., 2011. *Elements of Shipping*. 8th ed. London & N.Y.: Routledge.
2. Det Norske Veritas (DnV), 2010. Pathways to low carbon shipping, Abatement potential towards 2030. Hovik: DnV.
3. IMO, 2010. *Second IMO greenhouse gas study 2009*. London: CIP Books Limited.
4. Kim, G.H. et al., 2009. *Theory of ship financing* (in Korean). Seoul: Jeonyoungsa.
5. Korea International Trade Association (KITA), 2009. *Green Logistics Management Strategies* (in Korean). Seoul: KITA.
6. Organization for Economic Cooperation and Development (OECD), 2010. *Environmental and climate change issues in the shipbuilding industry*. OECD Council Working Party on Shipbuilding (WP6). Paris: OECD.
7. Stopford, M., 2009. *Maritime Economics*, 3rd ed. Abingdon & New York: Routledge.
8. The National Academy of Engineering of Korea (NAEK), 2010. *Policy proposal from the engineers* (in Korean). Seoul: NAEK.
9. The National Academy of Engineering of Korea (NAEK), 2010. *Process of technological development in Korea's major industries and challenge* (in Korean). Seoul: NAEK.
10. The National Academy of Engineering of Korea (NAEK), 2010. *Strategies to maintain highestcompetitiveness in shipbuilding and offshore industry* (in Korean). Seoul: NAEK.
11. UNCTAD (United Nations Conference on Trade and Development), 2010. *Review of Maritimetransport*, Geneva: UNCTAD.
12. Yoshida, K., 2010. Green ship: future environmentally friendly ships. In: *ISO TC 8 Seminar*. [Seminar note].

INDEX

A

B

C

E

F

G

H

I

J

K

L

M

N

O

P

Q

R

S

T

U

V

W

X ~ Z

****